BASIC OPERATIONAL AMPLIFIERS AND LINEAR INTEGRATED CIRCUITS

Second Edition

Thomas L. Floyd

David Buchla
Yuba College

Prentice Hall
Upper Saddle River, New Jersey   Columbus, Ohio

Cover Photo: Freeman/Mauritius/H. Armstrong Roberts
Editor: Scott Sambucci
Production Editor: Rex Davidson
Design Coordinator: Karrie Converse-Jones
Text Designer: John Edeen
Cover Designer: Dan Eckel
Production Manager: Patricia A. Tonneman
Illustrations: Jane Lopez
Marketing Manager: Ben Leonard

Library of Congress Cataloging-in-Publication Data

Floyd, Thomas L.
   Basic operational amplifiers and linear integrated
circuits /   Thomas L. Floyd, David Buchla.—2nd ed.
      p.   cm.
   Includes index.
   ISBN 0–13–082987-0
   1. Operational amplifiers. 2. Linear integrated circuits.
I. Buchla, David.   II. Title.
TK7871.58.06F565   1999
621.39'5—dc21
                                                    98-34031
                                                        CIP

© 1999 by Prentice-Hall, Inc.
Simon & Schuster/A Viacom Company
Upper Saddle River, New Jersey 07458

Earlier edition copyright © 1994 by Macmillan Publishing
Company.

Printed in the United States of America

ISBN: 0-13-082987-0

Prentice-Hall International (UK) Limited, *London*
Prentice-Hall of Australia Pty. Limited, *Sydney*
Prentice-Hall of Canada, Inc., *Toronto*
Prentice-Hall Hispanoamericana, S. A., *Mexico*
Prentice-Hall of India Private Limited, *New Delhi*
Prentice-Hall of Japan, Inc., *Tokyo*
Simon & Schuster Asia Pte. Ltd., *Singapore*
Editora Prentice-Hall do Brasil, Ltda., *Rio de Janeiro*

Thanks to our wives for their affectionate support. Like fine red wine, they get better with age.

Preface

Basic Operational Amplifiers and Linear Integrated Circuits, Second Edition, presents thorough coverage of operational amplifiers and other linear integrated circuits. Also, this textbook provides extensive troubleshooting and applications coverage with Test Bench sections that present circuits in a realistic printed-circuit board format.

Circuit currents in *Basic Operational Amplifiers and Linear Integrated Circuits,* Second Edition, are indicated by a meter notation rather than by directional arrows. This unique approach accomplishes two things. First, it eliminates the need to distinguish between conventional flow and electron flow because it indicates current direction by polarity signs, just as an actual ammeter does. Users can interpret current direction based on the meter polarity in accordance with their particular preference. Secondly, in addition to current direction, the meter notation provides relative magnitudes of the currents in a given circuit by the positions of the meter pointers.

Overview

The first chapter presents an introduction to analog electronics, analog signals, amplifiers, and troubleshooting. The remaining chapters cover analog integrated circuits as follows: Chapter 2 provides an introduction to operational amplifiers. Op-amp frequency response is covered in Chapter 3, and basic op-amp circuits (comparators, summing amplifiers, integrators, and differentiators) is the topic of Chapter 4. Active op-amp filters are covered in Chapter 5, and oscillators and timers are introduced in Chapter 6. Power supplies are covered in Chapter 7. Special amplifiers (instrumentation amplifiers, isolation amplifiers, operational transconductance amplifiers (OTAs), and log/antilog amplifiers) are introduced in Chapter 8. Communication circuits (AM and FM receivers, linear multipliers, mixers, and phase-locked loops) are studied in Chapter 9. Data conversion circuits such as analog switches, sample-and-hold circuits, digital-to-analog and analog-to-digital converters, and voltage-to-frequency and frequency-to-voltage converters are among the topics in Chapter 10. Finally, Chapter 11 covers various types of transducers and associated measurement circuits.

Features

Basic Operational Amplifiers and Linear Integrated Circuits, Second Edition, is innovative in four areas:

❑ Current in a circuit is indicated by a polarized meter symbol that allows the user to apply the direction of preference. Also, current meters show relative current magnitude in a circuit. See Figure P–1.

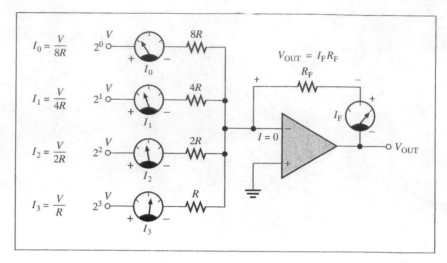

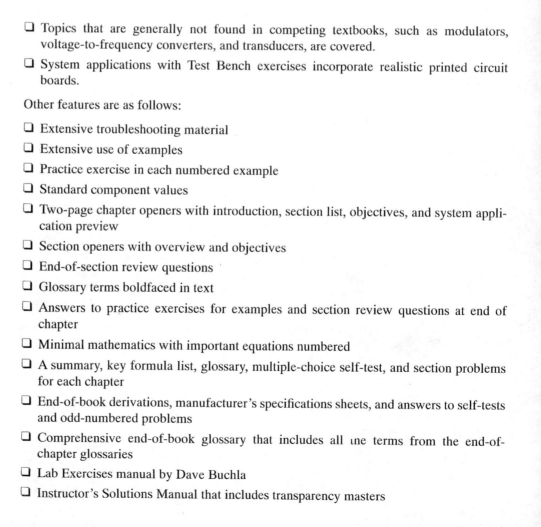

FIGURE P–1
Example of art showing meter symbols.

- ❏ Topics that are generally not found in competing textbooks, such as modulators, voltage-to-frequency converters, and transducers, are covered.
- ❏ System applications with Test Bench exercises incorporate realistic printed circuit boards.

Other features are as follows:

- ❏ Extensive troubleshooting material
- ❏ Extensive use of examples
- ❏ Practice exercise in each numbered example
- ❏ Standard component values
- ❏ Two-page chapter openers with introduction, section list, objectives, and system application preview
- ❏ Section openers with overview and objectives
- ❏ End-of-section review questions
- ❏ Glossary terms boldfaced in text
- ❏ Answers to practice exercises for examples and section review questions at end of chapter
- ❏ Minimal mathematics with important equations numbered
- ❏ A summary, key formula list, glossary, multiple-choice self-test, and section problems for each chapter
- ❏ End-of-book derivations, manufacturer's specifications sheets, and answers to self-tests and odd-numbered problems
- ❏ Comprehensive end-of-book glossary that includes all the terms from the end-of-chapter glossaries
- ❏ Lab Exercises manual by Dave Buchla
- ❏ Instructor's Solutions Manual that includes transparency masters

Chapter Pedagogy

Chapter Opener Each chapter begins with a two-page spread as indicated in Figure P–2.

Section Opener and Section Review Questions As illustrated in Figure P–3, each section within a chapter begins with an opening introduction and list of section objectives. Each section ends with a set of review questions that focus on key concepts. Answers to review questions are given at the end of the chapter.

Examples and Practice Exercises Worked-out examples are used to illustrate and clarify topics covered in the text. At the end of every example and within the example box is a practice exercise that either reinforces the example or focuses on a related topic. Answers to the practice exercises are given at the end of the chapter. This feature is illustrated in Figure P–4.

System Application As illustrated in Figure P–5, the last section of each chapter (except Chapter 1) is a system application of devices and circuits related to the chapter coverage. The system application is an optional feature which will not affect other topics if omitted. The variety of "systems" is intended to give the student an appreciation for the wide range of applications for electronic devices and to provide motivation to learn the basic concepts of each chapter.

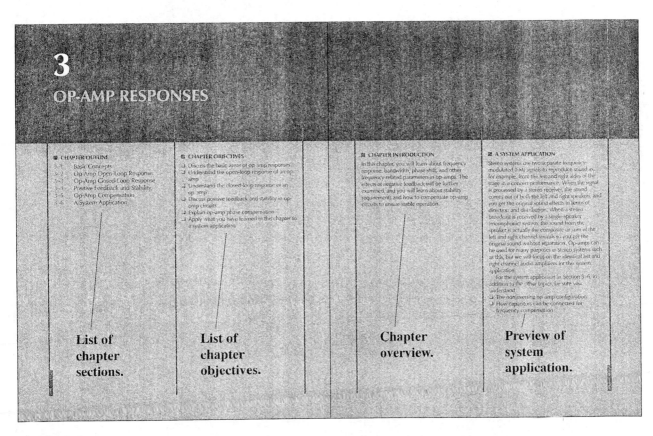

FIGURE P–2
Chapter opener.

Review questions end each section.

Introductory statements and a list of performance-based objectives begin each section.

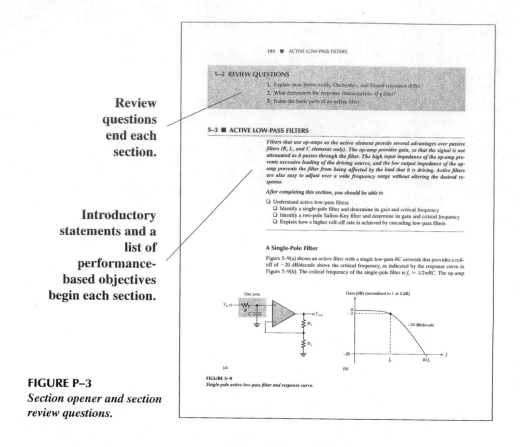

FIGURE P–3
Section opener and section review questions.

The system application sections can be used as follows:

❑ A part of each chapter for the purpose of relating devices to a realistic application and for establishing a useful purpose for devices covered. All or selected activities can be assigned and discussed in class or turned in for a grade.

❑ A separate out-of-class assignment to be turned in for extra credit.

❑ An in-class activity to promote and stimulate discussion and interaction among students and between students and instructor.

❑ An illustration to help answer the question that many students have: "Why do I need to know this?"

Chapter End Matter A summary, key formula list, glossary, self-test, and sectionalized problem sets are found at the end of each chapter. The answers to practice exercises and section review questions are also provided.

To the Student

Any career training requires hard work, and electronics is no exception. The best way to learn new material is by reading, thinking, and doing. This text is designed to help you along the way by providing an overview and objectives for each section, numerous worked-out examples, practice exercises, and review questions with answers.

**Examples
are contained
within a
ruled box.**

**Each example
contains an
exercise related
to the example.**

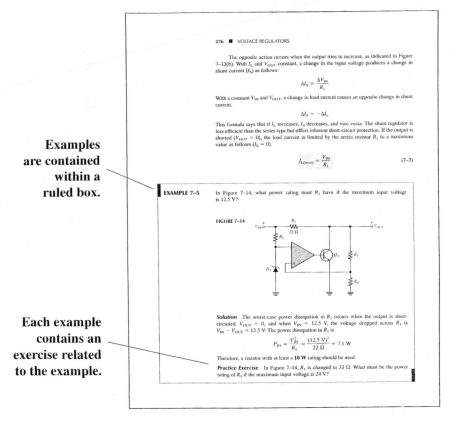

276 ■ VOLTAGE REGULATORS

The opposite action occurs when the output tries to increase, as indicated in Figure 7–13(b). With I_L and V_{OUT} constant, a change in the input voltage produces a change in shunt current (I_S) as follows:

$$\Delta I_S = \frac{\Delta V_{IN}}{R_1}$$

With a constant V_{IN} and V_{OUT}, a change in load current causes an opposite change in shunt current.

$$\Delta I_S = -\Delta I_L$$

This formula says that if I_L increases, I_S decreases, and vice versa. The shunt regulator is less efficient than the series type but offers inherent short-circuit protection. If the output is shorted ($V_{OUT} = 0$), the load current is limited by the series resistor R_1 to a maximum value as follows ($I_S = 0$).

$$I_{L(max)} = \frac{V_{IN}}{R_1} \qquad (7\text{–}7)$$

EXAMPLE 7–5 In Figure 7–14, what power rating must R_1 have if the maximum input voltage is 12.5 V?

FIGURE 7–14

Solution The worst-case power dissipation in R_1 occurs when the output is short-circuited. $V_{OUT} = 0$, and when $V_{IN} = 12.5$ V, the voltage dropped across R_1 is $V_{IN} - V_{OUT} = 12.5$ V. The power dissipation in R_1 is

$$P_{R1} = \frac{V_{R1}^2}{R_1} = \frac{(12.5\ V)^2}{22\ \Omega} = 7.1\ W$$

Therefore, a resistor with at least a **10 W** rating should be used.

Practice Exercise In Figure 7–14, R_1 is changed to 33 Ω. What must be the power rating of R_1 if the maximum input voltage is 24 V?

FIGURE P–4
Typical example and practice exercise.

Don't expect every concept to be crystal clear after a single reading. Read each section of the text carefully and think about what you have read. Work through the example problems step-by-step before trying the practice exercise that goes with the example. Sometimes more than one reading of a section will be necessary. After each section, check your understanding by answering the section review questions.

Review the chapter summary, glossary, and formula list. Take the multiple-choice self-test. Finally, work the problems at the end of the chapter. Check your answers to the self-test and the odd-numbered problems against those provided at the end of the book. Working problems is the most important way to check your comprehension and solidify concepts.

Milestones in Electronics

Before you begin your study of operational amplifiers and linear integrated circuits, let's briefly look at some of the important developments that led to electronics technology as we have today. The names of many of the early pioneers in electricity and electromagnetics still live on in terms of familiar units and quantities. Names such as Ohm, Ampere, Volta, Farad, Henry, Coulomb, Oersted, and Hertz are some of the better known examples. More widely known names such as Franklin and Edison are also significant in the history of electricity and electronics because of their tremendous contributions.

Section opener describes the assignment.

Realistic PC board provides visual information related to the assignment.

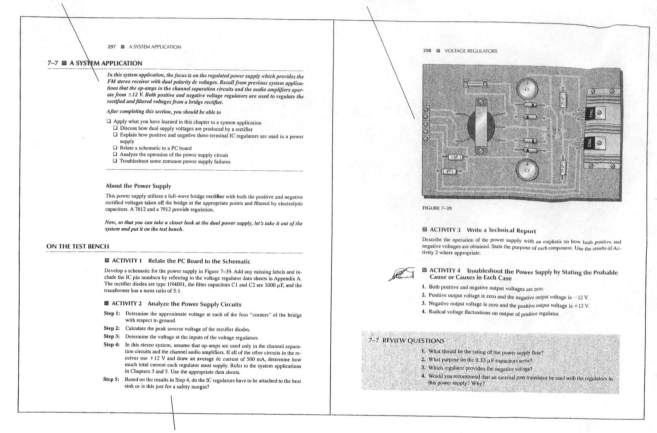

297 ■ A SYSTEM APPLICATION

7–7 ■ A SYSTEM APPLICATION

In this system application, the focus is on the regulated power supply which provides the FM stereo receiver with dual polarity dc voltages. Recall from previous system applications that the op-amps in the channel separation circuits and the audio amplifiers operate from ±12 V. Both positive and negative voltage regulators are used to regulate the rectified and filtered voltages from a bridge rectifier.

After completing this section, you should be able to

❑ Apply what you have learned in this chapter to a system application
  ❑ Discuss how dual supply voltages are produced by a rectifier
  ❑ Explain how positive and negative three-terminal IC regulators are used in a power supply
  ❑ Relate a schematic to a PC board
  ❑ Analyze the operation of the power supply circuit
  ❑ Troubleshoot some common power supply failures

About the Power Supply

This power supply utilizes a full-wave bridge rectifier with both the positive and negative rectified voltages taken off the bridge at the appropriate points and filtered by electrolytic capacitors. A 7812 and a 7912 provide regulation.

Now, so that you can take a closer look at the dual power supply, let's take it out of the system and put it on the test bench.

ON THE TEST BENCH

■ ACTIVITY 1   Relate the PC Board to the Schematic

Develop a schematic for the power supply in Figure 7–39. Add any missing labels and include the IC pin numbers by referring to the voltage regulator data sheets in Appendix A. The rectifier diodes are type 1N4001, the filter capacitors C1 and C2 are 1000 μF, and the transformer has a turns ratio of 5:1.

■ ACTIVITY 2   Analyze the Power Supply Circuits

Step 1: Determine the approximate voltage at each of the four "corners" of the bridge with respect to ground.

Step 2: Calculate the peak inverse voltage of the rectifier diodes.

Step 3: Determine the voltage at the inputs of the voltage regulators.

Step 4: In this stereo system, assume that op-amps are used only in the channel separation circuits and the channel audio amplifiers. If all of the other circuits in the receiver use +12 V and draw an average dc current of 500 mA, determine how much total current each regulator must supply. Refer to the system applications in Chapters 3 and 5. Use the appropriate data sheets.

Step 5: Based on the results in Step 4, do the IC regulators have to be attached to the heat sink or is this just for a safety margin?

298 ■ VOLTAGE REGULATORS

FIGURE 7–39

■ ACTIVITY 3   Write a Technical Report

Describe the operation of the power supply with an emphasis on how both positive and negative voltages are obtained. State the purpose of each component. Use the results of Activity 2 where appropriate.

■ ACTIVITY 4   Troubleshoot the Power Supply by Stating the Probable Cause or Causes in Each Case

1. Both positive and negative output voltages are zero.
2. Positive output voltage is zero and the negative output voltage is −12 V.
3. Negative output voltage is zero and the positive output voltage is +12 V.
4. Radical voltage fluctuations on output of positive regulator.

7–7 REVIEW QUESTIONS

1. What should be the rating of the power supply fuse?
2. What purpose do the 0.33 μF capacitors serve?
3. Which regulator provides the negative voltage?
4. Would you recommend that an external pass transistor be used with the regulators in this power supply? Why?

Steps instruct students to perform specific tasks.

FIGURE P–5
A typical system application.

The Beginning of Electronics   Early experiments with electronics involved electric currents in vacuum tubes. Heinrich Geissler (1814–1879) removed most of the air from a glass tube and found that the tube glowed when there was current through it. Later, Sir William Crookes (1832–1919) found the current in vacuum tubes seemed to consist of particles. Thomas Edison (1847–1931) experimented with carbon filament bulbs with plates and discovered that there was a current from the hot filament to a positively charged plate. He patented the idea but never used it.

Other early experimenters measured the properties of the particles that flowed in vacuum tubes. Sir Joseph Thompson (1856–1940) measured properties of these particles, later called *electrons*.

Although wireless telegraphic communication dates back to 1844, electronics is basically a 20th century concept that began with the invention of the vacuum tube amplifier. An early vacuum tube that allowed current in only one direction was constructed by John A. Fleming in 1904. Called the Fleming valve, it was the forerunner of vacuum tube diodes. In 1907, Lee deForest added a grid to the vacuum tube. The new device, called the audiotron, could amplify a weak signal. By adding the control element, deForest ushered in the electronics revolution. It was with an improved version of his device that made transcontinental telephone service and radios possible. In 1912, a radio amateur in San Jose, California, was regularly broadcasting music!

In 1921, the secretary of commerce, Herbert Hoover, issued the first license to a broadcast radio station; within two years over 600 licenses were issued. By the end of the 1920s radios were in many homes. A new type of radio, the superheterodyne radio, invented by Edwin Armstrong, solved problems with high-frequency communication. In 1923, Vladimir Zworykin, an American researcher, invented the first television picture tube, and in 1927 Philo T. Farnsworth applied for a patent for a complete television system.

The 1930s saw many developments in radio, including metal tubes, automatic gain control, "midgit sets," directional antennas, and more. Also started in this decade was the development of the first electronic computers. Modern computers trace their origins to the work of John Atanasoff at Iowa State University. Beginning in 1937, he envisioned a binary machine that could do complex mathematical work. By 1939, he and graduate student Clifford Berry had constructed a binary machine called ABC, (for Atanasoff-Berry Computer) that used vacuum tubes for logic and condensers (capacitors) for memory. In 1939, the magnetron, a microwave oscillator, was invented in Britain by Henry Boot and John Randall. In the same year, the klystron microwave tube was invented in America by Russell and Sigurd Varian.

During World War II, electronics developed rapidly. Radar and very high-frequency communication were made possible by the magnetron and klystron. Cathode ray tubes were improved for use in radar. Computer work continued during the war. By 1946, John von Neumann had developed the first stored program computer, the Eniac, at the University of Pennsylvania. The decade ended with one of the most important inventions ever, the transistor.

Solid-State Electronics   The crystal detectors used in early radios were the forerunners of modern solid-state devices. However, the era of solid-state electronics began with the invention of the transistor in 1947 at Bell Labs. The inventors were Walter Brattain, John Bardeen, and William Shockley, shown in Figure P–6. PC (printed circuit) boards were introduced in 1947, the year the transistor was invented. Commercial manufacturing of transistors began in Allentown, Pennsylvania, in 1951.

The most important invention of the 1950s was the integrated circuit. On September 12, 1958, Jack Kilby, at Texas Instruments, made the first integrated circuit. This invention literally created the modern computer age and brought about sweeping changes in medicine, communication, manufacturing, and the entertainment industry. Many billions of "chips"—as integrated circuits came to be called—have since been manufactured.

The 1960s saw the space race begin and spurred work on miniaturization and computers. The space race was the driving force behind the rapid changes in electronics that followed. The first successful "op-amp" was designed by Bob Widlar at Fairchild Semiconductor in 1965. Called the μA709, it was very successful but suffered from "latch-up" and other problems. Later, the most popular op-amp ever, the 741, was taking shape at

FIGURE P–6
The invention of the bipolar transistor.
Photo copyright by Bell Laboratories.
All rights reserved. Used with
permission.

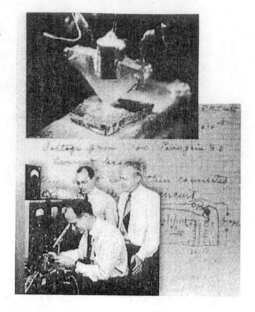

Fairchild. This op-amp became the industry standard and influenced design of op-amps for years to come.

By 1971, a new company that had been formed by a group from Fairchild introduced the first microprocessor. The company was Intel and the product was the 4004 chip, which had the same processing power as the Eniac computer. Later in that same year, Intel announced the first 8-bit processor, the 8008. In 1975, the first personal computer was introduced by Altair, and Popular Science magazine featured it on the cover of the January, 1975, issue. The 1970s also saw the introduction of the pocket calculator and new developments in optical integrated circuits.

By the 1980s, half of all U.S. homes were using cable hookups instead of television antennas. The reliability, speed, and miniaturization of electronics continued throughout the 1980s, including automated testing and calibrating of PC boards. The computer became a part of instrumentation and the virtual instrument was created. Computers became a standard tool on the workbench.

The 1990s saw a widespread application of the Internet. In 1993, there were 130 Web sites, and now there are millions. Companies scrambled to establish a home page and many of the early developments of radio broadcasting had parallels with the Internet. (The bean counters still want to know how it's going to make money!) In 1995, the FCC allocated spectrum space for a new service called Digital Audio Radio Service. Digital television standards were adopted in 1996 by the FCC for the nation's next generation of broadcast television. As the 20th century drew toward a close, historians could only breathe a sigh of relief. As one wag put it: "I'm all for new technologies, but I wish they'd let the old ones wear out first."

Acknowledgments

This textbook is the result of not only the authors' collaboration, but also the skills and efforts of all those at Prentice Hall who were involved in this project. We would particularly like to express our appreciation to Rex Davidson, Scott Sambucci, and Dave Garza. Lois Porter did another outstanding job of manuscript editing from beginning to end, and Jane Lopez deserves our admiration for her work on the art. Thanks also to Gary Snyder and Chuck Garbinski for their superb work in checking accuracy.

Tom Floyd

Dave Buchla

Contents

5 ■ ACTIVE FILTERS 182

6 ■ OSCILLATORS AND TIMERS 224

7 ■ VOLTAGE REGULATORS 262

8 ■ SPECIAL-PURPOSE AMPLIFIERS 308

1

BASIC CONCEPTS OF ANALOG CIRCUITS AND SIGNALS

Courtesy Hewlett-Packard Company

☑ CHAPTER OBJECTIVES

❑ Discuss the basic characteristics of analog electronics
❑ Describe analog signals
❑ Analyze signal sources
❑ Explain the characteristics of an amplifier
❑ Describe the process for troubleshooting an analog circuit

▣ CHAPTER INTRODUCTION

With the influence of computers and other digital devices, it's easy to overlook the fact that virtually all natural phenomena that we measure (for example, pressure, flow rate, and temperature) originate as analog signals. In electronics, transducers are used to convert these analog quantities into voltage or current. Usually amplification or other processing is required for these signals. Depending on the application, digital or analog techniques may be most efficient for processing. Analog circuits are found in nearly all power supplies, in many "real-time" applications (such as motor-speed controls), and in high-frequency communication systems. Digital processing is most effective when mathematical operations must be performed and has major advantages in reducing the noise inherent in processing analog signals. In short, the two sides of electronics (analog and digital) complement each other, and the competent technician needs to be knowledgeable of both.

1–1 ■ ANALOG ELECTRONICS

The field of electronics can be subdivided into various categories for study. The most basic division is to categorize signals between those that are represented by binary numbers (digital) and those that are represented by continuously variable quantities (analog). Digital electronics includes all arithmetic and logic operations such as performed in computers and calculators. Analog electronics includes virtually all other (nondigital) signals. Analog electronics includes signal-processing functions such as amplification, differentiation, and integration.

After completing this section, you should be able to

❑ Discuss the basic characteristics of analog electronics
   ❑ Contrast the characteristic curve for a linear component with that of a nonlinear component
   ❑ Explain what is meant by a characteristic curve
   ❑ Compare dc and ac resistance and explain how they differ
   ❑ Explain the difference between conventional current and electron flow

Linear Equations

In basic algebra, a linear equation is one that plots a straight line between the variables and is usually written in the following form:

$$y = mx + b$$

where y = the dependent variable
   x = the independent variable
   m = the slope
   b = the y-axis intercept

If the plot of the equation goes through the origin, then the y-axis intercept is zero, and the equation reduces to

$$y = mx$$

which has the same form as Ohm's law.

$$I = \frac{V}{R} \qquad\qquad (1\text{–}1)$$

As written here, the dependent variable in Ohm's law is current (I), the independent variable is voltage (V), and the slope is the reciprocal of resistance ($1/R$). Recall from your dc/ac course that this is simply the conductance, (G). By substitution, the linear form of Ohm's law is more obvious; that is,

$$I = GV$$

A **linear component** is one in which an increase in current is proportional to the applied voltage as given by Ohm's law. In general, a plot that shows the relationship between two variable properties of a device define a **characteristic curve.** For most electronic devices, a characteristic curve refers to a plot of the current, I, plotted as a function of voltage, V. For example, resistors have an IV characteristic described by the straight lines given in Figure 1–1. Notice that current is plotted on the y-axis because it is the dependent variable.

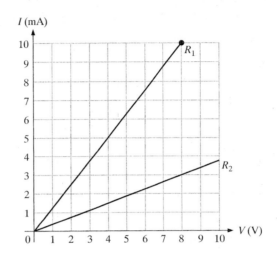

FIGURE 1–1
IV characteristic curve for two resistors.

EXAMPLE 1–1

Figure 1–1 shows the IV characteristic curve for two resistors. What are the conductance and resistance of R_1?

Solution   Find the conductance, G_1, by measuring the slope of the R_1 IV characteristic curve. The slope is the change in the y variable (written Δy) divided by the corresponding change in the x variable (written Δx).

$$\text{slope} = \frac{\Delta y}{\Delta x}$$

Choosing the point ($x = 8$ V, $y = 10$ mA) from Figure 1–1 and the origin, you can find the slope and therefore the conductance as

$$G_1 = \frac{10 \text{ mA} - 0 \text{ mA}}{8.0 \text{ V} - 0 \text{ V}} = \textbf{1.25 mS}$$

For a straight line, the slope is constant so you can use any two points to determine the conductance. The resistance is the reciprocal of the conductance.

$$R_1 = \frac{1}{G_1} = \frac{1}{1.25 \text{ mS}} = \textbf{0.8 k}\boldsymbol{\Omega}$$

Practice Exercise   Find the conductance and resistance of R_2.

AC Resistance

As you have seen, the graph of the characteristic curve for a resistor is a straight line that passes through the origin. The slope of the line is constant and represents the conductance of the resistor; the reciprocal of the slope represents resistance. Since the ratio of voltage to current is constant no matter where they are measured, the ratio is referred to as a *dc resistance.*

Many devices have a characteristic curve for which the current is not proportional to the voltage. These devices are nonlinear by nature but are included in the study of analog electronics because they take on a continuous range of input signals.

Figure 1–2 shows an *IV* characteristic curve for a diode, a nonlinear device used internally in op-amps and as an external component in circuits such as log amps (studied in Chapter 8). Generally, it is useful to define the resistance of a nonlinear analog device as a small change in voltage divided by the corresponding change in current, that is, $\Delta V/\Delta I$. The ratio of a small change in voltage divided by the corresponding small change in current is defined as the **ac resistance** of the device.

$$r_{ac} = \frac{\Delta V}{\Delta I}$$

This internal resistance (indicated with a lowercase italic *r*) is also called the *dynamic, small signal,* or *bulk resistance* of the device. The ac resistance depends on the particular point on the *IV* characteristic curve where the measurement is made.

FIGURE 1–2
An IV characteristic curve for a diode.

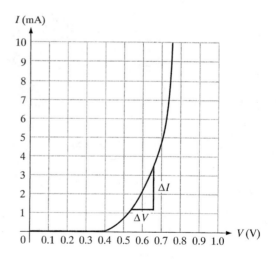

For the diode in Figure 1–2, the slope varies dramatically; the point where the ac resistance is measured needs to be specified with any measurement. For example, the slope at the point $x = 0.6$ V, $y = 2$ mA is found by computing the ratio of the change in current to the change in voltage as defined by the small triangle shown in the figure. The change in current, ΔI, is 3.4 mA − 1.2 mA = 2.2 mA and the change in voltage, ΔV, is 0.66 V − 0.54 V = 0.12 V. The ratio of $\Delta I/\Delta V$ is 2.2 mA/0.12 V = 18.3 mS. This represents the conductance at the specified point. The internal ac resistance is the reciprocal of this value:

$$r = \frac{1}{g} = \frac{1}{18.3 \text{ mS}} = 54.5 \text{ } \Omega$$

Conventional Current Versus Electron Flow

From your dc/ac circuits course, you know that current is the rate of flow of charge. The original definition of current was based on Benjamin Franklin's belief that electricity was an unseen substance that moved from positive to negative. *Conventional current* assumes for analysis purposes that current is out of the positive terminal of a voltage source, through the circuit, and into the negative terminal of the source. Engineers use this definition and many textbooks show current with arrows drawn with this viewpoint.

Today, it is known that in metallic conductors, the moving charge is actually negatively charged electrons. Electrons move from the negative to the positive point, just opposite to the defined direction of conventional current. The movement of electrons in a conductor is called *electron flow.* Many schools and textbooks show electron flow with current arrows drawn out of the negative terminal of a voltage source.

Unfortunately, the controversy between whether it is better to show conventional current or electron flow in representing circuit behavior has continued for many years and does not appear to be subsiding. It is not important which direction you use to form a mental picture of current. In practice, there is only one correct direction to connect a dc ammeter to make current measurements. Throughout this text, the proper polarity for dc meters is shown when appropriate. To avoid confusion, current paths are indicated with special meter symbols. In a given circuit, larger or smaller currents are indicated by the relative position of the meter needles.

1–1 REVIEW QUESTIONS

1. What is a characteristic curve for a component?
2. How does the characteristic curve for a large resistor compare to the curve for a smaller resistor?
3. What is the difference between dc resistance and ac resistance?

1–2 ■ ANALOG SIGNALS

A signal is any physical quantity that carries information. It can be an audible, visual, or other indication of information. In electronics, the term signal refers to the information that is carried by electrical waves, either in a conductor or as an electromagnetic field.

After completing this section, you should be able to

❑ Describe analog signals
  ❑ Compare an analog signal with a digital signal
  ❑ Define *sampling* and *quantizing*
  ❑ Apply the equation for a sinusoidal wave to find the instantaneous value of a voltage or current
  ❑ Find the peak, rms, or average value, given the equation for a sinusoidal wave
  ❑ Explain the difference between the time domain signal and the frequency-domain signal

Analog and Digital Signals

Signals can be classified as either continuous or discrete. A continuous signal changes smoothly, without interruption. A discrete signal can have only certain values. The terms *continuous* and *discrete* can be applied either to the amplitude or to the time characteristic of a signal.

In nature, most signals take on a continuous range of values within limits; such signals are referred to as **analog signals.** For example, consider a potentiometer that is used as a shaft encoder as shown in Figure 1–3(a). The output voltage can be continuously varied within the limit of the supply voltage, resulting in an analog signal that is related to the angular position of the shaft.

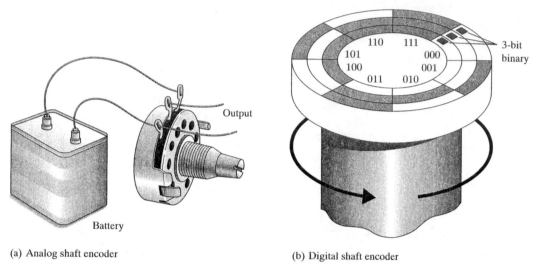

(a) Analog shaft encoder   (b) Digital shaft encoder

FIGURE 1–3
Analog and digital shaft encoders.

On the other hand, another type of encoder has a certain number of steps that can be selected as shown in Figure 1–3(b). When numbers are assigned to these steps, the result is called a **digital signal.**

Analog circuits are generally simple, have high speed and low cost, and can readily simulate natural phenomena. They are often used for operations such as performing linearizing functions, waveshaping, transforming voltage to current or current to voltage, multiplying, and mixing. By contrast, digital circuits have high noise immunity, no drift, and the ability to process data rapidly and to perform various calculations. In many electronic systems, a mix of analog and digital signals are required to optimize the overall system's performance or cost.

Many signals have their origin in a natural phenomenon such as a measurement of pressure or temperature. Transducer outputs are typically analog in nature; a microphone, for example, provides an analog signal to an amplifier. Frequently, the analog signal is converted to digital form for storing, processing, or transmitting.

Conversion from analog to digital form is accomplished by a two-step process: sampling and quantizing. **Sampling** is the process of breaking the analog waveform into time "slices" that approximate the original wave. This process always loses some information; however, the advantages of digital systems (noise reduction, digital storage, and processing)

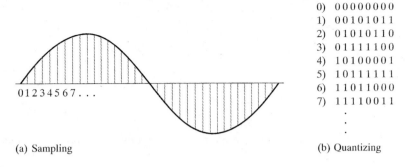

0) 00000000
1) 00101011
2) 01010110
3) 01111100
4) 10100001
5) 10111111
6) 11011000
7) 11110011
.
.
.

01234567...

(a) Sampling   (b) Quantizing

FIGURE 1–4
Digitizing an analog waveform.

outweigh the disadvantages. After sampling, the time slices are assigned a numeric value. This process, called **quantizing,** produces numbers that can be processed by digital computers or other digital circuits. Figure 1–4 illustrates the sampling and quantizing process.

Frequently, digital signals need to be converted back to their original analog form to be useful in their final application. For instance, the digitized sound on a CD must be converted to an analog signal and eventually back to sound by a loudspeaker.

Periodic Signals

To carry information, some property such as voltage or frequency of an electrical wave needs to vary. Frequently, an electrical signal repeats at a regular interval of time. Repeating waveforms are said to be **periodic.** The **period** *(T)* represents the time for a periodic wave to complete one cycle. A **cycle** is the complete sequence of values that a waveform exhibits before another identical pattern occurs. The period can be measured between any two corresponding points on successive cycles.

Periodic waveshapes are used extensively in electronics. Many practical electronic circuits such as oscillators generate periodic waves. Most oscillators are designed to produce a particular shaped waveform—either a sinusoidal wave or nonsinusoidal waves such as the square, rectangular, triangle, and sawtooth waves.

The most basic and important periodic waveform is the sinusoidal wave. Both the trigonometric sine and cosine functions have the shape of a sinusoidal wave. The term *sine wave* usually implies the trigonometric function, whereas the term *sinusoidal wave* means a waveform with the shape of a sine wave. A sinusoidal waveform is generated as the natural waveform from many ac generators and in radio waves. Sinusoidal waves are also present in physical phenomena from generation of laser light, the vibration of a tuning fork, or the motion of ocean waves.

A **vector** is any quantity that has both magnitude and direction. A sinusoidal curve can be generated by plotting the projection of the end point of a rotating vector that is turning with uniform circular motion, as illustrated in Figure 1–5. Successive revolutions of the point generate a periodic curve which can be expressed mathematically as

$$y(t) = A \sin(\omega t \pm \phi) \tag{1–2}$$

where $y(t)$ = vertical displacement of a point on the curve from the horizontal axis. The bracketed quantity (t) is an optional indicator, called *functional notation,*

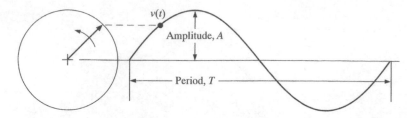

FIGURE 1–5
Generation of a sinusoidal waveform from the projection of a rotating vector.

to emphasize that the signals vary with time. Functional notation is frequently omitted when it isn't important to emphasize the time relationship but is introduced to familiarize you with the concept when it is shown.

A = amplitude. This is the maximum displacement from the horizontal axis.

ω = angular frequency of the rotating vector in radians per second.

t = time in seconds to a point on the curve.

ϕ = phase angle in radians. The **phase angle** is simply a fraction of a cycle that a waveform is shifted from a reference waveform of the same frequency. It is positive if the curve begins before $t = 0$ and is negative if the curve starts after $t = 0$.

Equation (1–2) illustrates that the sinusoidal wave can be defined in terms of three basic parameters. These are the frequency, amplitude, and phase angle.

Frequency and Period When the rotating vector has made one complete cycle, it has rotated through 2π radians. The number of complete cycles generated per second is called the **frequency.** Dividing the angular frequency (ω, in rad/s) of the rotating vector by the number of radians in one cycle (2π rad/cycle) gives the frequency in hertz.[1]

$$f\,(\text{Hz}) = \frac{\omega\,(\text{rad/s})}{2\pi\,(\text{rad/cycle})} \qquad (1\text{–}3)$$

One cycle per second is equal to 1 Hz. The frequency (f) of a periodic wave is the number of cycles in one second and the period (T) is the time for one cycle, so it is logical that the reciprocal of the frequency is the period and the reciprocal of the period is the frequency.

$$T = \frac{1}{f} \qquad (1\text{–}4)$$

and

$$f = \frac{1}{T} \qquad (1\text{–}5)$$

---

[1] The unit of frequency was cycles per second (cps) prior to 1960 but was renamed the hertz (abbreviated Hz) in honor of Heinrich Hertz, a German physicist who demonstrated radio waves. The old unit designation was more descriptive of the definition of frequency.

For example, if a signal repeats every 10 ms, then its period is 10 ms and its frequency is

$$f = \frac{1}{T} = \frac{1}{10 \text{ ms}} = 0.1 \text{ kHz}$$

Instantaneous Value of a Sinusoidal Wave   If the sinusoidal waveform shown in Figure 1–5 represents a voltage, Equation (1–2) is written

$$v(t) = V_p \sin(\omega t \pm \phi)$$

In this equation, $v(t)$ is a variable that represents the voltage. Since it changes as a function of time, it is often referred to as the *instantaneous voltage.*

Peak Value of a Sinusoidal Wave   The amplitude of a sinusoidal wave is the maximum displacement from the horizontal axis as shown in Figure 1–5. For a voltage waveform, the amplitude is called the peak voltage, V_p. When making voltage measurements with an oscilloscope, it is easier to measure the peak-to-peak voltage, V_{pp}. The peak-to-peak voltage is twice the peak value.

Average Value of a Sinusoidal Wave   During one cycle, a sinusoidal waveform has equal positive and negative excursions. Therefore, the mathematical definition of the average value of a sinusoidal waveform must be zero. However, the term *average value* is generally used to mean the average over a cycle without regard to the sign. That is, the average is usually computed by converting all negative values to positive values, then averaging. The average voltage is defined in terms of the peak voltage by the following equation:

$$V_{avg} = \frac{2V_p}{\pi}$$

Simplifying,

$$V_{avg} = 0.637V_p \qquad\qquad \textbf{(1–6)}$$

The average value is useful in certain practical problems. For example, if a rectified sinusoidal waveform is used to deposit material in an electroplating operation, the quantity of material deposited is related to the average current.

Effective Value (rms Value) of a Sinusoidal Wave   If you apply a dc voltage to a resistor, a steady amount of power is dissipated in the resistor and can be calculated using the following power law:

$$P = IV \qquad\qquad \textbf{(1–7)}$$

where V = dc voltage across the resistor (volts)
$\quad I$ = dc current in the resistor (amperes)
$\quad P$ = power dissipated (watts)

A sinusoidal waveform transfers maximum power at the peak excursions of the curve and no power at all at the instant the voltage crosses zero. In order to compare ac and dc voltages and currents, ac voltages and currents are defined in terms of the equivalent heating value of dc. This equivalent heating value is computed with calculus, and the result is called the rms (for *root-mean-square*) voltage or current. The rms voltage is related to the peak voltage by the following equation:

$$V_{rms} = 0.707V_p \qquad \qquad (1\text{--}8)$$

Likewise, the effective or rms current is

$$I_{rms} = 0.707I_p$$

EXAMPLE 1–2

A certain voltage waveform is described by the following equation:

$$v(t) = 15 \sin(600t)$$

(a) From this equation, determine the peak voltage and the average voltage. Give the angular frequency in rad/s.

(b) Find the instantaneous voltage at a time of 10 ms.

Solution

(a) The form of the equation is

$$y(t) = A \sin(\omega t)$$

The peak voltage is the same as the amplitude (A).

$$V_p = \textbf{15 V}$$

The average voltage is related to the peak voltage.

$$V_{avg} = 0.637V_p = 0.637(15 \text{ V}) = \textbf{9.56 V}$$

The radian frequency, ω, is **600 rad/s.**

(b) The instantaneous voltage at a time of 10 ms is

$$v(t) = 15 \sin(600t) = 15 \sin(600)(10 \text{ ms}) = \textbf{--4.19 V}$$

Note the negative value indicates that the waveform is below the axis at this point.

Practice Exercise Find the rms voltage, the frequency in hertz, and the period of the waveform described in the example.

Time-Domain Signals

Thus far, the signals you have looked at vary with time, and it is natural to associate time as the independent variable. Some instruments, such as the oscilloscope, are designed to record signals as a function of time. Time is therefore the independent variable. The values assigned to the independent variable are called the **domain.** Signals that have voltage, current, resistance, or other quantity vary as a function of time are called *time-domain* signals.

Frequency-Domain Signals

Sometimes it is useful to view a signal where frequency is represented on the horizontal axis and the signal amplitude (usually in logarithmic form) is plotted along the vertical axis. Since frequency is the independent variable, the instrument works in the *frequency domain,* and the plot of amplitude versus frequency is called a **spectrum.** The spectrum analyzer is an instrument used to view the spectrum of a signal. These instruments are extremely useful in radio frequency (RF) measurements for analyzing the frequency response of a circuit, testing for harmonic distortion, checking the percent modulation from transmitters, and many other applications.

You have seen how the sinusoidal wave can be defined in terms of three basic parameters. These are the amplitude, frequency, and phase angle. A continuous sinusoidal wave can be shown as a time-varying signal defined by these three parameters. The same sinusoidal wave can also be shown as a single line on a frequency spectrum. The frequency-domain representation gives information about the amplitude and frequency, but it does not show the phase angle. These two representations of a sinusoidal wave are compared in Figure 1–6. The height of the line on the spectrum is the amplitude of the sinusoidal wave.

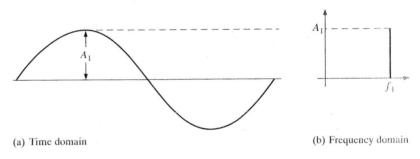

(a) Time domain

(b) Frequency domain

FIGURE 1–6
Time-domain and frequency-domain representations of a sinusoidal wave.

Harmonics   A nonsinusoidal repetitive waveform is composed of a fundamental frequency and harmonic frequencies. The fundamental frequency is the basic repetition rate of the waveform, and the **harmonics** are higher-frequency sinusoidal waves that are multiples of the fundamental. Interestingly, these multiples are all related to the fundamental by integers (whole numbers).

Odd harmonics are frequencies that are odd multiples of the fundamental frequency of a waveform. For example, a 1 kHz square wave consists of a fundamental of 1 kHz and odd harmonics of 3 kHz, 5 kHz, 7 kHz, and so on. The 3 kHz frequency in this case is called the third harmonic, the 5 kHz frequency is called the fifth harmonic, and so on.

Even harmonics are frequencies that are even multiples of the fundamental frequency. For example, if a certain wave has a fundamental of 200 Hz, the second harmonic is 400 Hz, the fourth harmonic is 800 Hz, and the sixth harmonic is 1200 Hz.

Any variation from a pure sinusoidal wave produces harmonics. A nonsinusoidal wave is a composite of the fundamental and certain harmonics. Some types of waveforms have only odd harmonics, some have only even harmonics, and some contain both. The shape of the wave is determined by its harmonic content. Generally, only the fundamental and the first few harmonics are important in determining the waveshape. For example, a square wave is formed from the fundamental and odd harmonics, as illustrated in Figure 1–7.

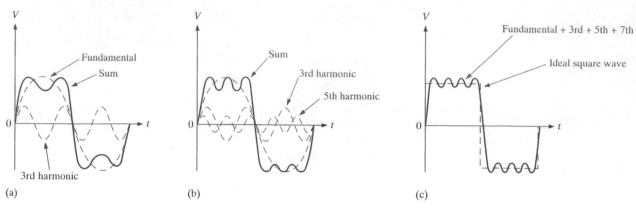

FIGURE 1–7
Odd harmonics combine to produce a square wave.

Fourier Series All periodic waves except the sinusoidal wave itself are complex waveforms composed of a series of sinusoidal waves. Jean Fourier, a French mathematician interested in problems of heat conduction, formed a mathematical series of trigonometry terms to describe periodic waves. This series is appropriately called the Fourier series.[2] With the Fourier series, one can mathematically determine the amplitude of each of the sinusoidal waves that compose a complex waveform.

The frequency spectrum developed by Fourier is often shown as an amplitude spectrum with units of voltage or power plotted on the *y*-axis plotted against Hz on the *x*-axis. Figure 1–8(a) illustrates the amplitude spectrum for several different periodic waveforms. Notice that all spectrums for periodic waves are depicted as lines located at harmonics of the fundamental frequency. These individual frequencies can be measured with a spectrum analyzer.

Nonperiodic signals such as speech, or other transient waveforms, can also be represented by a spectrum; however, the spectrum is no longer a series of lines as in the case of repetitive waves. Transient waveforms are computed by another method called the *Fourier transform*. The spectrum of a transient waveform contains a continuum of frequencies rather than just harmonically related components. A representative Fourier pair of signals for a nonrepetitive pulse are shown in Figure 1–8(b).

1–2 REVIEW QUESTIONS

1. What is the difference between an analog signal and a digital signal?

2. Describe the spectrum for a square wave.

3. How does the spectrum for a repetitive waveform differ from that of a nonrepetitive waveform?

---

[2] Although Fourier's work was significant and he was awarded a prize, his colleagues were uneasy about it. The famous mathematician, Legrange, argued in the French Academy of Science that Fourier's claim was impossible. For further information, see Scientific American, June 1989, p. 86.

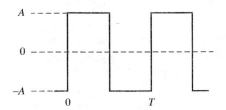
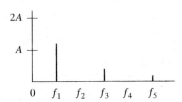

Square

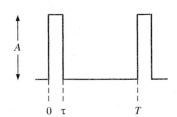
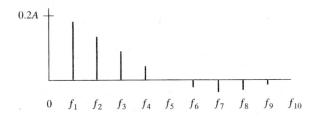

Rectangle

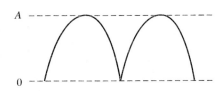
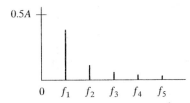

Full-wave rectified sinusoid

(a) Examples of time-domain and frequency-domain representations of repetitive waves

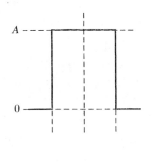
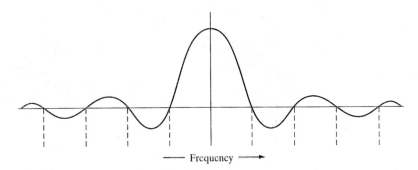

Time-domain representation   Frequency-domain representation

(b) Examples of the frequency spectrum of a nonrepetitive pulse waveform

FIGURE 1–8
Comparison of the frequency spectrum of repetitive and nonrepetitive waves.

1–3 ■ SIGNAL SOURCES

You may recall from basic electronics that Thevenin's theorem allows you to replace a complicated linear circuit with a single voltage source and a series resistance. The circuit is viewed from the standpoint of two output terminals. Likewise, Norton's theorem allows you to replace a complicated two-terminal, linear circuit with a single current source and a parallel resistance. These important theorems are useful for simplifying the analysis of a wide variety of circuits and should be thoroughly understood.

After completing this section, you should be able to

❑ Analyze signal sources
    ❑ Define two types of independent sources
    ❑ Draw a Thevenin or Norton equivalent circuit for a dc resistive circuit
    ❑ Show how to draw a load line for a Thevenin circuit
    ❑ Explain the meaning of Q-point
    ❑ Explain how a passive transducer can be modeled with a Thevenin equivalent circuit

Independent Sources

Signal sources can be defined in terms of either voltage or current and may be defined for either dc or ac. An ideal independent voltage source generates a voltage which does not depend on the load current. An ideal independent current source produces a current in the load which does not depend on the voltage across the load.

The value of an ideal independent source can be specified without regard to any other circuit parameter. Although a truly ideal source cannot be realized, in some cases, (such as a regulated power supply), it can be closely approximated. Actual sources can be modeled as consisting of an ideal source and a resistor (or other passive component for ac sources).

Thevenin's Theorem

Thevenin's theorem allows you to replace a complicated, two-terminal linear circuit with an ideal independent voltage source and a series resistance as illustrated in Figure 1–9. The source can be either a dc or ac source (a dc source is shown). **Thevenin's theorem** provides an equivalent circuit from the standpoint of the two output terminals. That is, the original circuit and the Thevenin circuit will produce exactly the same voltage and current in any load. Thevenin's theorem is particularly useful for analysis of linear circuits such as amplifiers, a topic that will be covered in Section 1–4.

FIGURE 1–9
Thevenin's equivalent for a dc circuit.

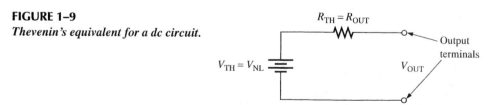

Only two quantities are needed to determine the equivalent Thevenin circuit—the Thevenin voltage and the Thevenin resistance. The Thevenin voltage, V_{TH}, is the open circuit (no load, NL) voltage from the original circuit. The Thevenin resistance, R_{TH}, is the resistance from the point of view of the output terminals with all voltage or current sources replaced by their internal resistance.

EXAMPLE 1–3

Find the equivalent Thevenin circuit for the dc circuit shown in Figure 1–10(a). The output terminals are represented by the open circles.

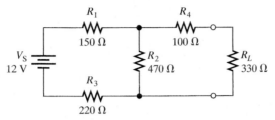

(a) Original circuit with load resistor, R_L

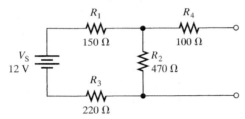 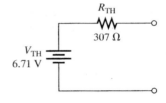

(b) Original circuit with load resistor, R_L, removed   (c) Thevenin equivalent of original circuit

FIGURE 1–10
Simplifying a circuit with Thevenin's theorem.

Solution   Find the Thevenin voltage by computing the voltage on the output terminals *as if the load were removed* as shown in Figure 1–10(b). With no load, there is no path for current through R_4. Therefore, there is no current and no voltage drop will appear across it. The output (Thevenin) voltage must be the same as the drop across R_2. Applying the voltage-divider rule for the equivalent series combination of R_1, R_2, and R_3, the voltage across R_2 is

$$V_{TH} = V_2 = V_S\left(\frac{R_2}{R_1 + R_2 + R_3}\right)$$

$$= 12\ V\left(\frac{470\ \Omega}{150\ \Omega + 470\ \Omega + 220\ \Omega}\right) = \textbf{6.71 V}$$

The Thevenin resistance is the resistance from the perspective of the output terminals with sources replaced with their internal resistance. The internal resistance of a voltage source is zero (ideally). Replacing the source with zero resistance places R_1 and R_3 in series and the combination in parallel with R_2. The equivalent resistance of these three resistors is in series with R_4. Thus, the Thevenin resistance for this circuit is

$$R_{TH} = [(R_1 + R_3) \parallel R_2] + R_4$$
$$= [(150\ \Omega + 220\ \Omega) \parallel 470\ \Omega] + 100\ \Omega = \textbf{307}\ \Omega$$

The Thevenin circuit is shown in Figure 1–10(c).

Practice Exercise   Use the Thevenin circuit to find the voltage across the 330 Ω load resistor.

Thevenin's theorem is a useful way of combining linear circuit elements to form an equivalent circuit that can be used to answer questions with respect to various loads. The requirement that the Thevenin circuit elements are linear places some restrictions on the use of Thevenin's theorem. In spite of this, if the circuit to be replaced is approximately linear, Thevenin's theorem may produce useful results. This is the case for many amplifier circuits that we will investigate later.

Norton's Theorem

Norton's theorem provides another equivalent circuit similar to the Thevenin equivalent circuit. Norton's equivalent circuit can also replace any two-terminal linear circuit with a reduced equivalent. Instead of a voltage source, the Norton equivalent circuit uses a current source in parallel with a resistance, as illustrated in Figure 1–11.

FIGURE 1–11
Norton circuit. The arrow in the current source symbol always points to the positive side of the source.

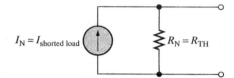

$I_N = I_{shorted\ load}$ $R_N = R_{TH}$

The magnitude of the Norton current source is found by replacing the load with a short and determining the current in the load. The Norton resistance is the same as the Thevenin resistance.

EXAMPLE 1–4   Find the equivalent Norton circuit for the dc circuit shown in Figure 1–12(a). The output terminals are represented by the open circles.

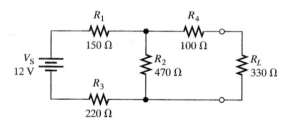

(a) Original circuit

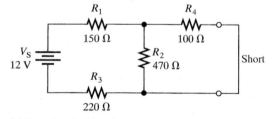

(b) R_L replaced with a short

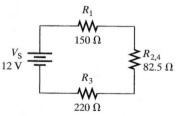

(c) R_2 and R_4 form an equivalent parallel resistor.

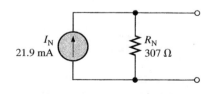

(d) The current in the short is equal to the Norton current.

FIGURE 1–12
Simplifying a circuit with Norton's theorem.

Solution   Find the Norton current by computing the current in the load *as if it were replaced by a short* as shown in Figure 1–12(b). The shorted load causes R_4 to appear in parallel with R_2 as shown in Figure 1–12(b). The total current in the equivalent circuit of Figure 1–12(c) can be found by applying Ohm's law to the total resistance.

$$I = \frac{V_S}{R_T} = \frac{12 \text{ V}}{R_1 + R_{2,4} + R_3} = \frac{12.0 \text{ V}}{452.5 \text{ } \Omega} = 26.5 \text{ mA}$$

The current (I_{SL}) in the shorted load is found by applying the current-divider rule to the R_2 and R_4 junction in the circuit of Figure 1–12(b).

$$I_{SL} = I_T \left(\frac{R_2}{R_2 + R_4} \right) = 26.5 \text{ mA} \left(\frac{470 \text{ } \Omega}{470 \text{ } \Omega + 100 \text{ } \Omega} \right) = \textbf{21.9 mA}$$

The current in the shorted load is the Norton current. The Norton resistance is equal to the Thevenin resistance, as found in Example 1–3. Notice that the Norton resistance is in parallel with the Norton current source. The equivalent circuit is shown in Figure 1–12(d).

Practice Exercise   Use Norton's theorem to find the voltage across the 330 Ω load resistor. Show that Norton's theorem gives the same result as Thevenin's theorem for this circuit (see Practice Exercise in Example 1–3).

Load Lines

An interesting way to obtain a "conceptual picture" of circuit operation is through the use of a load line for the circuit. Imagine a linear circuit that has an equivalent Thevenin circuit as shown in Figure 1–13. Let's see what happens if various loads are placed across the output terminals. First, assume there is a short (zero resistance). In this case, the voltage across the load is zero and the current is given by Ohm's law.

$$I_L = \frac{V_{TH}}{R_{TH}} = \frac{10 \text{ V}}{1.0 \text{ k}\Omega} = 10 \text{ mA}$$

Now assume the load is an open (infinite resistance). In this case, the load current is zero, and the voltage across the load is equal to the Thevenin voltage.

FIGURE 1–13

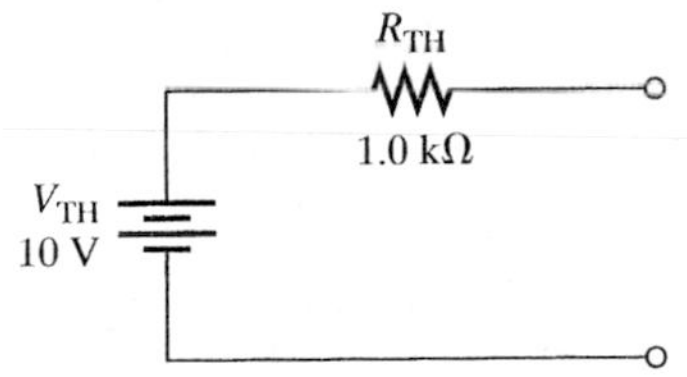

The two tested conditions represent the maximum and minimum current in the load. Table 1–1 shows the results of trying some more points to see what happens with different loads. Plotting the data as shown in Figure 1–14 establishes an *IV* curve for the Thevenin circuit. Because the circuit is a linear circuit, *any load that is placed across the output terminals falls onto the same straight line.* This line is called the **load line** for the circuit

TABLE 1–1
Various load conditions for the circuit in Figure 1–13.

R_L	V_L	I_L
0 Ω	0.0 V	10.0 mA
250 Ω	2.00 V	8.00 mA
500 Ω	3.33 V	6.67 mA
750 Ω	4.29 V	5.71 mA
1.0 kΩ	5.00 V	5.00 mA
2.0 kΩ	6.67 V	3.33 mA
4.0 kΩ	8.00 V	2.00 mA
open	10.0 V	0.00 mA

FIGURE 1–14
Load line for the circuit in Figure 1–13.

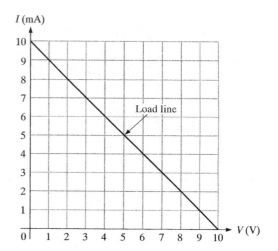

and describes the driving circuit (in this case, the Thevenin circuit), not the load itself. Since the load line is a straight line, the first two calculated conditions (a short and an open load) are all that are needed to establish it.

Before we leave the topic of load lines, consider one more idea. Recall that a resistor (or any other device) has its own characteristic that can be described by its *IV* curve. The characteristic curve for a resistor represents all of the possible operating points for the device, whereas the load line represents all of the possible operating points for the circuit. Combining these ideas, you can superimpose the *IV* curve for a resistor on the plot of the load line for the Thevenin circuit. The intersection of these two lines gives the operating point for the combination.

Figure 1–15(a) shows an 800 Ω load resistor added to the Thevenin circuit from Figure 1–13. The load line for the Thevenin circuit and the characteristic curve for resistor R_1 from Figure 1–1 are shown in Figure 1–15(b). R_1 now serves as a load resistor, R_L. The intersection of the two lines represents the operating point, or **quiescent point,** commonly referred to as the Q-point. Note that the load voltage (4.4 V) and load current (5.6 mA) can be read directly from the graph. This idea is frequently applied to circuits with discrete transistors and other devices to give a graphical tool for understanding circuit operation.

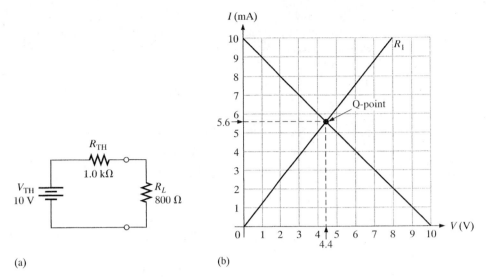

FIGURE 1–15
Load line and a resistor IV curve showing the Q-point.

Transducers

Analog circuits are frequently used in conjunction with a measurement that needs to be made. A **transducer** is a device that converts a physical quantity (such as position, pressure, or temperature) from one form to another; for electronic systems, input transducers convert a physical quantity to be measured into an electrical quantity (voltage, current, resistance). Transducers will be covered further in Chapter 11.

The signal from transducers is frequently very small, requiring amplification before being suitable for further processing. Passive transducers, such as strain gages, require a separate source of electrical power (called *excitation*) to perform their job. Others, such as thermocouples, are active transducers; they are self-generating devices that convert a small portion of the quantity to be measured into an electrical signal. Both passive and active transducers are often simplified to a Thevenin or Norton equivalent circuit for analysis.

In order to choose an appropriate amplifier, it is necessary to consider both the size of the source voltage and the size of the equivalent Thevenin or Norton resistance. When the equivalent resistance is very small, Thevenin's equivalent circuit is generally more useful because the circuit approximates an ideal voltage source. When the equivalent resistance is large, Norton's theorem is generally more useful because the circuit approximates an ideal current source. When the source resistance is very high, such as the case with a pH meter, a very high input impedance amplifier must be used. Other considerations, such as the frequency response of the system or the presence of noise, affect the choice of amplifier.

EXAMPLE 1–5

A piezoelectric crystal is used in a vibration monitor. Assume the output of the transducer should be a 60 mV rms sine wave with no load. When a technician connects an oscilloscope with a 10 MΩ input impedance across the output, the voltage is observed to be only 40 mV rms. Based on these observations, draw the Thevenin equivalent circuit for this transducer.

Solution The open circuit ac voltage is the Thevenin voltage; thus, $V_{th} = \textbf{60 mV}$. The Thevenin resistance can be found indirectly using the voltage-divider rule. The oscilloscope input impedance is considered the load resistance, R_L, in this case. The voltage across the load is

$$V_{R_L} = V_{th}\left(\frac{R_L}{R_L + R_{th}}\right)$$

Rearranging terms,

$$\frac{R_L + R_{th}}{R_L} = \frac{V_{th}}{V_{R_L}}$$

Now solving for R_{th} and substituting the given values,

$$R_{th} = R_L\left(\frac{V_{th}}{V_{R_L}} - 1\right) = 10 \text{ M}\Omega\left(\frac{60 \text{ mV}}{40 \text{ mV}} - 1\right) = \textbf{5.0 M}\boldsymbol{\Omega}$$

The equivalent transducer circuit is shown in Figure 1–16.

FIGURE 1–16

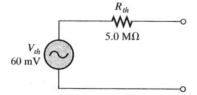

Practice Exercise Draw the Norton's equivalent circuit for the same transducer.

1–3 REVIEW QUESTIONS

1. What is an independent source?
2. What is the difference between a Thevenin and a Norton circuit?
3. What is the difference between a passive and an active transducer?

1–4 ■ AMPLIFIERS

Before processing, most signals require amplification. Amplification is simply increasing the magnitude of a signal (either voltage, current, or both) and is one of the most important operations in electronics. Other operations in the field of linear electronics include signal generation (oscillators), waveshaping, frequency conversion, modulation, and many other processes. In addition to strictly linear or strictly digital circuits, many electronic circuits involve a combination of digital and linear electronics. These include an important class of interfacing circuits that convert analog-to-digital and digital-to-analog. These circuits will be considered in Chapter 10.

After completing this section, you should be able to

❑ Explain the characteristics of an amplifier
  ❑ Write the equations for voltage gain and power gain
  ❑ Draw the transfer curve for an amplifier
  ❑ Show how an amplifier can be modeled as Thevenin or Norton equivalent circuits to represent the input circuit and the output circuit
  ❑ Describe how an amplifier can be formed by cascading stages
  ❑ Determine the loading effect of one amplifier stage on another
  ❑ Use a calculator to find the logarithm or antilog of a given number
  ❑ Compute decibel voltage and power gain for an amplifier or circuit

Linear Amplifiers

Linear circuits were discussed previously and that discussion can be extended to **amplifiers.** Linear amplifiers produce a magnified replica (**amplification**) of the input signal in order to produce a useful outcome (such as driving a loudspeaker). The concept of an *ideal amplifier* means that the amplifier introduces no noise or distortion to the signal; the output varies in time and replicates the input exactly. The ideas presented here apply to all amplifiers, including those made with discrete components and those made from integrated circuits.

Amplifiers are designed primarily to amplify either voltage or power. For a voltage amplifier, the output signal, $V_{out}(t)$, is proportional to the input signal, $V_{in}(t)$ and the ratio of output voltage to input voltage is voltage **gain.** To simplify the gain equation, you can omit the functional notation, (t), and simply show the ratio of the output signal voltage to the input signal voltage as

$$A_v = \frac{V_{out}}{V_{in}} \qquad (1-9)$$

where A_v = voltage gain
   V_{out} = output signal voltage
   V_{in} = input signal voltage

A useful way of looking at any circuit is to show the output for a given input. This plot, called a **transfer curve,** shows the response of the circuit. An ideal amplifier is characterized by a straight line that goes to infinity. For an actual linear amplifier, the transfer curve is a straight line until saturation is reached as shown in Figure 1–17. From this plot, the output voltage can be read for a given input voltage.

All amplifiers have certain limits, beyond which they no longer act as ideal. The output of the amplifier illustrated in Figure 1–17 eventually cannot follow the input; at this point the amplifier is no longer linear. Additionally, all amplifiers must operate from a source of energy, usually in the form of a dc power supply. Essentially, amplifiers convert some of this dc energy from the power supply into signal power. Thus, the output signal has larger power than the input signal. Frequently, block diagrams and other circuit representations omit the power supply, but it is understood to be present.

FIGURE 1–17
Transfer curve for a linear amplifier.

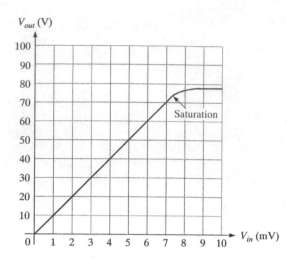

The Nonlinear Amplifier

Amplifiers are frequently used in situations where the output is not intended to be a replica of the input. These amplifiers form an important part of the field of analog electronics. They include two main categories: waveshaping and switching. A *waveshaping amplifier* is used to change the shape of a waveform. A *switching amplifier* produces a rectangular output from some other waveform. The input can be any waveform, for example, sinusoidal, triangle, or sawtooth. The rectangular output wave is generally used as a control signal for some digital applications.

EXAMPLE 1–6   The input and output signals for a linear amplifier are shown in Figure 1–18 and represent an oscilloscope display. What is the voltage gain of the amplifier?

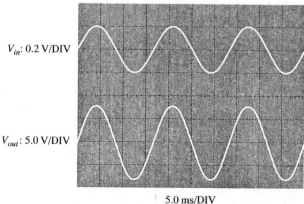

V_{in}: 0.2 V/DIV

V_{out}: 5.0 V/DIV

5.0 ms/DIV

FIGURE 1–18
Oscilloscope display.

Solution   The input signal is 2.0 divisions from peak-to peak.

$$V_{in} = 2.0 \text{ DIV} \times 0.2 \text{ V/DIV} = 0.4 \text{ V}$$

The output signal is 3.2 divisions from peak-to peak.

$$V_{out} = 3.2 \text{ DIV} \times 5.0 \text{ V/DIV} = 16 \text{ V}$$

$$A_v = \frac{V_{out}}{V_{in}} = \frac{16 \text{ V}}{0.4 \text{ V}} = \textbf{40}$$

Note that voltage gain is a ratio of voltages and therefore has no units. The answer is the same if rms or peak values had been used for both the input and output voltages.

Practice Exercise   The input to an amplifier is 20 mV. If the voltage gain is 300, what is the output signal?

Another gain parameter is power gain, A_p, defined as the ratio of the signal power out to the signal power in. Power is computed using rms values of voltage or current; however, power gain is a ratio so you can use any consistent units. Power gain, shown as a function of time, is given by the following equation:

$$A_p = \frac{P_{out}}{P_{in}} \qquad\qquad (1\text{--}10)$$

where A_p = power gain
P_{out} = power out
P_{in} = power in

Power can be expressed by any of the standard power relationships studied in basic electronics. For instance, given the voltage and current of the input and output signals, the power gain can be written

$$A_p = \frac{I_{out}V_{out}}{I_{in}V_{in}}$$

where I_{out} = output signal current to the load
I_{in} = input signal current

Power gain can also be expressed by substituting $P = V^2/R$ for the input and output power.

$$A_p = \left(\frac{V_{out}^2/R_L}{V_{in}^2/R_{in}}\right)$$

where R_L = load resistor
R_{in} = input resistance of the amplifier

The particular equation you choose depends on what you know.

Amplifier Model

An amplifier is a device that increases the magnitude of a signal for use by a load. Although op-amps and other amplifiers are complicated arrangements of transistors, resistors, and other components, a simplified description is all that is necessary when the requirement is to analyze the source and load behavior. The amplifier can be thought of as the interface between the

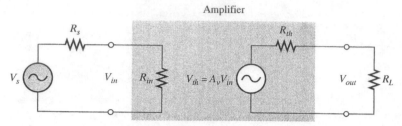

(a) Thevenin output circuit

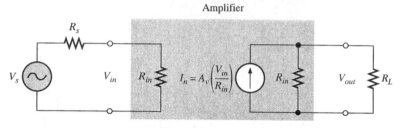

(b) Norton output circuit

FIGURE 1–19
Basic amplifier models showing the equivalent input resistance and dependent output circuits.

source and load, as shown in Figure 1–19(a) and 1–19(b). The concept of equivalent circuits, learned in basic electronics courses, can be applied to the more complicated case of an amplifier. By drawing an amplifier as an equivalent circuit, you can simplify equations related to its performance.

The input signal from a source is applied to the input terminals of the amplifier, and the output is taken from a second set of terminals. (Terminals are represented by circles on a schematic.) The amplifier's input terminals present an input resistance, R_{in}, to the source. This input resistance affects the input voltage to the amplifier because it forms a voltage divider with the source resistance.

The output of the amplifier can be drawn as either a Thevenin or Norton source, as shown in Figure 1–19. The magnitude of this source is dependent on the unloaded gain (A_v) and the input voltage; thus, the amplifier's output circuit (drawn as a Thevenin or Norton equivalent) is said to contain a *dependent* source. The value of a dependent source always depends on voltage or current elsewhere in the circuit.[3] The voltage or current values for the Thevenin and Norton cases are shown in Figure 1–19.

Cascaded Stages

The Thevenin and Norton models reduce an amplifier to its "bare-bones" for analysis purposes. In addition to considering the simplified model for source and load effects, the simplified model is also useful to analyze the internal loading when two or more stages are cascaded to form a single amplifier. Consider two stages cascaded as shown in Figure 1–20. The overall gain is affected by loading effects from each of the three loops. The loops are simple series circuits, so voltages can easily be calculated with the voltage-divider rule.

---

[3] The relationship between the dependent source and its reference cannot be broken. The superposition theorem, which allows sources to be treated separately, does not apply to dependent sources.

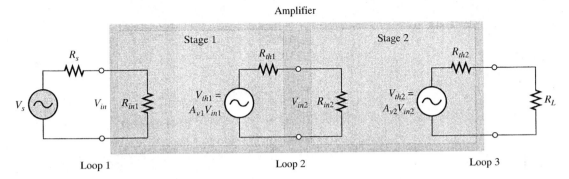

FIGURE 1–20
Cascaded stages in an amplifier.

EXAMPLE 1–7   Assume a transducer with a Thevenin (unloaded) source, V_s, of 10 mV and a Thevenin source resistance, R_s, of 50 kΩ is connected to a two-stage cascaded amplifier, as shown in Figure 1–21. Compute the voltage across a 1.0 kΩ load.

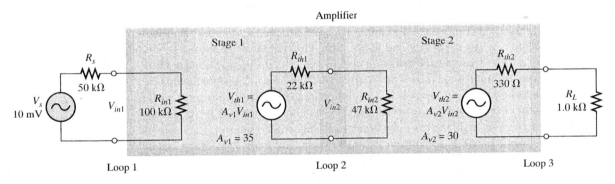

FIGURE 1–21
Two-stage cascaded amplifier.

Solution   Compute the input voltage to stage 1 from the voltage-divider rule applied to loop 1.

$$V_{in1} = V_s\left(\frac{R_{in1}}{R_{in1} + R_s}\right) = 10 \text{ mV}\left(\frac{100 \text{ k}\Omega}{100 \text{ k}\Omega + 50 \text{ k}\Omega}\right) = 6.67 \text{ mV}$$

The Thevenin voltage for stage 1 is

$$V_{th1} = A_{v1}V_{in1} = (35)(6.67 \text{ mV}) = 233 \text{ mV}$$

Compute the input voltage to stage 2 again from the voltage-divider rule, this time applied to loop 2.

$$V_{in2} = V_{th1}\left(\frac{R_{in2}}{R_{in2} + R_{th1}}\right) = 233 \text{ mV}\left(\frac{47 \text{ k}\Omega}{47 \text{ k}\Omega + 22 \text{ k}\Omega}\right) = 159 \text{ mV}$$

The Thevenin voltage for stage 2 is

$$V_{th2} = A_{v2}V_{in2} = (30)(159 \text{ mV}) = 4.77 \text{ V}$$

Apply the voltage-divider rule one more time to loop 3. The voltage across the 1.0 kΩ load is

$$V_{R_L} = V_{th2}\left(\frac{R_L}{R_L + R_{th2}}\right) = 4.77 \text{ V}\left(\frac{1.0 \text{ k}\Omega}{1.0 \text{ k}\Omega + 330 \text{ }\Omega}\right) = \mathbf{3.59 \text{ V}}$$

Practice Exercise   Assume a transducer with a Thevenin source voltage of 5.0 mV and a source resistance of 100 kΩ is connected to the same amplifier. Compute the voltage across the 1.0 kΩ load.

Logarithms

A widely used unit in electronics is the *decibel,* which is based on logarithms. Before defining the decibel, let's quickly review logarithms (sometimes called *logs*). A **logarithm** is simply an exponent. Consider the equation

$$y = b^x$$

The value of y is determined by the exponent of the base (b). The exponent, x, is said to be the logarithm of the number represented by the letter y.

Two bases are in common use—base ten and base e (discussed in mathematics courses). To distinguish the two, the abbreviation "log" is written to mean base ten, and the letters "ln" are written to mean base e. Base ten is standard for work with decibels. Thus, for base ten,

$$y = 10^x$$

Solving for x,

$$x = \log_{10}y$$

The subscript 10 can be omitted because it is implied by the abbreviation "log."

Logarithms are useful when you multiply or divide very large or small numbers. When two numbers written with exponents are multiplied, the exponents are simply added. That is,

$$10^x \times 10^y = 10^{x+y}$$

This is equivalent to writing

$$\log xy = \log x + \log y$$

This concept will be applied to problems involving multiple stages of amplification or attenuation.

EXAMPLE 1–8

(a) Determine the logarithm (base ten) for the numbers 2, 20, 200, and 2000.

(b) Find the numbers whose logarithms are 0.5, 1.5, and 2.5.

Solution

(a) Determine the logarithms by entering each number in a calculator and pressing the [log] key. The results are

$$\log 2 = \mathbf{0.30103} \qquad \log 20 = \mathbf{1.30103}$$
$$\log 200 = \mathbf{2.30103} \qquad \log 2000 = \mathbf{3.30103}$$

Notice that each factor-of-ten increase in y is an increase of 1.0 in the log.

(b) Find the number whose logarithm is a given value by entering the given value in a calculator and pressing the [10^x] function (or [INV] [log]). The results are

$$10^{0.5} = \mathbf{3.16228} \qquad 10^{1.5} = \mathbf{31.6228} \qquad 10^{2.5} = \mathbf{316.228}$$

Notice that each increase of 1 in x (the logarithm) is a factor-of-10 increase in the number.

Practice Exercise

(a) Find the logarithms for the numbers 0.04, 0.4, 4, and 40.

(b) What number has a logarithm of 4.8?

Decibel Power Ratios

Power ratios are often very large numbers. Early in the development of telephone communication systems, engineers devised the decibel as a means of describing large ratios of gain or attenuation (a signal reduction). The **decibel** (dB) is defined as 10 multiplied by the logarithmic ratio of the power gain.

$$dB = 10 \log\left(\frac{P_2}{P_1}\right) \tag{1–11}$$

where P_1 and P_2 are the two power levels being compared.

Previously, power gain was introduced and defined as the ratio of power delivered from an amplifier to the power supplied to the amplifier. To show power gain, A_p, as a decibel ratio, we use a prime in the abbreviation.

$$A_p' = 10 \log\left(\frac{P_{out}}{P_{in}}\right) \tag{1–12}$$

where A_p' = power gain expressed as a decibel ratio

P_{out} = power delivered to a load

P_{in} = power delivered to the amplifier

The decibel (dB) is a dimensionless quantity because it is a ratio. Any two power measurements with the same ratio are the same number of decibels. For example, the power ratio between 500 W and 1 W is 500:1, and the number of decibels this ratio represents is 27 dB. There is exactly the same number of decibels between 100 mW and 0.2 mW (500:1) or 27 dB. When the power ratio is less than 1, there is a power loss or **attenuation.** The decibel ratio is *positive* for power gain and *negative* for power loss.

One important power ratio is 2:1. This ratio is the defining power ratio for specifying the cutoff frequency of instruments, amplifiers, filters, and the like. By substituting into Equation (1–11), the dB equivalent of a 2:1 power ratio is

$$dB = 10 \log\left(\frac{P_2}{P_1}\right) = 10 \log\left(\frac{2}{1}\right) = 3.01 \text{ dB}$$

This result is usually rounded to 3 dB.

Since 3 dB represents a doubling of power, 6 dB represents another doubling of the original power (a power ratio of 4:1). Nine decibels represents an 8:1 ratio of power and so forth. If the ratio is reversed, the decibel result remains the same except for the sign.

$$dB = 10 \log\left(\frac{P_2}{P_1}\right) = 10 \log\left(\frac{1}{2}\right) = -3.01 \text{ dB}$$

The negative result indicates that P_2 is less than P_1.

Another useful ratio is 10:1. Since the log of 10 is 1, 10 dB equals a power ratio of 10:1. With this in mind, you can quickly estimate the overall gain (or attenuation) in certain situations. For example, if a signal is attenuated by 23 dB, it can be represented by two 10 dB attenuators and a 3 dB attenuator. Two 10 dB attenuators are a factor of 100 and another 3 dB represents another factor of 2 for an overall attenuation of 1:200.

EXAMPLE 1–9

Compute the overall power gain of the amplifier in Example 1–7. Express the answer as both power gain and decibel power gain.

Solution   The power delivered to the amplifier is

$$P_{in1} = \frac{V_{in1}^2}{R_{in1}} = \frac{(6.67 \text{ mV})^2}{100 \text{ k}\Omega} = 445 \text{ pW}$$

The power delivered to the load is

$$P_{out} = \frac{V_{R_L}^2}{R_L} = \frac{(3.59 \text{ V})^2}{1.0 \text{ k}\Omega} = 12.9 \text{ mW}$$

The power gain, A_p, is the ratio of P_{out}/P_{in1}.

$$A_p = \frac{P_{out}}{P_{in1}} = \frac{12.9 \text{ mW}}{445 \text{ pW}} = 29.0 \times 10^6$$

Expressed in dB,

$$A_p' = 10 \log 29.0 \times 10^6 = \textbf{74.6 dB}$$

29.3 [EXP] 9 [log] [×] 10 [=]

Practice Exercise   Compute the power gain (in dB) for an amplifier with an input power of 50 μW and a power delivered to the load of 4 W.

It is common in certain applications of electronics (microwave transmitters, for example) to combine several stages of gain or attenuation. When working with several stages of gain or attenuation, the total voltage gain is the product of the gains in absolute form.

$$A_{v(tot)} = A_{v1} \times A_{v2} \times \cdots \times A_{vn}$$

Decibel units are useful when combining these gains or losses because they involve just addition or subtraction. The algebraic addition of decibel quantities is equivalent to multiplication of the gains in absolute form.

$$A'_{v(tot)} = A'_{v1} = A'_{v2} + \cdots + A'_{vn}$$

EXAMPLE 1–10

Assume the transmitted power from a radar is 10 kW. A directional coupler (a device that samples the transmitted signal) has an output that represents -40 dB of attenuation. Two 3 dB attenuators are connected in series to this output, and the attenuated signal is terminated with a 50 Ω terminator (load resistor). What is the power dissipated in the terminator?

Solution

$$dB = 10 \log\left(\frac{P_2}{P_1}\right)$$

The transmitted power is attenuated by 46 dB (sum of the attenuators). Substituting,

$$-46 \text{ dB} = 10 \log\left(\frac{P_2}{10 \text{ kW}}\right)$$

Divide both sides by 10 and remove the log function.

$$10^{-4.6} = \frac{P_2}{10 \text{ kW}}$$

Therefore,

$$P_2 = 251 \text{ mW}$$

Practice Exercise   Assume one of the 3 dB attenuators is removed.
(a) What is the total attenuation?
(b) What is the new power dissipated in the terminator?

Although decibel power ratios are generally used to compare two power levels, they are occasionally used for absolute measurements when the reference power level is understood. Although different standard references are used depending on the application, the most common absolute measurement is the dBm. A **dBm** is the power level when the reference is understood to be 1 mW developed in some assumed load impedance. For radio frequency systems, this is commonly 50 Ω; for audio systems, it is generally 600 Ω. The dBm is defined as

$$dBm = 10 \log\left(\frac{P_2}{1 \text{ mW}}\right)$$

The dBm is commonly used to specify the output level of signal generators and is used in telecommunications to simplify the computation of power levels.

Decibel Voltage Ratios

Since power is given by the ratio of V^2/R, the decibel power ratio can be written.

$$dB = 10 \log\left(\frac{V_2^2/R_2}{V_1^2/R_1}\right)$$

where R_1, R_2 = resistances in which P_1 and P_2 are developed
V_1, V_2 = voltages across the resistances R_1 and R_2

If the resistances are equal, they cancel.

$$dB = 10 \log\left(\frac{V_2^2}{V_1^2}\right)$$

A property of logarithms is

$$\log x^2 = 2 \log x$$

Thus, the decimal voltage ratio is

$$dB = 20 \log\left(\frac{V_2}{V_1}\right)$$

When V_2 is the output voltage (V_{out}) and V_1 is the input voltage (V_{in}) for an amplifier, the equation defines the decibel voltage gain. By substitution,

$$A_v' = 20 \log\left(\frac{V_{out}}{V_{in}}\right) \tag{1-13}$$

where A_v' = voltage gain expressed as a decibel ratio
V_{out} = voltage delivered to a load
V_{in} = voltage delivered to the amplifier

Equation (1–13) gives the decibel voltage gain, a logarithmic ratio of amplitudes. The equation was originally derived from the decibel power equation when both the input and load resistances are the same (as in telephone systems).

Both the decibel voltage gain equation and decibel power gain equation give the same ratio if the input and load resistances are the same. However, it has become common practice to apply the decibel voltage equation to cases where the resistances are *not* the same. When the resistances are not equal, the two equations do not give the same result.[4]

In the case of decibel voltage gain, note that if the amplitudes have a ratio of 2:1, the decibel voltage ratio is very close to 6 dB (since $20 \log 2 = 6$). If the signal is attenuated by a factor of 2 (ratio = 1:2), the decibel voltage ratio is -6 (since $20 \log 1/2 = -6$). Another useful ratio is when the amplitudes have a 10:1 ratio; in this case, the decibel voltage ratio is 20 dB (since $20 \log 10 = 20$).

---

[4] The *IEEE Standard Dictionary of Electrical and Electronic Terms* recommends that a specific statement accompany this application of decibels to avoid confusion.

EXAMPLE 1–11   An amplifier with an input resistance of 200 kΩ drives a load resistance of 16 Ω. If the input voltage is 100 μV and the output voltage is 18 V, calculate the decibel power gain and the decibel voltage gain.

Solution   The power delivered to the amplifier is

$$P_{in} = \frac{V_{in}^2}{R_{in}} = \frac{(100 \; \mu V)^2}{200 \; k\Omega} = 5 \times 10^{-14} \; W$$

The output power (delivered to the load) is

$$P_{out} = \frac{V_{out}^2}{R_L} = \frac{(18 \; V)^2}{16 \; \Omega} = 20.25 \; W$$

The decibel power gain is

$$A_p' = 10 \log\left(\frac{P_{out}}{P_{in}}\right) = 10 \log\left(\frac{20.25 \; W}{5 \times 10^{-14} \; W}\right) = \textbf{146 dB}$$

The decibel voltage gain is

$$A_v' = 20 \log\left(\frac{V_{out}}{V_{in}}\right) = 20 \log\left(\frac{18 \; V}{100 \; \mu V}\right) = \textbf{105 dB}$$

Practice Exercise   A video amplifier with an input resistance of 75 Ω drives a load of 75 Ω.
(a) How do the power gain and voltage gains compare?
(b) If the input voltage is 20 mV and the output voltage is 1.0 V, what is the decibel voltage gain?

1–4 REVIEW QUESTIONS

1. What is an ideal amplifier?
2. What is a dependent source?
3. What is a decibel?

1–5 ■ TROUBLESHOOTING ANALOG CIRCUITS

Technicians must diagnose and repair malfunctioning circuits or systems. Troubleshooting is the application of logical thinking to correct the malfunctioning circuit or system. Troubleshooting skills will be emphasized throughout the text.

After completing this section, you should be able to

❑ Describe the process for troubleshooting an analog circuit
  ❑ Explain what is meant by *half-splitting*
  ❑ Cite basic rules for replacing a part in a printed circuit (PC) board
  ❑ Describe basic bench test equipment for troubleshooting

Analysis, Planning, and Measuring

When troubleshooting any circuit, the first step is to analyze the clues (symptoms) of a failure. The analysis can begin by determining the answer to several questions: Has the circuit ever worked? If so, under what conditions did it fail? What are the symptoms of a failure? What are possible causes of this failure? The process of asking these questions is part of the analysis of a problem.

After analyzing the clues, the second step in the troubleshooting process is forming a logical plan for troubleshooting. A lot of time can be saved by planning the process. As part of this plan, you must have a working understanding of the circuit you are troubleshooting. Take the time to review schematics, operating instructions, or other pertinent information if you are not certain how the circuit should operate. It may turn out that the failure was that of the operator, not the circuit! A schematic with proper voltages or waveforms marked at various test points is particularly useful for troubleshooting.

Logical thinking is the most important tool of troubleshooting but rarely can solve the problem by itself. The third step is to narrow the possible failures by making carefully thought out measurements. These measurements usually confirm the direction you are taking in solving the problem or point to a new direction. Occasionally, you may find a totally unexpected result!

The thinking process that is part of analysis, planning, and measuring is best illustrated with an example. Suppose you have a string of 16 decorative lamps connected in series to a 120 V source as shown in Figure 1–22. Assume that this circuit worked at one time and stopped after moving it to a new location. When plugged in, the lamps fail to turn on. How would you go about finding the trouble?

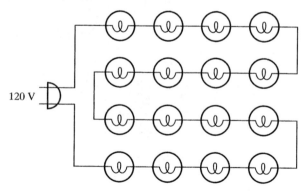

FIGURE 1–22
A series of lights. Is one of them open?

You might think like this: Since the circuit worked before it was moved, the problem could be that there is no voltage at this location. Or perhaps the wiring was loose and pulled apart when moved. It's possible a bulb burned out or became loose. This reasoning has considered possible causes and failures that could have occurred. The fact that the circuit was once working eliminates the possibility that the original circuit may have been incorrectly wired. In a series circuit, the possibility of two open paths occurring together is unlikely. You have analyzed the problem and now you are ready to plan the troubleshooting approach.

The first part of your plan is to measure (or test) for voltage at the new location. If voltage is present, then the problem is in the light string. If voltage is not present, check the circuit breakers at the input panel to the building. Before resetting breakers, you should think about why a breaker may be tripped.

The second part of your plan assumes voltage is present and the string is bad. You can disconnect power from the string and make resistance checks to begin isolating the problem. Alternatively, you could apply power to the string and measure voltage at various points. The decision whether to measure resistance or voltage is a toss-up and can be made based on the ease of making the test. Seldom is a troubleshooting plan developed so completely that all possible contingencies are included. The troubleshooter will frequently need to modify the plan as tests are made. You are ready to make measurements.

Suppose you have a digital multimeter (DMM) handy. You check the voltage at the source and find 120 V present. Now you have eliminated one possibility (no voltage). You know the problem is in the string, so you proceed with the second part of your plan. You might think: Since I have voltage across the entire string, and apparently no current in the circuit (since no bulb is on), there is almost certainly an open in the path—either a bulb or a connection. To eliminate testing each bulb, you decide to break the circuit in the middle and to check the *resistance* of each half of the circuit.

Now you are using logical thinking to reduce the effort needed. The technique you are using is a common troubleshooting procedure called *half-splitting*. By measuring the resistance of half the bulbs at once, you can reduce the effort required to find the open. Continuing along these lines, by half-splitting again, will lead to the solution in a few tests.

Unfortunately, most troubleshooting is more difficult than this example. However, analysis and planning are important for effective troubleshooting. As measurements are made, the plan is modified; the experienced troubleshooter narrows the search by fitting the symptoms and measurements into a possible cause.

Soldering

When repairing circuit boards, sooner or later the technician will need to replace a soldered part. When you replace any part, it is important to be able to remove the old part without damaging the board by excessive force or heat. Transfer of heat for removal of a part is facilitated with a chisel tip (as opposed to a conical tip) on the soldering iron.

Before installing a new part, the area must be clean. Old solder should be completely removed without exposing adjacent devices to excess heat. A degreasing cleaner or alcohol is suggested for cleaning (remember—solder won't stick to a dirty board!). Solder must be a resin core type (acid solder is never used in electronic circuits and shouldn't even be on your workbench!). Solder is applied to the joint (not to the iron). As the solder cools, it must be kept still. A good solder connection is a smooth, shiny one and the solder *flows* into the printed circuit trace. A poor solder connection looks dull. During repair, it is possible for excessive solder to short together two parts or two pins on an integrated circuit (this rarely happens when boards are machine soldered). This is called a solder bridge, and the technician must be alert for this type of error when repairing boards. After the repair is completed, any flux must be removed from the board with alcohol or other cleaner.

Basic Test Equipment

The ability to troubleshoot effectively requires the technician to have a set of test equipment available and to be familiar with the operation of the instruments. Some representative test instruments for troubleshooting are shown in Figure 1–23. No one instrument is best for all situations, so it is important to understand the limitations of the test equipment at hand. All electronic measuring instruments become part of the circuit they are measuring and thus affect the measurement itself (an effect called *instrument loading*). In addition, instruments are specified for a range of frequencies and must be properly calibrated if readings are to be trusted. An expert troubleshooter must consider these effects when making electronic measurements.

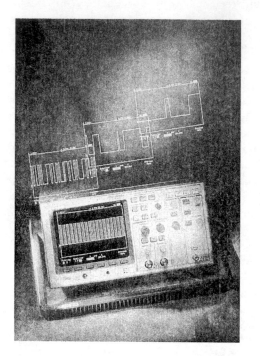

(a) Oscilloscope

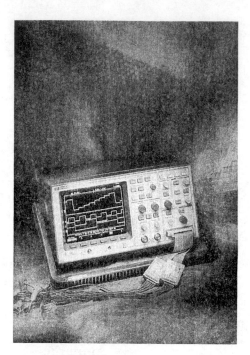

(b) Mixed signal oscilloscope

(c) Triple-output power supply

FIGURE 1–23
Test instruments. (Photos courtesy of Hewlett-Packard Co.).

For general-purpose troubleshooting of analog circuits, all technicians need access to an oscilloscope and a DMM. The oscilloscope needs to be a good two-channel scope, fast enough to spot noise or ringing when it occurs. A set of switchable probes, with the ability to switch between ×1 and ×10 is useful for looking at large or small signals. (Note that in the ×1 position, the scope loses bandwidth.)

The DMM is a general-purpose meter that has the advantage of very high input impedance but may have error if used in circuits with frequencies above a few kilohertz. Many new DMMs offer special features, such as continuity testing and diode checking, and may include capacitance, inductance, and frequency measurements. While DMMs are excellent test instruments, the VOM (volt-ohm-milliammeter) has some advantages (for example, spotting trends faster than a digital meter). Although generally not as accurate as a DMM, a VOM has very small capacitance to ground, and it is isolated from the line voltage. Also, because a VOM is a passive device, it will not tend to inject noise into a circuit under test.

Many times the circuit under test needs to have a test signal injected to simulate operation in a system. The circuit's response is then observed with a scope or other instrument. This type of testing is called *stimulus-response testing* and is commonly used when a portion of a complete system is tested. For general-purpose troubleshooting, the function generator is used as the stimulus instrument. All function generators have a sine wave, square wave, and triangle wave output; the frequency range varies widely, from a low frequency of 1 μHz to a high of 50 MHz (or more) depending on the generator. Higher-quality function generators offer the user a choice of other waveforms (pulses and ramps, for example) and may have triggered or gated outputs as well as other features.

The basic function generator waveforms (sine, square, and triangle) are used in many tests of electronic circuits and equipment. A common application of a function generator is to inject a sine wave into a circuit to check the circuit's response. The signal is capacitively coupled to the circuit to avoid upsetting the bias network; the response is observed on an oscilloscope. With a sine wave, it is easy to ascertain if the circuit is operating properly by checking the amplitude and shape of the sine wave at various points or to look for possible troubles such as high-frequency oscillation.

A common test for wide-band amplifiers is to inject a square wave into a circuit to test the frequency response. Recall that a square wave consists of the fundamental frequency and an infinite number of odd harmonics (as discussed in Section 1–2). The square wave is applied to the input of the test circuit and the output is monitored. The shape of the output square wave indicates if specific frequencies are selectively attenuated.

Figure 1–24 illustrates square wave distortions due to selective attenuation of low or high frequencies. A good amplifier should show a high-quality replica of the input. If the square wave sags, as in Figure 1–24(b), low frequencies are not being passed properly by the circuit. The rising edge contains mostly higher-frequency harmonics. If the square wave rolls over before reaching the peak, as in Figure 1–24(c), high frequencies are being attenuated. The rise time of the square wave is an indirect measurement of the bandwidth of the circuit.

For testing dc voltages or providing power to a circuit under test, a multiple output power supply, with both positive and negative outputs, is necessary. The outputs should be variable from 0 to 15 V. A separate low voltage supply is also handy for powering logic circuits or as a dc source for analog circuits.

For certain situations and applications, there are specialized measuring instruments designed for the application. Some of this specialized equipment is designed for a specific frequency range or for a specific application, so they won't be discussed here. If available,

FIGURE 1–24
Square-wave response of wide-band amplifiers.

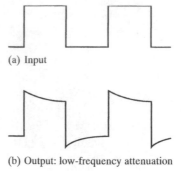

(a) Input

(b) Output: low-frequency attenuation

(c) Output: high-frequency attenuation

a digital storage scope can be of help. It has some particular advantages for troubleshooting because it can be used to store and compare waveforms from a known good unit or to capture a failure that occurs intermittently. It also has the ability to display events that occur before and after the trigger event, a feature that is invaluable with intermittent problems.

A complete list of "nice to have" accessories could be quite long indeed, but another handy set of instruments is a pulser and pulse tracer. These tools are useful for tracing a short such as onc from the power supply to ground. The pulser stimulates the circuit with a series of very short pulses. The current tracer can follow the path of the current and lead right to the short. These tools are useful for both digital and analog circuits.

Other Troubleshooting Materials

In general, some materials that are useful for general-purpose troubleshooting that fall under the "must have" category include the following:

❑ A basic set of hand tools for electronics, including long-nose pliers, diagonal wire cutters, wire strippers, screwdrivers (especially jeweler's screwdrivers), and a small flashlight.

❑ Soldering and desoldering tools, including solder wick and a magnifying glass for inspecting work or looking for hairline cracks, solder splashes, or other problems.

❑ A collection of spare parts (resistors, capacitors, transistors, diodes, switches, ICs). In this category, you will also need extra clip leads, cables with various connectors, banana to alligator converters, heat shrink, and the like.

❑ A capacitor and a resistor substitution box. This is a useful tool for various tests such as changing the time constant in a circuit under test.

❑ A hair dryer and freeze spray for testing thermal effects of a circuit.

❑ A static safe wrist strap (and static-free work station, if possible) to prevent damaging static-sensitive circuits.

1–5 REVIEW QUESTIONS

1. What is the first step in troubleshooting a circuit?
2. What is meant by half-splitting?
3. Name two ways that an instrument can load a circuit under test.

■ SUMMARY

- A linear component is one in which an increase in current is proportional to the applied voltage.

- An analog signal takes on a continuous range of values within limits. A digital signal is a discrete signal that can have only certain values. Many circuits use a combination of analog and digital circuits.

- Waveforms that repeat in a certain interval of time are said to be periodic. A cycle is the complete sequence of values that a waveform exhibits before an identical pattern occurs. The period is the time interval for one cycle.

- Signals that have voltage, current, resistance, or other quantity vary as a function of time are called time-domain signals. When the frequency is made the independent variable, the result is a frequency-domain signal. Any signal can be observed in either the time domain or the frequency domain.

- Thevenin's theorem replaces a complicated, two-terminal, linear circuit with an ideal independent voltage source and a series resistance. The Thevenin circuit is equivalent to the original circuit for any load that is connected to the output terminals.

- Norton's theorem replaces a complicated, two-terminal, linear circuit with an ideal independent current source and a parallel resistance. The Norton circuit is equivalent to the original circuit for any load that is connected to the output terminals.

- A transducer is a device that converts a physical quantity from one form to another; for electronic systems, input transducers convert a physical quantity to an electrical quantity (voltage, current, resistance).

- An ideal amplifier increases the magnitude of an input signal in order to produce a useful outcome. For a voltage amplifier, the output signal, $v_{out}(t)$, is proportional to the input signal, $v_{in}(t)$. The ratio of the output voltage to the input voltage is called the voltage gain, A_v.

- The decibel is a dimensionless number that is ten times the logarithmic ratio of two powers. Decibel gains and losses are combined by algebraic addition.

- Troubleshooting begins with analyzing the symptoms of a failure; then forming a logical plan. Carefully thought-out measurements are made to narrow the search for the cause of the failure. These measurements may modify or change the plan.

- For general-purpose troubleshooting, a reasonable fast, two-channel oscilloscope and a DMM are the principal measuring instruments. The most common stimulus instruments are a function generator and a regulated power supply.

■ GLOSSARY

These terms are included in the end-of-book glossary.

AC resistance   The ratio of a small change in voltage divided by a corresponding change in current for a given device; also called *dynamic, small-signal,* or *bulk* resistance.

Amplification   The process of producing a larger voltage, current, or power using a smaller input signal as a "pattern."

Amplifier   An electronic circuit having the capability of amplification and designed specifically for that purpose.

Analog signal   A signal that can take on a continuous range of values within certain limits.

Attenuation   The reduction in the level of power, current, or voltage.

Characteristic curve   A plot which shows the relationship between two variable properties of a device. For most electronic devices, a characteristic curve refers to a plot of the current, I, plotted as a function of voltage, V.

Cycle   The complete sequence of values that a waveform exhibits before another identical pattern occurs.

dBm   Decibel power level when the reference is understood to be 1 mW (see *Decibel*).

Decibel   A dimensionless quantity that is 10 times the logarithm of a power ratio or 20 times the logarithm of a voltage ratio.

Digital signal   A noncontinuous signal that has discrete numerical values assigned to specific steps.

Domain   The values assigned to the independent variable. For example, frequency or time are typically used as the independent variable for plotting signals.

Frequency   The number of repetitions per unit of time for a periodic waveform.

Gain   The amount of amplification. Gain is a ratio of an output quantity to an input quantity (e.g., voltage gain is the ratio of the output voltage to the input voltage).

Harmonics   Higher-frequency sinusoidal waves that are integer multiples of a fundamental frequency.

Linear component   A component in which an increase in current is proportional to the applied voltage.

Load line   A straight line plotted on a current versus voltage plot that represents all possible operating points for an external circuit.

Logarithm   In mathematics, the logarithm of a number is the power to which a base must be raised to give that number.

Norton's theorem   An equivalent circuit that replaces a complicated two-terminal linear circuit with a single current source and a parallel resistance.

Period (T)   The time for one cycle of a repeating wave.

Periodic   A waveform that repeats at regular intervals.

Phase angle (in radians)   The fraction of a cycle that a waveform is shifted from a reference waveform of the same frequency.

Quantizing   The process of assigning numbers to sampled data.

Quiescent point   The point on a load line that represents the current and voltage conditions for a circuit with no signal (also called *operating* or *Q-point*). It is the intersection of a device characteristic curve with a load line.

Sampling   The process of breaking the analog waveform into time "slices" that approximate the original wave.

Spectrum   A plot of amplitude versus frequency for a signal.

Thevenin's theorem   An equivalent circuit that replaces a complicated two-terminal linear circuit with a single voltage source and a series resistance.

Transducer   A device that converts a physical quantity from one form to another; for example, a microphone converts sound into voltage.

Transfer curve   A plot of the output of a circuit or system for a given input.

Vector   Any quantity that has both magnitude and direction.

■ KEY FORMULAS

(1–1)	$I = \dfrac{V}{R}$	Ohm's law
(1–2)	$y(t) = A \sin(\omega t \pm \phi)$	Instantaneous value of a sinusoidal wave
(1–3)	$f\,(\text{Hz}) = \dfrac{\omega\,(\text{rad/s})}{2\pi\,(\text{rad/cycle})}$	Conversion from radian frequency (rad/s) to hertz (Hz)
(1–4)	$T = \dfrac{1}{f}$	Conversion from frequency to period
(1–5)	$f = \dfrac{1}{T}$	Conversion from period to frequency
(1–6)	$V_{avg} = 0.637 V_p$	Conversion from peak voltage to average voltage for a sinusoidal wave
(1–7)	$P = IV$	Power law
(1–8)	$V_{rms} = 0.707 V_p$	Conversion from peak voltage to rms voltage for a sinusoidal wave
(1–9)	$A_v = \dfrac{V_{out}}{V_{in}}$	Voltage gain
(1–10)	$A_p = \dfrac{P_{out}}{P_{in}}$	Power gain
(1–11)	$\text{dB} = 10 \log\left(\dfrac{P_2}{P_1}\right)$	Definition of the decibel
(1–12)	$A'_p = 10 \log\left(\dfrac{P_{out}}{P_{in}}\right)$	Decibel power gain
(1–13)	$A'_v = 20 \log\left(\dfrac{V_{out}}{V_{in}}\right)$	Decibel voltage gain

■ SELF-TEST

1. The graph of a linear equation
   (a) always has a constant slope   (b) always goes through the origin
   (c) must have a positive slope   (d) all of the above
   (e) none of the above

2. AC resistance is defined as
   (a) voltage divided by current
   (b) a change in voltage divided by a corresponding change in current
   (c) current divided by voltage
   (d) a change in current divided by a corresponding change in voltage

3. A discrete signal
   (a) changes smoothly   (b) can take on any value
   (c) is the same thing as an analog signal   (d) none of the above
   (e) all of the above

4. The process of assigning numeric values to a signal is called
   (a) sampling   (b) multiplexing   (c) quantizing   (d) digitizing

5. The reciprocal of the repetition time of a periodic signal is the
   (a) frequency   (b) angular frequency   (c) period   (d) amplitude

6. If a sinusoidal wave has a peak amplitude of 10 V, the rms voltage is
   (a) 0.707 V   (b) 6.37 V   (c) 7.07 V   (d) 20 V

7. If a sinusoidal wave has a peak-to-peak amplitude of 325 V, the rms voltage is
   (a) 103 V   (b) 115 V   (c) 162.5 V   (d) 460 V

8. Assume the equation for a sinusoidal wave is $v(t) = 200 \sin(500t)$. The peak voltage is
   (a) 100 V   (b) 200 V   (c) 400 V   (d) 500 V

9. A harmonic is
   (a) an integer multiple of a fundamental frequency
   (b) an unwanted signal that adds noise to a system
   (c) a transient signal
   (d) a pulse

10. A Thevenin circuit consists of a
    (a) current source in parallel with a resistor   (b) current source in series with a resistor
    (c) voltage source in parallel with a resistor   (d) voltage source in series with a resistor

11. A Norton circuit consists of a
    (a) current source in parallel with a resistor   (b) current source in series with a resistor
    (c) voltage source in parallel with a resistor   (d) voltage source in series with a resistor

12. A load line is a plot that describes
    (a) the IV characteristic curve for a load resistor   (b) a driving circuit
    (c) both of the above   (d) none of the above

13. The intersection of an IV curve with the load line is called the
    (a) transfer curve   (b) transition point   (c) load point   (d) Q-point

14. Assume a certain amount of power is attenuated by 20 dB. This is a factor of
    (a) 10   (b) 20   (c) 100   (d) 200

15. Assume an amplifier has a decibel voltage gain of 100 dB. The output will be larger than the input by a factor of
    (a) 100   (b) 1000   (c) 10,000   (d) 100,000

16. An important rule for soldering is
    (a) always use a good acid-based solder
    (b) always apply solder directly to the iron, never to the parts being soldered
    (c) wiggle the solder joint as it cools to strengthen it
    (d) all of the above
    (e) none of the above

■ PROBLEMS

SECTION 1–1   Analog Electronics

1. What is the conductance of a 22 kΩ resistor?

2. How does the ac resistance of a diode change as the voltage increases?

3. Compute the ac resistance of the diode in Figure 1–2 at the point $V = 0.7$ V, $I = 5.0$ mA.

4. Sketch the shape of an IV curve for a device that has a decreasing ac resistance as voltage increases.

SECTION 1–2   Analog Signals

5. Assume a sinusoidal wave is described by the equation $v(t) = 100 \sin(200t + 0.52)$.
   (a) From this expression, determine the peak voltage, the average voltage, and the angular frequency in rad/s.
   (b) Find the instantaneous voltage at a time of 2.0 ms. (Reminder: the angles are in radians in this equation).

6. Determine the frequency (in Hz) and the period (in s) for the sinusoidal wave described in Problem 5.

7. An oscilloscope shows a wave repeating every 27 μs. What is the frequency of the wave?

8. A DMM indicates the rms value of a sinusoidal wave. If a DMM indicates a sinusoidal wave is 3.5 V, what peak-to-peak voltage would you expect to observe on an oscilloscope?

9. The ratio of the rms voltage to the average voltage for any wave is called the *form factor* (used occasionally to convert meter readings). What is the form factor for a sinusoidal wave?

10. What is the fifth harmonic of a 500 Hz triangular wave?

11. What is the only type of harmonics found in a square wave?

SECTION 1–3   Analog Sources

12. Draw the Thevenin equivalent circuit for the circuit shown in Figure 1–25. Show values on your drawing.

13. Assume a 1.0 kΩ, 2.7 kΩ and 3.6 kΩ load resistor are connected, one at a time, across the output terminals of the circuit in Figure 1–25. Determine the voltage across each load.

14. Draw the Norton equivalent circuit for the circuit shown in Figure 1–25. Show values on your drawing.

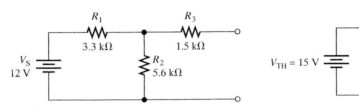

FIGURE 1–25   FIGURE 1–26

15. Draw a graph showing the load line for the Thevenin circuit shown in Figure 1–26. On the same graph, show the *IV* curve for a 150 kΩ resistor. Show the Q-point on your plot.

16. Assume the output of a transducer is a 10 mV ac signal with no load and drops to 5 mV when a 100 kΩ load is connected to the outputs. Based on these observations, draw the Thevenin equivalent circuit for this transducer.

17. Draw the Norton equivalent circuit for the transducer circuit described in Problem 16.

SECTION 1–4   Amplifiers

18. For the amplifier described by the transfer curve in Figure 1–17, what is the voltage gain in the linear region? What is the largest output voltage before saturation?

19. The input to an amplifier is 80 μV. If the voltage gain of the amplifier is 50,000, what is the output signal?

20. Assume a transducer with a Thevenin (unloaded) voltage of 5.0 mV and a Thevenin resistance of 20 kΩ is connected to a two-stage cascaded amplifier with the following specifications:

$R_{in1} = 50\ \text{k}\Omega$
A_{v1} (unloaded) $= 50$
$R_{th1} = 5\ \text{k}\Omega$
$R_{in2} = 10\ \text{k}\Omega$
A_{v2} (unloaded) $= 40$
$R_{th2} = 1.0\ \text{k}\Omega$

Draw the amplifier model and compute the voltage across a 2.0 kΩ load.

21. Compute the decibel voltage gain for the amplifier in Problem 20.

22. Compute the decibel power gain for the amplifier in Problem 20.

23. Assume you want to attenuate the voltage from a signal generator by a factor of 1000. What is the decibel attenuation required?

24. **(a)** What is the power dissipated in a 50 Ω load resistor when 20 V is across the load?
   (b) Express your answer in dBm.

25. A certain instrument is limited to 2 watts of input power dissipated in its internal 50 Ω input resistor.
   (a) How much attenuation (in dB) is required in order to connect a 20 W source to the instrument?
   (b) What is the maximum allowable voltage at the input?

SECTION 1–5   Troubleshooting Analog Circuits

26. Figure 1–27 shows a small system consisting of four microphones connected to a two-channel amplifier through a selector switch (SW1). Either the A set or the B set of microphones is selected and amplified. The output of the amplifier is connected to two speakers. Power to the amplifier is supplied by a single power supply that furnishes dc voltages to the amplifier and two batteries that provide power to two each of the microphones as shown.

   Assume no sound is heard when the system is plugged in and turned on. Outline a basic troubleshooting plan by indicating the tests you would make to isolate the trouble to either the power supply, amplifier, a microphone, microphone battery, switch, speaker, or other fault.

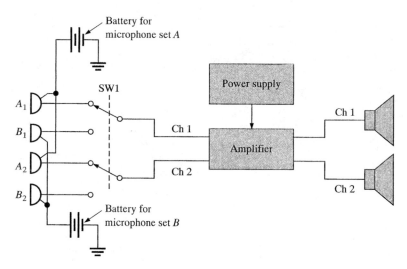

FIGURE 1–27
A small system consisting of a two-channel amplifier and four microphones.

27. For the system described in Problem 26, outline a basic troubleshooting plan for the case where Channel 1 operates normally but no sound is heard from Channel 2. Indicate the tests you would make to isolate the trouble. (Can you think of a method to do half-splitting?)

28. What information is obtained when a square wave calibration signal is used as the input to an oscilloscope?

29. How can you protect a static-sensitive circuit from damage when you are working on it?

30. Cite two important advantages of a digital storage oscilloscope over an analog oscilloscope.

ANSWERS TO REVIEW QUESTIONS

Section 1–1

1. A graph of the relationship between current and voltage for the component
2. The slope of curve is lower for larger resistors.
3. DC resistance is the voltage divided by the current. AC resistance is the *change* in voltage divided by the *change* in current.

Section 1–2

1. An analog signal takes on a continuous range of values; a digital signal represents information that has a discrete number of codes.
2. The spectrum is a line spectrum with a single line at the fundamental and other lines at the odd harmonic frequencies. See Figure 1–8(a).
3. The spectrum for the repetitive waveform is a line spectrum; the spectrum for the nonrepetitive waveform is a continuous spectrum.

Section 1–3

1. An independent source is a voltage or current source that can be specified without regard to any other circuit parameter.
2. A Thevenin circuit consists of a series voltage source and resistor that duplicates the performance of a more complicated circuit for any given load. A Norton circuit consists of a parallel current source and resistor that duplicates the performance of a more complicated circuit for any given load.
3. Passive transducers require a separate source of electrical power; active transducers are self-generating devices.

Section 1–4

1. An ideal amplifier is one that introduces no noise or distortion to the signal; the output varies in time and replicates the input exactly.
2. A dependent source is one whose value depends on voltage or current elsewhere in the circuit.
3. A decibel is a dimensionless number that is 10 multiplied by the logarithmic ratio of two powers.

Section 1–5

1. Analyzing the symptoms of a failure by asking questions: Has the circuit ever worked? If so, under what conditions did it fail? What are the symptoms of a failure? What are possible causes of this failure?
2. Half-splitting divides a troubleshooting problem into halves and determines which half of the circuit is likely to have the problem.
3. Resistive or capacitive loading

■ **ANSWERS TO PRACTICE EXERCISES FOR EXAMPLES**

1–1 $G_2 = 375$ mS, $R_2 = 2.67$ kΩ

1–2 $V_{rms} = 10.6$ V, $f = 95$ Hz, $T = 10.5$ ms

1–3 $V_{R_L} = 3.48$ V

1–4 $V_{R_L} = 3.48$ V

1–5 See Figure 1–28.

FIGURE 1–28

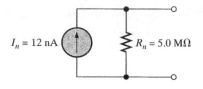

$I_n = 12$ nA $R_n = 5.0$ MΩ

1–6 6 V

1–7 1.34 V

1–8 (a) $\log 0.04 = -1.398$ (b) $10^{4.8} = 63{,}096$
$\log 0.4 = -0.398$
$\log 4.0 = 0.602$
$\log 40 = 1.602$

1–9 49 dB

1–10 (a) -43 dB   (b) 503 mW

1–11 (a) The decibel power gain is one-half the decibel voltage gain.   (b) 34 dB

2

OPERATIONAL AMPLIFIERS

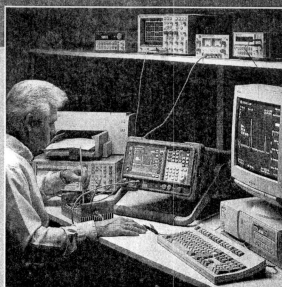

Courtesy Hewlett-Packard Company

■ CHAPTER OBJECTIVES

☐ Describe the basic op-amp and its characteristics
☐ Discuss the differential amplifier and its operation
☐ Discuss several op-amp parameters
☐ Explain negative feedback in op-amp circuits
☐ Analyze three op-amp configurations
☐ Describe impedances of the three op-amp configurations
☐ Troubleshoot op-amp circuits
☐ Apply what you have learned in this chapter to a system application

In this chapter, you are introduced to a general-purpose IC, the operational amplifier (op-amp), which is the most versatile and widely used of all linear integrated circuits. Although the op-amp is made up of many resistors, diodes, and transistors, it is treated as a single device. This means that you will be concerned with what the circuit does more from an external viewpoint than from an internal, component-level viewpoint.

In medical laboratories, an instrument known as a spectrophotometer is used to analyze chemicals in solutions by determining how much absorption of light occurs over a range of wavelengths. A basic system is shown in Figure 2–35. Light is passed through a prism; and as the light source and prism are pivoted, different wavelengths of visible light pass through the slit. The wavelength coming through the slit at a given pivot angle passes through the solution and is detected by a photocell. The op-amp circuit is used to amplify the output of the photocell and send the signal to a processor and display instrument. Since every chemical and compound absorbs light in a different way, the output of the spectro-photometer can be used to accurately identify the contents of the solution.

For the system application in Section 2–8, in addition to the other topics, be sure you understand

☐ The functions of the inputs and outputs of an op-amp

☐ How an op-amp works

☐ The pin configurations of an op amp

2–1 ■ INTRODUCTION TO OPERATIONAL AMPLIFIERS

Early operational amplifiers (op-amps) were used primarily to perform mathematical operations such as addition, subtraction, integration, and differentiation, thus the term operational. These early devices were constructed with vacuum tubes and worked with high voltages. Today, op-amps are linear integrated circuits that use relatively low supply voltages and are reliable and inexpensive.

After completing this section, you should be able to

❑ Describe the basic op-amp and its characteristics
    ❑ Recognize the op-amp symbol
    ❑ Identify the terminals on op-amp packages
    ❑ Describe the ideal op-amp
    ❑ Describe the practical op-amp

Symbol and Terminals

The standard **operational amplifier** (op-amp) symbol is shown in Figure 2–1(a). It has two input terminals, the inverting ($-$) input and the noninverting ($+$) input, and one output terminal. The typical op-amp operates with two dc supply voltages, one positive and the other negative, as shown in Figure 2–1(b). Usually these dc voltage terminals are left off the schematic symbol for simplicity but are always understood to be there. Some typical op-amp IC packages are shown in Figure 2–1(c).

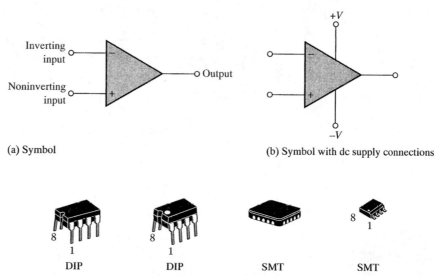

(a) Symbol

(b) Symbol with dc supply connections

DIP   DIP   SMT   SMT

(c) Typical packages. Looking from the top, pin 1 always is to the left of the notch or dot on the Dual In-line Package (DIP) and Surface-Mount Technology (SMT) packages.

FIGURE 2–1
Op-amp symbols and packages.

The Ideal Op-Amp

To illustrate what an op-amp is, let's consider its *ideal* characteristics. A practical op-amp, of course, falls short of these ideal standards, but it is much easier to understand and analyze the device from an ideal point of view.

First, the ideal op-amp has *infinite voltage gain* and an *infinite input impedance* (open), so that it does not load the driving source. Finally, it has a *zero output impedance*. These characteristics are illustrated in Figure 2–2. The input voltage V_{in} appears between the two input terminals, and the output voltage is $A_v V_{in}$, as indicated by the internal voltage source symbol. The concept of infinite input impedance is a particularly valuable analysis tool for the various op-amp configurations, which will be discussed in Section 2–5.

FIGURE 2–2
Ideal op-amp representation.

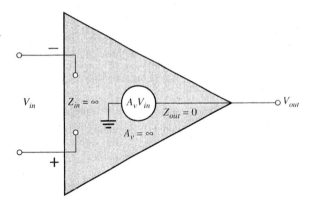

The Practical Op-Amp

Although modern integrated circuit (IC) op-amps approach parameter values that can be treated as ideal in many cases, no practical op-amp can be ideal. Any device has limitations, and the IC op-amp is no exception. Op-amps have both voltage and current limitations. Peak-to-peak output voltage, for example, is usually limited to slightly less than the difference between the two supply voltages. Output current is also limited by internal restrictions such as power dissipation and component ratings.

Characteristics of a practical op-amp are *high voltage gain, high input impedance, low output impedance,* and *wide bandwidth.* Some of these are illustrated in Figure 2–3.

FIGURE 2–3
Practical op-amp representation.

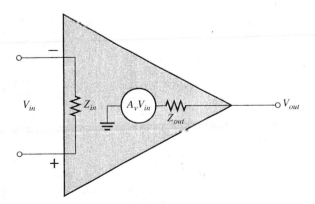

2–1 REVIEW QUESTIONS

1. What are the connections to a basic op-amp?
2. Describe some of the characteristics of a practical op-amp.

2–2 ■ THE DIFFERENTIAL AMPLIFIER

The op-amp, in its basic form, typically consists of two or more differential amplifier stages. Because the differential amplifier (diff-amp) is fundamental to the op-amp's internal operation, it is useful to have a basic understanding of this type of circuit.

After completing this section, you should be able to

❑ Discuss the differential amplifier and its operation
   ❑ Explain single-ended input operation
   ❑ Explain differential-input operation
   ❑ Explain common-mode operation
   ❑ Define *common-mode rejection ratio*
   ❑ Discuss the use of differential amplifiers in op-amps

A basic **differential amplifier** (diff-amp) circuit and its symbol are shown in Figure 2–4. The diff-amp stages that make up part of the op-amp provide high voltage gain and common-mode rejection (defined later in this section). Notice that the differential amplifier has two outputs where the op-amp has only one output.

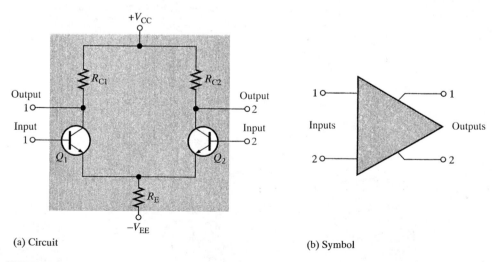

(a) Circuit   (b) Symbol

FIGURE 2–4
Basic differential amplifier.

Basic Operation

Although an op-amp typically has more than one differential amplifier **stage,** we will use a single diff-amp to illustrate the basic operation. The following discussion is in relation to Figure 2–5 and consists of a basic dc analysis of the diff-amp's operation.

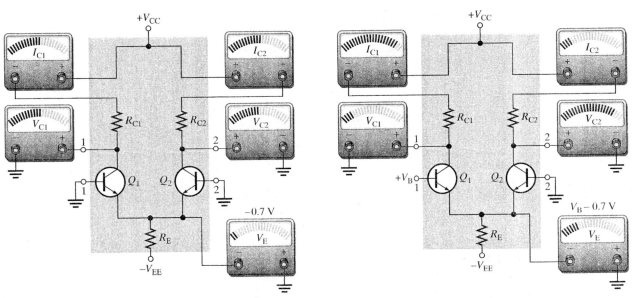

(a) Both inputs grounded

(b) Bias voltage on input 1 with input 2 grounded

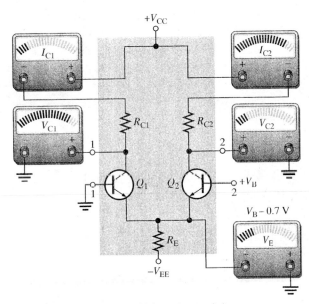

(c) Bias voltage on input 2 with input 1 grounded

FIGURE 2–5

Basic operation of a differential amplifier (ground is zero volts) showing relative changes in currents and voltages.

First, when both inputs are grounded (0 V), the emitters are at -0.7 V, as indicated in Figure 2–5(a). It is assumed that the transistors, Q_1 and Q_2, are identically matched by careful process control during manufacturing so that their dc emitter currents are the same when there is no input signal. Thus,

$$I_{E1} = I_{E2}$$

Since both emitter currents combine through R_E,

$$I_{E1} = I_{E2} = \frac{I_{R_E}}{2}$$

where

$$I_{R_E} = \frac{V_E - V_{EE}}{R_E}$$

Based on the approximation that $I_C \cong I_E$, it can be stated that

$$I_{C1} = I_{C2} \cong \frac{I_{R_E}}{2}$$

Since both collector currents and both collector resistors are equal (when the input voltage is zero),

$$V_{C1} = V_{C2} = V_{CC} - I_{C1}R_{C1}$$

This condition is illustrated in Figure 2–5(a).

Next, input 2 is left grounded, and a positive bias voltage is applied to input 1, as shown in Figure 2–5(b). The positive voltage on the base of Q_1 increases I_{C1} and raises the emitter voltage to

$$V_E = V_B - 0.7 \text{ V}$$

This action reduces the forward bias (V_{BE}) of Q_2 because its base is held at 0 V (ground), thus causing I_{C2} to decrease as indicated in part (b) of the diagram. The net result is that the increase in I_{C1} causes a decrease in V_{C1}, and the decrease in I_{C2} causes an increase in V_{C2}, as shown.

Finally, input 1 is grounded and a positive bias voltage is applied to input 2, as shown in Figure 2–5(c). The positive bias voltage causes Q_2 to conduct more, thus increasing I_{C2}. Also, the emitter voltage is raised. This reduces the forward bias of Q_1, since its base is held at ground, and causes I_{C1} to decrease. The result is that the increase in I_{C2} produces a decrease in V_{C2}, and the decrease in I_{C1} causes V_{C1} to increase, as shown.

Modes of Signal Operation

Single-Ended Input   When a diff-amp is operated in this mode, one input is grounded and the signal voltage is applied only to the other input, as shown in Figure 2–6. In the case where the signal voltage is applied to input 1 as in part (a), an inverted, amplified signal voltage appears at output 1 as shown. Also, a signal voltage appears in phase at the emitter of Q_1. Since the emitters of Q_1 and Q_2 are common, the emitter signal becomes an input to Q_2, which functions as a common-base amplifier. The signal is amplified by Q_2 and appears, noninverted, at output 2. This action is illustrated in part (a).

In the case where the signal is applied to input 2 with input 1 grounded, as in Figure 2–6(b), an inverted, amplified signal voltage appears at output 2. In this situation, Q_1 acts as a common-base amplifier, and a noninverted, amplified signal appears at output 1. This action is illustrated in part (b) of the figure.

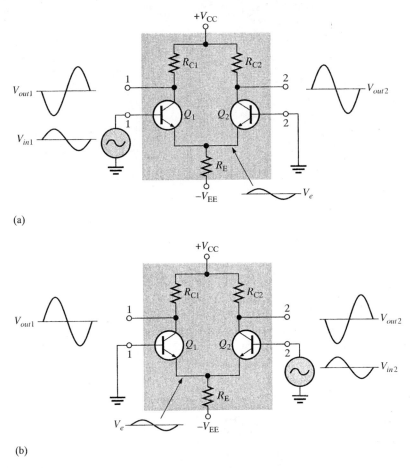

(a)

(b)

FIGURE 2–6
Single-ended input operation of a differential amplifier.

Differential Input   In this mode, two opposite-polarity (out-of-phase) signals are applied to the inputs, as shown in Figure 2–7(a). This type of operation is also referred to as *double-ended.* Each input affects the outputs, as you will see in the following discussion.

Figure 2–7(b) shows the output signals due to the signal on input 1 acting alone as a single-ended input. Figure 2–7(c) shows the output signals due to the signal on input 2 acting alone as a single-ended input. Notice in parts (b) and (c) that the signals on output 1 are of the same polarity. The same is also true for output 2. By superimposing both output 1 signals and both output 2 signals, you get the total differential operation, as shown in Figure 2–7(d).

Common-Mode Input   One of the most important aspects of the operation of a differential amplifier can be seen by considering the **common-mode** condition where two signal voltages of the same phase, frequency, and amplitude are applied to the two inputs, as shown in Figure 2–8(a). Again, by considering each input signal as acting alone, the basic operation can be understood.

Figure 2–8(b) shows the output signals due to the signal on only input 1, and Figure 2–8(c) shows the output signals due to the signal on only input 2. Notice that the corresponding signals on output 1 are of the opposite polarity, and so are the ones on output 2. When the input signals are applied to both inputs, the outputs are superimposed and they cancel, resulting in a zero output voltage, as shown in Figure 2–8(d).

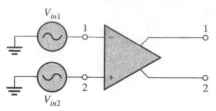

(a) Differential inputs

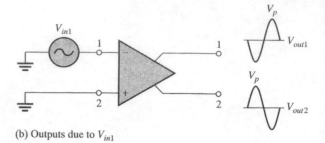

(b) Outputs due to V_{in1}

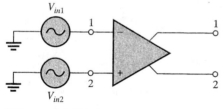

(c) Outputs due to V_{in2}

(d) Total outputs due to differential inputs

FIGURE 2–7
Differential operation of a differential amplifier.

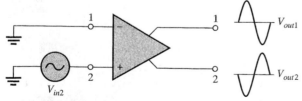

(a) Common-mode inputs

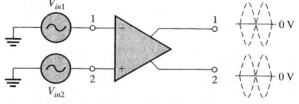

(b) Outputs due to V_{in1}

(c) Outputs due to V_{in2}

(d) Outputs cancel when common-mode signals are applied. Output signals of equal amplitude but opposite phase cancel, producing 0 V on each output.

FIGURE 2–8
Common-mode operation of a differential amplifier.

This action is called *common-mode rejection*. Its importance lies in the situation where an unwanted signal appears commonly on both diff-amp inputs. Common-mode rejection means that this unwanted signal will not appear on the outputs to distort the desired signal. Common-mode signals (noise) generally are the result of the pick-up of radiated energy on the input lines, from adjacent lines, or the 60 Hz power line, or other sources.

Common-Mode Rejection Ratio

Desired signals appear on only one input or with opposite polarities on both input lines. These desired signals are amplified and appear on the outputs as previously discussed. Unwanted signals (noise) appearing with the same polarity on both input lines are essentially cancelled by the diff-amp and do not appear on the outputs. The measure of an amplifier's ability to reject common-mode signals is a parameter called the **common-mode rejection ratio (CMRR).**

Ideally, a differential amplifier provides a very high gain for desired signals (single-ended or differential) and zero gain for common-mode signals. Practical diff-amps, however, do exhibit a very small common-mode gain (usually much less than 1), while providing a high differential voltage gain (usually several thousand). The higher the differential gain with respect to the common-mode gain, the better the performance of the diff-amp in terms of rejection of common-mode signals. This suggests that a good measure of the diff-amp's performance in rejecting unwanted common-mode signals is the ratio of the differential gain $A_{v(d)}$ to the common-mode gain, A_{cm}. This ratio is the common-mode rejection ratio, CMRR.

$$CMRR = \frac{A_{v(d)}}{A_{cm}} \tag{2-1}$$

The higher the CMRR, the better. A very high value of CMRR means that the differential gain $A_{v(d)}$ is high and the common-mode gain A_{cm} is low.

The CMRR is often expressed in decibels (dB) as

$$CMRR' = 20 \log\left(\frac{A_{v(d)}}{A_{cm}}\right) \tag{2-2}$$

EXAMPLE 2–1 A certain diff-amp has a differential voltage gain of 2000 and a common-mode gain of 0.2. Determine the CMRR and express it in decibels.

Solution $A_{v(d)} = 2000$, and $A_{cm} = 0.2$. Therefore,

$$CMRR = \frac{A_{v(d)}}{A_{cm}} = \frac{2000}{0.2} = \mathbf{10,000}$$

Expressed in dB,

$$CMRR' = 20 \log(10,000) = \mathbf{80\ dB}$$

Practice Exercise Determine the CMRR and express it in dB for an amplifier with a differential voltage gain of 8500 and a common-mode gain of 0.25.

A CMRR of 10,000, for example, means that the desired input signal (differential) is amplified 10,000 times more than the unwanted noise (common-mode). So, as an example, if the amplitudes of the differential input signal and the common-mode noise are equal, the desired signal will appear on the output 10,000 times greater in amplitude than the noise. Thus, the noise or interference has been essentially eliminated.

Example 2–2 illustrates further the idea of common-mode rejection and the general signal operation of the differential amplifier.

EXAMPLE 2–2

The differential amplifier shown in Figure 2–9 has a differential voltage gain of 2500 and a CMRR of 30,000. In part (a), a single-ended input signal of 500 μV rms is applied. At the same time, a 1 V, 60 Hz common-mode interference signal appears on both inputs as a result of radiated pick-up from the ac power system. In part (b), differential input signals of 500 μV rms each are applied to the inputs. The common-mode interference is the same as in part (a).

(a) Determine the common-mode gain.
(b) Express the CMRR in dB.
(c) Determine the rms output signal for Figure 2–9(a) and (b).
(d) Determine the rms interference voltage on the output.

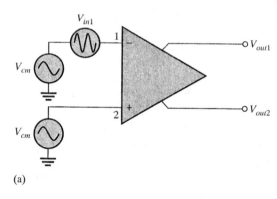

(a)

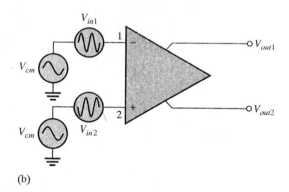

(b)

FIGURE 2–9

Solution

(a) $\text{CMRR} = \dfrac{A_{v(d)}}{A_{cm}}$. Therefore,

$$A_{cm} = \frac{A_{v(d)}}{\text{CMRR}} = \frac{2500}{30,000} = \textbf{0.083}$$

(b) $\text{CMRR}' = 20 \log(30,000) = \textbf{89.5 dB}$

(c) In Figure 2–9(a), the differential input voltage is the difference between the voltage on input 1 and that on input 2. Since input 2 is grounded, its voltage is zero. Therefore,

$$V_{in(d)} = V_{in1} - V_{in2} = 500 \text{ μV} - 0 \text{ V} = 500 \text{ μV}$$

The output signal voltage in this case is taken at output 1.

$$V_{out1} = A_{v(d)}V_{in(d)} = (2500)(500 \text{ μV}) = \textbf{1.25 V rms}$$

In Figure 2–9(b), the differential input voltage is the difference between the two opposite-polarity, 500 μV signals.

$$V_{in(d)} = V_{in1} - V_{in2} = 500 \text{ μV} - (-500 \text{ μV}) = 1000 \text{ μV} = 1 \text{ mV}$$

The output voltage signal is

$$V_{out1} = A_{v(d)}V_{in(d)} = (2500)(1 \text{ mV}) = \textbf{2.5 V rms}$$

This shows that a differential input (two opposite-polarity signals) results in a gain that is double that for a single-ended input.

(d) The common-mode input is 1 V rms. The common-mode gain A_{cm} is 0.083. The interference (common-mode) voltage on the output is, therefore,

$$A_{cm} = \frac{V_{out(cm)}}{V_{in(cm)}}$$

$$V_{out(cm)} = A_{cm}V_{in(cm)} = (0.083)(1 \text{ V}) = \textbf{0.083 V}$$

Practice Exercise   The amplifier in Figure 2–9 has a differential voltage gain of 4200 and a CMRR of 25,000. For the same single-ended and differential input signals as described in the example: **(a)** Find A_{cm}. **(b)** Express the CMRR in dB. **(c)** Determine the rms output signal for parts (a) and (b) of the figure. **(d)** Determine the rms interference (common-mode) voltage appearing on the output.

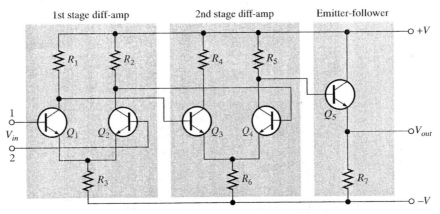

(a) Circuit

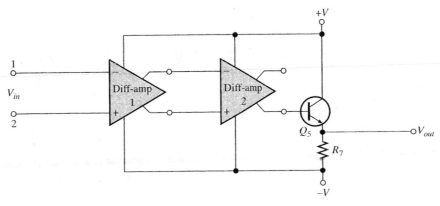

(b) Block diagram

FIGURE 2–10
Simplified internal circuitry of a basic op-amp.

A Simple Op-Amp Arrangement

Figure 2–10, on the previous page, shows two differential amplifier (diff-amp) stages and an emitter-follower connected to form a simple op-amp. The first stage can be used with a single-ended or a differential input. The differential outputs of the first stage are directly coupled into the differential inputs of the second stage. The output of the second stage is single-ended to drive an emitter-follower to achieve a low output impedance. Both differential stages together provide a high voltage gain and a high CMRR.

2–2 REVIEW QUESTIONS

1. Distinguish between differential and single-ended inputs.
2. Define *common-mode rejection?*
3. For a given value of differential gain, does a higher CMRR result in a higher or lower common-mode gain?

2–3 ■ OP-AMP DATA SHEET PARAMETERS

In this section, several important op-amp parameters are defined. (These are listed in the objectives that follow.) Also several IC op-amps are compared in terms of these parameters.

After completing this section, you should be able to

❑ Discuss several op-amp parameters
   ❑ Define *input offset voltage*
   ❑ Discuss input offset voltage drift with temperature
   ❑ Define *input bias current*
   ❑ Define *input impedance*
   ❑ Define *input offset current*
   ❑ Define *output impedance*
   ❑ Discuss common-mode input voltage range
   ❑ Discuss open-loop voltage gain
   ❑ Define *common-mode rejection ratio*
   ❑ Define *slew rate*
   ❑ Discuss frequency response
   ❑ Compare the parameters of several types of IC op-amps

Input Offset Voltage

The ideal op-amp produces zero volts out for zero volts in. In a practical op-amp, however, a small dc voltage, $V_{OUT(error)}$, appears at the output when no differential input voltage is applied. Its primary cause is a slight mismatch of the base-emitter voltages of the differential input stage of an op-amp, as illustrated in Figure 2–11(a).

The output voltage of the differential input stage is expressed as

$$V_{OUT(error)} = I_{C2}R_C - I_{C1}R_C$$

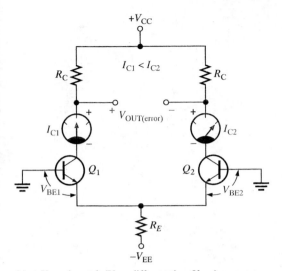

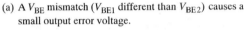

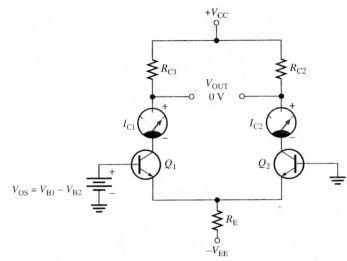

(a) A V_{BE} mismatch (V_{BE1} different than V_{BE2}) causes a small output error voltage.

(b) The input offset voltage is the difference in the voltage between the inputs that is necessary to eliminate the output error voltage (makes $V_{OUT} = 0$).

FIGURE 2–11
Illustration of input offset voltage, V_{OS}.

A small difference in the base-emitter voltages of Q_1 and Q_2 causes a small difference in the collector currents. This results in a nonzero value of V_{OUT}, which is the error voltage. (The collector resistors are equal.)

As specified on an op-amp data sheet, the *input offset voltage* (V_{OS}) is the differential dc voltage required between the inputs to force the differential output to zero volts. V_{OS} is demonstrated in Figure 2–11(b). Typical values of input offset voltage are in the range of 2 mV or less. In the ideal case, it is 0 V.

Input Offset Voltage Drift with Temperature

The *input offset voltage drift* is a parameter related to V_{OS} that specifies how much change occurs in the input offset voltage for each degree change in temperature. Typical values range anywhere from about 5 μV per degree Celsius to about 50 μV per degree Celsius. Usually, an op-amp with a higher nominal value of input offset voltage exhibits a higher drift.

Input Bias Current

You have seen that the input terminals of a bipolar differential amplifier are the transistor bases and, therefore, the input currents are the base currents.

The *input bias current* is the dc current required by the inputs of the amplifier to properly operate the first stage. By definition, the input bias current is the *average* of both input currents and is calculated as follows:

$$I_{BIAS} = \frac{I_1 + I_2}{2}$$

The concept of input bias current is illustrated in Figure 2–12.

FIGURE 2–12
Input bias current is the average of the two op-amp input currents.

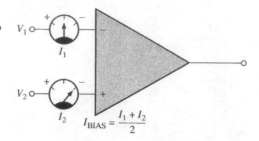

Input Impedance

Two basic ways of specifying the input impedance of an op-amp are the differential and the common mode. The *differential input impedance* is the total resistance between the inverting and the noninverting inputs, as illustrated in Figure 2–13(a). Differential input impedance is measured by determining the change in bias current for a given change in differential input voltage. The *common-mode input impedance* is the resistance between each input and ground and is measured by determining the change in bias current for a given change in common-mode input voltage. It is depicted in Figure 2–13(b).

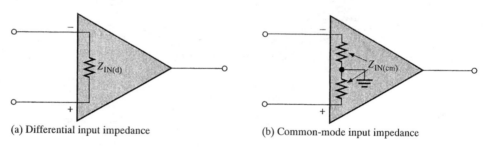

(a) Differential input impedance   (b) Common-mode input impedance

FIGURE 2–13
Op-amp input impedance.

Input Offset Current

Ideally, the two input bias currents are equal, and thus their difference is zero. In a practical op-amp, however, the bias currents are not exactly equal.

The *input offset current*, I_{OS}, is the difference of the input bias currents, expressed as an absolute value.

$$I_{OS} = |I_1 - I_2|$$

Actual magnitudes of offset current are usually at least an order of magnitude (ten times) less than the bias current. In many applications, the offset current can be neglected. However, high-gain, high-input impedance amplifiers should have as little I_{OS} as possible because the difference in currents through large input resistances develops a substantial offset voltage, as shown in Figure 2–14.

The offset voltage developed by the input offset current is

$$V_{OS} = I_1 R_{in} - I_2 R_{in} = (I_1 - I_2)R_{in} = I_{OS}R_{in}$$

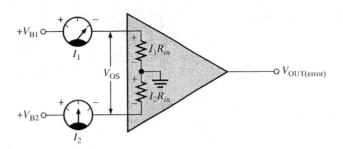

FIGURE 2–14
Effect of input offset current.

The error created by I_{OS} is amplified by the gain A_v of the op-amp and appears in the output as

$$V_{OUT(error)} = A_v I_{OS} R_{in}$$

A change in offset current with temperature affects the error voltage. Values of temperature coefficient for the offset current in the range of 0.5 nA per degree Celsius are common.

Output Impedance

Output impedance is the resistance viewed from the output terminal of the op-amp, as indicated in Figure 2–15.

FIGURE 2–15
Op-amp output impedance.

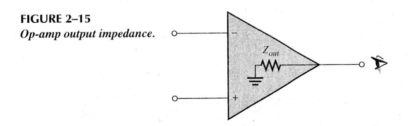

Common-Mode Input Voltage Range

All op-amps have limitations on the range of voltages over which they will operate. The *common-mode input voltage range* is the range of input voltages which, when applied to both inputs, will not cause clipping or other output distortion. Many op-amps have common-mode ranges of no more than ±10 V with dc supply voltages of ±15 V, while in others the output can go as high as the supply voltages (this is called rail-to-rail).

Open-Loop Voltage Gain

The **open loop voltage gain,** A_{ol} of an op-amp is the internal voltage gain of the device and represents the ratio of output voltage to input voltage when there are no external components. The open-loop voltage gain is set entirely by the internal design. Open-loop voltage gain can range up to 200,000 and is *not a well-controlled parameter.* Data sheets often refer to the open-loop voltage gain as the *large-signal voltage gain.*

Common-Mode Rejection Ratio

The *common-mode rejection ratio* (CMRR), as discussed in conjunction with the diff-amp, is a measure of an op-amp's ability to reject common-mode signals. An infinite value of CMRR means that the output is zero when the same signal is applied to both inputs (common-mode).

An infinite CMRR is never achieved in practice, but a good op-amp does have a very high value of CMRR. As previously mentioned, common-mode signals are undesired interference voltages such as 60 Hz power-supply ripple and noise voltages due to pick-up of radiated energy. A high CMRR enables the op-amp to virtually eliminate these interference signals from the output.

The accepted definition of CMRR for an op-amp is the open-loop voltage gain (A_{ol}) divided by the common-mode gain.

$$\text{CMRR} = \frac{A_{ol}}{A_{cm}} \tag{2-3}$$

It is commonly expressed in decibels as follows:

$$\text{CMRR}' = 20 \log\left(\frac{A_{ol}}{A_{cm}}\right) \tag{2-4}$$

EXAMPLE 2–3   A certain op-amp has an open-loop voltage gain of 100,000 and a common-mode gain of 0.25. Determine the CMRR and express it in decibels.

Solution
$$\text{CMRR} = \frac{A_{ol}}{A_{cm}} = \frac{100,000}{0.25} = \mathbf{400,000}$$
$$\text{CMRR}' = 20 \log(400,000) = \mathbf{112\ dB}$$

Practice Exercise   If a particular op-amp has a CMRR′ of 90 dB and a common-mode gain of 0.4, what is the open-loop voltage gain?

Slew Rate

The maximum rate of change of the output voltage in response to a step input voltage is the **slew rate** of an op-amp. The slew rate is dependent upon the high-frequency response of the amplifier stages within the op-amp.

Slew rate is measured with an op-amp connected as shown in Figure 2–16(a). This particular op-amp connection is a unity-gain, noninverting configuration which will be discussed later. It gives a worst-case (slowest) slew rate. Recall that the high-frequency components of a voltage step are contained in the rising edge and that the upper critical frequency of an amplifier limits its response to a step input. The lower the upper critical frequency is, the more gradual the slope on the output for a step input.

A pulse is applied to the input as shown in Figure 2–16(b), and the ideal output voltage is measured as indicated. The width of the input pulse must be sufficient to allow the output to "slew" from its lower limit to its upper limit, as shown. As you can see, a certain

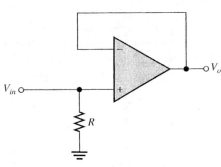

(a) Test circuit

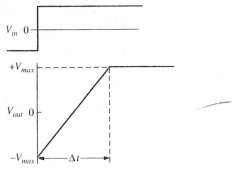

(b) Step input voltage and the resulting output voltage

FIGURE 2–16
Slew rate measurement.

time interval, Δt, is required for the output voltage to go from its lower limit $-V_{max}$ to its upper limit $+V_{max}$, once the input step is applied. The slew rate is expressed as

$$\text{Slew rate} = \frac{\Delta V_{out}}{\Delta t} \qquad (2\text{–}5)$$

where $\Delta V_{out} = +V_{max} - (-V_{max})$. The unit of slew rate is volts per microsecond (V/μs).

EXAMPLE 2–4

The output voltage of a certain op-amp appears as shown in Figure 2–17 in response to a step input. Determine the slew rate.

FIGURE 2–17

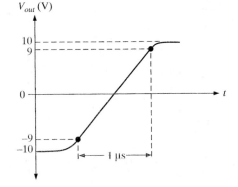

Solution The output goes from the lower to the upper limit in 1 μs. Since this response is not ideal, the limits are taken at the 90% points, as indicated, So, the upper limit is $+9$ V and the lower limit is -9 V. The slew rate is

$$\text{Slew rate} = \frac{\Delta V}{\Delta t} = \frac{+9\,\text{V} - (-9\,\text{V})}{1\ \mu s} = \textbf{18 V/μs}$$

Practice Exercise When a pulse is applied to an op-amp, the output voltage goes from -8 V to $+7$ V in 0.75 μs. What is the slew rate?

Frequency Response

The internal amplifier stages that make up an op-amp have voltage gains limited by junction capacitances. Although the differential amplifiers used in op-amps are somewhat different from the basic amplifiers discussed, the same principles apply. An op-amp has no internal coupling capacitors, however; therefore, the low-frequency response extends down to dc (0 Hz). Frequency-related characteristics of op-amps will be discussed in the next chapter.

Comparison of Op-Amp Parameters

Table 2–1 provides a comparison of values of some of the parameters just described for several common IC op-amps. Any values not listed were not given on the manufacturer's data sheet.

TABLE 2–1

Op-amp	Input offset voltage (mV) (max)	Input bias current (nA) (max)	Input impedance (MΩ) (min)	Open-loop gain (typ)	Slew rate (V/μs) (typ)	CMRR' (dB) (min)	Comment
LM741C	6	500	0.3	200,000	0.5	70	Industry standard
LM101A	7.5	250	1.5	160,000	—	80	General-purpose
OP113E	0.075	600	—	2,400,000	1.2	100	Low noise, low drift
OP177A	0.01	1.5	26	12,000,000	0.3	130	Ultra precision
OP184E	0.065	350	—	240,000	2.4	60	Precision, rail-to-rail*
AD8009AR	5	150	—	—	5500	50	$BW = 700$ MHz, ultra fast, low distortion, current feedback
AD8041A	7	2000	0.16	56,000	160	74	$BW = 160$ MHz, rail-to-rail
AD8055A	5	1200	10	3500	1400	82	Very fast voltage feedback

*Rail-to-rail means that the output voltage can go as high as the supply voltages.

Other Features

Most available op-amps have three important features: short-circuit protection, no latch-up, and input offset nulling. Short-circuit protection keeps the circuit from being damaged if the output becomes shorted, and the no latch-up feature prevents the op-amp from hanging up in one output state (high or low voltage level) under certain input conditions. Input offset nulling is achieved by an external potentiometer that sets the output voltage at precisely zero with zero input.

2–3   REVIEW QUESTIONS

1. List ten or more op-amp parameters.
2. Which two parameters, not including frequency response, are frequency dependent?

2–4   ■   NEGATIVE FEEDBACK

Negative feedback is one of the most useful concepts in electronics, particularly in op-amp applications. Negative feedback is the process whereby a portion of the output voltage of an amplifier is returned to the input with a phase angle that opposes (or subtracts from) the input signal.

After completing this section, you should be able to

❑ Explain negative feedback in op-amp circuits
   ❑ Describe the effects of negative feedback
   ❑ Discuss why negative feedback is used

Negative feedback is illustrated in Figure 2–18. The inverting ($-$) input effectively makes the feedback signal 180° out of phase with the input signal. The op-amp has extremely high gain and amplifies the *difference* in the signals applied to the inverting and noninverting inputs. A very tiny difference in these two signals is all the op-amp needs to produce the required output. *When negative feedback is present, the noninverting and inverting inputs are nearly identical.* This concept can help you figure out what signal to expect in many op-amp circuits.

FIGURE 2–18
Illustration of negative feedback.

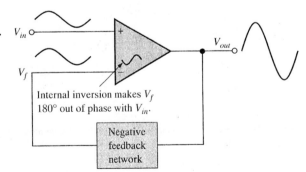

Internal inversion makes V_f 180° out of phase with V_{in}.

Now let's review how negative feedback works and why the signals at the inverting and noninverting terminals are nearly identical when negative feedback is used. Assume a 1.0 V input signal is applied to the noninverting terminal and the open-loop gain of the op-amp is 100,000. The amplifier responds to the voltage at its noninverting input terminal and moves the output toward saturation. Immediately, a fraction of this output is returned to the inverting terminal through the feedback path. But if the feedback signal ever reaches 1.0 V, there is nothing left for the op-amp to amplify! Thus, the feedback signal tries (but never quite succeeds) in matching the input signal. The gain is controlled by the amount of feedback used. When you are troubleshooting an op-amp circuit with negative feedback present, remember that the two inputs will look identical on a scope but in fact are very slightly different.

Now suppose something happens that reduces the internal gain of the op-amp. This causes the output signal to drop a small amount, returning a smaller signal to the inverting input via the feedback path. This means the difference between the signals is larger than it was. The output increases, compensating for the original drop in gain. The net change in the output is so small, it can hardly be measured. The main point is that any variation in the amplifier is immediately compensated for by the negative feedback, resulting in a very stable, predictable output.

Why Use Negative Feedback?

As you have seen, the inherent open-loop gain of a typical op-amp is very high (usually greater than 100,000). Therefore, an extremely small difference in the two input voltages drives the op-amp into its saturated output states. In fact, even the input offset voltage of the op-amp can drive it into saturation. For example, assume $V_{in} = 1$ mV and $A_{ol} = 100,000$. Then,

$$V_{in}A_{ol} = (1 \text{ mV})(100,000) = 100 \text{ V}$$

Since the output level of an op-amp can never reach 100 V, it is driven into saturation and the output is limited to its maximum output levels, as illustrated in Figure 2–19 for both a positive and a negative input voltage of 1 mV.

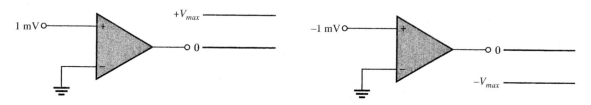

FIGURE 2–19
Without negative feedback, an extremely small difference in the two input voltages drives the op-amp to its output limits and it becomes nonlinear.

The usefulness of an op-amp operated in this manner is severely restricted and is generally limited to comparator applications (to be studied in Chapter 4). With negative feedback, the overall closed-loop voltage gain (A_{cl}) can be reduced and controlled so that the op-amp can function as a linear amplifier. In addition to providing a controlled, stable voltage gain, negative feedback also provides for control of the input and output impedances and amplifier bandwidth. Table 2–2 summarizes the general effects of negative feedback on op-amp performance.

TABLE 2–2

	Voltage Gain	Input Z	Output Z	Bandwidth
Without negative feedback	A_{ol} is too high for linear amplifier applications	Relatively high (see Table 2–1)	Relatively low	Relatively narrow (because the gain is so high)
With negative feedback	A_{cl} is set to desired value by the feedback network	Can be increased or reduced to a desired value depending on type of circuit	Can be reduced to a desired value	Significantly wider

2–4 REVIEW QUESTIONS

1. What are the benefits of negative feedback in an op-amp circuit?
2. Why is it necessary to reduce the gain of an op-amp from its open-loop value?
3. When troubleshooting an op-amp circuit in which negative feedback is present, what do you expect to observe on the input terminals?

2–5 ■ OP-AMP CONFIGURATIONS WITH NEGATIVE FEEDBACK

In this section, we will discuss three basic ways in which an op-amp can be connected using negative feedback to stabilize the gain and increase frequency response. As mentioned, the extremely high open-loop gain of an op-amp creates an unstable situation because a small noise voltage on the input can be amplified to a point where the amplifier is driven out of its linear region. Also, unwanted oscillations can occur. In addition, the open-loop gain parameter of an op-amp can vary greatly from one device to the next. Negative feedback takes a portion of the output and applies it back out of phase with the input, creating an effective reduction in gain. This closed-loop gain is usually much less than the open-loop gain and independent of it.

After completing this section, you should be able to

❑ Analyze three op-amp configurations
   ❑ Identify the noninverting amplifier configuration
   ❑ Determine the voltage gain of a noninverting amplifier
   ❑ Identify the voltage-follower configuration
   ❑ Identify the inverting amplifier configuration
   ❑ Determine the voltage gain of an inverting amplifier

Closed-Loop Voltage Gain, A_{cl}

The **closed-loop voltage gain** is the voltage gain of an op-amp with negative feedback. The amplifier configuration consists of the op-amp and an external feedback network that connects the output to the inverting input. The closed-loop voltage gain is then determined by the component values in the feedback network and can be precisely controlled by them.

Noninverting Amplifier

An op-amp connected in a **closed-loop** configuration as a **noninverting amplifier** is shown in Figure 2–20. The input signal is applied to the noninverting (+) input. A portion of the output is applied back to the inverting (−) input through the feedback network. This constitutes negative feedback. The feedback fraction, B, is the portion of the output returned to the inverting input and determines the gain of the amplifier as you will see. This smaller feedback voltage, V_f, can be written

$$V_f = BV_{out}$$

FIGURE 2–20
Noninverting amplifier.

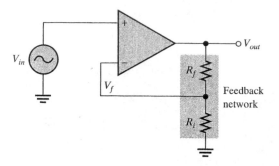

The differential voltage, V_{diff}, between the op-amp's input terminals is illustrated in Figure 2–21 and can be expressed as

$$V_{diff} = V_{in} - V_f$$

This input differential voltage is forced to be very small as a result of the negative feedback and the high open-loop gain, A_{ol}. Therefore, a close approximation is

$$V_{in} \cong V_f$$

By substitution,

$$V_{in} \cong BV_{out}$$

Rearranging,

$$\frac{V_{out}}{V_{in}} \cong \frac{1}{B}$$

FIGURE 2–21
Differential input, $V_{in} - V_f$.

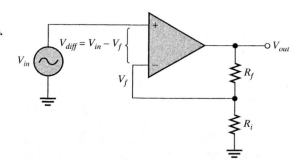

The ratio of the output voltage to the input voltage is the closed-loop gain. This result shows that the closed-loop gain for the noninverting amplifier, $A_{cl(\text{NI})}$, is approximately

$$A_{cl(\text{NI})} = \frac{V_{out}}{V_{in}} \cong \frac{1}{B}$$

The feedback fraction is determined by R_f and R_i, which form a voltage-divider network. The fraction of the output voltage, V_{out}, that is returned to the inverting input is found by applying the voltage-divider rule to the feedback network.

$$V_{in} \cong BV_{out} \cong \left(\frac{R_i}{R_f + R_i}\right)V_{out}$$

Rearranging,

$$\frac{V_{out}}{V_{in}} = \left(\frac{R_f + R_i}{R_i}\right)$$

which can be expressed as follows:

$$A_{cl(\text{NI})} = \frac{R_f}{R_i} + 1 \qquad (2\text{--}6)$$

Equation (2–6) shows that the closed-loop voltage gain, $A_{cl(\text{NI})}$, of the noninverting (NI) amplifier is not dependent on the op-amp's open-loop gain but can be set by selecting values of R_i and R_f. This equation is based on the assumption that the open-loop gain is very high compared to the ratio of the feedback resistors, causing the input differential voltage, V_{diff}, to be very small. In nearly all practical circuits, this is an excellent assumption.

For those rare cases where a more exact equation is necessary, the output voltage can be expressed as

$$V_{out} = V_{in}\left(\frac{A_{ol}}{1 + A_{ol}B}\right)$$

The following formula gives the exact solution of the closed-loop gain:

$$A_{cl(\text{NI})} = \frac{V_{out}}{V_{in}} = \left(\frac{A_{ol}}{1 + A_{ol}B}\right)$$

EXAMPLE 2–5   Determine the closed-loop voltage gain of the amplifier in Figure 2–22.

FIGURE 2–22

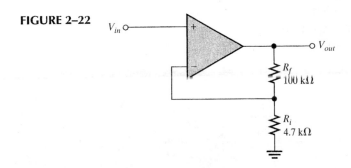

Solution This is a noninverting op-amp configuration. Therefore, the closed-loop voltage gain is

$$A_{cl(\text{NI})} = \frac{R_f}{R_i} + 1 = \frac{100\,\text{k}\Omega}{4.7\,\text{k}\Omega} + 1 = \mathbf{22.3}$$

Practice Exercise If R_f in Figure 2–22 is increased to 150 kΩ, determine the closed-loop gain.

Voltage-Follower The **voltage-follower** configuration is a special case of the noninverting amplifier where all of the output voltage is fed back to the inverting input by a straight connection, as shown in Figure 2–23. As you can see, the straight feedback connection has a voltage gain of approximately 1. The closed-loop voltage gain of a noninverting amplifier is $1/B$ as previously derived. Since $B = 1$, the closed-loop gain of the voltage-follower is

$$A_{cl(\text{VF})} = 1 \qquad\qquad (2\text{–}7)$$

The most important features of the voltage-follower configuration are its very high input impedance and its very low output impedance. These features make it a nearly ideal buffer amplifier for interfacing high-impedance sources and low-impedance loads. This is discussed further in Section 2–6.

FIGURE 2–23
Op-amp voltage-follower.

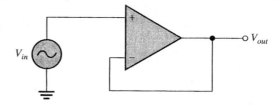

Inverting Amplifier

An op-amp connected as an **inverting amplifier** with a controlled amount of voltage gain is shown in Figure 2–24. The input signal is applied through a series input resistor (R_i) to the inverting input. Also, the output is fed back through R_f to the inverting input. The noninverting input is grounded.

FIGURE 2–24
Inverting amplifier.

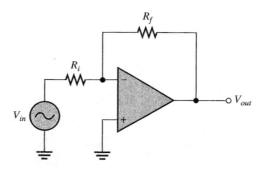

At this point, the ideal op-amp parameters mentioned earlier are useful in simplifying the analysis of this circuit. In particular, the concept of infinite input impedance is of great value. An infinite input impedance implies that there is *no* current out of the inverting input. If there is no current through the input impedance, then there must be *no* voltage drop between the inverting and noninverting inputs. This means that the voltage at the inverting (−) input is zero because the noninverting (+) input is grounded. This zero voltage at the inverting input terminal is referred to as *virtual ground*. This condition is illustrated in Figure 2–25(a).

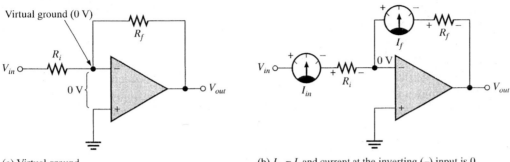

(a) Virtual ground

(b) $I_{in} = I_f$ and current at the inverting (−) input is 0.

FIGURE 2–25
Virtual ground concept and closed-loop voltage gain development for the inverting amplifier.

Since there is no current at the inverting input, the current through R_i and the current through R_f are equal, as shown in Figure 2–25(b).

$$I_{in} = I_f$$

The voltage across R_i equals V_{in} because of virtual ground on the other side of the resistor. Therefore,

$$I_{in} = \frac{V_{in}}{R_i}$$

Also, the voltage across R_f equals $-V_{out}$ because of virtual ground, and therefore,

$$I_f = \frac{-V_{out}}{R_f}$$

Since $I_f = I_{in}$,

$$\frac{-V_{out}}{R_f} = \frac{V_{in}}{R_i}$$

Rearranging the terms,

$$\frac{V_{out}}{V_{in}} = -\frac{R_f}{R_i}$$

Of course, V_{out}/V_{in} is the overall gain of the inverting amplifier.

$$A_{cl(I)} = -\frac{R_f}{R_i}$$

(2–8)

Equation (2–8) shows that the closed-loop voltage gain $A_{cl(I)}$ of the inverting amplifier is the ratio of the feedback resistance R_f to the input resistance R_i. *The closed-loop gain is independent of the op-amp's internal open-loop gain.* Thus, the negative feedback stabilizes the voltage gain. The negative sign indicates inversion.

EXAMPLE 2–6

Given the op-amp configuration in Figure 2–26, determine the value of R_f required to produce a closed-loop voltage gain of -100.

FIGURE 2–26

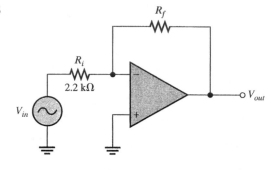

Solution   Knowing that $R_i = 2.2 \text{ k}\Omega$ and $A_{cl(I)} = -100$, calculate R_f as follows:

$$A_{cl(I)} = -\frac{R_f}{R_i}$$

$$R_f = -A_{cl(I)}R_i = -(-100)(2.2 \text{ k}\Omega) = \textbf{220 k}\boldsymbol{\Omega}$$

Practice Exercise
(a) If R_i is changed to $2.7 \text{ k}\Omega$ in Figure 2–26, what value of R_f is required to produce a closed-loop gain of -25?
(b) If R_f failed open, what would you expect to see at the output?

2–5 REVIEW QUESTIONS

1. What is the main purpose of negative feedback?
2. The closed-loop voltage gain of each of the op-amp configurations discussed is dependent on the internal open-loop voltage gain of the op-amp. (True or False)
3. The attenuation of the negative feedback network of a noninverting op-amp configuration is 0.02. What is the closed-loop gain of the amplifier?

2–6 ■ OP-AMP IMPEDANCES

In this section, you will see how a negative feedback connection affects the input and output impedances of an op-amp. The effects on both inverting and noninverting amplifiers are examined.

After completing this section, you should be able to

❑ Describe impedances of the three op-amp configurations
    ❑ Determine input and output impedances of a noninverting amplifier
    ❑ Determine input and output impedances of a voltage-follower
    ❑ Determine input and output impedances of an inverting amplifier

Input Impedance of the Noninverting Amplifier

Recall that negative feedback causes the feedback voltage, V_f, to nearly equal the input voltage, V_{in}. The difference between the input and feedback voltage, V_{diff}, is approximately zero, and ideally, can be assumed to have this value. This assumption implies that the input signal current to the op-amp is also zero. Since the input impedance is the ratio of input voltage to input current, the input impedance of a noninverting amplifier is

$$Z_{in} = \frac{V_{in}}{I_{in}} \cong \frac{V_{in}}{0} = \text{infinity } (\infty)$$

For many practical circuits, this assumption is good for obtaining a basic idea of the operation. A more exact analysis takes into account the fact that the input signal current is not zero.

The exact input impedance of this op-amp configuration is developed with the aid of Figure 2–27. For this analysis, a small differential voltage, V_{diff}, is assumed to exist between the two inputs, as indicated. This means that you cannot assume the op-amp's input impedance to be infinite or the input current to be zero. The input voltage can be expressed as

$$V_{in} = V_{diff} + V_f$$

FIGURE 2–27

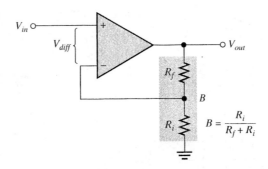

Substituting BV_{out} for V_f,

$$V_{in} = V_{diff} + BV_{out}$$

Since $V_{out} \cong A_{ol}V_{diff}$ (A_{ol} is the open-loop gain of the op-amp),

$$V_{in} = V_{diff} + A_{ol}BV_{diff} = (1 + A_{ol}B)V_{diff}$$

Because $V_{diff} = I_{in}Z_{in}$,

$$V_{in} = (1 + A_{ol}B)I_{in}Z_{in}$$

where Z_{in} is the open-loop input impedance of the op-amp (without feedback connections).

$$\frac{V_{in}}{I_{in}} = (1 + A_{ol}B)Z_{in}$$

V_{in}/I_{in} is the overall input impedance of the closed-loop noninverting configuration.

$$Z_{in(NI)} = (1 + A_{ol}B)Z_{in} \qquad (2\text{--}9)$$

This equation shows that the input impedance of this amplifier configuration with negative feedback is much greater than the internal input impedance of the op-amp itself (without feedback).

Output Impedance of the Noninverting Amplifier

In addition to the input impedance, negative feedback also produces an advantage for the output impedance of an op-amp. The output impedance of an amplifier without feedback is relatively low. With feedback, the output impedance is even lower. For many applications, the assumption that the output impedance with feedback is zero will produce sufficient accuracy. That is,

$$Z_{out(NI)} \cong 0$$

An exact analysis to find the output impedance with feedback is developed with the aid of Figure 2–28. By applying Kirchhoff's law to the output circuit,

$$V_{out} = A_{ol}V_{diff} - Z_{out}I_{out}$$

The differential input voltage is $V_{in} - V_f$; so, by assuming that $A_{ol}V_{diff} >> Z_{out}I_{out}$, the output voltage can be expressed as

$$V_{out} \cong A_{ol}(V_{in} - V_f)$$

Substituting BV_{out} for V_f,

$$V_{out} \cong A_{ol}(V_{in} - BV_{out})$$

Remember, B is the attenuation of the negative feedback network. Expanding, factoring and rearranging terms,

$$A_{ol}V_{in} \cong V_{out} + A_{ol}BV_{out} = (1 + A_{ol}B)V_{out}$$

FIGURE 2–28

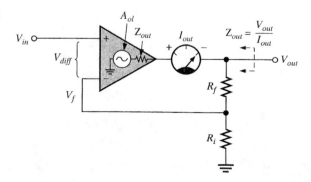

Since the output impedance of the noninverting configuration is $Z_{out(NI)} = V_{out}/I_{out}$, you can substitute $I_{out}Z_{out(NI)}$ for V_{out}; therefore,

$$A_{ol}V_{in} = (1 + A_{ol}B)I_{out}Z_{out(NI)}$$

Dividing both sides of the above expression by I_{out} yields

$$\frac{A_{ol}V_{in}}{I_{out}} = (1 + A_{ol}B)Z_{out(NI)}$$

The term on the left is the internal output impedance of the op-amp (Z_{out}) because, without feedback, $A_{ol}V_{in} = V_{out}$. Therefore,

$$Z_{out} = (1 + A_{ol}B)Z_{out(NI)}$$

Thus,

$$Z_{out(NI)} = \frac{Z_{out}}{1 + A_{ol}B} \qquad (2\text{-}10)$$

This equation shows that the output impedance of this amplifier configuration with negative feedback is much less than the internal output impedance of the op-amp itself (without feedback) because it is divided by the factor $1 + A_{ol}B$.

EXAMPLE 2–7

(a) Determine the input and output impedances of the amplifier in Figure 2–29. The op-amp data sheet gives $Z_{in} = 2$ MΩ, $Z_{out} = 75$ Ω, and $A_{ol} = 200,000$.

(b) Find the closed-loop voltage gain.

FIGURE 2–29

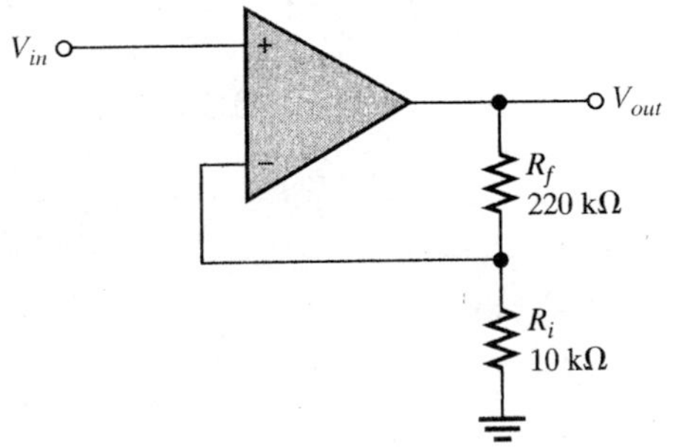

Solution

(a) The attenuation, B, of the feedback network is

$$B = \frac{R_i}{R_i + R_f} = \frac{10\ \text{k}\Omega}{230\ \text{k}\Omega} = 0.0435$$

$$Z_{in(NI)} = (1 + A_{ol}B)Z_{in} = [1 + (200,000)(0.0435)](2\ \text{M}\Omega)$$
$$= (1 + 8700)(2\ \text{M}\Omega) - 17,402\ \text{M}\Omega - 17.4\ \text{G}\Omega$$

$$Z_{out(NI)} = \frac{Z_{out}}{1 + A_{ol}B} = \frac{75\ \Omega}{1 + 8700} = 0.0086\ \Omega = \textbf{8.6 m}\Omega$$

(b) $$A_{cl(NI)} = \frac{1}{B} = \frac{1}{0.0435} \cong \textbf{23}$$

Practice Exercise

(a) Determine the input and output impedances in Figure 2–29 for op-amp data sheet values of $Z_{in} = 3.5$ MΩ, $Z_{out} = 82$ Ω, and $A_{ol} = 135,000$.

(b) Find A_{cl}.

Voltage-Follower Impedances

Since the voltage-follower is a special case of the noninverting configuration, the same impedance formulas are used with $B = 1$.

$$Z_{in(VF)} = (1 + A_{ol})Z_{in} \tag{2–11}$$

$$Z_{out(VF)} = \frac{Z_{out}}{1 + A_{ol}} \tag{2–12}$$

As you can see, the voltage-follower input impedance is greater for a given A_{ol} and Z_{in} than for the noninverting configuration with the voltage-divider feedback network. Also, its output impedance is much smaller because B is normally much smaller than 1 for a noninverting configuration.

EXAMPLE 2–8

The same op-amp as in Example 2–7 is used in a voltage-follower configuration. Determine the input and output impedances.

Solution Since $B = 1$,

$$Z_{in(VF)} = (1 + A_{ol})Z_{in} = (1 + 200,000)(2 \text{ MΩ}) = \mathbf{400 \text{ GΩ}}$$

$$Z_{out(VF)} = \frac{Z_{out}}{1 + A_{ol}} = \frac{75 \text{ Ω}}{1 + 200,000} = \mathbf{375 \text{ μΩ}}$$

Notice that $Z_{in(VF)}$ is much greater than $Z_{in(NI)}$, and $Z_{out(VF)}$ is much less than $Z_{out(NI)}$ from Example 2–7.

Practice Exercise If the op-amp in this example is replaced with one having a higher open-loop gain, how are the input and output impedances affected?

Impedances of the Inverting Amplifier

The input impedance of this op-amp configuration is developed with the aid of Figure 2–30. Because both the input signal and the negative feedback are applied, through resistors, to the inverting terminal, Miller's theorem can be applied to this configuration. According to Miller's theorem, the effective input impedance of an amplifier with a feedback resistor from output to input as in Figure 2–30 is

$$Z_{in(Miller)} = \frac{R_f}{1 + A_{ol}}$$

and

$$Z_{out(Miller)} = \left(\frac{A_{ol}}{1 + A_{ol}}\right)R_f$$

FIGURE 2–30
Inverting amplifier.

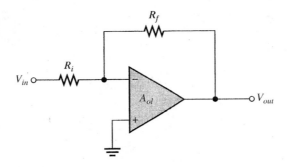

Applying Miller's theorem to the circuit of Figure 2–30, you get the equivalent circuit of Figure 2–31. As indicated, the Miller input impedance appears in parallel with the internal input impedance of the op-amp, and R_i appears in series with this as follows:

$$Z_{in(I)} = R_i + \frac{R_f}{1 + A_{ol}} \parallel Z_{in}$$

FIGURE 2–31
Miller equivalent for the inverting amplifier in Figure 2–30.

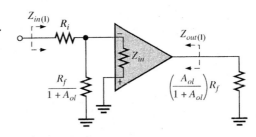

Typically, $R_f/(1 + A_{ol})$ is much less than the Z_{in} of an open-loop op-amp; also, $A_{ol} \gg 1$. So the previous equation simplifies to

$$Z_{in(I)} \cong R_i + \frac{R_f}{A_{ol}}$$

Since R_i appears in series with R_f/A_{ol} and if $R_i \gg R_f/A_{ol}$, $Z_{in(I)}$ reduces to

$$Z_{in(I)} \cong R_i \qquad\qquad (2\text{–}13)$$

The Miller output impedance appears in parallel with Z_{out} of the op-amp.

$$Z_{out(I)} = \left(\frac{A_{ol}}{1 + A_{ol}}\right)R_f \parallel Z_{out}$$

Normally $A_{ol} \gg 1$ and $R_f \gg Z_{out}$, so $Z_{out(I)}$ simplifies to

$$Z_{out(I)} \cong Z_{out} \qquad\qquad (2\text{–}14)$$

EXAMPLE 2–9   Find the values of the input and output impedances in Figure 2–32. Also, determine the closed-loop voltage gain. The op-amp has the following parameters: $A_{ol} = 50{,}000$; $Z_{in} = 4\ \text{M}\Omega$; and $Z_{out} = 50\ \Omega$.

FIGURE 2–32

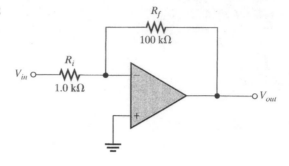

Solution

$$Z_{in(I)} \cong R_i = \textbf{1.0 k}\boldsymbol{\Omega}$$

$$Z_{out(I)} \cong Z_{out} = \textbf{50 }\boldsymbol{\Omega}$$

$$A_{cl(I)} = -\frac{R_f}{R_i} = -\frac{100 \text{ k}\Omega}{1.0 \text{ k}\Omega} = \textbf{-100}$$

Practice Exercise   Determine the input and output impedances and the closed-loop voltage gain in Figure 2–32. The op-amp parameters and circuit values are as follows: $A_{ol} = 100{,}000$; $Z_{in} = 5$ MΩ; $Z_{out} = 75$ Ω; $R_i = 560$ Ω; and $R_f = 82$ kΩ.

2–6 REVIEW QUESTIONS

1. How does the input impedance of a noninverting amplifier configuration compare to the input impedance of the op-amp itself?
2. When an op-amp is connected in a voltage-follower configuration, does the input impedance increase or decrease?
3. Given that $R_f = 100$ kΩ; $R_i = 2.0$ kΩ; $A_{ol} = 120{,}000$; $Z_{in} = 2$ MΩ; and $Z_{out} = 60$ Ω, what are $Z_{in(I)}$ and $Z_{out(I)}$ for an inverting amplifier configuration?

2–7 ■ TROUBLESHOOTING

As a technician, you will no doubt encounter situations in which an op-amp or its associated circuitry has malfunctioned. The op-amp is a complex integrated circuit with many types of internal failures possible. However, since you cannot troubleshoot the op-amp internally, you treat it as a single device with only a few connections to it. If it fails, you replace it just as you would a resistor, capacitor, or transistor.

After completing this section, you should be able to

❑ Troubleshoot op-amp circuits
   ❑ Analyze faults in a noninverting amplifier
   ❑ Analyze faults in a voltage-follower
   ❑ Analyze faults in an inverting amplifier

In op-amp configurations, there are only a few components that can fail. Both inverting and noninverting amplifiers have a feedback resistor, R_f, and an input resistor, R_i. Depending on the circuit, a load resistor, bypass capacitors, or a voltage compensation resistor may also be present. Any of these components can appear to be open or appear to be shorted. An open is not always due to the component itself, but may be due to a poor solder connection or a bent pin on the op-amp. Likewise, a short circuit may be due to a solder bridge. Of course, the op-amp itself can fail. Let's examine the basic configurations, considering only the feedback and input resistor failure modes and associated symptoms.

Faults in the Noninverting Amplifier

The first thing to do when you suspect a faulty circuit is to check for the proper power supply voltage. *The positive and negative supply voltages should be measured on the op-amp's pins* with respect to a nearby circuit ground. If either voltage is missing or incorrect, trace the power connections back toward the supply before making other checks. Check that the ground path is not open, giving a misleading power supply reading. If you have verified the supply voltages and ground path, possible faults with the basic amplifier are as follows.

Open Feedback Resistor If the feedback resistor, R_f, in Figure 2–33 opens, the op-amp is operating with its very high open-loop gain, which causes the input signal to drive the device into nonlinear operation and results in a severely clipped output signal as shown in part (a).

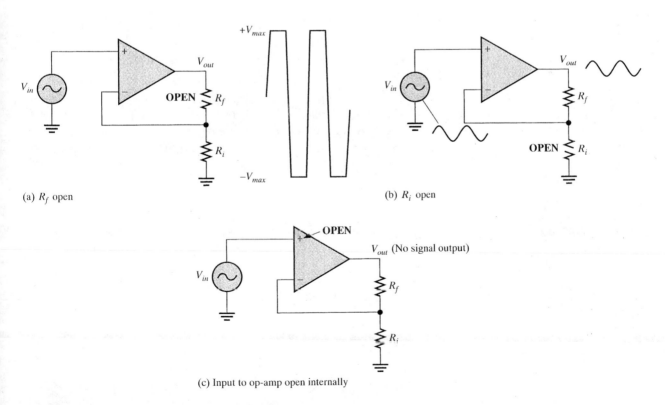

(a) R_f open

(b) R_i open

(c) Input to op-amp open internally

FIGURE 2–33

Faults in the noninverting amplifier.

Open Input Resistor In this case, you still have a closed-loop configuration. But, since R_i is open and effectively equal to infinity, ∞, the closed-loop gain from Equation (2–6) is

$$A_{cl(\text{NI})} = \frac{R_f}{R_i} + 1 = \frac{R_f}{\infty} + 1 = 0 + 1 = 1$$

This shows that the amplifier acts like a voltage-follower. You would observe an output signal that is the same as the input, as indicated in Figure 2–33(b).

Internally Open Noninverting Op-Amp Input In this situation, because the input voltage is not applied to the op-amp, the output is zero. This is indicated in Figure 2–33(c).

Other Op-Amp Faults In general, an internal failure will result in a loss or distortion of the output signal. The best approach is to first make sure that there are no external failures or faulty conditions. If everything else is good, then the op-amp must be bad.

Faults in the Voltage-Follower

The voltage-follower is a special case of the noninverting amplifier. Except for a bad power supply, a bad op-amp, or an open or short at a connection, about the only thing that can happen in a voltage-follower circuit is an open feedback loop. This would have the same effect as an open feedback resistor as previously discussed.

Faults in the Inverting Amplifier

Power Supply As in the case of the noninverting amplifier, the power supply voltages should be checked first. Power supply voltages should be checked on the op-amp's pins with respect to a nearby ground.

Open Feedback Resistor If R_f opens as indicated in Figure 2–34(a), the input signal still feeds through the input resistor and is amplified by the high open-loop gain of the op-amp. This forces the device to be driven into nonlinear operation, and you will see an output something like that shown. This is the same result as in the noninverting configuration.

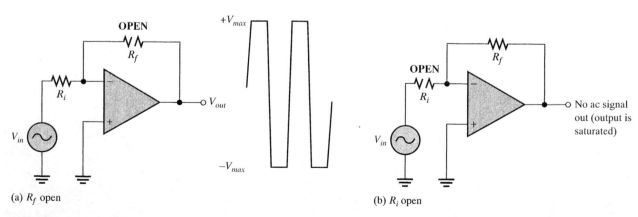

(a) R_f open

(b) R_i open

FIGURE 2–34
Faults in the inverting amplifier.

Open Input Resistor This prevents the input signal from getting to the op-amp input, so there will be no output signal, as indicated in Figure 2–34(b).

Failures in the op-amp itself have the same effects as previously discussed for the noninverting amplifier.

2–7 REVIEW QUESTIONS

1. If you notice that the op-amp output is saturated, what should you check first?
2. If there is no op-amp output signal when there is a verified input signal, what should you check first?

2–8 ■ A SYSTEM APPLICATION

The spectrophotometer system presented at the beginning of this chapter combines light optics with electronics to analyze the chemical makeup of various solutions. This type of system is common in medical laboratories as well as many other areas. It is another example of a mixed system in which electronic circuits interface with other types of systems, such as mechanical and optical, to accomplish a specific function. When you are a technician or technologist in industry, you will probably be working with different types of mixed systems from time to time.

After completing this section, you should be able to

❑ Apply what you have learned in this chapter to a system application
   ❑ Understand the role of electronics in a mixed system
   ❑ See how an op-amp is used in the system
   ❑ See how an electronic circuit interfaces with an optical device
   ❑ Translate between a printed circuit board and a schematic
   ❑ Troubleshoot some common system problems

A Brief Description of the System

The light source shown in Figure 2–35 produces a beam of visible light containing a wide spectrum of wavelengths. Each component wavelength in the beam of light is refracted at a different angle by the prism as indicated. Depending on the angle of the platform as set by the pivot angle controller, a certain wavelength passes through the narrow slit and is transmitted through the solution under analysis. By precisely pivoting the light source and prism, a selected wavelength can be transmitted. Every chemical and compound absorbs different wavelengths of light in different ways, so the resulting light coming through the solution has a unique "signature" that can be used to define the chemicals in the solution.

The photocell on the circuit board produces a voltage that is proportional to the amount of light and wavelength. The op-amp circuit amplifies the photocell output and sends the resulting signal to the processing and display unit where the type of chemical(s) in the solution is identified. The focus of this system application is the photocell/amplifier circuit board.

FIGURE 2–35

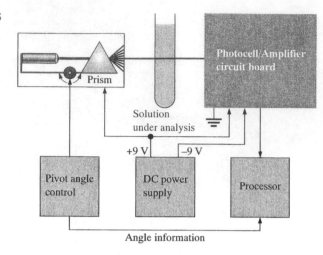

Now, so that you can take a closer look at the photocell/amplifier circuit board, let's take it out of the system and put it on the test bench.

ON THE TEST BENCH

■ ACTIVITY 1 Relate the PC Board to the Schematic

Develop a complete schematic by carefully following the conductive traces on the PC board shown in Figure 2–36 to see how the components are interconnected. Two interconnecting traces are on the reverse side of the board and are indicated as slightly darker trace. Refer to the chapter material or the 741 data sheet for the pin layout. This op-amp is housed in a surface-mount SO-8 package. A pad to which no component lead is connected represents a feedthrough to the other side.

FIGURE 2–36

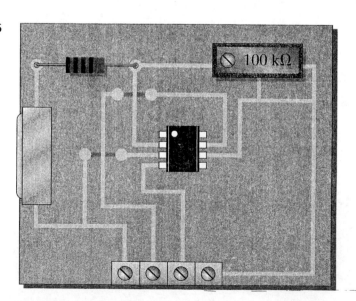

■ ACTIVITY 2   Analyze the Circuit

Step 1:   Determine the resistance value to which the feedback rheostat must be adjusted for a voltage gain of 50.

Step 2:   Assume the maximum linear output of the op-amp is 1 V less than the dc supply voltage. Determine the voltage gain required and the value to which the feedback resistor must be set to achieve the maximum linear output. The maximum voltage from the photocell is 0.5 V.

Step 3:   The system light source produces wavelengths ranging from 400 nm to 700 nm, which is approximately the full range of visible light from violet to red. Determine the op-amp output voltage over this range of wavelengths in 50 nm intervals and plot a graph of the results. Refer to the photocell response characteristic in Figure 2–37.

FIGURE 2–37
Photocell response curve.

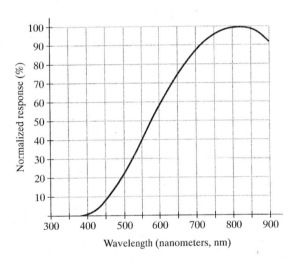

■ ACTIVITY 3   Write a Technical Report

Describe the circuit operation. Be sure to identify the type of op-amp circuit configuration and explain the purpose of the potentiometer. Use the results of Activity 2 to specify the performance of the circuit.

■ ACTIVITY 4   Troubleshoot the Photocell/Amplifier Circuit for Each of the Following Symptoms By Stating the Probable Cause or Causes

1. No voltage at the op-amp output
2. Output of op-amp stays at approximately −9 V.
3. A small dc voltage on the op-amp output under no-light conditions
4. Zero output voltage as light source is pivoted with verified photocell output voltage

2–8 REVIEW QUESTIONS

1. What is the purpose of the 100 kΩ potentiometer on the circuit board?
2. Explain why the light source and prism must be pivoted.

■ SUMMARY

- The basic op-amp has three terminals not including power and ground: inverting (−) input, noninverting (+) input, and output.
- Most op-amps require both a positive and a negative dc supply voltage.
- The ideal (perfect) op-amp has infinite input impedance, zero output impedance, infinite open-loop voltage gain, infinite bandwidth, and infinite CMRR.
- A good practical op-amp has high input impedance, low output impedance, high open-loop voltage gain, and a wide bandwidth.
- A differential amplifier is normally used for the input stage of an op-amp.
- A differential input voltage appears between the inverting and noninverting inputs of a differential amplifier.
- A single-ended input voltage appears between one input and ground (with the other input grounded).
- A differential output voltage appears between two output terminals of a diff-amp.
- A single-ended output voltage appears between the output and ground of a diff-amp.
- Common mode occurs when equal in-phase voltages are applied to both input terminals.
- Input offset voltage produces an output error voltage (with no input voltage).
- Input bias current also produces an output error voltage (with no input voltage).
- Input offset current is the difference between the two bias currents.
- Open-loop voltage gain is the gain of an op-amp with no external feedback connections.
- Closed-loop voltage gain is the gain of an op-amp with external feedback.
- The common-mode rejection ratio (CMRR) is a measure of an op-amp's ability to reject common-mode inputs.
- Slew rate is the rate in volts per microsecond at which the output voltage of an op-amp can change in response to a step input.
- Figure 2–38 shows the op-amp symbol and the three basic op-amp configurations.

FIGURE 2–38

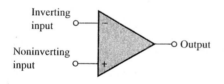

(a) Basic op-amp symbol

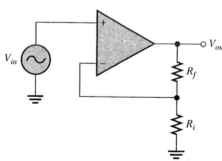

(b) Noninverting amplifier

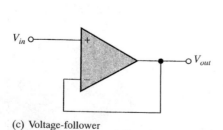

(c) Voltage-follower

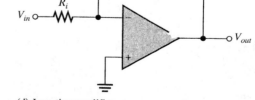

(d) Inverting amplifier

- All op-amp configurations listed use negative feedback. Negative feedback occurs when a portion of the output voltage is connected back to the inverting input such that it subtracts from the input voltage, thus reducing the voltage gain but increasing the stability and bandwidth.
- A noninverting amplifier configuration has a higher input impedance and a lower output impedance than the op-amp itself (without feedback).
- An inverting amplifier configuration has an input impedance approximately equal to the input resistor R_i and an output impedance approximately equal to the output impedance of the op-amp itself.
- The voltage-follower has the highest input impedance and the lowest output impedance of the three configurations.

■ GLOSSARY

These terms are included in the end-of-book glossary.

Closed-loop   An op-amp configuration in which the output is connected back to the input through a feedback circuit.

Closed-loop voltage gain, A_{cl}   The net voltage gain of an amplifier when negative feedback is included.

Common mode   A condition characterized by the presence of the same signal on both op-amp inputs.

Common-mode rejection ratio (CMRR)   The ratio of open-loop gain to common-mode gain; a measure of an op-amp's ability to reject common-mode signals.

Differential amplifier (diff-amp)   An amplifier that produces an output voltage proportional to the difference of the two input voltages.

Inverting amplifier   An op-amp closed-loop configuration in which the input signal is applied to the inverting input.

Negative feedback   The process of returning a portion of the output signal to the input of an amplifier such that it is out of phase with the input signal.

Noninverting amplifier   An op-amp closed-loop configuration in which the input signal is applied to the noninverting input.

Open-loop voltage gain, A_{ol}   The voltage gain of an amplifier without external feedback.

Operational amplifier (op-amp)   A type of amplifier that has very high voltage gain, very high input impedance, very low output impedance, and good rejection of common-mode signals.

Slew rate   The rate of change of the output voltage of an op-amp in response to a step input.

Stage   Each transistor in a multistage amplifier that amplifies a signal.

Voltage-follower   A closed-loop, noninverting op-amp with a voltage gain of 1.

■ KEY FORMULAS

Differential Amplifiers

(2–1) $CMRR = \dfrac{A_{v(d)}}{A_{cm}}$ Common-mode rejection ratio (diff-amp)

(2–2) $CMRR' = 20 \log\left(\dfrac{A_{v(d)}}{A_{cm}}\right)$ Common-mode rejection ratio (dB) (diff-amp)

Op-Amp Parameters

$$(2\text{-}3) \qquad \text{CMRR} = \frac{A_{ol}}{A_{cm}} \qquad\qquad \text{Common-mode rejection ratio (op-amp)}$$

$$(2\text{-}4) \qquad \text{CMRR}' = 20 \log\!\left(\frac{A_{ol}}{A_{cm}}\right) \qquad \text{Common-mode rejection ratio (dB) (op-amp)}$$

$$(2\text{-}5) \qquad \text{Slew rate} = \frac{\Delta V_{out}}{\Delta t} \qquad\qquad \text{Slew rate}$$

Op-Amp Configurations

$$(2\text{-}6) \qquad A_{cl(NI)} = \frac{R_f}{R_i} + 1 \qquad\qquad \text{Voltage gain (noninverting)}$$

$$(2\text{-}7) \qquad A_{cl(VF)} = 1 \qquad\qquad\qquad \text{Voltage gain (voltage-follower)}$$

$$(2\text{-}8) \qquad A_{cl(I)} = -\frac{R_f}{R_i} \qquad\qquad \text{Voltage gain (inverting)}$$

Op-Amp Impedances

$$(2\text{-}9) \qquad Z_{in(NI)} = (1 + A_{ol}B)Z_{in} \qquad \text{Input impedance (noninverting)}$$

$$(2\text{-}10) \qquad Z_{out(NI)} = \frac{Z_{out}}{1 + A_{ol}B} \qquad \text{Output impedance (noninverting)}$$

$$(2\text{-}11) \qquad Z_{in(VF)} = (1 + A_{ol})Z_{in} \qquad \text{Input impedance (voltage-follower)}$$

$$(2\text{-}12) \qquad Z_{out(VF)} = \frac{Z_{out}}{1 + A_{ol}} \qquad \text{Output impedance (voltage-follower)}$$

$$(2\text{-}13) \qquad Z_{in(I)} \cong R_i \qquad\qquad \text{Input impedance (inverting)}$$

$$(2\text{-}14) \qquad Z_{out(I)} \cong Z_{out} \qquad\qquad \text{Output impedance (inverting)}$$

$$B = \frac{R_i}{R_i + R_f}$$

■ SELF-TEST

1. An integrated circuit (IC) op-amp has
   (a) two inputs and two outputs   (b) one input and one output
   (c) two inputs and one output

2. Which of the following characteristics does not necessarily apply to an op-amp?
   (a) High gain   (b) Low power
   (c) High input impedance   (d) Low output impedance

3. A differential amplifier
   (a) is part of an op-amp   (b) has one input and one output
   (c) has two outputs   (d) answers (a) and (c)

4. When a differential amplifier is operated single-ended,
   (a) the output is grounded
   (b) one input is grounded and a signal is applied to the other
   (c) both inputs are connected together
   (d) the output is not inverted

5. In the differential mode,
   (a) opposite polarity signals are applied to the inputs   (b) the gain is 1
   (c) the outputs are different amplitudes   (d) only one supply voltage is used

6. In the common mode,
   (a) both inputs are grounded
   (b) the outputs are connected together
   (c) an identical signal appears on both inputs
   (d) the output signals are in phase

7. Common-mode gain is
   (a) very high   (b) very low   (c) always unity   (d) unpredictable

8. Differential gain is
   (a) very high
   (b) very low
   (c) dependent on the input voltage
   (d) about 100

9. If $A_{v(d)} = 3500$ and $A_{cm} = 0.35$, the CMRR is
   (a) 1225   (b) 10,000   (c) 80 dB   (d) answers (b) and (c)

10. With zero volts on both inputs, an op-amp ideally should have an output
    (a) equal to the positive supply voltage
    (b) equal to the negative supply voltage
    (c) equal to zero
    (d) equal to the CMRR

11. Of the values listed, the most realistic value for open-loop gain of an op-amp is
    (a) 1   (b) 2000   (c) 80 dB   (d) 100,000

12. A certain op-amp has bias currents of 50 μA and 49.3 μA. The input offset current is
    (a) 700 nA   (b) 99.3 μA   (c) 49.65 μA   (d) none of these

13. The output of a particular op-amp increases 8 V in 12 μs. The slew rate is
    (a) 96 V/μs   (b) 0.67 V/μs   (c) 1.5 V/μs   (d) none of these

14. For an op-amp with negative feedback, the output is
    (a) equal to the input
    (b) increased
    (c) fed back to the inverting input
    (d) fed back to the noninverting input

15. The use of negative feedback
    (a) reduces the voltage gain of an op-amp
    (b) makes the op-amp oscillate
    (c) makes linear operation possible
    (d) answers (a) and (c)

16. Negative feedback
    (a) increases the input and output impedances
    (b) increases the input impedance and the bandwidth
    (c) decreases the output impedance and the bandwidth
    (d) does not affect impedances or bandwidth

17. A certain noninverting amplifier has an R_i of 1.0 kΩ and an R_f of 100 kΩ. The closed-loop gain is
    (a) 100,000   (b) 1000   (c) 101   (d) 100

18. If the feedback resistor in Question 17 is open, the voltage gain
    (a) increases   (b) decreases   (c) is not affected   (d) depends on R_i

19. A certain inverting amplifier has a closed-loop gain of 25. The op-amp has an open-loop gain of 100,000. If another op-amp with an open-loop gain of 200,000 is substituted in the configuration, the closed-loop gain
    (a) doubles   (b) drops to 12.5   (c) remains at 25   (d) increases slightly

20. A voltage-follower
    (a) has a gain of one
    (b) is noninverting
    (c) has no feedback resistor
    (d) has all of these

■ PROBLEMS

SECTION 2–1 Introduction to Operational Amplifiers

1. Compare a practical op-amp to the ideal.

2. Two IC op-amps are available to you. Their characteristics are listed below. Choose the one you think is more desirable.
   Op-amp 1: $Z_{in} = 5$ MΩ, $Z_{out} = 100$ Ω, $A_{ol} = 50,000$
   Op-amp 2: $Z_{in} = 10$ MΩ, $Z_{out} = 75$ Ω, $A_{ol} = 150,000$

SECTION 2–2 The Differential Amplifier

3. Identify the type of input and output configuration for each basic differential amplifier in Figure 2–39.

4. The dc base voltages in Figure 2–40 are zero. Using your knowledge of transistor analysis, determine the dc differential output voltage. Assume that for Q_1, $I_C/I_E = 0.98$ and for Q_2, $I_C/I_E = 0.975$.

5. Identify the quantity being measured by each meter in Figure 2–41.

6. A differential amplifier stage has collector resistors of 5.1 kΩ each. If $I_{C1} = 1.35$ mA and $I_{C2} = 1.29$ mA, what is the differential output voltage?

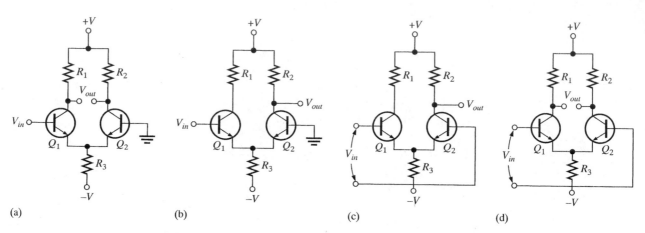

(a)   (b)   (c)   (d)

FIGURE 2–39

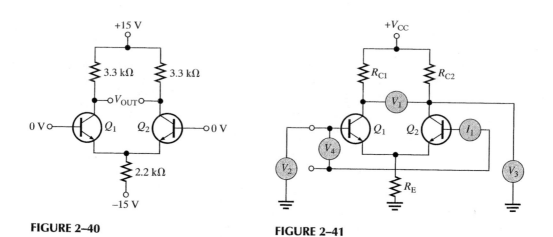

FIGURE 2–40   **FIGURE 2–41**

SECTION 2–3 Op-Amp Data Sheet Parameters

7. Determine the bias current, I_{BIAS}, given that the input currents to an op-amp are 8.3 μA and 7.9 μA.

8. Distinguish between input bias current and input offset current, and then calculate the input offset current in Problem 7.

9. A certain op-amp has a CMRR of 250,000. Convert this to dB.

10. The open-loop gain of a certain op-amp is 175,000. Its common-mode gain is 0.18. Determine the CMRR in dB.

11. An op-amp data sheet specifies a CMRR of 300,000 and an A_{ol} of 90,000. What is the common-mode gain?

12. Figure 2–42 shows the output voltage of an op-amp in response to a step input. What is the slew rate?

FIGURE 2–42

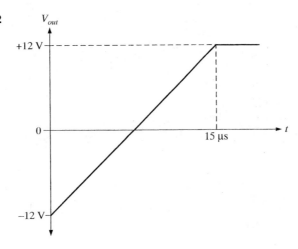

13. How long does it take the output voltage of an op-amp to go from -10 V to $+10$ V, if the slew rate is 0.5 V/µs?

SECTION 2–5   Op-Amp Configurations with Negative Feedback

14. Identify each of the op-amp configurations in Figure 2–43.

15. A noninverting amplifier has an R_i of 1.0 kΩ and an R_f of 100 kΩ. Determine V_f and B if $V_{out} = 5$ V.

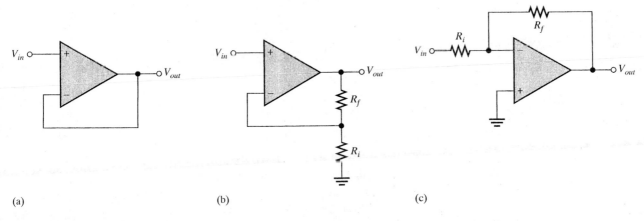

(a)   (b)   (c)

FIGURE 2–43

FIGURE 2–44

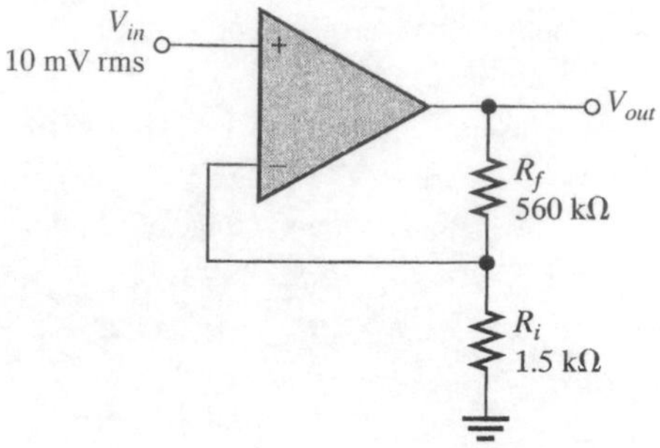

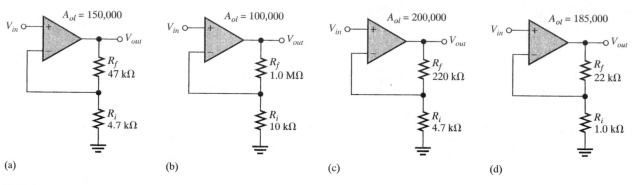

FIGURE 2–45

(a)   (b)   (c)   (d)

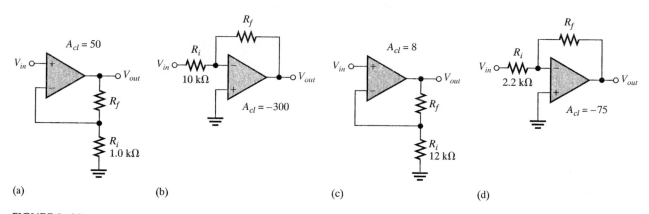

(a)   (b)   (c)   (d)

FIGURE 2–46

16. For the amplifier in Figure 2–44, determine the following:
   (a) $A_{cl(NI)}$ **(b)** V_{out} **(c)** V_f

17. Determine the closed-loop gain of each amplifier in Figure 2–45.

18. Find the value of R_f that will produce the indicated closed-loop gain in each amplifier in Figure 2–46.

19. Find the gain of each amplifier in Figure 2–47.

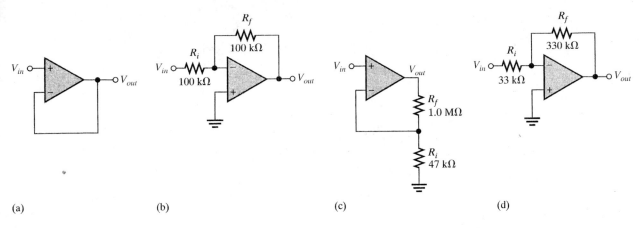

(a) (b) (c) (d)

FIGURE 2–47

FIGURE 2–48

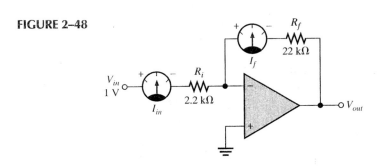

20. If a signal voltage of 10 mV rms is applied to each amplifier in Figure 2–47, what are the output voltages and what is their phase relationship with inputs?

21. Determine the approximate values for each of the following quantities in Figure 2–48.
(a) I_{in} (b) I_f (c) V_{out} (d) Closed-loop gain

SECTION 2–6 Op-Amp Impedances

22. Determine the input and output impedances for each amplifier configuration in Figure 2–49.

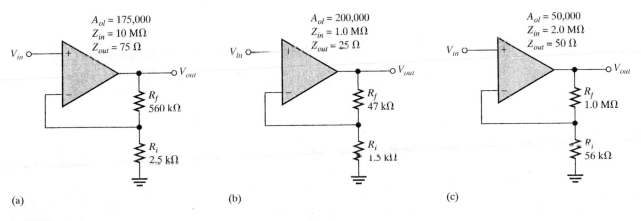

(a) (b) (c)

FIGURE 2–49

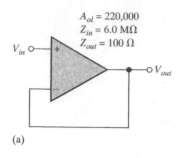

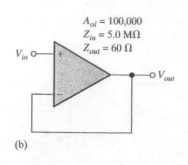

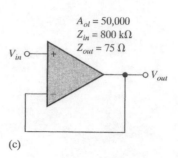

(a)

(b)

(c)

FIGURE 2–50

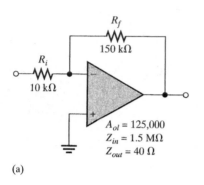

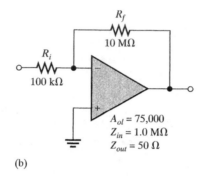

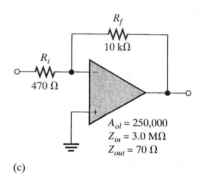

(a)

(b)

(c)

FIGURE 2–51

23. Repeat Problem 22 for each circuit in Figure 2–50.

24. Repeat Problem 22 for each circuit in Figure 2–51.

 SECTION 2–7   Troubleshooting

25. Determine the most likely fault(s) for each of the following symptoms in Figure 2–52 with a 100 mV signal applied.

(a) No output signal.

(b) Output severely clipped on both positive and negative swings.

FIGURE 2–52

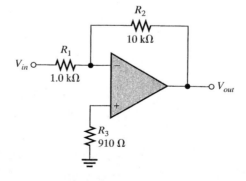

FIGURE 2–53

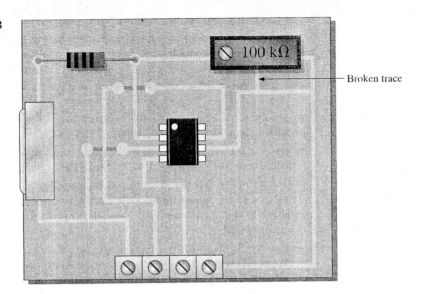

Broken trace

26. Determine the effect on the output if the circuit in Figure 2–52 has the following fault (one fault at a time).
   (a) Output pin is shorted to the inverting input.
   (b) R_3 is open.
   (c) R_3 is 10 kΩ instead of 910 Ω.
   (d) R_1 and R_2 are swapped.

27. On the circuit board in Figure 2–53, what happens if the middle lead (wiper) of the 100 kΩ potentiometer is broken?

■ ANSWERS TO REVIEW QUESTIONS

Section 2–1

1. Inverting input, noninverting input, output, positive and negative supply voltages

2. A practical op-amp has high input impedance, low output impedance, high voltage gain, and wide bandwidth.

Section 2–2

1. Differential input is between two input terminals. Single-ended input is from one input terminal to ground (with other input grounded).

2. Common-mode rejection is the ability of an op-amp to produce very little output when the same signal is applied to both inputs.

3. A higher CMRR results in a lower common-mode gain.

Section 2–3

1. Input bias current, input offset voltage, drift, input offset current, input impedance, output impedance, common-mode input voltage range, CMRR, open-loop voltage gain, slew rate, frequency response

2. Slew rate and voltage gain are both frequency dependent.

Section 2–4

1. Negative feedback provides a stable controlled voltage gain, control of input and output impedances, and wider bandwidth.

2. The open-loop gain is so high that a very small signal on the input will drive the op-amp into saturation.

3. Both inputs will be the same.

Section 2–5

1. The main purpose of negative feedback is to stabilize the gain.

2. False

3. $A_{cl} = 1/0.02 = 50$

Section 2–6

1. The noninverting configuration has a higher Z_{in} than the op-amp alone.

2. Z_{in} increases in a voltage-follower.

3. $Z_{in(I)} \cong R_i = 2.0$ kΩ, $Z_{out(I)} \cong Z_{out} = 60$ Ω.

Section 2–7

1. Check power supply voltages with respect to ground. Verify ground connections. Check for an open feedback resistor.

2. Verify power supply voltages and ground leads. For inverting amplifiers, check for open R_i. For noninverting amplifiers, check that V_{in} is actually on (+) pin; if so, check (−) pin for identical signal.

Section 2–8

1. The 100 kΩ potentiometer is the feedback resistor.

2. The light source and prism must be pivoted to allow different wavelengths of light to pass through the slit.

■ **ANSWERS TO PRACTICE EXERCISES FOR EXAMPLES**

2–1 34,000; 90.6 dB

2–2 **(a)** 0.168   **(b)** 87.96 dB   **(c)** 2.1 V rms, 4.2 V rms   **(d)** 0.168V

2–3 12,649

2–4 20 V/μs

2–5 32.9

2–6 **(a)** 67.5 kΩ **(b)** The amplifier would have an open-loop gain producing a square wave.

2–7 **(a)** 20.6 GΩ, 14 mΩ **(b)** 23

2–8 Input Z increases, output Z decreases.

2–9 $Z_{in(I)} = 560$ Ω, $Z_{out(I)} = 75$ Ω, $A_{cl} = -146$

3

OP-AMP RESPONSES

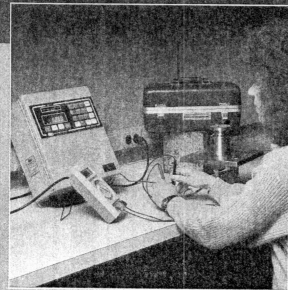

Courtesy Yuba College

■ CHAPTER OBJECTIVES

▢ Discuss the basic areas of op-amp responses
▢ Understand the open-loop response of an op-amp
▢ Understand the closed-loop response of an op-amp
▢ Discuss positive feedback and stability in op-amp circuits
▢ Explain op-amp phase compensation
▢ Apply what you have learned in this chapter to a system application

In this chapter, you will learn about frequency response, bandwidth, phase shift, and other frequency-related parameters in op-amps. The effects of negative feedback will be further examined, and you will learn about stability requirements and how to compensate op-amp circuits to ensure stable operation.

Stereo systems use two separate frequency-modulated (FM) signals to reproduce sound as, for example, from the left and right sides of the stage in a concert performance. When the signal is processed by a stereo receiver, the sound comes out of both the left and right speakers, and you get the original sound effects in terms of direction and distribution. When a stereo broadcast is received by a single-speaker (monophonic) system, the sound from the speaker is actually the composite or sum of the left and right channel sounds so you get the original sound without separation. Op-amps can be used for many purposes in stereo systems such as this, but we will focus on the identical left and right channel audio amplifiers for this system application.

For the system application in Section 3–6, in addition to the other topics, be sure you understand
☐ The noninverting op-amp configuration
☐ How capacitors can be connected for frequency compensation

3–1 ■ BASIC CONCEPTS

Chapter 2 demonstrated how closed-loop voltage gains of the basic op-amp configurations are determined, and the distinction between open-loop voltage gain and closed-loop voltage gain was established. Because of the importance of these two different types of voltage gain, the definitions are restated in this section.

After completing this section, you should be able to

❑ Discuss the basic areas of op-amp responses
   ❑ Explain open-loop gain
   ❑ Explain closed-loop gain
   ❑ Discuss the frequency dependency of gain
   ❑ Explain the open-loop bandwidth
   ❑ Explain the unity-gain bandwidth
   ❑ Determine phase shift

Open-Loop Gain

The *open-loop gain* (A_{ol}) of an op-amp is the internal voltage gain of the device and represents the ratio of output voltage to input voltage, as indicated in Figure 3–1(a). Notice that there are no external components, so the open-loop gain is set entirely by the internal design. Open-loop voltage gain varies widely for different op-amps. Table 2–1 listed the open-loop gain for some representative op-amps. Data sheets often refer to the open-loop gain as the *large-signal voltage gain.*

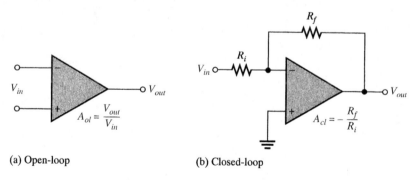

(a) Open-loop   (b) Closed-loop

FIGURE 3–1
Open-loop and closed-loop op-amp configurations.

Closed-Loop Gain

The *closed-loop gain* (A_{cl}) is the voltage gain of an op-amp with external feedback. The amplifier configuration consists of the op-amp and an external negative feedback network that connects the output to the inverting ($-$) input. The closed-loop gain is determined by the external component values, as illustrated in Figure 3–1(b) for an inverting amplifier configuration. The closed-loop gain can be precisely controlled by external component values.

The Gain Is Frequency Dependent

In Chapter 2, all of the gain expressions applied to the midrange gain and were considered independent of the frequency. The midrange open-loop gain of an op-amp extends from zero frequency (dc) up to a critical frequency at which the gain is 3 dB less than the midrange value. The difference here is that op-amps are dc amplifiers (no capacitive coupling between stages), and therefore, there is no lower critical frequency. This means that the midrange gain extends down to zero frequency (dc), and dc voltages are amplified the same as midrange signal frequencies.

An open-loop response curve (Bode plot) for a certain op-amp is shown in Figure 3–2. Most op-amp data sheets show this type of curve or specify the midrange open-loop gain. Notice that the curve rolls off (decreases) at −20 dB per decade (−6 dB per octave). The midrange gain is 200,000, which is 106 dB, and the critical (cutoff) frequency is approximately 10 Hz.

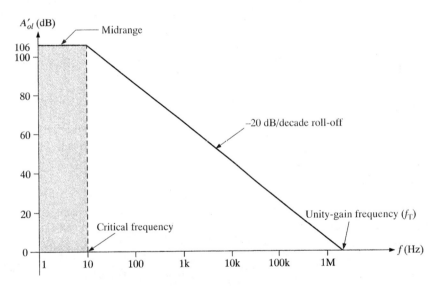

FIGURE 3–2
Ideal plot of open-loop voltage gain versus frequency for typical op-amp. The frequency scale is logarithmic.

3 dB Open-Loop Bandwidth

The **bandwidth** of an ac amplifier is the frequency range between the points where the gain is 3 dB less than the midrange gain. In general, the bandwidth equals the upper critical frequency (f_{cu}) minus the lower critical frequency (f_{cl}).

$$BW = f_{cu} - f_{cl}$$

Since f_{cl} for an op-amp is zero, the bandwidth is simply equal to the upper critical frequency.

$$BW = f_{cu} \qquad\qquad (3\text{–}1)$$

From now on, we will refer to f_{cu} as simply f_c; and we will use open-loop (*ol*) or closed-loop (*cl*) subscript designators. For example, $f_{c(ol)}$ is the open-loop upper critical frequency and $f_{c(cl)}$ is the closed-loop upper critical frequency.

Unity-Gain Bandwidth

Notice in Figure 3–2 that the gain steadily decreases to a point where it is equal to one (0 dB). The value of the frequency at which this unity gain occurs is the *unity-gain bandwidth*.

Gain-Versus-Frequency Analysis

The *RC* lag (low-pass) networks within an op-amp are responsible for the roll-off in gain as the frequency increases. From basic ac circuit theory, the attenuation of an *RC* lag network, such as in Figure 3–3, is expressed as

$$\frac{V_{out}}{V_{in}} = \frac{X_C}{\sqrt{R^2 + X_C^2}}$$

Dividing both the numerator and denominator to the right of the equal sign by X_C,

$$\frac{V_{out}}{V_{in}} = \frac{1}{\sqrt{1 + R^2/X_C^2}}$$

FIGURE 3–3
RC lag network.

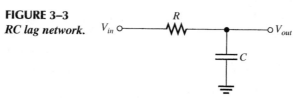

The critical frequency of an *RC* network is

$$f_c = \frac{1}{2\pi RC}$$

Dividing both sides by f gives

$$\frac{f_c}{f} = \frac{1}{2\pi RCf} = \frac{1}{(2\pi fC)R}$$

Since $X_C = 1/(2\pi fC)$, the above expression can be written as

$$\frac{f_c}{f} = \frac{X_C}{R}$$

Substituting this result into the second equation produces the following expression for the attenuation of an *RC* lag network:

$$\frac{V_{out}}{V_{in}} = \frac{1}{\sqrt{1 + f^2/f_c^2}}$$

FIGURE 3–4
Op-amp represented by gain element and internal RC network.

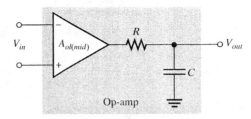

If an op-amp is represented by a voltage gain element with a gain of $A_{ol(mid)}$ and a single RC lag network, as shown in Figure 3–4, then the total open-loop gain of the op-amp is the product of the midrange open-loop gain $A_{ol(mid)}$ and the attenuation of the RC network.

$$A_{ol} = \frac{A_{ol(mid)}}{\sqrt{1 + f^2/f_c^2}} \tag{3–2}$$

As you can see from Equation (3–2), the open-loop gain equals the midrange value when the signal frequency f is much less than the critical frequency f_c and drops off as the frequency increases. Since f_c is part of the open-loop response of an op-amp, we will refer to it as $f_{c(ol)}$.

The following example demonstrates how the open-loop gain decreases as the frequency increases above $f_{c(ol)}$.

EXAMPLE 3–1

Determine A_{ol} for the following values of f. Assume $f_{c(ol)} = 100$ Hz and $A_{ol(mid)} = 100,000$.

(a) $f = 0$ Hz   **(b)** $f = 10$ Hz   **(c)** $f = 100$ Hz   **(d)** $f = 1000$ Hz

Solution

(a) $A_{ol} = \dfrac{A_{ol(mid)}}{\sqrt{1 + f^2/f_{c(ol)}^2}} = \dfrac{100,000}{\sqrt{1 + 0}} = \mathbf{100,000}$

(b) $A_{ol} = \dfrac{100,000}{\sqrt{1 + (0.1)^2}} = \mathbf{99,500}$

(c) $A_{ol} = \dfrac{100,000}{\sqrt{1 + (1)^2}} = \dfrac{100,000}{\sqrt{2}} = \mathbf{70,700}$

(d) $A_{ol} = \dfrac{100,000}{\sqrt{1 + (10)^2}} = \mathbf{9950}$

Practice Exercise Find A_{ol} for the following frequencies. Assume $f_{c(ol)} = 200$ Hz and $A_{ol(mid)} = 80,000$.

(a) $f = 2$ Hz   **(b)** $f = 10$ Hz   **(c)** $f = 2500$ Hz

Phase Shift

As you know, an *RC* network causes a propagation delay from input to output, thus creating a **phase shift** between the input signal and the output signal. An *RC* lag network such as found in an op-amp stage causes the output signal voltage to lag the input, as shown in Figure 3–5. From basic ac circuit theory, the phase shift, φ, is

$$\phi = -\tan^{-1}\left(\frac{R}{X_C}\right)$$

Since $R/X_C = f/f_c$,

$$\phi = -\tan^{-1}\left(\frac{f}{f_c}\right) \tag{3–3}$$

The negative sign indicates that the output lags the input. This equation shows that the phase shift increases with frequency and approaches $-90°$ as f becomes much greater than f_c.

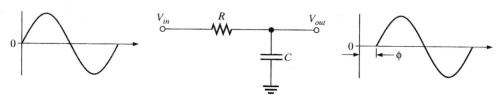

FIGURE 3–5
Output voltage lags input voltage.

EXAMPLE 3–2

Calculate the phase shift for an *RC* lag network for each of the following frequencies, and then plot the curve of phase shift versus frequency. Assume $f_c = 100$ Hz.

(a) $f = 1$ Hz   **(b)** $f = 10$ Hz   **(c)** $f = 100$ Hz
(d) $f = 1000$ Hz   **(e)** $f = 10$ kHz

Solution

(a) $\phi = -\tan^{-1}\left(\dfrac{f}{f_c}\right) = -\tan^{-1}\left(\dfrac{1\text{ Hz}}{100\text{ Hz}}\right) = \mathbf{-0.6°}$

(b) $\phi = -\tan^{-1}\left(\dfrac{10\text{ Hz}}{100\text{ Hz}}\right) = \mathbf{-5.7°}$

(c) $\phi = -\tan^{-1}\left(\dfrac{100\text{ Hz}}{100\text{ Hz}}\right) = \mathbf{-45.0°}$

(d) $\phi = -\tan^{-1}\left(\dfrac{1000\text{ Hz}}{100\text{ Hz}}\right) = \mathbf{-84.3°}$

(e) $\phi = -\tan^{-1}\left(\dfrac{10\text{ kHz}}{100\text{ Hz}}\right) = \mathbf{-89.4°}$

The phase shift versus frequency curve is plotted in Figure 3–6. Note that the frequency axis is logarithmic.

FIGURE 3–6

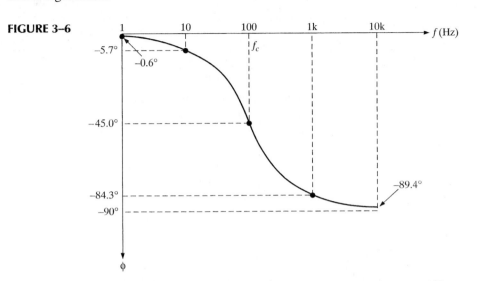

Practice Exercise At what frequency, in this example, is the phase shift $-60°$?

3–1 REVIEW QUESTIONS

1. How do the open-loop gain and the closed-loop gain of an op-amp differ?
2. The upper critical frequency of a particular op-amp is 100 Hz. What is its open-loop 3 dB bandwidth?
3. Does the open-loop gain increase or decrease with frequency above the critical frequency?

3–2 ■ OP-AMP OPEN-LOOP RESPONSE

In this section, you will learn about the open-loop frequency response and the open-loop phase response of an op-amp. Open-loop responses relate to an op-amp with no external feedback. The frequency response indicates how the voltage gain changes with frequency, and the phase response indicates how the phase shift between the input and output signal changes with frequency. The open-loop gain, like the β of a transistor, varies greatly from one device to the next of the same type.

After completing this section, you should be able to

❑ Understand the open-loop response of an op-amp
   ❑ Discuss how internal stages affect the overall response
   ❑ Discuss critical frequencies and roll-off rates
   ❑ Determine overall phase response

Frequency Response

In Section 3–1, an op-amp was assumed to have a constant roll-off of -20 dB/decade above its critical frequency. For a large number of op-amps, this is indeed the case. Op-amps that have a constant -20 dB/decade roll-off from f_c to unity gain are called *compensated op-amps*. A compensated op-amp has only one RC network that determines its frequency characteristic. Thus, the roll-off rate is the same as that of a basic RC network.

For some op-amp circuits, the situation is more complicated. The frequency response may be determined by several internal stages, where each stage has its own critical frequency. As a result, the overall response is affected by more than one cascaded stage and the overall response is a composite of the individual responses. An op-amp that has more than one critical frequency is called an *uncompensated op-amp*.

Uncompensated op-amps require careful attention to the feedback network to avoid oscillation. As an example, a three-stage op-amp is represented in Figure 3–7(a), and the frequency response of each stage is shown in Figure 3–7(b). As you know, dB gains are added so that the total op-amp frequency response is as shown in Figure 3–7(c). Since the roll-off rates are additive, the total roll-off rate increases by -20 dB/decade (-6 dB/octave) as each critical frequency is reached.

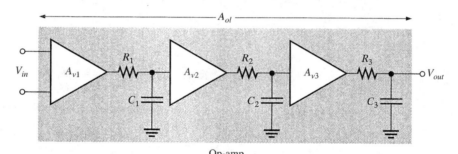

(a) Representation of an op-amp with three internal stages

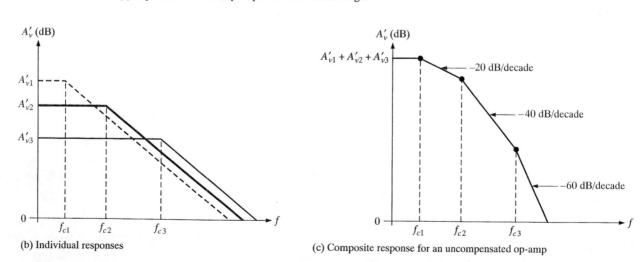

(b) Individual responses

(c) Composite response for an uncompensated op-amp

FIGURE 3–7
Op-amp open-loop frequency response.

Phase Response

In a multistage amplifier, each stage contributes to the total phase lag. As you have seen, each RC lag network can produce up to a $-90°$ phase shift. Since each stage in an op-amp includes an RC lag network, a three-stage op-amp, for example, can have a maximum phase lag of $-270°$. Also, the phase lag of each stage is less than $-45°$ when the frequency is below the critical frequency, equal to $-45°$ at the critical frequency, and greater than $-45°$ when the frequency is above the critical frequency. The phase lags of the stages of an op-amp are added to produce a total phase lag, according to the following formula for three stages:

$$\phi_{tot} = -\tan^{-1}\left(\frac{f}{f_{c1}}\right) - \tan^{-1}\left(\frac{f}{f_{c2}}\right) - \tan^{-1}\left(\frac{f}{f_{c3}}\right)$$

EXAMPLE 3–3

A certain op-amp has three internal amplifier stages with the following gains and critical frequencies:

Stage 1: $A'_{v1} = 40$ dB, $f_{c1} = 2000$ Hz
Stage 2: $A'_{v2} = 32$ dB, $f_{c2} = 40$ kHz
Stage 3: $A'_{v3} = 20$ dB, $f_{c3} = 150$ kHz

Determine the open-loop midrange dB gain and the total phase lag when $f = f_{c1}$.

Solution

$$A'_{ol(mid)} = A'_{v1} + A'_{v2} + A'_{v3} = 40 \text{ dB} + 32 \text{ dB} + 20 \text{ dB} = \mathbf{92 \text{ dB}}$$

$$\phi_{tot} = -\tan^{-1}\left(\frac{f}{f_{c1}}\right) - \tan^{-1}\left(\frac{f}{f_{c2}}\right) - \tan^{-1}\left(\frac{f}{f_{c3}}\right)$$

$$= -\tan^{-1}(1) - \tan^{-1}\left(\frac{2}{40}\right) - \tan^{-1}\left(\frac{2}{150}\right)$$

$$= -45° - 2.86° - 0.76° = \mathbf{-48.6°}$$

Practice Exercise The internal stages of a two-stage amplifier have the following characteristics: $A'_{v1} = 50$ dB, $A'_{v2} = 25$ dB, $f_{c1} = 1500$ Hz, and $f_{c2} = 3000$ Hz. Determine the open-loop midrange gain in dB and the total phase lag when $f = f_{c1}$.

3–2 REVIEW QUESTIONS

1. If the individual stage gains of an op-amp are 20 dB and 30 dB, what is the total gain in dB?
2. If the individual phase lags are $-49°$ and $-5.2°$, what is the total phase lag?

3–3 ■ OP-AMP CLOSED-LOOP RESPONSE

Op-amps are normally used in a closed-loop configuration with negative feedback in order to achieve precise control of the gain and bandwidth. In this section, you will see how feedback affects the gain and frequency response of an op-amp.

After completing this section, you should be able to

❑ Understand the closed-loop response of an op-amp
  ❑ Determine the closed-loop gain
  ❑ Explain the effect of negative feedback on bandwidth
  ❑ Explain gain-bandwidth product

Recall from Chapter 2 that midrange gain is reduced by negative feedback, as indicated by the following closed-loop gain expressions for the three configurations previously covered. For the noninverting amplifier,

$$A_{cl(\text{NI})} = \frac{R_f}{R_i} + 1$$

For the voltage-follower,

$$A_{cl(\text{VF})} \cong 1$$

For the inverting amplifier,

$$A_{cl(\text{I})} \cong -\frac{R_f}{R_i}$$

Effect of Negative Feedback on Bandwidth

You know how negative feedback affects the gain; now you will learn how it affects the amplifier's bandwidth. The closed-loop critical frequency of an op-amp is

$$f_{c(cl)} = f_{c(ol)}(1 + BA_{ol(mid)}) \tag{3–4}$$

This expression shows that the closed-loop critical frequency, $f_{c(cl)}$, is higher than the open-loop critical frequency $f_{c(ol)}$ by the factor $1 + BA_{ol(mid)}$. Recall that B is the feedback attenuation, $R_i/(R_i + R_f)$. A derivation of Equation (3–4) can be found in Appendix B.

Since $f_{c(cl)}$ equals the bandwidth for the closed-loop amplifier, the bandwidth is also increased by the same factor.

$$BW_{cl} = BW_{ol}(1 + BA_{ol(mid)}) \tag{3–5}$$

EXAMPLE 3–4

A certain amplifier has an open-loop midrange gain of 150,000 and an open-loop 3 dB bandwidth of 200 Hz. The attenuation of the feedback loop is 0.002. What is the closed-loop bandwidth?

Solution

$$BW_{cl} = BW_{ol}(1 + BA_{ol(mid)}) = 200 \text{ Hz}[1 + (0.002)(150,000)] = \mathbf{60.2 \text{ kHz}}$$

Practice Exercise If $A_{ol(mid)} = 200{,}000$ and $B = 0.05$, what is the closed loop bandwidth?

Figure 3–8 graphically illustrates the concept of closed-loop response for a compensated op-amp. When the open-loop gain of an op-amp is reduced by negative feedback, the bandwidth is increased. The closed-loop gain is independent of the open-loop gain up to the point of intersection of the two gain curves. This point of intersection is the critical frequency, $f_{c(cl)}$, for the closed-loop response. Notice that the closed-loop gain has the same roll-off rate as the open-loop gain, beyond the closed-loop critical frequency.

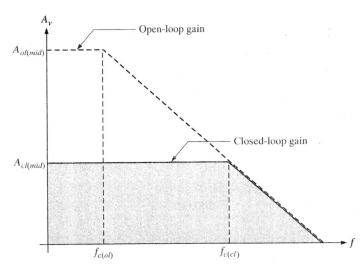

FIGURE 3–8
Closed-loop gain compared to open-loop gain.

Gain-Bandwidth Product

An increase in closed-loop gain causes a decrease in the bandwidth and vice versa, such that *the product of gain and bandwidth is a constant.* This is true as long as the roll-off rate is a fixed -20 dB/decade. If you let A_{cl} represent the gain of any of the closed-loop configurations and $f_{c(cl)}$ represent the closed-loop critical frequency (same as the bandwidth), then

$$A_{cl}f_{c(cl)} = A_{ol}f_{c(ol)}$$

The gain-bandwidth product is always equal to the frequency at which the op-amp's open-loop gain is unity (unity-gain bandwidth).[1]

$$A_{cl}f_{c(cl)} = \text{unity-gain bandwidth} \tag{3–6}$$

---

[1] Technically speaking, this equation is true only for noninverting configurations.

EXAMPLE 3–5   Determine the bandwidth of each of the amplifiers in Figure 3–9. Both op-amps have an open-loop gain of 100 dB and a unity-gain bandwidth of 3 MHz.

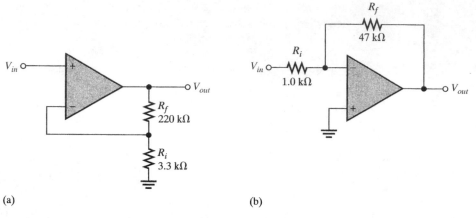

(a)   (b)

FIGURE 3–9

Solution

(a) For the noninverting amplifier in Figure 3–9(a), the closed-loop gain is

$$A_{cl(NI)} = \frac{R_f}{R_i} + 1 = \frac{220 \text{ k}\Omega}{3.3 \text{ k}\Omega} + 1 = 67.7$$

Use Equation (3–6) and solve for $f_{c(cl)}$ (where $f_{c(cl)} = BW_{cl}$).

$$f_{c(cl)} = BW_{cl} = \frac{\text{unity-gain } BW}{A_{cl}}$$

$$BW_{cl} = \frac{3 \text{ MHz}}{67.7} = \textbf{44.3 kHz}$$

(b) For the inverting amplifier in Figure 3–9(b), the closed-loop gain is

$$A_{cl(I)} = -\frac{R_f}{R_i} = -\frac{47 \text{ k}\Omega}{1.0 \text{ k}\Omega} = -47$$

The closed-loop bandwidth is

$$BW_{cl} = \frac{3 \text{ MHz}}{47} = \textbf{63.8 kHz}$$

Practice Exercise   Determine the bandwidth of each of the amplifiers in Figure 3–9. Both op-amps have an A'_{ol} of 90 dB and a unity-gain bandwidth of 2 MHz.

3–4 ■ POSITIVE FEEDBACK AND STABILITY

Stability is a very important consideration when using op-amps. Stable operation means that the op-amp does not oscillate under any condition. Instability produces oscillations, which are unwanted voltage swings on the output when there is no signal present on the input, or in response to noise or transient voltages on the input.

After completing this section, you should be able to

❑ Discuss positive feedback and stability in op-amp circuits
    ❑ Define *positive feedback*
    ❑ Define *loop gain*
    ❑ Define *phase margin* and discuss its importance
    ❑ Determine if an op-amp circuit is stable
    ❑ Summarize the criteria for stability

Positive Feedback

To understand stability, you must first examine instability and its causes. As you know, with negative feedback, the signal fed back to the input of an amplifier is out of phase with the input signal, thus subtracting from it and effectively reducing the voltage gain. As long as the feedback is negative, the amplifier is stable.

When the signal fed back from output to input is in phase with the input signal, a positive feedback condition exists and the amplifier can oscillate. That is, **positive feedback** occurs when the total phase shift through the op-amp and feedback network is 360°, which is equivalent to no phase shift (0°).

Loop Gain

For instability to occur, (a) there must be positive feedback, and (b) the loop gain of the closed-loop amplifier must be greater than 1. The **loop gain** of a closed loop amplifier is defined to be the op-amp's open-loop gain times the attenuation of the feedback network.

$$\text{Loop gain} = A_{ol}B \qquad\qquad (3\text{–}7)$$

Phase Margin

Notice that for each amplifier configuration in Figure 3–10, the feedback loop is connected to the inverting input. There is an inherent phase shift of 180° because of the *inversion* between input and output. Additional phase shift (ϕ_{tot}) is produced by the *RC* lag networks (not shown) within the amplifier. So, the total phase shift around the loop is $180° + \phi_{tot}$.

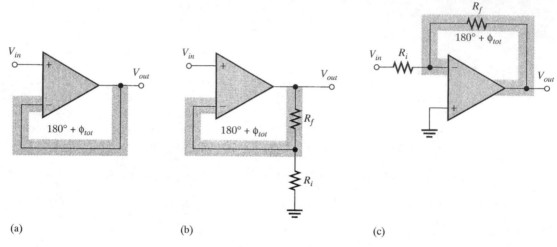

FIGURE 3–10
Feedback-loop phase shift.

The **phase margin,** ϕ_{pm}, is the amount of additional phase shift required to make the total phase shift around the loop 360°. (360° is equivalent to 0°.)

$$180° + \phi_{tot} + \phi_{pm} = 360°$$

$$\phi_{pm} = 180° - |\phi_{tot}| \tag{3–8}$$

If the phase margin is positive, the total phase shift is less than 360° and the amplifier is stable. If the phase margin is zero or negative, then the amplifier is potentially unstable because the signal fed back can be in phase with the input. As you can see from Equation (3–8), when the total lag network phase shift ϕ_{tot} equals or exceeds 180°, then the phase margin is 0° or negative and an unstable condition exists, which would cause the amplifier to oscillate.

Stability Analysis

Since most op-amp configurations use a loop gain greater than 1 ($A_{ol}B > 1$), the criteria for stability are based on the phase angle of the internal lag networks. As previously mentioned, operational amplifiers are composed of multiple stages, each of which has a critical frequency. For compensated op-amps, only one critical frequency is dominant, and stability due to the feedback is not a problem. Stability problems generally manifest themselves as unwanted oscillations. Feedback stability occurs near the unity-gain frequency for the op-amp.

To illustrate the concept of feedback **stability,** we will use an uncompensated three-stage op-amp with an open-loop response as shown in the Bode plot of Figure 3–11. For this case, there are three different critical frequencies, which indicate three internal RC lag networks. At the first critical frequency, f_{c1}, the gain begins rolling off at -20 dB/decade; when the second critical frequency, f_{c2}, is reached, the gain decreases at -40 dB/decade; and when the third critical frequency, f_{c3}, is reached, the gain drops at -60 dB/decade.

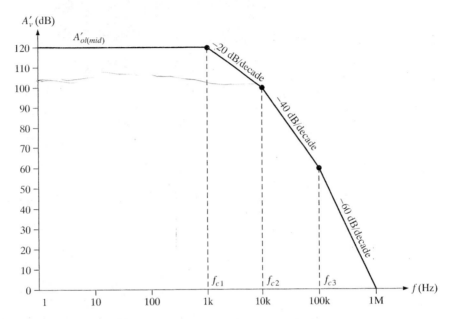

FIGURE 3–11
Bode plot of example of three-stage op-amp response.

To analyze an uncompensated closed-loop amplifier for stability, the phase margin must be determined. A positive phase margin will indicate that the amplifier is stable for a given value of closed-loop gain. Three example cases will be considered in order to demonstrate the conditions for instability.

Case 1   The closed-loop gain intersects the open-loop response on the -20 dB/decade slope, as shown in Figure 3–12. The midrange closed-loop gain is 106 dB, and the closed-loop critical frequency is 5 kHz. If we assume that the amplifier is not operated out of its midrange, the maximum phase shift for the 106 dB amplifier occurs at the highest midrange frequency (in this case, 5 kHz). The total phase shift at this frequency due to the three lag networks is calculated as follows:

$$\phi_{tot} = -\tan^{-1}\left(\frac{f}{f_{c1}}\right) - \tan^{-1}\left(\frac{f}{f_{c2}}\right) - \tan^{-1}\left(\frac{f}{f_{c3}}\right)$$

where $f = 5$ kHz, $f_{c1} = 1$ kHz, $f_{c2} = 10$ kHz, and $f_{c3} = 100$ kHz. Therefore,

$$\phi_{tot} = -\tan^{-1}\left(\frac{5 \text{ kHz}}{1 \text{ kHz}}\right) - \tan^{-1}\left(\frac{5 \text{ kHz}}{10 \text{ kHz}}\right) - \tan^{-1}\left(\frac{5 \text{ kHz}}{100 \text{ kHz}}\right)$$

$$= -78.7° - 26.6° - 2.9° = -108.1°$$

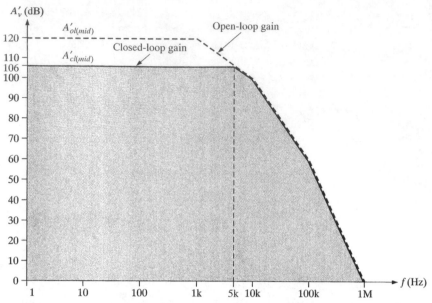

FIGURE 3–12
Case where closed-loop gain intersects open-loop gain on −20 dB/decade slope (stable operation).

The phase margin, ϕ_{pm}, is

$$\phi_{pm} = 180° - |\phi_{tot}| = 180° - 108.1° = +71.9°$$

The phase margin is positive, so the amplifier is stable for all frequencies in its midrange. In general, an amplifier is stable for all midrange frequencies if its closed-loop gain intersects the open-loop response curve on a −20 dB/decade slope.

Case 2 The closed-loop gain is lowered to where it intersects the open-loop response on the −40 dB/decade slope, as shown in Figure 3–13. The midrange closed-loop gain in this case is 72 dB, and the closed-loop critical frequency is approximately 30 kHz. The total phase shift at $f = 30$ kHz due to the three lag networks is calculated as follows:

$$\phi_{tot} = -\tan^{-1}\left(\frac{30\ \text{kHz}}{1\ \text{kHz}}\right) - \tan^{-1}\left(\frac{30\ \text{kHz}}{10\ \text{kHz}}\right) - \tan^{-1}\left(\frac{30\ \text{kHz}}{100\ \text{kHz}}\right)$$
$$= -88.1° - 71.6° - 16.7° = -176.4°$$

The phase margin is

$$\phi_{pm} = 180° - 176.4° = +3.6°$$

The phase margin is positive, so the amplifier is still stable for frequencies in its midrange, but a very slight increase in frequency above f_c would cause it to oscillate. Therefore, it is marginally stable and may oscillate due to other paths. It is very close to instability because instability occurs where $\phi_{pm} = 0°$. As a general rule, a minimum 45° phase margin is recommended to avoid marginal conditions.

Case 3 The closed-loop gain is further decreased until it intersects the open-loop response on the −60 dB/decade slope, as shown in Figure 3–14. The midrange closed-loop

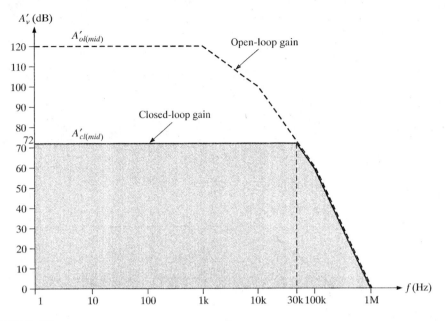

FIGURE 3–13

Case where closed-loop gain intersects open-loop gain on −40 dB/decade slope (marginally stable operation).

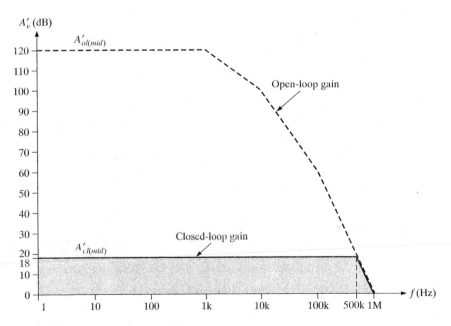

FIGURE 3–14

Case where closed-loop gain intersects open loop gain on −60 dB/decade slope (unstable operation).

gain in this case is 18 dB, and the closed-loop critical frequency is 500 kHz. The total phase shift at $f = 500$ kHz due to the three lag networks is

$$\phi_{tot} = -\tan^{-1}\left(\frac{500 \text{ kHz}}{1 \text{ kHz}}\right) - \tan^{-1}\left(\frac{500 \text{ kHz}}{10 \text{ kHz}}\right) - \tan^{-1}\left(\frac{500 \text{ kHz}}{100 \text{ kHz}}\right)$$

$$= -89.9° - 88.9° - 78.7° = -257.5°$$

The phase margin is

$$\phi_{pm} = 180° - 257.5° = -77.5°$$

Here the phase margin is negative and the amplifier is unstable at the upper end of its midrange.

Summary of Stability Criteria The stability analysis of the three example cases has demonstrated that an amplifier's closed-loop gain must intersect the open-loop gain curve on a -20 dB/decade slope to ensure stability for all of its midrange frequencies. If the closed-loop gain is lowered to a value that intersects on a -40 dB/decade slope, then marginal stability or complete instability can occur. In the previous situations (Cases 1, 2, and 3), the closed-loop gain should be greater than 72 dB.

If the closed-loop gain intersects the open-loop response on a -60 dB/decade slope, instability will definitely occur at some frequency within the amplifier's midrange unless a specially designed feedback network is used. Therefore, to ensure stability for all of the midrange frequencies, an op-amp must be operated at a closed-loop gain such that the roll-off rate beginning at its dominant critical frequency does not exceed -20 dB/decade.

Troubleshooting Unwanted Oscillations

The stability problems mentioned in this section can be brought under control, even in the case of a negative phase margin (Case 3), by specially designed feedback networks. A lead network in the feedback path can be used to increase the phase margin and thus increase the stability. In some cases, a complicated feedback network with an amplifier or other active element is added to a design to increase stability.

Not all stability problems are due to the feedback network. If oscillations are not near the unity-gain frequency of the op-amp, the feedback loop is probably not the culprit. Causes of oscillations can include the presence of an external feedback path, a grounding problem, or an extraneous noise signal coupled into the power supply lines. When oscillations are a problem, a simple test is to increase the gain and see if they disappear. (This means the closed-loop gain will intersect the open-loop gain at a higher point.) If the oscillations persist, the problem may be something other than a negative phase margin.

To eliminate unwanted oscillations, check ground paths (try to use single-point grounding), add bypass capacitors to the supply voltages, and try to eliminate extraneous capacitive coupling paths to the input. A coupling path may not be obvious but can be due to a protoboard, especially if it has no ground plane, or may be caused by long leads in the circuit (remember that wires have capacitance). Power supply noise can produce feedback in the amplifier that can result in oscillations. At low frequencies, a simple bypass capacitor (1 μF to 10 μF tantalum) may be all that is necessary to solve the problem. At high frequencies, a single bypass capacitor may have a self-resonance, requiring the addition of a secondary bypass capacitor.

Occasionally, oscillations are due to interference from nearby sources and may require shielding. It is also possible to induce oscillations when a low-level signal shares a common ground path with a high-level signal or because of long leads in the circuit layout.

Try reconstructing the circuit with shorter leads, paying attention to ground paths and making sure a ground plane is present, if possible.

3–4 REVIEW QUESTIONS

1. Under what feedback condition can an amplifier oscillate?
2. How much can the phase shift of an amplifier's internal *RC* network be before instability occurs? What is the phase margin at the point where instability begins?
3. What is the maximum roll-off rate of the open-loop gain of an op-amp for which the device will still be stable?

3–5 ■ OP-AMP COMPENSATION

The last section demonstrated that instability can occur when an op-amp's response has roll-off rates exceeding −20 dB/decade and the op-amp is operated in a closed-loop configuration having a gain curve that intersects a higher roll-off rate portion of the open-loop response. In situations like those examined in the last section, the closed-loop voltage gain is restricted to very high values. In many applications, lower values of closed-loop gain are necessary or desirable. To allow op-amps to be operated at low closed-loop gain, phase lag compensation is required.

After completing this section, you should be able to

❑ Explain op-amp phase compensation
   ❑ Describe phase-lag compensation
   ❑ Explain a compensating circuit
   ❑ Apply single-capacitor compensation
   ❑ Apply feedforward compensation

Phase Lag Compensation

As you have seen, the cause of instability is excessive phase shift through an op-amp's internal lag networks. When these phase shifts equal or exceed 180°, the amplifier can oscillate. **Compensation** is used to either eliminate open-loop roll-off rates greater than −20 dB/decade or extend the −20 dB/decade rate to a lower gain. These concepts are illustrated in Figure 3–15.

Compensating Network

There are two basic methods of compensation for integrated circuit op-amps: internal and external. In either case an *RC* network is added. The basic compensating action is as follows. Consider first the *RC* network shown in Figure 3–16(a). At low frequencies where the reactance of the compensating capacitor, X_{C_c}, is extremely large, the output voltage approximately equals the input voltage. When the frequency reaches its critical value, $f_c = 1/[2\pi(R_1 + R_2)C_c]$, the output voltage decreases at −20 dB/decade. This roll-off rate continues until $X_{C_c} \cong 0$, at which point the output voltage levels off to a value determined

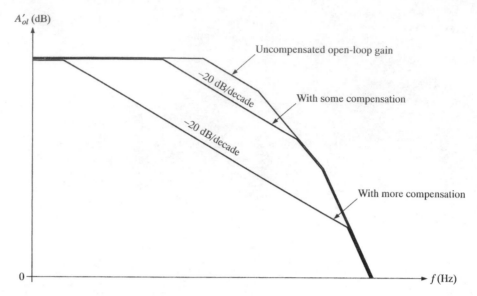

FIGURE 3–15

Bode plot illustrating effect of phase compensation on open-loop gain of typical op-amp.

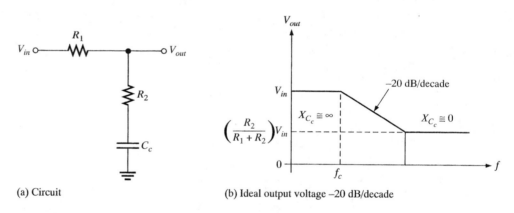

(a) Circuit   (b) Ideal output voltage –20 dB/decade

FIGURE 3–16

Basic compensating network action.

by R_1 and R_2, as indicated in Figure 3–16(b). This is the principle used in the phase compensation of an op-amp.

To see how a compensating network changes the open-loop response of an op-amp, refer to Figure 3–17. This diagram represents a two-stage op-amp. The individual stages are within the blue-shaded blocks along with the associated lag networks. A compensating network is shown connected at point A on the output of stage 1.

The critical frequency of the compensating network is set to a value less than the dominant (lowest) critical frequency of the internal lag networks. This causes the −20 dB/decade roll-off to begin at the compensating network's critical frequency. The roll-off of the compensating network continues up to the critical frequency of the dominant lag network. At this point, the response of the compensating network levels off, and the

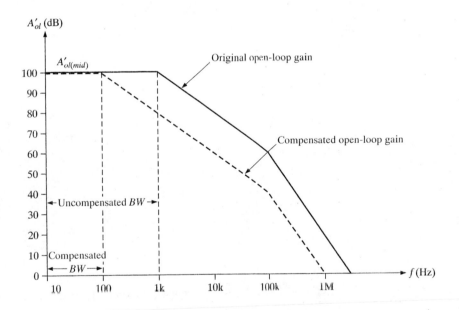

FIGURE 3–17
Representation of op-amp with compensation.

−20 dB/decade roll-off of the dominant lag network takes over. The net result is a shift of the open-loop response to the left, thus reducing the bandwidth, as shown in Figure 3–18. The response curve of the compensating network is shown in proper relation to the overall open-loop response.

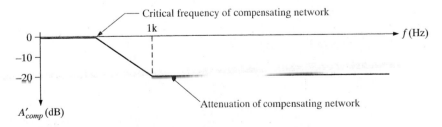

FIGURE 3–18
Example of compensated op-amp frequency response.

EXAMPLE 3–6 A certain op-amp has the open-loop response in Figure 3–19. As you can see, the lowest closed-loop gain for which stability is assured is approximately 40 dB (where the closed-loop gain line still intersects the −20 dB/decade slope). In a particular application, a 20 dB closed-loop gain is required.

(a) Determine the critical frequency for the compensating network.
(b) Sketch the ideal response curve for the compensating network.
(c) Sketch the total ideal compensated open-loop response.

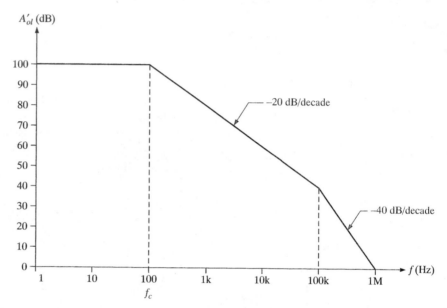

FIGURE 3–19
Original open-loop response.

Solution

(a) The gain must be dropped so that the −20 dB/decade roll-off extends down to 20 dB rather than to 40 dB. To achieve this, the midrange open-loop gain must be made to roll off a decade sooner. Therefore, the critical frequency of the compensating network must be 10 Hz.

(b) The roll-off of the compensating network must end at 100 Hz, as shown in Figure 3–20(a).

(c) The total open-loop response resulting from compensation is shown in Figure 3–20(b).

FIGURE 3–20

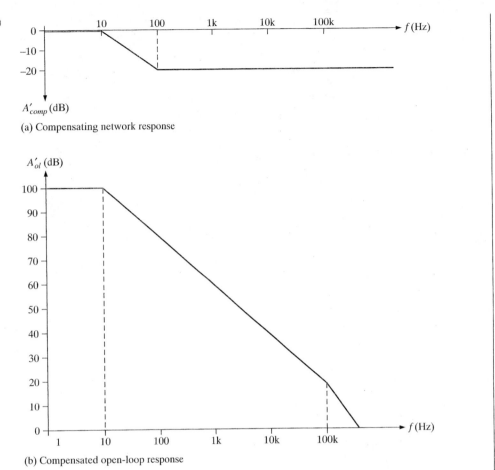

(a) Compensating network response

(b) Compensated open-loop response

Practice Exercise   In this example, what is the uncompensated bandwidth? What is the compensated bandwidth?

Extent of Compensation

A larger compensating capacitor will cause the open-loop roll-off to begin at a lower frequency and thus extend the −20 dB/decade roll-off to lower gain levels, as shown in Figure 3–21(a). With a sufficiently large compensating capacitor, an op-amp can be made unconditionally stable, as illustrated in Figure 3–21(b), where the −20 dB/decade slope is extended all the way down to unity gain. This is normally the case when internal compensation is provided by the manufacturer. An internally, fully compensated op-amp can be used for any value of closed-loop gain and remain stable. The 741 is an example of an internally fully compensated device.

A disadvantage of internally fully compensated op amps is that bandwidth is sacrificed; thus the slew rate is decreased. Therefore, many IC op-amps have provisions for external compensation. Figure 3–22 shows typical package layouts of an LM101A op-amp with pins available for external compensation with a small capacitor. With provisions for external connections, just enough compensation can be used for a given application without sacrificing more performance than necessary.

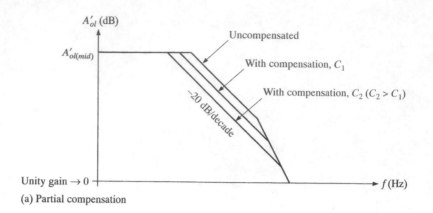

(a) Partial compensation

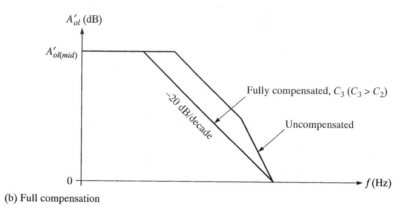

(b) Full compensation

FIGURE 3–21
Extent of compensation.

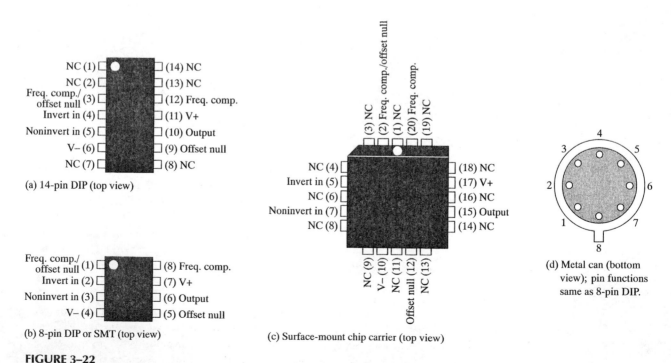

(a) 14-pin DIP (top view)

(b) 8-pin DIP or SMT (top view)

(c) Surface-mount chip carrier (top view)

(d) Metal can (bottom view); pin functions same as 8-pin DIP.

FIGURE 3–22
Typical op-amp packages.

Single-Capacitor Compensation

As an example of compensating an IC op-amp, a capacitor C_1 is connected to pins 1 and 8 of an LM101A in an inverting amplifier configuration, as shown in Figure 3–23(a). Part (b) of the figure shows the open-loop frequency response curves for two values of C_1. The 3 pF compensating capacitor produces a unity-gain bandwidth approaching 10 MHz. Notice that the −20 dB/decade slope extends to a very low gain value. When C_1 is increased ten times to 30 pF, the bandwidth is reduced by a factor of ten. Notice that the −20 dB/decade slope now extends through unity gain.

When the op-amp is used in a closed-loop configuration, as in Figure 3–23(c), the useful frequency range depends on the compensating capacitor. For example, with a closed-loop gain of 40 dB as shown in part (c), the bandwidth is approximately 10 kHz for $C_1 = 30$ pF and increases to approximately 100 kHz when C_1 is decreased to 3 pF.

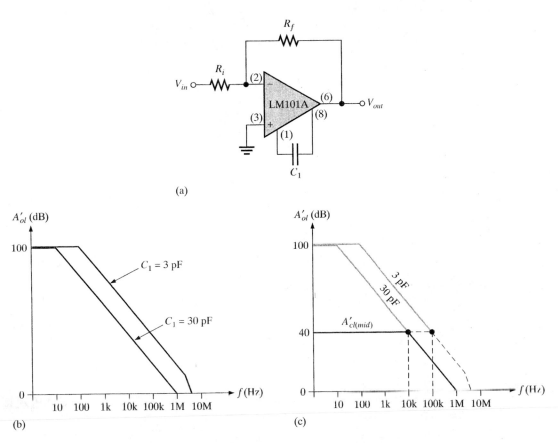

FIGURE 3–23

Example of single-capacitor compensation of an LM101A op-amp.

Feedforward Compensation

Another method of phase compensation is called **feedforward.** This type of compensation results in less bandwidth reduction than the method previously discussed. The basic concept is to bypass the internal input stage of the op-amp at high frequencies and drive the higher-frequency second stage, as shown in Figure 3–24.

FIGURE 3–24
Feedforward compensation showing high-frequency bypassing of first stage.

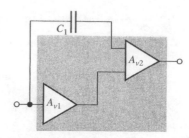

Feedforward compensation of an LM101A is shown in Figure 3–25(a). The feedforward capacitor C_1 is connected from the inverting input to the compensating terminal. A small capacitor is needed across R_f to ensure stability. The Bode plot in Figure 3–25(b) shows the feedforward compensated response and the standard compensated response that was discussed previously. The use of feedforward compensation is restricted to the inverting amplifier configuration. Other compensation methods are also used. Often, recommendations are provided by the manufacturer on the data sheet.

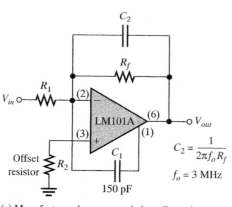

$$C_2 = \frac{1}{2\pi f_o R_f}$$

$$f_o = 3 \text{ MHz}$$

(a) Manufacturers' recommended configuration

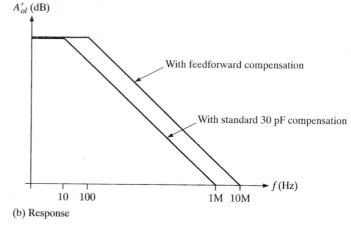

(b) Response

FIGURE 3–25
Feedforward compensation of an LM101A op-amp and the response curves.

3–5 REVIEW QUESTIONS

1. What is the purpose of phase compensation?

2. What is the main difference between internal and external compensation?

3. When you compensate an amplifier, does the bandwidth increase or decrease?

3–6 ■ A SYSTEM APPLICATION

In this system application, we are focusing on the two audio amplifier boards in an FM stereo receiver. Both boards are identical except one is for the left channel sound and the other for the right channel sound. This circuit is a good example of a mixed use of an integrated circuit and discrete components.

After completing this section, you should be able to

❑ Apply what you have learned in this chapter to a system application
    ❑ See how an op-amp is used as an audio amplifier
    ❑ Identify the functions of various components on the board
    ❑ Analyze the circuit's operation
    ❑ Translate between a printed circuit board and a schematic
    ❑ Troubleshoot some common amplifier failures

A Brief Description of the System

Some general information about the stereo system might be helpful before you concentrate on the audio amplifiers. When an FM stereo broadcast is received by a standard single-speaker system, the output to the speaker is equal to the sum of the left plus the right channel audio, so you get the original sound without separation. When a stereo receiver is used, the full stereo effect is reproduced by the two speakers. Stereo FM signals are transmitted on a carrier frequency of 88 MHz to 108 MHz. The complete stereo signal consists of three modulating signals. These are the sum of the left and right channel audio, the difference of the left and right channel audio, and a pilot subcarrier. These three signals are detected and are used to separate out the left and right channel audio by special circuits. The channel audio amplifiers then amplify each signal equally and drive the speakers. It is not necessary for you to understand this process for the purposes of this system application, although you may be interested in doing further study in this area on your own.

The two channel audio amplifiers are identical, so we will look at only one. The op-amp serves basically as a preamplifier that drives the power amplifier stage.

Now, so that you can take a closer look at one of the audio amplifier boards, let's take one out of the system and put it on the test bench.

ON THE TEST BENCH

■ ACTIVITY 1 Relate the PC Board to the Schematic

The schematic for the audio amplifier board in Figure 3–26 is shown in Figure 3–27. Using this schematic, locate and label each component on the PC board. The board has several feedthrough pads for connections that are on the back side. Backside traces are shown as darker lines. Compare the board to the schematic.

■ ACTIVITY 2 Analyze the Circuit

Step 1: Determine the midrange voltage gain.

Step 2: Determine the lower critical frequency. Given that the upper critical frequency is 15 kHz, what is the bandwidth?

Step 3: Determine the maximum peak-to-peak input voltage that can be applied without producing a distorted output signal. Assume that the maximum output peaks are 1 V less than the supply voltages.

■ ACTIVITY 3 Write a Technical Report

Describe the overall operation of the circuit and the function of each component. In discussing the general operation and basic purpose of each component, make sure you identify the negative feedback loop, the type of op-amp configuration, which components de-

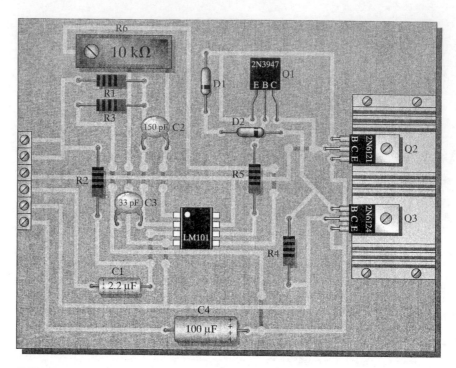

FIGURE 3–26

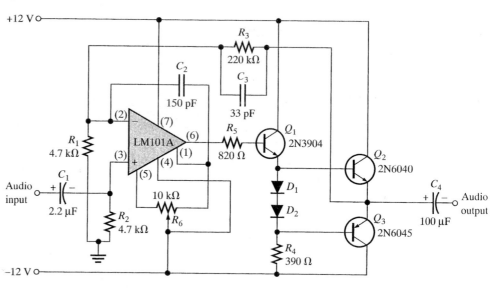

FIGURE 3–27

termine the voltage gain, which components set the lower critical frequency, and the purpose of each of the capacitors. Use the results of Activity 2 when appropriate.

■ **ACTIVITY 4   Troubleshoot the Audio Amplifier PC Boards for Each of the Following Problems by Stating the Probable Cause or Causes in Each Case**

1. No final output signal when there is a verified input signal.
2. The positive half-cycle of the output voltage is severely distorted or missing.
3. Output severely clipped on both positive and negative cycles.

3–6 REVIEW QUESTIONS

1. How can the lower critical frequency of the amplifier be reduced?
2. Which transistors form the class B power amplifier?
3. What is the purpose of Q_1 and what type of circuit is it?
4. Calculate the power that would be delivered to an 8 Ω speaker for the maximum voltage output from Activity 2.

■ SUMMARY

- Open-loop gain is the voltage gain of an op-amp without feedback.
- Closed-loop gain is the voltage gain of an op-amp with negative feedback.
- The closed-loop gain is always less than the open-loop gain.
- The midrange gain of an op-amp extends down to dc.
- Above the critical frequency, the gain of an op-amp decreases.
- The internal RC lag networks that are inherently part of the amplifier stages cause the gain to roll off as frequency goes up.
- The internal RC lag networks also cause a phase shift between input and output signals.
- Negative feedback lowers the gain and increases the bandwidth.
- The product of gain and bandwidth is constant for a compensated op-amp.
- The gain-bandwidth product equals the frequency at which unity voltage gain occurs.
- Positive feedback occurs when the total phase shift through the op-amp (including 180° inversion) and feedback network is 0° (equivalent to 360°) or more.
- The phase margin is the amount of additional phase shift required to make the total phase shift around the loop 360°.
- When the closed-loop gain of an op-amp intersects the open-loop response curve on a −20 dB/decade (−6 dB/octave) slope, the amplifier is stable.
- When the closed-loop gain intersects the open-loop response curve on a slope greater than −20 dB/decade, the amplifier can be either marginally stable or unstable.
- A minimum phase margin of 45° is recommended to provide a sufficient safety factor for stable operation.

- A fully compensated op-amp has a -20 dB/decade roll-off all the way down to unity gain.
- Compensation reduces bandwidth and increases slew rate.
- Internally compensated op-amps such as the 741 are available. These are usually fully compensated with a large sacrifice in bandwidth.
- Externally compensated op-amps such as LM101A are available. External compensating networks can be connected to specified pins, and the compensation can be tailored to a specific application. In this way, bandwidth and slew rate are not degraded more than necessary.

■ GLOSSARY

These terms are included in the end-of-book glossary.

Bandwidth   The range of frequencies between the lower critical frequency and the upper critical frequency.

Compensation   The process of modifying the roll-off rate of an amplifier to ensure stability.

Feedforward   A method of frequency compensation in op-amp circuits.

Loop gain   An op-amp's open-loop voltage gain times the attenuation of the feedback network.

Phase shift   The relative angular displacement of a time-varying function relative to a reference.

Phase margin   The difference between the total phase shift through an amplifier and 180°; the additional amount of phase shift that can be allowed before instability occurs.

Positive feedback   The return of a portion of the output signal to the input such that it reinforces the output. This output signal is in phase with the input signal.

Stability   A condition in which an amplifier circuit does not oscillate.

■ KEY FORMULAS

(3–1) $BW = f_{cu}$ Op-amp bandwidth

(3–2) $A_{ol} = \dfrac{A_{ol(mid)}}{\sqrt{1 + f^2/f_c^2}}$ Open-loop gain

(3–3) $\phi = -\tan^{-1}\left(\dfrac{f}{f_c}\right)$ RC phase shift

(3–4) $f_{c(cl)} = f_{c(ol)}(1 + BA_{ol(mid)})$ Closed-loop critical frequency

(3–5) $BW_{cl} = BW_{ol}(1 + BA_{ol(mid)})$ Closed-loop bandwidth

(3–6) $A_{cl}f_{c(cl)} = $ unity-gain bandwidth

(3–7)   Loop gain $= A_{ol}B$

(3–8) $\phi_{pm} = 180° - |\phi_{tot}|$ Phase margin

■ SELF-TEST

1. The open-loop gain of an op-amp is always
   (a) less than the closed-loop gain
   (b) equal to the closed-loop gain
   (c) greater than the closed-loop gain
   (d) a very stable and constant quantity for a given type of op-amp

2. The bandwidth of an ac amplifier having a lower critical frequency of 1 kHz and an upper critical frequency of 10 kHz is
   (a) 1 kHz   (b) 9 kHz   (c) 10 kHz   (d) 11 kHz

3. The bandwidth of a dc amplifier having an upper critical frequency of 100 kHz is
   (a) 100 kHz   (b) unknown   (c) infinity   (d) 0 kHz

4. The midrange open-loop gain of an op-amp
   (a) extends from the lower critical frequency to the upper critical frequency
   (b) extends from 0 Hz to the upper critical frequency
   (c) rolls off at −20 dB/decade beginning at 0 Hz
   (d) answers (b) and (c)

5. The frequency at which the open-loop gain is equal to one is called
   (a) the upper critical frequency   (b) the cutoff frequency
   (c) the notch frequency   (d) the unity-gain frequency

6. Phase shift through an op-amp is caused by
   (a) the internal RC networks   (b) the external RC networks
   (c) the gain roll-off   (d) negative feedback

7. Each RC network in an op-amp
   (a) causes the gain to roll off at −6 dB/octave
   (b) causes the gain to roll off at −20 dB/decade
   (c) reduces the midrange gain by 3 dB
   (d) answers (a) and (b)

8. When negative feedback is used, the gain-bandwidth product of an op-amp
   (a) increases   (b) decreases   (c) stays the same   (d) fluctuates

9. If a certain noninverting op-amp has a midrange open-loop gain of 200,000 and a unity-gain frequency of 5 MHz, the gain-bandwidth product is
   (a) 200,000 Hz   (b) 5,000,000 Hz
   (c) 1×10^{12} Hz   (d) not determinable from the information

10. If a certain noninverting op-amp has a closed-loop gain of 20 and an upper critical frequency of
    10 MHz, the gain-bandwidth product is
    (a) 200 MHz   (b) 10 MHz   (c) the unity-gain frequency   (d) answers (a) and (c)

11. Positive feedback occurs when
    (a) the output signal is fed back to the input in-phase with the input signal
    (b) the output signal is fed back to the input out-of-phase with the input signal
    (c) the total phase shift through the op-amp and feedback network is 360°
    (d) answers (a) and (c)

12. For a closed-loop op-amp circuit to be unstable,
    (a) there must be positive feedback
    (b) the loop gain must be greater than 1
    (c) the loop gain must be less than 1
    (d) answers (a) and (b)

13. The amount of additional phase shift required to make the total phase shift around a closed loop
    equal to zero is called
    (a) the unity-gain phase shift   (b) phase margin
    (c) phase lag   (d) phase bandwidth

14. For a given value of closed-loop gain, a positive phase margin indicates
    (a) an unstable condition   (b) too much phase shift
    (c) a stable condition   (d) nothing

15. The purpose of phase-lag compensation is to
    (a) make the op-amp stable at very high values of gain
    (b) make the op-amp stable at low values of gain
    (c) reduce the unity-gain frequency
    (d) increase the bandwidth

■ PROBLEMS

SECTION 3–1 Basic Concepts

1. The midrange open-loop gain of a certain op-amp is 120 dB. Negative feedback reduces this gain by 50 dB. What is the closed-loop gain?

2. The upper critical frequency of an op-amp's open-loop response is 200 Hz. If the midrange gain is 175,000, what is the ideal gain at 200 Hz? What is the actual gain? What is the op-amp's open-loop bandwidth?

3. An RC lag network has a critical frequency of 5 kHz. If the resistance value is 1.0 kΩ, what is X_C when $f = 3$ kHz?

4. Determine the attenuation of an RC lag network with $f_c = 12$ kHz for each of the following frequencies.
   (a) 1 kHz   (b) 5 kHz   (c) 12 kHz   (d) 20 kHz   (e) 100 kHz

5. The midrange open-loop gain of a certain op-amp is 80,000. If the open-loop critical frequency is 1 kHz, what is the open-loop gain at each of the following frequencies?
   (a) 100 Hz   (b) 1 kHz   (c) 10 kHz   (d) 1 MHz

6. Determine the phase shift through each network in Figure 3–28 at a frequency of 2 kHz.

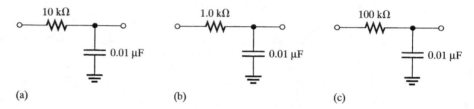

(a)   (b)   (c)

FIGURE 3–28

7. An RC lag network has a critical frequency of 8.5 kHz. Determine the phase for each frequency and plot a graph of its phase angle versus frequency.
   (a) 100 Hz   (b) 400 Hz   (c) 850 Hz
   (d) 8.5 kHz   (e) 25 kHz   (f) 85 kHz

SECTION 3–2 Op-Amp Open-Loop Response

8. A certain op-amp has three internal amplifier stages with midrange gains of 30 dB, 40 dB, and 20 dB. Each stage also has a critical frequency associated with it as follows: $f_{c1} = 600$ Hz, $f_{c2} = 50$ kHz, and $f_{c3} = 200$ kHz.
   (a) What is the midrange open-loop gain of the op-amp, expressed in dB?
   (b) What is the total phase shift through the amplifier, including inversion, when the signal frequency is 10 kHz?

9. What is the gain roll-off rate in Problem 8 between the following frequencies?
   (a) 0 Hz and 600 Hz   (b) 600 Hz and 50 kHz
   (c) 50 kHz and 200 kHz   (d) 200 kHz and 1 MHz

SECTION 3–3 Op-Amp Closed-Loop Response

10. Determine the midrange gain in dB of each amplifier in Figure 3–29. Are these open-loop or closed-loop gains?

11. A certain amplifier has an open-loop gain in midrange of 180,000 and an open-loop critical frequency of 1500 Hz. If the attenuation of the feedback path is 0.015, what is the closed-loop bandwidth?

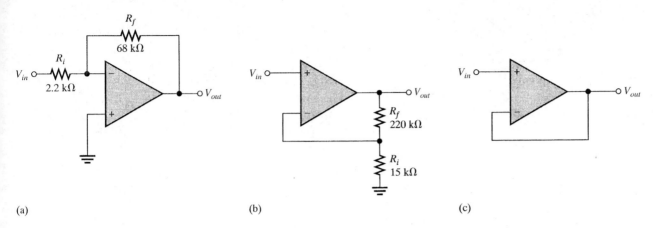

(a)

(b)

(c)

FIGURE 3–29

12. Given that $f_{c(ol)}$ = 750 Hz, A'_{ol} = 89 dB, and $f_{c(cl)}$ = 5.5 kHz, determine the closed-loop gain in dB.

13. What is the unity-gain bandwidth in Problem 12?

14. For each amplifier in Figure 3–30, determine the closed-loop gain and bandwidth. The op-amps in each circuit exhibit an open-loop gain of 125 dB and a unity-gain bandwidth of 2.8 MHz.

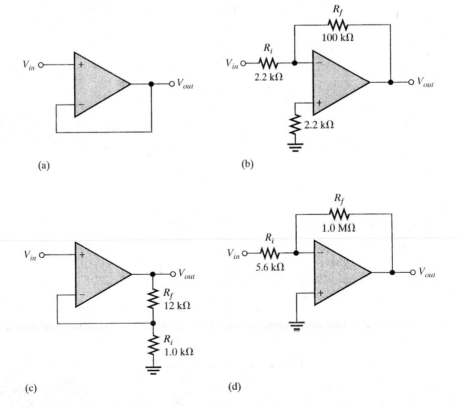

(a)

(b)

(c)

(d)

FIGURE 3–30

FIGURE 3-31

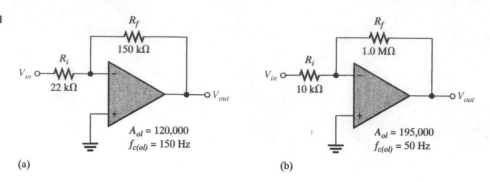

(a)   (b)

15. Which of the amplifiers in Figure 3–31 has the smaller bandwidth?

SECTION 3–4 Positive Feedback and Stability

16. It has been determined that the op-amp circuit in Figure 3–32 has three internal critical frequencies as follows: 1.2 kHz, 50 kHz, 250 kHz. If the midrange open-loop gain is 100 dB, is the amplifier configuration stable, marginally stable, or unstable?

17. Determine the phase margin for each value of phase lag.
   (a) 30°   **(b)** 60°   **(c)** 120°   **(d)** 180°   **(e)** 210°

18. A certain op-amp has the following internal critical frequencies in its open-loop response: 125 Hz, 25 kHz, and 180 kHz. What is the total phase shift through the amplifier when the signal frequency is 50 kHz?

19. Each graph in Figure 3–33 shows both the open-loop and the closed-loop response of a particular op-amp configuration. Analyze each case for stability.

FIGURE 3–32

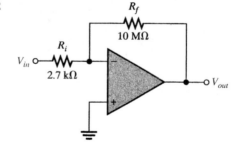

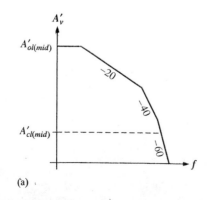

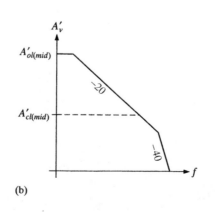

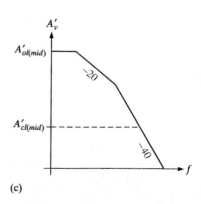

(a)   (b)   (c)

FIGURE 3–33

SECTION 3–5   Op-Amp Compensation

20. A certain operational amplifier has an open-loop response curve as shown in Figure 3–34. A particular application requires a 30 dB closed-loop midrange gain. In order to achieve a 30 dB gain, compensation must be added because the 30 dB line intersects the uncompensated open-loop gain on the −40 dB/decade slope and, therefore, stability is not assured.

    (a) Find the critical frequency of the compensating network such that the −20 dB/decade slope is lowered to a point where it intersects the 30 dB gain line.

    (b) Sketch the ideal response curve for the compensating network.

    (c) Sketch the total ideal compensated open-loop response.

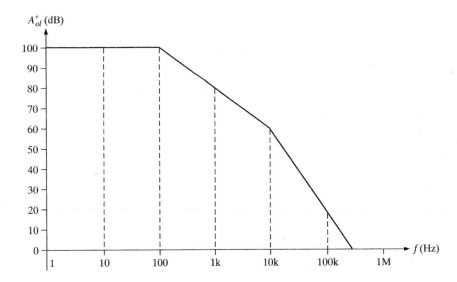

FIGURE 3–34

21. The open-loop gain of a certain op-amp rolls off at −20 dB/decade, beginning at $f = 250$ Hz. This roll-off rate extends down to a gain of 60 dB. If a 40 dB closed-loop gain is required, what is the critical frequency for the compensating network?

22. Repeat Problem 21 for a closed-loop gain of 20 dB.

■ ANSWERS TO REVIEW QUESTIONS

Section 3–1

1. Open-loop gain is without feedback, and closed-loop gain is with negative feedback. Open-loop gain is larger.

2. $BW = 100$ Hz

3. A_{ol} decreases.

Section 3–2

1. $A'_{v(tot)} = 20$ dB $+ 30$ dB $= 50$ dB

2. $\phi_{tot} = -49° + (-5.2°) = -54.2°$

Section 3–3

1. Yes, A_{cl} is always less than A_{ol}.
2. $BW = 3{,}000 \text{ kHz}/60 = 50 \text{ kHz}$
3. unity-gain $BW = 3{,}000 \text{ kHz}/1 = 3 \text{ MHz}$

Section 3–4

1. Positive feedback
2. 180°, 0°
3. -20 dB/decade (-6 dB/octave)

Section 3–5

1. Phase compensation increases the phase margin at a given frequency.
2. Internal compensation is full compensation; external compensation can be tailored to maximize bandwidth.
3. Bandwidth decreases.

Section 3–6

1. f_{cl} can be reduced by increasing C_1 or R_2.
2. Q_2 and Q_3
3. Q_1 is an emitter-follower buffer stage.
4. $P_{avg} = V_{out(rms)}^2 / R_L = (8.4)^2 / 8 = 8.8 \text{ W}$

■ ANSWERS TO PRACTICE EXERCISES FOR EXAMPLES

3–1 (a) 80,000   (b) 79,900   (c) 6400
3–2 173 Hz
3–3 75 dB; $-71.6°$
3–4 2.00 MHz
3–5 (a) 29.6 kHz   (b) 42.6 kHz
3–6 100 Hz; 10 Hz

4

BASIC OP-AMP CIRCUITS

Courtesy Hewlett-Packard Company

■ CHAPTER OBJECTIVES

☐ Understand the operation of several basic comparator circuits
☐ Understand the operation of several types of summing amplifiers
☐ Understand the operation of integrators and differentiators
☐ Understand the operation of several special op-amp circuits
☐ Troubleshoot basic op-amp circuits
☐ Apply what you have learned in this chapter to a system application

In the last two chapters, you learned about the principles, operation, and characteristics of the operational amplifier. Op-amps are used in such a wide variety of applications that it is impossible to cover all of them in one chapter, or even in one book. Therefore, in this chapter, we will examine some of the more fundamental applications to illustrate how versatile the op-amp is and to give you a foundation in basic op-amp circuits.

This system application illustrates a very interesting application of three types of op-amp circuits that will be studied in this chapter: the summing amplifier, integrator, and comparator. The system diagram in Figure 4–47 shows one basic type of analog-to-digital converter that takes an audio input, such as voice or music, and converts it to binary codes that can be recorded digitally. Analog-to-digital converters are covered thoroughly in Chapter 10.

Op-amps play a key role in this system, and we will be focusing on the analog-to-digital converter board to see how these circuits are used in a representative application. The digital circuits are discussed just enough to allow you to understand what the overall system does. You do not need to have a background in digital circuits for our purposes here. However, this particular system application points out the fact, again, that many systems include combinations of both analog and digital circuits.

For the system application in Section 4–6, in addition to the other topics, be sure you understand

☐ How a summing amplifier works
☐ How an integrator works
☐ How a comparator works

4–1 ■ COMPARATORS

Operational amplifiers are often used as nonlinear devices to compare the amplitude of one voltage with another. In this application, the op-amp is used in the open-loop configuration, with the input voltage on one input and a reference voltage on the other.

❑ Understand the operation of several basic comparator circuits
    ❑ Describe the operation of a zero-level detector
    ❑ Describe the operation of a nonzero-level detector
    ❑ Discuss how input noise affects comparator operation
    ❑ Define *hysteresis*
    ❑ Explain how hysteresis reduces noise effects
    ❑ Describe a Schmitt trigger circuit
    ❑ Describe the operation of bounded comparators
    ❑ Describe the operation of a window comparator
    ❑ Discuss two comparator applications including analog-to-digital conversion

Zero-Level Detection

One application of an op-amp used as a **comparator** is to determine when an input voltage exceeds a certain level. Figure 4–1(a) shows a zero-level detector. Notice that the inverting ($-$) input is grounded to produce a zero level and that the input signal voltage is applied to the noninverting ($+$) input. Because of the high open-loop voltage gain, a very small difference voltage between the two inputs drives the amplifier into saturation, causing the output voltage to go to its limit.

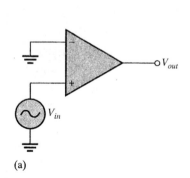

(a)

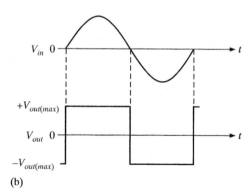

(b)

FIGURE 4–1
The op-amp as a zero-level detector.

For example, consider an op-amp having $A_{ol} = 100,000$. A voltage difference of only 0.25 mV between the inputs could produce an output voltage of $(0.25 \text{ mV})(100,000) = 25$ V *if* the op-amp were capable. However, since most op-amps have output voltage limitations of ± 15 V or less, the device would be driven into saturation. For many comparison applications, special op-amp comparators are selected. These ICs are generally uncompensated to maximize speed. In less stringent applications, a general-purpose op-amp works nicely as a comparator.

Figure 4–1(b) shows the result of a sinusoidal input voltage applied to the noninverting input of the zero-level detector. When the sine wave is negative, the output is at its maximum negative level. When the sine wave crosses 0, the amplifier is driven to its opposite state and the output goes to its maximum positive level, as shown. As you can see, the zero-level detector can be used as a squaring circuit to produce a square wave from a sine wave.

Nonzero-Level Detection

The zero-level detector in Figure 4–1 can be modified to detect voltages other than zero by connecting a fixed reference voltage to the inverting ($-$) input, as shown in Figure 4–2(a). A more practical arrangement is shown in Figure 4–2(b) using a voltage divider to set the reference voltage as follows:

$$V_{REF} = \frac{R_2}{R_1 + R_2}(+V) \qquad (4\text{–}1)$$

where $+V$ is the positive op-amp supply voltage. The circuit in Figure 4–2(c) uses a zener diode to set the reference voltage ($V_{REF} = V_Z$). As long as the input voltage V_{in} is less than

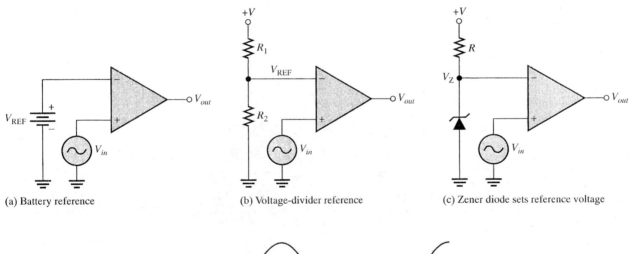

(a) Battery reference   (b) Voltage-divider reference   (c) Zener diode sets reference voltage

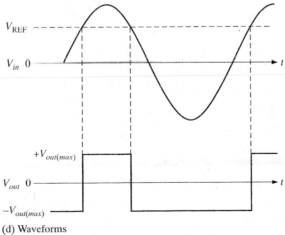

(d) Waveforms

FIGURE 4–2
Nonzero-level detectors.

V_{REF}, the output remains at the maximum negative level. When the input voltage exceeds the reference voltage, the output goes to its maximum positive state, as shown in Figure 4–2(d) with a sinusoidal input voltage.

EXAMPLE 4–1

The input signal in Figure 4–3(a) is applied to the comparator circuit in Figure 4–3(b). Make a sketch of the output showing its proper relationship to the input signal. Assume the maximum output levels of the op-amp are ±12 V.

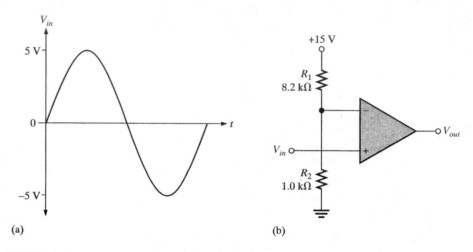

(a)   (b)

FIGURE 4–3

Solution   The reference voltage is set by R_1 and R_2 as follows:

$$V_{REF} = \frac{R_2}{R_1 + R_2}(+V) = \frac{1.0 \text{ k}\Omega}{8.2 \text{ k}\Omega + 1.0 \text{ k}\Omega}(+15 \text{ V}) = 1.63 \text{ V}$$

As shown in Figure 4–4, each time the input exceeds +1.63 V, the output voltage switches to its +12 V level, and each time the input goes below +1.63 V, the output switches back to its −12 V level.

FIGURE 4–4

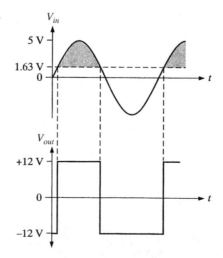

Practice Exercise   Determine the reference voltage in Figure 4–3 if $R_1 = 22$ kΩ and $R_2 = 3.3$ kΩ.

Effects of Input Noise on Comparator Operation

In many practical situations, unwanted voltage or current fluctuations (**noise**) may appear on the input line. This noise voltage becomes superimposed on the input voltage, as shown in Figure 4–5, and can cause a comparator to erratically switch output states.

FIGURE 4–5
Sine wave with superimposed noise.

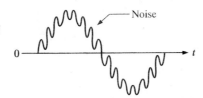

In order to understand the potential effects of noise voltage, consider a low-frequency sinusoidal voltage applied to the noninverting (+) input of an op-amp comparator used as a zero-level detector, as shown in Figure 4–6(a). Part (b) of the figure shows the input sine wave plus noise and the resulting output. As you can see, when the sine wave approaches 0, the fluctuations due to noise cause the total input to vary above and below 0 several times, thus producing an erratic output voltage.

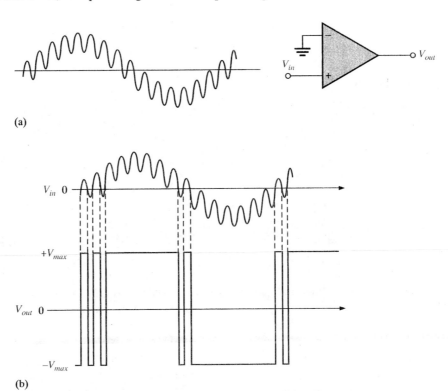

FIGURE 4–6
Effects of noise on comparator circuit.

Reducing Noise Effects with Hysteresis

An erratic output voltage caused by noise on the input occurs because the op-amp comparator switches from its negative output state to its positive output state at the same input voltage level that causes it to switch in the opposite direction, from positive to negative. This unstable condition occurs when the input voltage hovers around the reference voltage, and any small noise fluctuations cause the comparator to switch first one way and then the other.

In order to make the comparator less sensitive to noise, a technique incorporating positive feedback, called **hysteresis,** can be used. Basically, hysteresis means that there is a higher reference level when the input voltage goes from a lower to higher value than when it goes from a higher to a lower value. A good example of hysteresis is a common household thermostat that turns the furnace on at one temperature and off at another.

The two reference levels are referred to as the upper trigger point (UTP) and the lower trigger point (LTP). This two-level hysteresis is established with a positive feedback arrangement, as shown in Figure 4–7. Notice that the noninverting (+) input is connected to a resistive voltage divider such that a portion of the output voltage is fed back to the input. The input signal is applied to the inverting (−) input in this case.

FIGURE 4–7
Comparator with positive feedback for hysteresis.

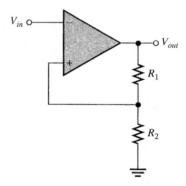

The basic operation of the comparator with hysteresis is as follows and is illustrated in Figure 4–8. Assume that the output voltage is at its positive maximum, $+V_{out(max)}$. The voltage fed back to the noninverting input is V_{UTP} and is expressed as

$$V_{\text{UTP}} = \frac{R_2}{R_1 + R_2}[+V_{out(max)}]$$

When the input voltage V_{in} exceeds V_{UTP}, the output voltage drops to its negative maximum, $-V_{out(max)}$. Now the voltage fed back to the noninverting input is V_{LTP} and is expressed as

$$V_{\text{LTP}} = \frac{R_2}{R_1 + R_2}[-V_{out(max)}]$$

The input voltage must now fall below V_{LTP} before the device will switch back to its other voltage level. This means that a small amount of noise voltage has no effect on the output, as illustrated by Figure 4–8.

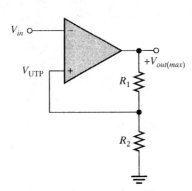

(a) Output at the maximum positive voltage

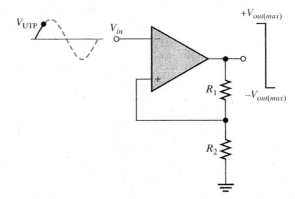

(b) Input exceeds UTP; output switches from the maximum positive voltage to the maximum negative voltage.

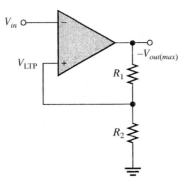

(c) Output at the maximum negative voltage

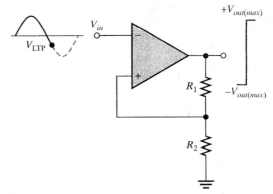

(d) Input goes below LTP; output switches from maximum negative voltage back to maximum positive voltage.

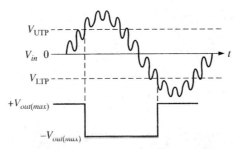

(e) Device triggers only once when UTP or LTP is reached; thus, there is immunity to noise that is riding on the input signal.

FIGURE 4–8

Operation of a comparator with hysteresis.

A comparator with hysteresis is sometimes known as a **Schmitt trigger.** The amount of hysteresis is defined by the difference of the two trigger levels.

$$V_{HYS} = V_{UTP} - V_{LTP}$$ (4–2)

EXAMPLE 4–2

Determine the upper and lower trigger points and the hysteresis for the comparator circuit in Figure 4–9. Assume that $+V_{out(max)} = +5$ V and $-V_{out(max)} = -5$ V.

FIGURE 4–9

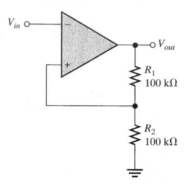

Solution

$$V_{UTP} = \frac{R_2}{R_1 + R_2}[+V_{out(max)}] = 0.5(5 \text{ V}) = \textbf{+2.5 V}$$

$$V_{LTP} = \frac{R_2}{R_1 + R_2}[-V_{out(max)}] = 0.5(-5 \text{ V}) = \textbf{−2.5 V}$$

$$V_{HYS} = V_{UTP} - V_{LTP} = 2.5 \text{ V} - (-2.5 \text{ V}) = \textbf{5 V}$$

Practice Exercise Determine the upper and lower trigger points and the hysteresis in Figure 4–9 for $R_1 = 68$ kΩ and $R_2 = 82$ kΩ. The maximum output voltage levels are ±7 V.

Output Bounding

In some applications, it is necessary to limit the output voltage levels of a comparator to a value less than that provided by the saturated op-amp. A single zener diode can be used as shown in Figure 4–10 to limit the output voltage to the zener voltage in one direction and to the forward diode drop in the other. This process of limiting the output range is called **bounding.**

FIGURE 4–10
Comparator with output bounding.

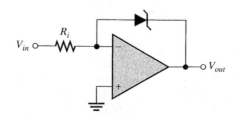

The operation is as follows. Since the anode of the zener is connected to the inverting (−) input, it is at virtual ground (≅0 V). Therefore, when the output voltage reaches a positive value equal to the zener voltage, it limits at that value, as illustrated in Figure 4–11. When the output switches negative, the zener acts as a regular diode and becomes forward-biased at 0.7 V, limiting the negative output voltage to this value, as shown. Turning the zener around limits the output voltage in the opposite direction.

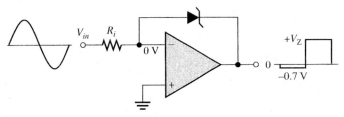

(a) Bounded at a positive value

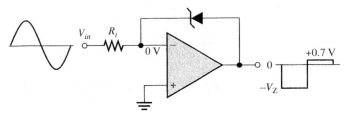

(b) Bounded at a negative value

FIGURE 4–11
Operation of a bounded comparator.

Two zener diodes arranged as in Figure 4–12 limit the output voltage to the zener voltage plus the forward voltage drop (0.7 V) of the forward-biased zener, both positively and negatively, as shown in Figure 4–12.

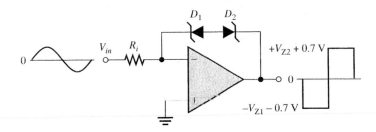

FIGURE 4–12
Double-bounded comparator.

EXAMPLE 4–3

Determine the output voltage waveform for Figure 4–13.

Solution   This comparator has both hysteresis and zener bounding.
The voltage across D_1 and D_2 in either direction is 4.7 V + 0.7 V = 5.4 V. This is because one zener is always forward-biased with a drop of 0.7 V when the other one is in breakdown.

FIGURE 4–13

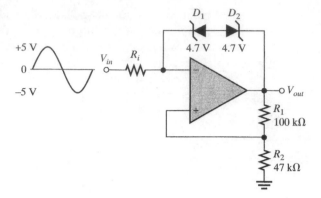

The voltage at the inverting $(-)$ op-amp input is $V_{out} \pm 5.4$ V. Since the differential voltage is negligible, the voltage at the noninverting $(+)$ op-amp input is also approximately $V_{out} \pm 5.4$ V. Thus,

$$V_{R1} = V_{out} - (V_{out} \pm 5.4 \text{ V}) = \pm 5.4 \text{ V}$$

$$I_{R1} = \frac{V_{R1}}{R_1} = \frac{\pm 5.4 \text{ V}}{100 \text{ k}\Omega} = \pm 54 \text{ } \mu\text{A}$$

Since the current at the noninverting input is negligible,

$$I_{R2} = I_{R1} = \pm 54 \text{ } \mu\text{A}$$
$$V_{R2} = R_2 I_{R2} = (47 \text{ k}\Omega)(\pm 54 \text{ } \mu\text{A}) = \pm 2.54 \text{ V}$$
$$V_{out} = V_{R1} + V_{R2} = \pm 5.4 \text{ V} \pm 2.54 \text{ V} = \pm 7.94 \text{ V}$$

The upper trigger point (UTP) and the lower trigger point (LTP) are as follows:

$$V_{\text{UTP}} = \left(\frac{R_2}{R_1 + R_2}\right)(+V_{out}) = \left(\frac{47 \text{ k}\Omega}{147 \text{ k}\Omega}\right)(+7.94 \text{ V}) = +2.54 \text{ V}$$

$$V_{\text{LTP}} = \left(\frac{R_2}{R_1 + R_2}\right)(-V_{out}) = \left(\frac{47 \text{ k}\Omega}{147 \text{ k}\Omega}\right)(-7.94 \text{ V}) = -2.54 \text{ V}$$

The output waveform for the given input voltage is shown in Figure 4–14.

FIGURE 4–14

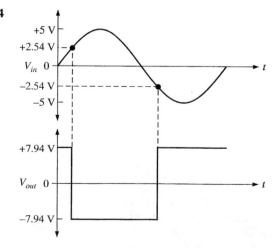

Practice Exercise Determine the upper and lower trigger points for Figure 4–13 if $R_1 = 150 \text{ k}\Omega$, $R_2 = 68 \text{ k}\Omega$, and the zener diodes are 3.3 V devices.

Window Comparator

Two individual op-amp comparators arranged as in Figure 4–15 form what is known as a *window comparator*. This circuit detects when an input voltage is between two limits, an upper and a lower, called the "window."

FIGURE 4–15
A basic window comparator.

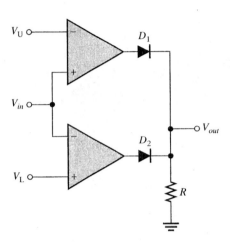

The upper and lower limits are set by reference voltages designated V_U and V_L. These voltages can be established with voltage dividers, zener diodes, or any type of voltage source. As long as V_{in} is within the window (less than V_U and greater than V_L), the output of each comparator is at its low saturated level. Under this condition, both diodes are reverse-biased and V_{out} is held at zero by the resistor to ground. When V_{in} goes above V_U or below V_L, the output of the associated comparator goes to its high saturated level. This action forward-biases the diode and produces a high-level V_{out}. This is illustrated in Figure 4–16 with V_{in} varying arbitrarily.

FIGURE 4–16
Example of window comparator operation.

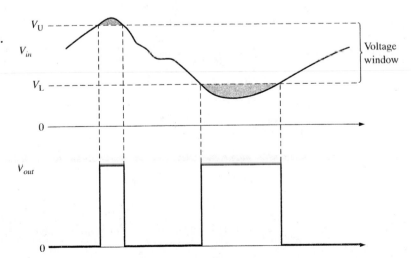

A Comparator Application: Over-Temperature Sensing Circuit

Figure 4–17 shows an op-amp comparator used in a precision over-temperature sensing circuit to determine when the temperature reaches a certain critical value. The circuit consists of a Wheatstone bridge with the op-amp used to detect when the bridge is balanced. One leg of the bridge contains a thermistor (R_1), which is a temperature-sensing resistor with a negative temperature coefficient (its resistance decreases as temperature increases and vice versa). The potentiometer (R_2) is set at a value equal to the resistance of the thermistor at the critical temperature. At normal temperatures (below critical), R_1 is greater than R_2, thus creating an unbalanced condition that drives the op-amp to its low saturated output level and keeps transistor Q_1 off.

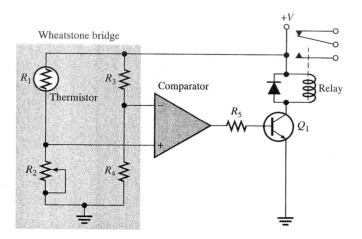

FIGURE 4–17
An over-temperature sensing circuit.

As the temperature increases, the resistance of the thermistor decreases. When the temperature reaches the critical value, R_1 becomes equal to R_2, and the bridge becomes balanced (since $R_3 = R_4$). At this point the op-amp switches to its high saturated output level, turning Q_1 on. This energizes the relay, which can be used to activate an alarm or initiate an appropriate response to the over-temperature condition.

A Comparator Application: Analog-to-Digital (A/D) Conversion

A/D conversion is a common interfacing process often used when a linear *analog* system must provide inputs to a *digital* system. Many methods for A/D conversion are available and some of these will be covered thoroughly in Chapter 10. However, in this discussion, only one type is used to demonstrate the concept.

The *simultaneous,* or *flash,* method of A/D conversion uses parallel comparators to compare the linear input signal with various reference voltages developed by a voltage divider. When the input voltage exceeds the reference voltage for a given comparator, a high level is produced on that comparator's output. Figure 4–18 shows an analog-to-digital converter (ADC) that produces three-digit binary numbers on its output, which represent the values of the analog input voltage as it changes. This converter requires seven compara-

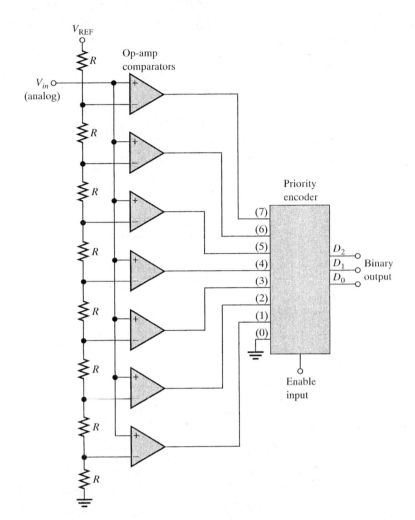

FIGURE 4–18
A simultaneous (flash) analog-to-digital converter (ADC) using op-amps as comparators.

tors. In general, $2^n - 1$ comparators are required for conversion to an n-digit binary number. The large number of comparators necessary for a reasonably sized binary number is one of the drawbacks of this type of ADC. Its chief advantage is that it provides a fast conversion time.

The reference voltage for each comparator is set by the resistive voltage-divider network and V_{REF}. The output of each comparator is connected to an input of the priority encoder. The *priority encoder* is a digital device that produces a binary number representing the highest-value input.

The encoder *samples* its input when a pulse occurs on the enable line (sampling pulse), and a three-digit binary number proportional to the value of the analog input signal appears on the encoder's outputs. The sampling rate determines the accuracy with which the sequence of binary numbers represents the changing input signal. The more samples taken in a given unit of time, the more accurately the analog signal is represented in digital form.

4–1 REVIEW QUESTIONS

1. What is the reference voltage for each comparator in Figure 4–19?

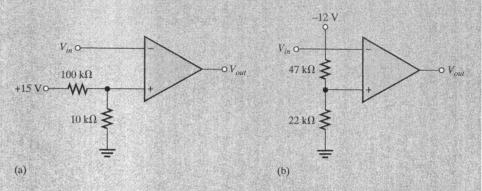

(a)   (b)

FIGURE 4–19

2. What is the purpose of hysteresis in a comparator?

3. Define the term *bounding* in relation to a comparator's output.

4–2 ■ SUMMING AMPLIFIERS

The summing amplifier is a variation of the inverting op-amp configuration covered in Chapter 2. The summing amplifier has two or more inputs, and its output voltage is proportional to the negative of the algebraic sum of its input voltages. In this section, you will see how a summing amplifier works, and you will learn about the averaging amplifier and the scaling amplifier, which are variations of the basic summing amplifier.

After completing this section, you should be able to

❑ Understand the operation of several types of summing amplifiers
    ❑ Describe the operation of a unity-gain summing amplifier
    ❑ Discuss how to achieve any specified gain greater than unity
    ❑ Describe the operation of an averaging amplifier
    ❑ Describe the operation of a scaling adder
    ❑ Discuss a scaling adder used as a digital-to-analog converter

Summing Amplifier with Unity Gain

A two-input summing amplifier is shown in Figure 4–20, but any number of inputs can be used.

    The operation of the circuit and derivation of the output expression are as follows. Two voltages, V_{IN1} and V_{IN2}, are applied to the inputs and produce currents I_1 and I_2, as shown. From the concepts of infinite input impedance and virtual ground, the voltage at the inverting ($-$) input of the op-amp is approximately 0 V, and therefore there is no current at the input. This means that both input currents I_1 and I_2 combine at this summing point and form the total current, which is through R_f, as indicated.

$$I_T = I_1 + I_2$$

FIGURE 4–20
Two-input inverting summing amplifier.

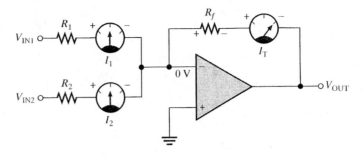

Since $V_{OUT} = -I_T R_f$, the following steps apply.

$$V_{OUT} = -(I_1 + I_2)R_f = -\left(\frac{V_{IN1}}{R_1} + \frac{V_{IN2}}{R_2}\right)R_f$$

If all three of the resistors are equal in value ($R_1 = R_2 = R_f = R$), then

$$V_{OUT} = -\left(\frac{V_{IN1}}{R} + \frac{V_{IN2}}{R}\right)R = -(V_{IN1} + V_{IN2})$$

The previous equation shows that the output voltage is the sum of the two input voltages. A general expression is given in Equation (4–3) for a summing amplifier with n inputs, as shown in Figure 4–21 where all resistors are equal in value.

$$V_{OUT} = -(V_{IN1} + V_{IN2} + \cdots + V_{INn}) \tag{4–3}$$

FIGURE 4–21
Summing amplifier with n inputs.

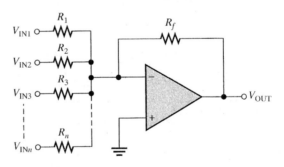

EXAMPLE 4–4 Determine the output voltage in Figure 4–22.

FIGURE 4–22

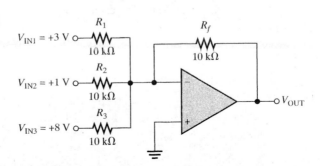

Solution

$$V_{OUT} = -(V_{IN1} + V_{IN2} + V_{IN3}) = -(3 \text{ V} + 1 \text{ V} + 8 \text{ V}) = -12 \text{ V}$$

Practice Exercise If a fourth input of +0.5 V is added to Figure 4–22 with a 10 kΩ resistor, what is the output voltage?

Summing Amplifier with Gain Greater Than Unity

When R_f is larger than the input resistors, the amplifier has a gain of R_f/R, where R is the value of each input resistor. The general expression for the output is

$$V_{OUT} = -\frac{R_f}{R}(V_{IN1} + V_{IN2} + \cdots + V_{INn}) \qquad (4\text{–}4)$$

As you can see, the output is the sum of all the input voltages multiplied by a constant determined by the ratio R_f/R.

EXAMPLE 4–5   Determine the output voltage for the summing amplifier in Figure 4–23.

FIGURE 4–23

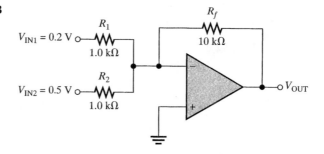

Solution $R_f = 10$ kΩ, and $R = R_1 = R_2 = 1.0$ kΩ. Therefore,

$$V_{OUT} = -\frac{R_f}{R}(V_{IN1} + V_{IN2}) = -\frac{10 \text{ k}\Omega}{1.0 \text{ k}\Omega}(0.2 \text{ V} + 0.5 \text{ V}) = -10(0.7 \text{ V}) = -7 \text{ V}$$

Practice Exercise   Determine the output voltage in Figure 4–23 if the two input resistors are 2.2 kΩ and the feedback resistor is 18 kΩ.

Averaging Amplifier

A summing amplifier can be made to produce the mathematical average of the input voltages. This is done by setting the ratio R_f/R equal to the reciprocal of the number of inputs (n); that is, $R_f/R = 1/n$.

You obtain the average of several numbers by first adding the numbers and then dividing by the quantity of numbers you have. Examination of Equation (4–4) and a little thought will convince you that a summing amplifier will do this. The next example illustrates this idea.

EXAMPLE 4–6

Show that the amplifier in Figure 4–24 produces an output whose magnitude is the mathematical average of the input voltages.

FIGURE 4–24

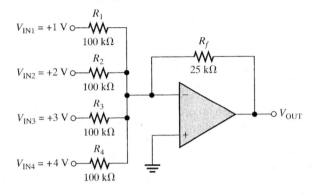

Solution   Since the input resistors are equal, $R = 100 \text{ k}\Omega$. The output voltage is

$$V_{OUT} = -\frac{R_f}{R}(V_{IN1} + V_{IN2} + V_{IN3} + V_{IN4})$$

$$= -\frac{25 \text{ k}\Omega}{100 \text{ k}\Omega}(1 \text{ V} + 2 \text{ V} + 3 \text{ V} + 4 \text{ V}) = -\frac{1}{4}(10 \text{ V}) = -2.5 \text{ V}$$

A simple calculation shows that the average of the input values is the same magnitude as V_{OUT} but of opposite sign.

$$V_{IN(avg)} = \frac{1 \text{ V} + 2 \text{ V} + 3 \text{ V} + 4 \text{ V}}{4} = \frac{10 \text{ V}}{4} = 2.5 \text{ V}$$

Practice Exercise   Specify the changes required in the averaging amplifier in Figure 4–24 in order to handle five inputs.

Scaling Adder

A different weight can be assigned to each input of a summing amplifier by simply adjusting the values of the input resistors. As you have seen, the output voltage can be expressed as

$$V_{OUT} = -\left(\frac{R_f}{R_1}V_{IN1} + \frac{R_f}{R_2}V_{IN2} + \cdots + \frac{R_f}{R_n}V_{INn}\right) \tag{4–5}$$

The weight of a particular input is set by the ratio of R_f to the resistance for that input. For example, if an input voltage is to have a weight of 1, then $R = R_f$. Or, if a weight of 0.5 is required, $R = 2R_f$. The smaller the value of R, the greater the weight, and vice versa.

EXAMPLE 4–7

Determine the weight of each input voltage for the scaling adder in Figure 4–25 and find the output voltage.

FIGURE 4–25

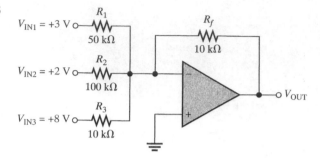

Solution

Weight of input 1: $\dfrac{R_f}{R_1} = \dfrac{10\ k\Omega}{50\ k\Omega} = \mathbf{0.2}$

Weight of input 2: $\dfrac{R_f}{R_2} = \dfrac{10\ k\Omega}{100\ k\Omega} = \mathbf{0.1}$

Weight of input 3: $\dfrac{R_f}{R_3} = \dfrac{10\ k\Omega}{10\ k\Omega} = \mathbf{1}$

The output voltage is

$$V_{OUT} = -\left(\frac{R_f}{R_1}V_{IN1} + \frac{R_f}{R_2}V_{IN2} + \frac{R_f}{R_3}V_{IN3}\right)$$

$$= -[0.2(3\ V) + 0.1(2\ V) + 1(8\ V)] = -(0.6\ V + 0.2\ V + 8\ V) = \mathbf{-8.8\ V}$$

Practice Exercise Determine the weight of each input voltage in Figure 4–25 if $R_1 = 22\ k\Omega$, $R_2 = 82\ k\Omega$, $R_3 = 56\ k\Omega$, and $R_f = 10\ k\Omega$. Also find V_{OUT}.

A Scaling Adder Application: Digital-to-Analog (D/A) Conversion

D/A conversion is an important interface process for converting digital signals to analog (linear) signals. An example is a voice signal that is digitized for storage, processing, or transmission and must be changed back into an approximation of the original audio signal in order to drive a speaker. Digital-to-analog converters will be covered thoroughly in Chapter 10.

One method of D/A conversion uses a scaling adder with input resistor values that represent the binary weights of the digital input code. Figure 4–26 shows a four-digit digital-to-analog converter (DAC) of this type (called a *binary-weighted resistor DAC*). The switch symbols represent transistor switches for applying each of the four binary digits to the inputs.

The inverting (−) input is at virtual ground, so that the output voltage is proportional to the current through the feedback resistor R_f (sum of input currents). The lowest-value resistor R corresponds to the highest weighted binary input (2^3). All of the other resistors are multiples of R and correspond to the binary weights 2^2, 2^1, and 2^0.

FIGURE 4–26

A scaling adder as a four-digit digital-to-analog converter (DAC).

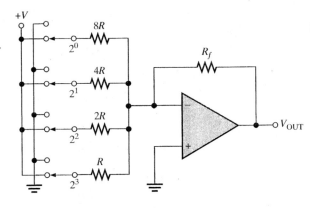

4–2 REVIEW QUESTIONS

1. Define *summing point.*
2. What is the value of R_f/R for a five-input averaging amplifier?
3. A certain scaling adder has two inputs, one having twice the weight of the other. If the resistor value for the lower weighted input is 10 kΩ, what is the value of the other input resistor?

4–3 ■ INTEGRATORS AND DIFFERENTIATORS

An op-amp integrator simulates mathematical integration, which is basically a summing process that determines the total area under the curve of a function. An op-amp differentiator simulates mathematical differentiation, which is a process of determining the instantaneous rate of change of a function. The integrators and differentiators shown in this section are idealized to show basic principles. Practical integrators often have an additional resistor or other circuitry in parallel with the feedback capacitor to prevent saturation. Practical differentiators may include a series resistor to reduce high frequency noise.

After completing this section, you should be able to

❑ Understand the operation of integrators and differentiators
   ❑ Identify an integrator
   ❑ Discuss how a capacitor charges
   ❑ Determine the rate of change of an integrator's output
   ❑ Identify a differentiator
   ❑ Determine the output voltage of a differentiator

The Op-Amp Integrator

A basic **integrator** is shown in Figure 4–27. Notice that the feedback element is a capacitor that forms an *RC* circuit with the input resistor.

FIGURE 4–27
An op-amp integrator.

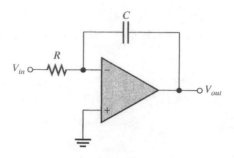

How a Capacitor Charges   To understand how the integrator works, it is important to review how a capacitor charges. Recall that the charge Q on a capacitor is proportional to the charging current (I_C) and the time (t).

$$Q = I_C t$$

Also, in terms of the voltage, the charge on a capacitor is

$$Q = CV_C$$

From these two relationships, the capacitor voltage can be expressed as

$$V_C = \left(\frac{I_C}{C}\right)t$$

This expression is an equation for a straight line which begins at zero with a constant slope of I_C/C. (Remember from algebra that the general formula for a straight line is $y = mx + b$. In this case, $y = V_C, m = I_C/C, x = t,$ and $b = 0$.)

Recall that the capacitor voltage in a simple RC network is not linear but is exponential. This is because the charging current continuously decreases as the capacitor charges and causes the rate of change of the voltage to continuously decrease. The key thing about using an op-amp with an RC network to form an integrator is that the capacitor's charging current is made constant, thus producing a straight-line (linear) voltage rather than an exponential voltage. Now let's see why this is true.

In Figure 4–28, the inverting input of the op-amp is at virtual ground (0 V), so the voltage across R_i equals V_{in}. Therefore, the input current is

$$I_{in} = \frac{V_{in}}{R_i}$$

If V_{in} is a constant voltage, then I_{in} is also a constant because the inverting input always remains at 0 V, keeping a constant voltage across R_i. Because of the very high input

FIGURE 4–28
Currents in an integrator.

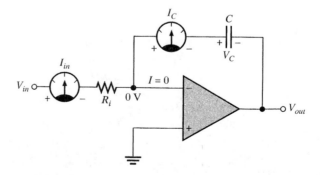

impedance of the op-amp, there is negligible current at the inverting input. This makes all of the input current charge the capacitor, so

$$I_C = I_{in}$$

The Capacitor Voltage   Since I_{in} is constant, so is I_C. The constant I_C charges the capacitor linearly and produces a linear voltage across C. The positive side of the capacitor is held at 0 V by the virtual ground of the op-amp. The voltage on the negative side of the capacitor decreases linearly from zero as the capacitor charges, as shown in Figure 4–29. This voltage is called a *negative ramp*.

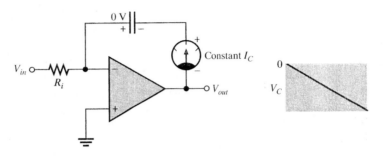

FIGURE 4–29
A linear ramp voltage is produced across C by the constant charging current.

The Output Voltage V_{out} is the same as the voltage on the negative side of the capacitor. When a constant input voltage in the form of a step or pulse (a pulse has a constant amplitude when high) is applied, the output ramp decreases negatively until the op-amp saturates at its maximum negative level. This is indicated in Figure 4–30.

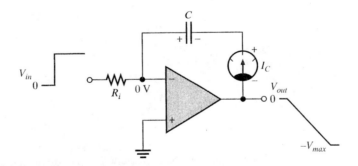

FIGURE 4–30
A constant input voltage produces a ramp on the output of the integrator.

Rate of Change of the Output   The rate at which the capacitor charges, and therefore the slope of the output ramp, is set by the ratio I_C/C, as you have seen. Since $I_C = V_{in}/R_i$, the rate of change or slope of the integrator's output voltage is

$$\frac{\Delta V_{out}}{\Delta t} = -\frac{V_{in}}{R_i C} \tag{4–6}$$

Integrators are especially useful in triangular-wave generators as you will see in Chapter 6.

EXAMPLE 4–8 (a) Determine the rate of change of the output voltage in response to the first input pulse in a pulse waveform, as shown for the integrator in Figure 4–31(a). The output voltage is initially zero.

(b) Describe the output after the first pulse. Draw the output waveform.

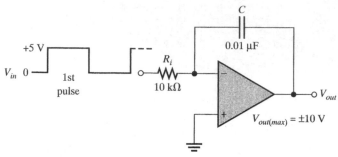

(a)

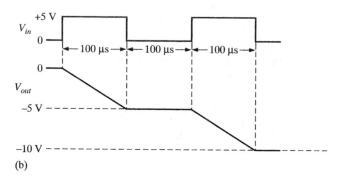

(b)

FIGURE 4–31

Solution

(a) The rate of change of the output voltage during the time that the input pulse is high is

$$\frac{\Delta V_{out}}{\Delta t} = -\frac{V_{in}}{R_i C} = -\frac{5 \text{ V}}{(10 \text{ k}\Omega)(0.01 \text{ }\mu\text{F})} = -50 \text{ kV/s} = \mathbf{-50 \text{ mV/}\mu\text{s}}$$

(b) The rate of change was found to be -50 mV/μs in part (a). When the input is at $+5$ V, the output is a negative-going ramp. When the input is at 0 V, the output is a constant level. In 100 μs, the voltage decreases.

$$\Delta V_{out} = (-50 \text{ mV/}\mu\text{s})(100 \text{ }\mu\text{s}) = \mathbf{-5 \text{ V}}$$

Therefore, the negative-going ramp reaches -5 V at the end of the pulse. The output voltage then remains constant at -5 V for the time that the input is zero. On the next pulse, the output again is a negative-going ramp that reaches -10 V. Since this is the maximum limit, the output remains at -10 V as long as pulses are applied. The waveforms are shown in Figure 4–31(b).

Practice Exercise Modify the integrator in Figure 4–31 to make the output change from 0 to -5 V in 50 μs with the same input.

The Op-Amp Differentiator

A basic **differentiator** is shown in Figure 4–32. Notice how the placement of the capacitor and resistor differ from that in the integrator. The capacitor is now the input element. A differentiator produces an output that is proportional to the rate of change of the input voltage.

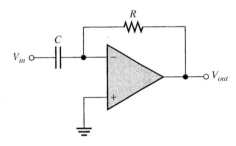

FIGURE 4–32
An op-amp differentiator.

To see how the differentiator works, let's apply a positive-going ramp voltage to the input as indicated in Figure 4–33. In this case, $I_C = I_{in}$ and the voltage across the capacitor is equal to V_{in} at all times ($V_C = V_{in}$) because of virtual ground on the inverting input.

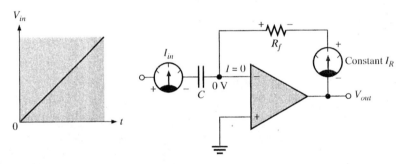

FIGURE 4–33
A differentiator with a ramp input.

From the basic formula, which is $V_C = (I_C/C)t$,

$$I_C = \left(\frac{V_C}{t}\right)C$$

Since the current at the inverting input is negligible, $I_R = I_C$. Both currents are constant because the slope of the capacitor voltage (V_C/t) is constant. The output voltage is also constant and equal to the voltage across R_f because one side of the feedback resistor is always 0 V (virtual ground).

$$V_{out} = I_R R_f = I_C R_f$$

$$V_{out} = -\left(\frac{V_C}{t}\right)R_f C \tag{4–7}$$

The output is negative when the input is a positive-going ramp and positive when the input is a negative-going ramp, as illustrated in Figure 4–34. During the positive slope of the input, the capacitor is charging from the input source with constant current through the feed-

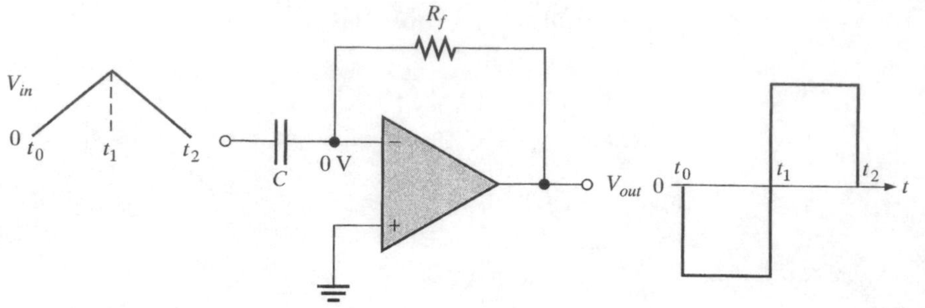

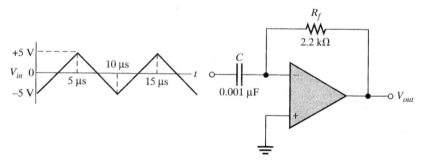

FIGURE 4–34
Output of a differentiator with a series of positive and negative ramps (triangle wave) on the input.

back resistor. During the negative slope of the input, the constant current is in the opposite direction because the capacitor is discharging.

Notice in Equation (4–7) that the term V_C/t is the slope of the input. If the slope increases, V_{out} becomes more negative. If the slope decreases, V_{out} becomes more positive. So, the output voltage is proportional to the negative slope (rate of change) of the input. The constant of proportionality is the time constant, R_fC.

EXAMPLE 4–9

Determine the output voltage of the op-amp differentiator in Figure 4–35 for the triangular-wave input shown.

FIGURE 4–35

Solution   Starting at $t = 0$, the input voltage is a positive-going ramp ranging from -5 V to $+5$ V (a $+10$ V change) in 5 μs. Then it changes to a negative-going ramp ranging from $+5$ V to -5 V (a -10 V change) in 5 μs.

Substituting into Equation (4–7), the output voltage for the positive-going ramp is

$$V_{out} = -\left(\frac{V_C}{t}\right)R_fC = -\left(\frac{10\text{ V}}{5\text{ μs}}\right)(2.2\text{ k}\Omega)(0.001\text{ μF}) = \mathbf{-4.4\text{ V}}$$

The output voltage for the negative-going ramp is calculated the same way.

$$V_{out} = -\left(\frac{V_C}{t}\right)R_fC = -\left(\frac{-10\text{ V}}{5\text{ μs}}\right)(2.2\text{ k}\Omega)(0.001\text{ μF}) = \mathbf{+4.4\text{ V}}$$

Finally, the output voltage waveform is graphed relative to the input as shown in Figure 4–36.

FIGURE 4–36

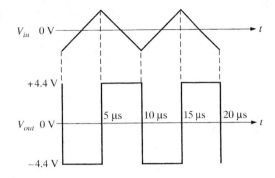

Practice Exercise What would the output voltage be if the feedback resistor in Figure 4–35 is changed to 3.3 kΩ?

4–3 REVIEW QUESTIONS

1. What is the feedback element in an op-amp integrator?
2. For a constant input voltage to an integrator, why is the voltage across the capacitor linear?
3. What is the feedback element in an op-amp differentiator?
4. How is the output of a differentiator related to the input?

4–4 ■ CONVERTERS AND OTHER OP-AMP CIRCUITS

This section introduces a few more op-amp circuits that represent basic applications of the op-amp. You will learn about the constant-current source, the current-to-voltage converter, the voltage-to-current converter, and the peak detector. This is, of course, not a comprehensive coverage of all possible op-amp circuits but is intended only to introduce you to some common and basic uses.

After completing this section, you should be able to

❑ Understand the operation of several special op-amp circuits
    ❑ Identify and explain the operation of an op-amp constant-current source
    ❑ Identify and explain the operation of an op-amp current-to-voltage converter
    ❑ Identify and explain the operation of an op-amp voltage-to-current converter
    ❑ Explain how an op-amp can be used as a peak detector

Constant-Current Source

A constant-current source delivers a load current that remains constant when the load resistance changes. Figure 4–37 shows a basic circuit in which a stable voltage source (V_{in}) provides a constant current (I_i) through the input resistor (R_i). Since the inverting

FIGURE 4–37
A basic constant-current source.

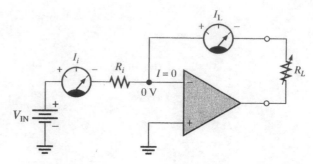

input of the op-amp is at virtual ground (0 V), the value of I_i is determined by V_{IN} and R_i as

$$I_i = \frac{V_{IN}}{R_i}$$

Now, since the internal input impedance of the op-amp is extremely high (ideally infinite), practically all of I_i is through R_L, which is connected in the feedback path. Since $I_i = I_L$,

$$I_L = \frac{V_{IN}}{R_i} \qquad\qquad (4\text{–}8)$$

If R_L changes, I_L remains constant as long as V_{IN} and R_i are held constant.

Current-to-Voltage Converter

A current-to-voltage converter converts a variable input current to a proportional output voltage. A basic circuit that accomplishes this is shown in Figure 4–38(a). Since practically all of I_i is through the feedback path, the voltage dropped across R_f is I_iR_f. Because the left side of R_f is at virtual ground (0 V), the output voltage equals the voltage across R_f, which is proportional to I_i.

$$V_{OUT} = I_iR_f \qquad\qquad (4\text{–}9)$$

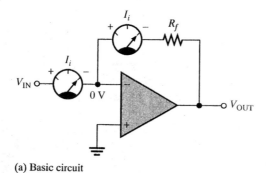

(a) Basic circuit

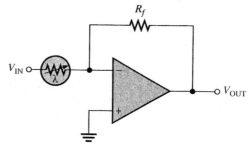

(b) Circuit for sensing light level and converting
it to a proportional output voltage

FIGURE 4–38
Current-to-voltage converter.

A specific application of this circuit is illustrated in Figure 4–38(b), where a photoconductive cell is used to sense changes in light level. As the amount of light changes, the current through the photoconductive cell varies because of the cell's change in resistance. This change in resistance produces a proportional change in the output voltage ($\Delta V_{OUT} = \Delta I_i R_f$).

Voltage-to-Current Converter

A basic voltage-to-current converter is shown in Figure 4–39. This circuit is used in applications where it is necessary to have an output (load) current that is controlled by an input voltage.

Neglecting the input offset voltage, both inverting and noninverting input terminals of the op-amp are at the same voltage, V_{IN}. Therefore, the voltage across R_1 equals V_{IN}. Since there is negligible current for the inverting input, the current through R_1 is the same as the current through R_L; thus,

$$I_L = \frac{V_{IN}}{R_1} \tag{4–10}$$

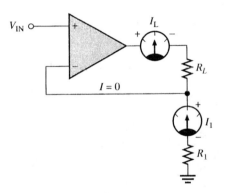

FIGURE 4–39
Voltage-to-current converter.

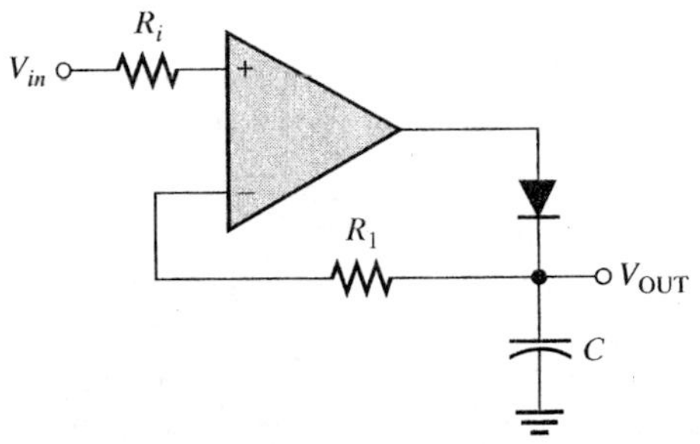

FIGURE 4–40
A basic peak detector.

Peak Detector

An interesting application of the op-amp is in a peak detector circuit such as the one shown in Figure 4–40. In this case the op-amp is used as a comparator. The purpose of this circuit is to detect the peak of the input voltage and store that peak voltage on a capacitor. For example, this circuit can be used to detect and store the maximum value of a voltage surge; this value can then be measured at the output with a voltmeter or recording device. The basic operation is as follows. When a positive voltage is applied to the noninverting input of the op-amp through R_i, the high-level output voltage of the op-amp forward-biases the diode and charges the capacitor. The capacitor continues to charge until its voltage reaches a value equal to the input voltage and thus both op-amp inputs are at the same voltage. At this point, the op-amp comparator switches, and its output goes to the low level. The diode is now reverse-biased, and the capacitor stops charging. It has reached a voltage equal to the peak of V_{in} and will hold this voltage until the charge eventually leaks off. If a greater input peak occurs, the capacitor charges to the new peak.

4–4 REVIEW QUESTIONS

1. For the constant-current source in Figure 4–37, the input reference voltage is 6.8 V and R_i is 10 kΩ. What value of constant current does the circuit supply to a 1 kΩ load? To a 5 kΩ load?
2. What element determines the constant of proportionality that relates input current to output voltage in the current-to-voltage converter?

4–5 ▪ TROUBLESHOOTING

Although integrated circuit op-amps are extremely reliable and trouble-free, failures do occur from time to time. One type of internal failure mode is a condition where the op-amp output is "stuck" in a saturated state resulting in a constant high or constant low level, regardless of the input. Also, external component failures will produce various types of failure modes in op-amp circuits. Some examples are presented in this section.

After completing this section, you should be able to

❑ Troubleshoot basic op-amp circuits
   ❑ Identify failures in comparator circuits
   ❑ Identify failures in summing amplifiers

Figure 4–41 illustrates an internal failure of a comparator circuit that results in a "stuck" output.

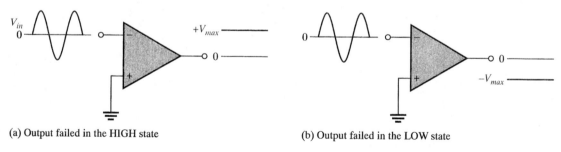

(a) Output failed in the HIGH state   (b) Output failed in the LOW state

FIGURE 4–41
Internal comparator failures typically result in the output being "stuck" in the HIGH or LOW state.

Symptoms of External Component Failures in Comparator Circuits

A comparator with zener-bounding and hysteresis is shown in Figure 4–42. In addition to a failure of the op-amp itself, a zener diode or one of the resistors could be faulty. For example, suppose one of the zener diodes opens. This effectively eliminates both zeners, and the circuit operates as an unbounded comparator, as indicated in Figure 4–43(a). With a shorted diode, the output is limited to the zener voltage (bounded) only in one direction depending on which diode remains operational, as illustrated in Figure 4–43(b). In the other direction, the output is held at the forward diode voltage.

FIGURE 4–42
A bounded comparator.

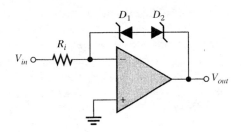

FIGURE 4–43
Examples of comparator circuit failures and their effects.

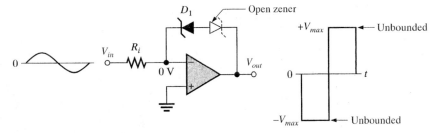

(a) The effect of an open zener

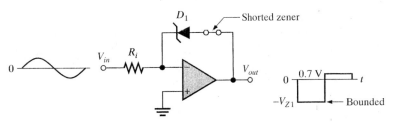

(b) The effect of a shorted zener

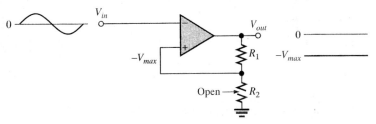

(c) Open R_2 causes output to "stick" in one state (either high or low)

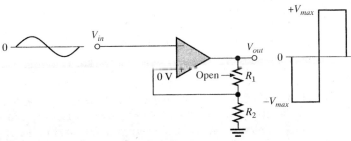

(d) Open R_1 forces the circuit to operate as a zero-level detector

Recall that R_1 and R_2 set the UTP and LTP for the hysteresis comparator. Now, suppose that R_2 opens. Essentially all of the output voltage is fed back to the noninverting input, and, since the input voltage will never exceed the output, the device will remain in one of its saturated states. This symptom can also indicate a faulty op-amp, as mentioned before. Now, assume that R_1 opens. This leaves the noninverting input near ground potential and causes the circuit to operate as a zero-level detector. These conditions are shown in parts (c) and (d) of Figure 4–43.

EXAMPLE 4–10   One channel of a dual-trace oscilloscope is connected to the comparator output and the other channel to the input signal, as shown in Figure 4–44. From the observed waveforms, determine if the circuit is operating properly, and if not, what the most likely failure is.

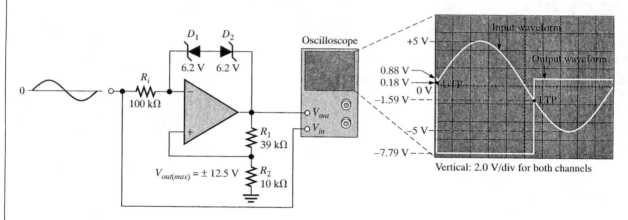

FIGURE 4–44

Solution   The output should be limited to ± 8.67 V. However, the positive maximum is $+0.88$ V and the negative maximum is -7.79 V. This indicates that D_2 is shorted. Refer to Example 4–3 for analysis of the bounded comparator.

Practice Exercise   What would the output voltage look like if D_1 shorted rather than D_2?

Symptoms of Component Failures in Summing Amplifiers

If one of the input resistors in a unity-gain summing amplifier opens, the output will be less than the normal value by the amount of the voltage applied to the open input. Stated another way, the output will be the sum of the remaining input voltages.

If the summing amplifier has a nonunity gain, an open input resistor causes the output to be less than normal by an amount equal to the gain times the voltage at the open input.

EXAMPLE 4–11   (a) What is the normal output voltage in Figure 4–45?

(b) What is the output voltage if R_2 opens?

(c) What happens if R_5 opens?

FIGURE 4–45

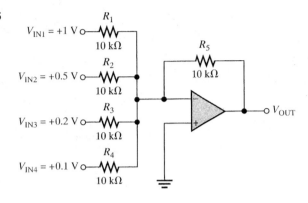

Solution

(a) $V_{OUT} = V_{IN1} + V_{IN2} + \cdots + V_{INn}) = -(1\,\text{V} + 0.5\,\text{V} + 0.2\,\text{V} + 0.1\,\text{V}) = \mathbf{-1.8\,V}$

(b) $V_{OUT} = -(1\,\text{V} + 0.2\,\text{V} + 0.1\,\text{V}) = \mathbf{-1.3\,V}$

(c) If R_5 opens, the circuit becomes a comparator and the output goes to $-V_{max}$.

Practice Exercise   In Figure 4–45, $R_5 = 47\,\text{k}\Omega$. What is the output voltage if R_1 opens?

As another example, let's look at an averaging amplifier. An open input resistor will result in an output voltage that is the average of all the inputs with the open input averaged in as a zero.

EXAMPLE 4–12   (a) What is the normal output voltage for the averaging amplifier in Figure 4–46?

(b) If R_4 opens, what is the output voltage? What does the output voltage represent?

FIGURE 4–46

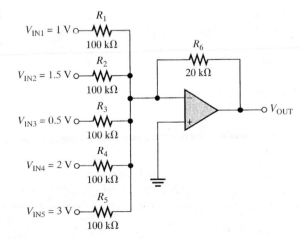

Solution Since the input resistors are equal, $R = 100$ kΩ. $R_f = R_6$.

(a) $V_{OUT} = -\dfrac{R_f}{R}(V_{IN1} + V_{IN2} + \cdots + V_{INn})$

$= -\dfrac{20 \text{ k}\Omega}{100 \text{ k}\Omega}(1 \text{ V} + 1.5 \text{ V} + 0.5 \text{ V} + 2 \text{ V} + 3 \text{ V}) = -0.2(8 \text{ V}) = \mathbf{-1.6 \text{ V}}$

(b) $V_{OUT} = -\dfrac{20 \text{ k}\Omega}{100 \text{ k}\Omega}(1 \text{ V} + 1.5 \text{ V} + 0.5 + 3 \text{ V}) = -0.2(6 \text{ V}) = \mathbf{-1.2 \text{ V}}$

The 1.2 V result is the average of five voltages with the 2 V input replaced by 0 V. Notice that the output is not the average of the four remaining input voltages.

Practice Exercise If R_4 is open, as was the case in this example, what would you have to do to make the output equal to the average of the remaining four input voltages?

4–5 REVIEW QUESTIONS

1. Describe one type of internal op-amp failure.
2. If a certain malfunction is attributable to more than one possible component failure, what would you do to isolate the problem?

4–6 ■ A SYSTEM APPLICATION

The system presented at the beginning of this chapter is a dual-slope analog-to-digital converter (ADC). This is one of several methods for A/D conversion. You were introduced to another type, called a simultaneous (or flash) ADC as an example in the chapter. The topic of data conversion including both ADCs and DACs is covered thoroughly in Chapter 10. Although A/D conversion is used for many purposes, in this particular application the converter is used to change an audio signal into digital form for recording. Although many parts of this system are digital, you will focus on the analog-to-digital converter board, which includes op-amps used in several types of circuits that you have learned about in this chapter.

After completing this section, you should be able to

❑ Apply what you have learned in this chapter to a system application
    ❑ Describe one way in which a summing amplifier is used
    ❑ State how the integrator is a key element in A/D conversion
    ❑ Explain how a comparator is used
    ❑ Translate between a printed circuit board and a schematic
    ❑ Troubleshoot some common system problems

A Brief Description of the System

The dual-slope ADC in Figure 4–47 accepts an audio signal and converts it to a series of digital codes for the purpose of recording. The audio signal voltage is applied to the sample-and-hold circuit. (Sample-and-hold circuits are covered in detail in Chapter 10.) At fixed intervals, sample pulses cause the amplitude at that point on the audio waveform to be converted to proportional dc levels that are then processed by the rest of the circuits and represented by a series of digital codes.

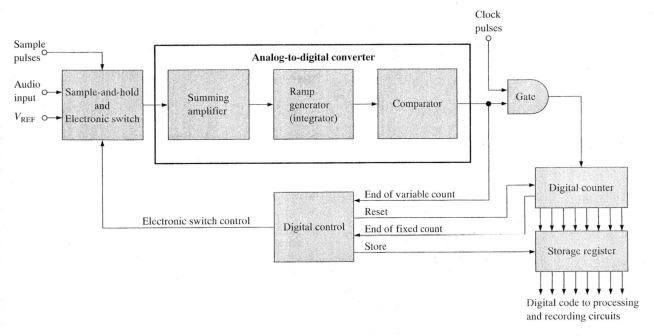

FIGURE 4–47
Basic dual-slope analog-to-digital converter.

The sample pulses occur at a much higher rate than the audio frequency so that a sufficient number of points on the audio waveform are sampled and converted to obtain an accurate digital representation of the audio signal. A rough approximation of the sampling process is illustrated in Figure 4–48. As the frequency of the sample pulses increases relative to the audio frequency, an increasingly accurate representation is achieved.

The summing amplifier has only one input active at a time. For example, when the audio input is switched in, the reference voltage input is zero and vice versa.

During the time between each sample pulse, the dc level from the sample-and-hold circuit is switched electronically into the summing amplifier on the ADC board. The output of the summing amplifier goes to the ramp generator which is an integrator circuit. At the same time, the digital counter starts counting up from zero. During the fixed time interval of the counting sequence, the integrator (ramp generator) produces a positive-going ramp voltage whose slope depends on the level of the sampled audio voltage. At the end of the fixed time interval, the ramp voltage at the output of the integrator has reached a voltage that is proportional to the sampled audio voltage. At this time, the digital control logic switches from the sample-and-hold input to the negative dc reference voltage input and resets the digital counter to zero.

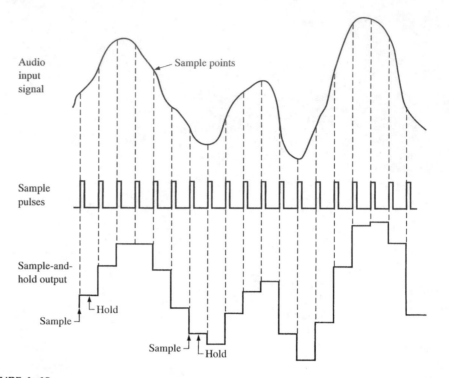

FIGURE 4–48

Sample-and-hold process. The sample-and-hold output is a rough approximation of the audio voltage for purposes of illustration. As the frequency of the sample pulses is increased, an increasingly accurate representation is achieved.

The summing amplifier applies this negative dc reference to the integrator input, which starts a negative-going ramp on the output. This ramp voltage has a slope that is fixed by the value V_{REF}. At the same time, the digital counter begins to count up again and will continue to count up until the negative-going ramp output of the integrator reaches zero volts.

At this point, the comparator switches to its negative saturated output voltage and disables the gate so that there are no additional clock pulses to the counter. At this time, the digital code in the counter is proportional to the time that it took for the negative-going ramp at the integrator output to reach zero and it will vary for each different sampled value.

Recall that the negative-going ramp started at a positive voltage that was dependent on the sampled value of the audio signal. Therefore, the digital code in the counter is also proportional to, and represents, the amplitude of the sampled audio voltage. This code is then shifted out to the register and then processed and recorded.

This process is repeated many times during a typical audio cycle. The result is a sequence of digital codes that represent the audio voltage amplitude as it varies with time. Figure 4–49 illustrates this for several sampled values. As mentioned, you will focus on the ADC board, which contains the summing amplifier, integrator, and comparator.

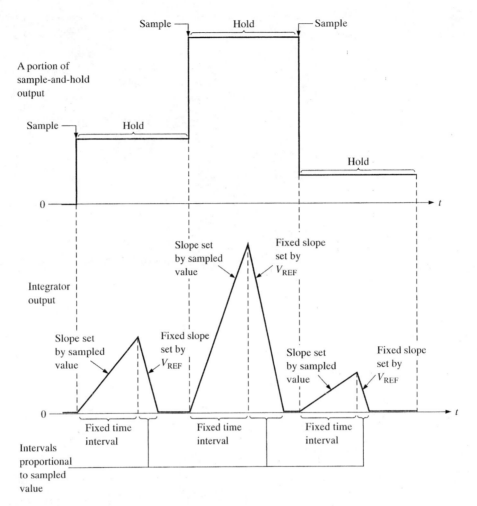

FIGURE 4–49
During the fixed-time interval of the positive-going ramp, the sampled audio input is applied to the integrator. During the variable-time interval of the negative-going ramp, the reference voltage is applied to the integrator. The counter sets the fixed-time interval and is then reset. Another count begins during the variable interval and the code in the counter at the end of this interval represents the sampled value.

Now, so you can take a closer look at the ADC board, let's take it out of the system and put it on the test bench.

ON THE TEST BENCH

■ ACTIVITY 1   Relate the PC Board to the Schematic

Identify each component on the circuit board in Figure 4–50 using the schematic in Figure 4–51. Also, identify each input and output on the board. There are several backside connections with corresponding feedthrough pads. Locate and identify the components associated with the summing amplifier, the integrator, and the comparator. Note that the two 10 kΩ potentiometers are for nulling the output offset voltage.

FIGURE 4–50
Analog-to-digital (ADC) board.

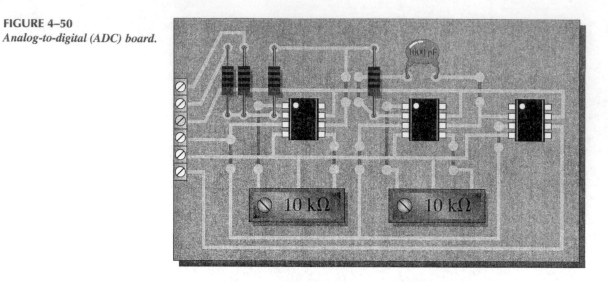

FIGURE 4–51
Schematic of the ADC.

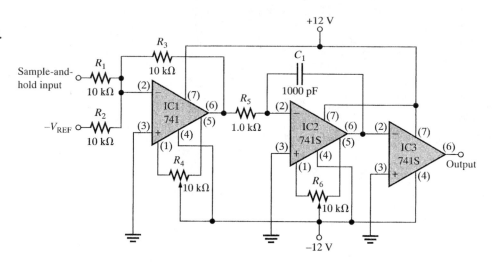

■ ACTIVITY 2   Analyze the Circuit

Step 1: Determine the gain of the summing amplifier.

Step 2: Determine the slope of the integrator ramp in volts per microsecond when a sampled audio voltage of $+2$ V is applied.

Step 3: Determine the slope of the integrator ramp in volts per microsecond when the reference voltage of -8 V is applied.

Step 4: Given that the reference voltage is -8 V and the fixed-time interval of the negative-going slope is 1 μs, sketch the dual-slope output of the integrator when an instantaneous audio voltage of $+3$ V is sampled.

Step 5: Assuming that the maximum audio voltage to be sampled is $+6$ V, determine the maximum audio frequency that can be sampled by this particular system if there are to be 100 samples per cycle. What is the sample pulse rate in this case?

▨ ACTIVITY 3   Write a Technical Report

Discuss the detailed operation of the ADC board circuitry and explain how it interfaces with the overall system. Discuss the purpose of each component on the circuit board.

▨ ACTIVITY 4   Troubleshoot the Circuit for Each of the Following Problems by Stating the Probable Cause or Causes

1. Zero volts on the output of IC1 when there are voltages on the sampled audio input and on the reference voltage input.
2. IC1 goes back and forth between its saturated states as the positive audio voltage and the negative reference voltage are alternately switched in.
3. The inverting input of IC1 never goes negative.
4. The output of IC2 stays at zero volts under normal operating conditions.

4–6 REVIEW QUESTIONS

1. Identify the summing amplifier, the integrator, and the comparator by IC number.
2. The 741S op-amps are high slew-rate devices with a minimum slew rate of 10 V/μs. Why are these used in this application?
3. What is the purpose of R_4 and R_6 in the circuit of Figure 4–51?
4. What type of output voltage does the ADC board produce and how is it used in the system?
5. If a sample pulse rate of 500 kHz is used, how long does each sampled audio voltage remain on the input to the ADC board?
6. Although IC1 is connected in the form of a summing amplifier, it does not actually perform a summing operation in this application. Why?

▨ SUMMARY

- In an op-amp comparator, when the input voltage exceeds a specified reference voltage, the output changes state.
- Hysteresis gives an op-amp noise immunity.
- A comparator switches to one state when the input reaches the upper trigger point (UTP) and back to the other state when the input drops below the lower trigger point (LTP).
- The difference between the UTP and the LTP is the hysteresis voltage.
- Bounding limits the output amplitude of a comparator.
- The output voltage of a summing amplifier is proportional to the sum of the input voltages.
- An averaging amplifier is a summing amplifier with a closed-loop gain equal to the reciprocal of the number of inputs.
- In a scaling adder, a different weight can be assigned to each input, thus making the input contribute more or contribute less to the output.
- Integration is a mathematical process for determining the area under a curve.
- Integration of a step produces a ramp with a slope proportional to the amplitude.
- Differentiation is a mathematical process for determining the rate of change of a function.
- Differentiation of a ramp produces a step with an amplitude proportional to the slope.

■ GLOSSARY

These terms are included in the end-of-book glossary.

A/D conversion   A process whereby information in analog form is converted into digital form.

Bounding   The process of limiting the output range of an amplifier or other circuit.

Comparator   A circuit which compares two input voltages and produces an output in either of two states indicating the greater than or less than relationship of the inputs.

D/A conversion   The process of converting a sequence of digital codes to an analog form.

Differentiator   A circuit that produces an inverted output which approximates the rate of change of the input function.

Hysteresis   Characteristic of a circuit in which two different trigger levels create an effect or lag in the switching action.

Integrator   A circuit that produces an inverted output which approximates the area under the curve of the input function.

Noise   An unwanted voltage or current fluctuation.

Schmitt trigger   A comparator with hysteresis.

■ KEY FORMULAS

Comparators

$$(4\text{–}1) \qquad V_{\text{REF}} = \frac{R_2}{R_1 + R_2}(+V) \qquad\qquad \text{Comparator reference}$$

$$(4\text{–}2) \qquad V_{\text{HYS}} = V_{\text{UTP}} - V_{\text{LTP}} \qquad\qquad \text{Hysteresis voltage}$$

Summing Amplifier

$$(4\text{–}3) \qquad V_{\text{OUT}} = -(V_{\text{IN1}} + V_{\text{IN2}} + \cdots + V_{\text{IN}n}) \qquad\qquad n\text{-input adder}$$

$$(4\text{–}4) \qquad V_{\text{OUT}} = -\frac{R_f}{R}(V_{\text{IN1}} + V_{\text{IN2}} + \cdots + V_{\text{IN}n}) \qquad\qquad \text{Scaling adder with gain}$$

$$(4\text{–}5) \qquad V_{\text{OUT}} = -\left(\frac{R_f}{R_1}V_{\text{IN1}} + \frac{R_f}{R_2}V_{\text{IN2}} + \cdots + \frac{R_f}{R_n}V_{\text{IN}n}\right) \qquad\qquad \text{Scaling adder}$$

Integrator and Differentiator

$$(4\text{–}6) \qquad \frac{\Delta V_{out}}{\Delta t} = -\frac{V_{in}}{R_i C} \qquad\qquad \text{Integrator output rate of change}$$

$$(4\text{–}7) \qquad V_{out} = -\left(\frac{V_C}{t}\right)R_f C \qquad\qquad \text{Differentiator output voltage with ramp input}$$

Miscellaneous

$$(4\text{–}8) \qquad I_{\text{L}} = \frac{V_{\text{IN}}}{R_i} \qquad\qquad \text{Constant-current source}$$

$$(4\text{–}9) \qquad V_{\text{OUT}} = I_i R_f \qquad\qquad \text{Current-to-voltage converter}$$

$$(4\text{–}10) \qquad I_{\text{L}} = \frac{V_{in}}{R_1} \qquad\qquad \text{Voltage-to-current converter}$$

■ SELF-TEST

1. In a zero-level detector, the output changes state when the input
   (a) is positive   (b) is negative
   (c) crosses zero   (d) has a zero rate of change

2. The zero-level detector is one application of a
   (a) comparator   (b) differentiator
   (c) summing amplifier   (d) diode

3. Noise on the input of a comparator can cause the output to
   (a) hang up in one state
   (b) go to zero
   (c) change back and forth erratically between two states
   (d) produce the amplified noise signal

4. The effects of noise can be reduced by
   (a) lowering the supply voltage   (b) using positive feedback
   (c) using negative feedback   (d) using hysteresis
   (e) answers (b) and (d)

5. A comparator with hysteresis
   (a) has one trigger point   (b) has two trigger points
   (c) has a variable trigger point   (d) is like a magnetic circuit

6. In a comparator with hysteresis,
   (a) a bias voltage is applied between the two inputs
   (b) only one supply voltage is used
   (c) a portion of the output is fed back to the inverting input
   (d) a portion of the output is fed back to the noninverting input

7. Using output bounding in a comparator
   (a) makes it faster   (b) keeps the output positive
   (c) limits the output levels   (d) stabilizes the output

8. A window comparator detects when
   (a) the input is between two specified limits
   (b) the input is not changing
   (c) the input is changing too fast
   (d) the amount of light exceeds a certain value

9. A summing amplifier can have
   (a) only one input   (b) only two inputs
   (c) any number of inputs

10. If the voltage gain for each input of a summing amplifier with a 4.7 kΩ feedback resistor is unity, the input resistors must have a value of
    (a) 4.7 kΩ
    (b) 4.7 kΩ divided by the number of inputs
    (c) 4.7 kΩ times the number of inputs

11. An averaging amplifier has five inputs. The ratio R_f/R_{in} must be
    (a) 5   (b) 0.2   (c) 1

12. In a scaling adder, the input resistors are
    (a) all the same value
    (b) all of different values
    (c) each proportional to the weight of its inputs
    (d) related by a factor of two

13. In an integrator, the feedback element is a
    (a) resistor   (b) capacitor
    (c) zener diode   (d) voltage divider

14. For a step input, the output of an integrator is a
   (a) pulse   (b) triangular waveform
   (c) spike   (d) ramp

15. The rate of change of an integrator's output voltage in response to a step input is set by
   (a) the RC time constant   (b) the amplitude of the step input
   (c) the current through the capacitor   (d) all of these

16. In a differentiator, the feedback element is a
   (a) resistor   (b) capacitor
   (c) zener diode   (d) voltage divider

17. The output of a differentiator is proportional to
   (a) the RC time constant   (b) the rate at which the input is changing
   (c) the amplitude of the input   (d) answers (a) and (b)

18. When you apply a triangular waveform to the input of a differentiator, the output is
   (a) a dc level   (b) an inverted triangular waveform
   (c) a square waveform   (d) the first harmonic of the triangular waveform

---

■ PROBLEMS

SECTION 4–1   Comparators

1. A certain op-amp has an open-loop gain of 80,000. The maximum saturated output levels of this particular device are ± 12 V when the dc supply voltages are ± 15 V. If a differential voltage of 0.15 mV rms is applied between the inputs, what is the peak-to-peak value of the output?

2. Determine the output level (maximum positive or maximum negative) for each comparator in Figure 4–52.

3. Calculate the V_{UTP} and V_{LTP} in Figure 4–53. $V_{out(max)} = -10$ V.

4. What is the hysteresis voltage in Figure 4–53?

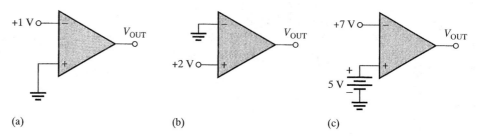

(a)   (b)   (c)

FIGURE 4–52

FIGURE 4–53

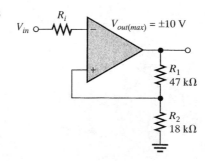

5. Sketch the output voltage waveform for each circuit in Figure 4–54 with respect to the input. Show voltage levels.

6. Determine the hysteresis voltage for each comparator in Figure 4–55. The maximum output levels are ±11 V.

7. A 6.2 V zener diode is connected from the output to the inverting input in Figure 4–53 with the cathode at the output. What are the positive and negative output levels?

8. Determine the output voltage waveform in Figure 4–56.

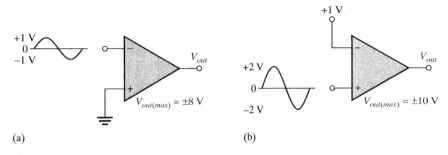

(a)   (b)

FIGURE 4–54

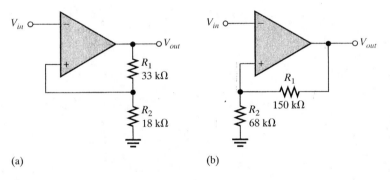

(a)   (b)

FIGURE 4–55

FIGURE 4–56

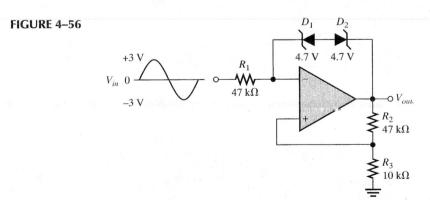

SECTION 4–2 Summing Amplifiers

9. Determine the output voltage for each circuit in Figure 4–57.

10. Refer to Figure 4–58. Determine the following:
   (a) V_{R1} and V_{R2}
   (b) Current through R_f
   (c) V_{OUT}

11. Find the value of R_f necessary to produce an output that is five times the sum of the inputs in Figure 4–58.

12. Design a summing amplifier that will average eight input voltages. Use input resistances of 10 kΩ each.

13. Find the output voltage when the input voltages shown in Figure 4–59 are applied to the scaling adder. What is the current through R_f?

14. Determine the values of the input resistors required in a six-input scaling adder so that the lowest weighted input is 1 and each successive input has a weight twice the previous one. Use $R_f =$ 100 kΩ.

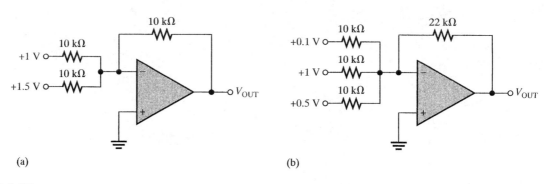

(a)   (b)

FIGURE 4–57

FIGURE 4–58

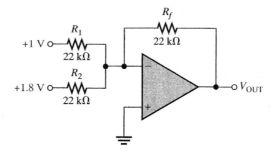

FIGURE 4–59

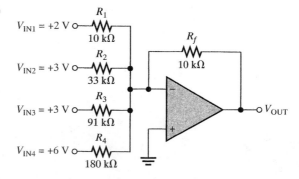

SECTION 4–3   Integrators and Differentiators

15. Determine the rate of change of the output voltage in response to the step input to the integrator in Figure 4–60.

16. A triangular waveform is applied to the input of the circuit in Figure 4–61 as shown. Determine what the output should be and sketch its waveform in relation to the input.

17. What is the magnitude of the capacitor current in Problem 16?

18. A triangular waveform with a peak-to-peak voltage of 2 V and a period of 1 ms is applied to the differentiator in Figure 4–62(a). What is the output voltage?

19. Beginning in position 1 in Figure 4–62(b), the switch is thrown into position 2 and held there for 10 ms, then back to position 1 for 10 ms, and so forth. Sketch the resulting output waveform. The saturated output levels of the op-amp are ±12 V.

FIGURE 4–60

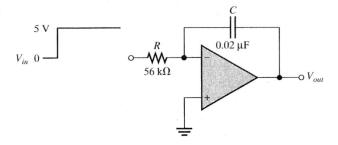

FIGURE 4–61

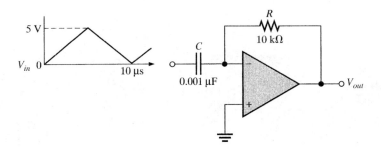

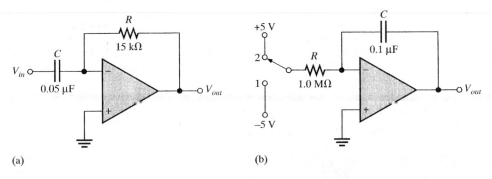

(a)   (b)

FIGURE 4–62

SECTION 4–4 Converters and Other Op-Amp Circuits

20. Determine the load current in each circuit of Figure 4–63. (Hint: Thevenize the circuit to the left of R_i.)

21. Devise a circuit for remotely sensing temperature and producing a proportional voltage that can then be converted to digital form for display. A thermistor can be used as the temperature-sensing element.

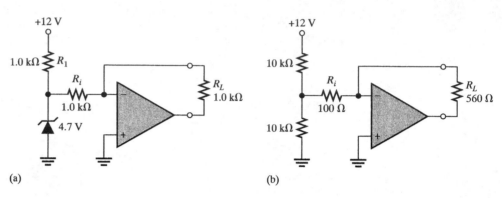

(a) (b)

FIGURE 4–63

SECTION 4–5 Troubleshooting

22. The waveforms given in Figure 4–64(a) are observed at the indicated points in Figure 4–64(b). Is the circuit operating properly? If not, what is a likely fault?

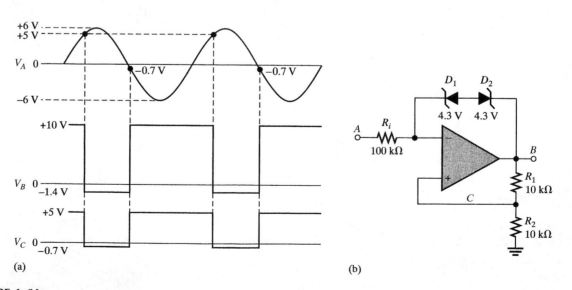

(a) (b)

FIGURE 4–64

23. The waveforms shown for the window comparator in Figure 4–65 are measured. Determine if the output waveform is correct and, if not, specify the possible fault(s).

24. The sequences of voltage levels shown in Figure 4–66 are applied to the summing amplifier and the indicated output is observed. First, determine if this output is correct. If it is not correct, determine the fault.

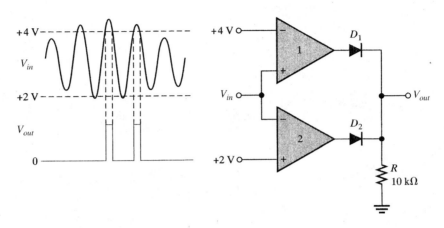

FIGURE 4–65

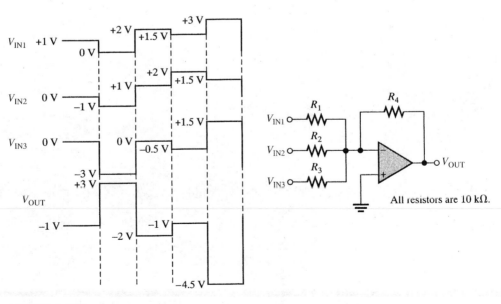

FIGURE 4–66

25. The given ramp voltages are applied to the op-amp circuit in Figure 4–67. Is the given output correct? If it isn't, what is the problem?

26. The ADC board, shown in Figure 4–68 in the system application, has just come off the assembly line and a pass/fail test indicates that it doesn't work. The board now comes to you for troubleshooting. What is the very first thing you should do? Can you isolate the problem(s) by this first step in this case?

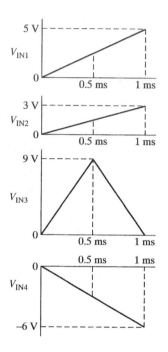

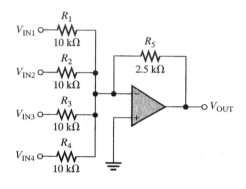

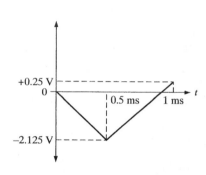

FIGURE 4–67

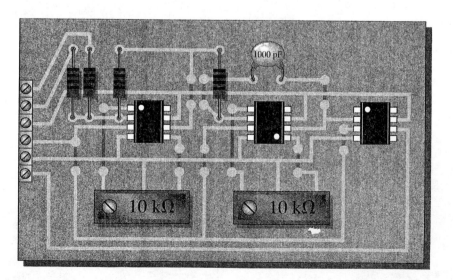

FIGURE 4–68

■ ANSWERS TO REVIEW QUESTIONS

Section 4–1

1. (a) $V = (10 \text{ k}\Omega/110 \text{ k}\Omega)15 \text{ V} = 1.36 \text{ V}$
   (b) $V = 22 \text{ k}\Omega/69 \text{ k}\Omega)(-12 \text{ V}) = -3.83 \text{ V}$

2. Hysteresis makes the comparator noise-free.

3. Bounding limits the output amplitude to a specified level.

Section 4–2

1. The summing point is the point where the input resistors are commonly connected.

2. $R_f/R = 1/5 = 0.2$

3. $5 \text{ k}\Omega$

Section 4–3

1. The feedback element in an integrator is a capacitor.

2. The capacitor voltage is linear because the capacitor current is constant.

3. The feedback element in a differentiator is a resistor.

4. The output of a differentiator is proportional to the rate of change of the input.

Section 4–4

1. $I_L = 6.8 \text{ V}/10 \text{ k}\Omega = 0.68 \text{ mA}$; same value to $5 \text{ k}\Omega$ load.

2. The feedback resistor is the constant of proportionality.

Section 4–5

1. An op-amp can fail with a shorted output.

2. Replace suspected components one by one.

Section 4–6

1. Summing amplifier—IC1, integrator—IC2, comparator—IC3

2. A high slew-rate op-amp is used in the integrator to avoid slew-rate limitation of the output ramps. One is used as a comparator, to achieve a fast switching time.

3. R_4 and R_6 are for eliminating output offset (nulling).

4. The board output is the comparator output. The transition of the comparator output from its positive state to its negative state notifies the control logic of the end of the variable-time interval.

5. $1/500 \text{ kHz} = 2 \text{ } \mu\text{s}$

6. Because of the electronic switch, only one input voltage at a time is actually applied.

■ ANSWERS TO PRACTICE EXERCISES FOR EXAMPLES

4–1 1.96 V

4–2 +3.83 V, −3.83 V, $V_{HYS} = 7.65$ V

4–3 +1.81 V, −1.81 V

4–4 −12.5 V

4–5 −5.73 V

4–6 Changes require an additional 100 kΩ input resistor and a change of R_f to 20 kΩ.

4–7 0.45, 0.12, 0.18; $V_{OUT} = -3.03$ V

4–8 Change C to 5000 pF or change R to 5.0 kΩ.

4–9 Same waveform with an amplitude of 6.6 V

4–10 A pulse from −0.88 V to +7.79 V

4–11 −3.76 V

4–12 Change R_6 to 25 kΩ.

5

ACTIVE FILTERS

Courtesy Sunsweet Growers

■ **CHAPTER OBJECTIVES**

- Describe the gain-versus-frequency responses of the basic filters
- Describe the three basic filter response characteristics and other filter parameters
- Understand active low-pass filters
- Understand active high-pass filters
- Understand active band-pass filters
- Understand active band-stop filters
- Discuss two methods for measuring frequency response
- Apply what you have learned in this chapter to a system application

In this chapter, active filters used for signal processing are introduced. Filters are circuits that are capable of passing input signals with certain selected frequencies through to the output while rejecting signals with other frequencies. This property is called *selectivity*.

Filters use active devices such as transistors or op-amps and passive *RC* networks. The active devices provide voltage gain and the passive networks provide frequency selectivity. In terms of general response, there are four basic categories of active filters: low-pass, high-pass, band-pass, and band-stop. In this chapter, you will study active filters using op-amps and *RC* networks.

In Chapter 3, you worked with an FM stereo multiplex receiver, concentrating on the audio amplifiers. In this chapter, we again look at this same system, but this time our focus is on the filters in the left and right channel separation circuits in which several types of active filters are used. The FM stereo multiplex signal that is received is quite complex, and it is beyond the scope of our coverage to investigate the reasons it is transmitted in such a way. It is interesting, however, to see how filters, such as the ones studied in this chapter can be used to separate out the audio signals that go to the left and right speakers.

For the system application in Section 5–8, in addition to the other topics, be sure you understand

Filter responses

How low-pass and band-pass filters work

5–1 ■ BASIC FILTER RESPONSES

Filters are usually categorized by the manner in which the output voltage varies with the frequency of the input voltage. The categories of active filters are low-pass, high-pass, band-pass, and band-stop. We will examine each of these general responses in this section.

After completing this section, you should be able to

❑ Describe the gain-versus-frequency responses of the basic filters
    ❑ Explain the low-pass response
    ❑ Determine the critical frequency and bandwidth of a low-pass filter
    ❑ Explain the high-pass response
    ❑ Determine the critical frequency of a high-pass filter
    ❑ Explain the band-pass response
    ❑ Explain the significance of the quality factor
    ❑ Determine the critical frequency, bandwidth, quality factor, and damping factor of a band-pass filter
    ❑ Explain the band-stop response

Low-Pass Filter Response

A **filter** is a circuit that passes certain frequencies and attenuates or rejects all other frequencies. The **passband** of a filter is the region of frequencies that are allowed to pass through the filter with minimum attenuation (usually defined as less than -3 dB of attenuation). The **critical frequency**, f_c, (also called the *cutoff frequency*) defines the end of the passband and is normally specified at the point where the response drops -3 dB (70.7%) from the passband response. Following the passband is a region called *the transition region* that leads into a region called the *stopband*. There is no precise point between the transition region and the stopband.

A **low-pass filter** is one that passes frequencies from dc to f_c and significantly attenuates all other frequencies. The passband of the ideal low-pass filter is shown in the blue shaded area of Figure 5–1(a); the response drops to zero at frequencies beyond the passband. This ideal response is sometimes referred to as a "brick-wall" because nothing gets through beyond the wall. The bandwidth of an ideal low-pass filter is equal to f_c.

$$BW = f_c \qquad\qquad (5\text{–}1)$$

The ideal response shown in Figure 5–1(a) is not attainable by any practical filter. Actual filter responses depend on the number of **poles,** a term used with filters to describe the number of bypass circuits contained in the filter.[1] The most basic low-pass filter is a simple *RC* network consisting of just one resistor and one capacitor; the output is taken across the capacitor as shown in Figure 5–1(b). This basic *RC* filter has a single pole and it rolls off at -20 dB/decade beyond the critical frequency. The actual response is indicated by the blue line in Figure 5–1(a). The response is plotted on a standard log-log plot that is used for filters to show details of the curve as the gain drops. Notice that the gain drops off slowly until the frequency is at the critical frequency; after this, the gain drops rapidly.

---

[1] A pole is used to describe certain complex mathematical characteristics of the transfer function for the filter.

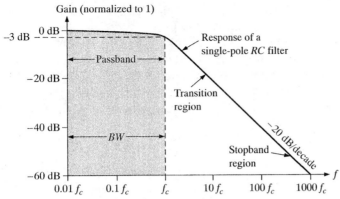

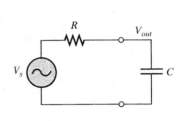

(a) Comparison of an ideal low-pass filter response with actual response

(b) Basic low-pass circuit

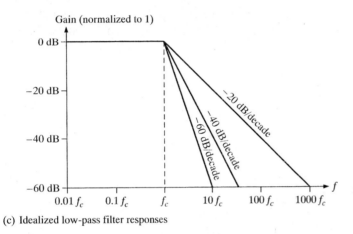

(c) Idealized low-pass filter responses

FIGURE 5–1
Low-pass filter responses.

The −20 dB/decade roll-off rate for the gain of a basic *RC* filter means that at a frequency of $10f_c$, the output will be −20 dB (10%) of the input. This rather gentle roll-off is not a particularly good filter characteristic because too much of the unwanted frequencies (beyond the passband) are allowed through the filter.

The critical frequency of the simple low-pass *RC* filter occurs when $X_C = R$, where

$$f_c = \frac{1}{2\pi RC}$$

Recall from your basic dc/ac course that the output at the critical frequency is 70.7% of the input. This response is equivalent to an attenuation of −3 dB.

Figure 5–1(c) illustrates several idealized low-pass response curves including the basic one pole response (−20 dB/decade). The approximations show a *flat* response to the cutoff frequency and a roll-off at a constant rate after the cutoff frequency. Actual filters do not have a perfectly flat response to the cutoff frequency but have dropped to −3 dB at this point as described previously.

In order to produce a filter that has a steeper transition region, (and hence form a more effective filter), it is necessary to add additional circuitry to the basic filter. Responses that are steeper than −20 dB/decade in the transition region cannot be obtained by simply cascading identical *RC* stages (due to loading effects). However, by combining an op-amp with frequency-selective feedback networks, filters can be designed with roll-off rates of −40, −60, or more dB/decade. Filters that include one or more op-amps in the design are called **active filters.** These filters can optimize the roll-off rate or other attribute (such as phase response) with a particular filter design. In general, the more poles the filter uses, the steeper its transition region will be. The exact response depends on the type of filter and the number of poles.

High-Pass Filter Response

A **high-pass filter** is one that significantly attenuates or rejects all frequencies below f_c and passes all frequencies above f_c. The critical frequency is, again, the frequency at which the output is 70.7% of the input (or −3 dB) as shown in Figure 5–2(a). The ideal response, indicated by the shaded area, has an instantaneous drop at f_c, which, of course, is not achievable. Ideally, the passband of a high-pass filter is all frequencies above the critical frequency. The high-frequency response of practical circuits is limited by the op-amp or other components that make up the filter.

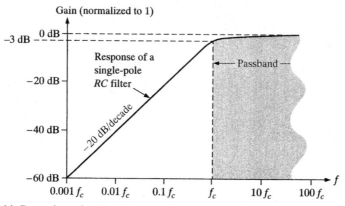

(a) Comparison of an ideal high-pass filter response with actual response

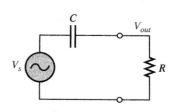

(b) Basic high-pass circuit

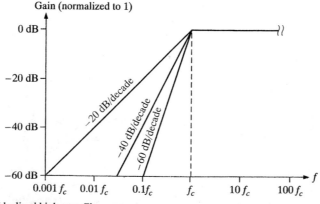

(c) Idealized high-pass filter responses

FIGURE 5–2
High-pass filter responses.

A simple *RC* network consisting of a single resistor and capacitor can be configured as a high-pass filter by taking the output across the resistor as shown in Figure 5–2(b). As in the case of the low-pass filter, the basic *RC* network has a roll-off rate of -20 dB/decade as indicated by the blue line in Figure 5–2(a). Also, the critical frequency for the basic high-pass filter occurs when $X_C = R$, where

$$f_c = \frac{1}{2\pi RC}$$

Figure 5–2(c) illustrates several idealized high-pass response curves including the basic one pole response (-20 dB/decade) for a basic *RC* network. As in the case of the low-pass filter, the approximations show a *flat* response to the cutoff frequency and a roll-off at a constant rate after the cutoff frequency. Actual high-pass filters do not have the perfectly flat response indicated or the precise roll-off rate shown. Responses that are steeper than -20 dB/decade in the transition region are also possible with active high-pass filters; the particular response depends on the type of filter and the number of poles.

Band-Pass Filter Response

A **band-pass filter** passes all signals lying within a band between a lower-frequency limit and an upper-frequency limit and essentially rejects all other frequencies that are outside this specific band. A generalized band-pass response curve is shown in Figure 5–3. The *bandwidth (BW)* is defined as the difference between the upper critical frequency (f_{c2}) and the lower critical frequency (f_{c1}).

$$BW = f_{c2} - f_{c1} \qquad \text{(5–2)}$$

The critical frequencies are the points at which the response curve is 70.7% of its maximum. These critical frequencies are also called *3 dB frequencies*. The frequency about which the passband is centered is called the *center frequency, f_0*, defined as the geometric mean of the critical frequencies.

$$f_0 = \sqrt{f_{c1}f_{c2}} \qquad \text{(5–3)}$$

FIGURE 5–3
General band-pass response curve.

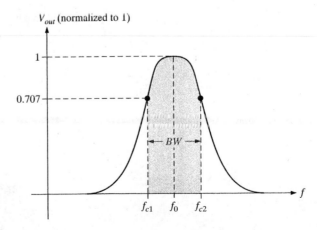

Quality Factor The **quality factor** (*Q*) of a band-pass filter is defined as the ratio of the center frequency to the bandwidth.

$$Q = \frac{f_0}{BW} \qquad\qquad (5\text{--}4)$$

The value of Q is an indication of the selectivity of a band-pass filter. The higher the value of Q, the narrower the bandwidth and the better the selectivity for a given value of f_0. Band-pass filters are sometimes classified as narrow-band ($Q > 10$) or wide-band ($Q < 10$). The Q can also be expressed in terms of the damping factor *(DF)* of the filter as

$$Q = \frac{1}{DF}$$

You will study the damping factor in Section 5–2.

EXAMPLE 5–1

A certain band-pass filter has a center frequency of 15 kHz and a bandwidth of 1 kHz. Determine the Q and classify the filter as narrow-band or wide-band.

Solution
$$Q = \frac{f_0}{BW} = \frac{15 \text{ kHz}}{1 \text{ kHz}} = \mathbf{15}$$

Because $Q > 10$, this is a **narrow-band** filter.

Practice Exercise If the Q of the filter is doubled, what will the bandwidth be?

Band-Stop Filter Response

Another category of active filter is the **band-stop filter,** also known as the *notch, band-reject,* or *band-elimination filter.* A general response curve for a band-stop filter is shown in Figure 5–4. Notice that the bandwidth is the band of frequencies between the 3 dB points, just as in the case of the band-pass filter response. You can think of the operation as opposite to that of the band-pass filter because frequencies within a certain bandwidth are rejected, and frequencies outside the bandwidth are passed.

FIGURE 5–4
General band-stop filter response.

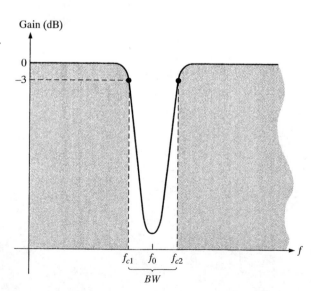

5–1 REVIEW QUESTIONS

1. What determines the bandwidth of a low-pass filter?
2. What limits the bandwidth of an active high-pass filter?
3. How are the Q and the bandwidth of a band-pass filter related? Explain how the selectivity is affected by the Q of a filter.

5–2 ■ FILTER RESPONSE CHARACTERISTICS

Each type of filter (low-pass, high-pass, band-pass, or band-stop) can be tailored by circuit component values to have either a Butterworth, Chebyshev, or Bessel characteristic. Each of these characteristics is identified by the shape of the response curve, and each has an advantage in certain applications.

After completing this section, you should be able to

❑ Describe the three basic filter response characteristics and other filter parameters
  ❑ Describe the Butterworth characteristic
  ❑ Describe the Chebyshev characteristic
  ❑ Describe the Bessel characteristic
  ❑ Define *damping factor* and discuss its significance
  ❑ Calculate the damping factor of a filter
  ❑ Discuss the order of a filter and its effect on the roll-off rate

Butterworth, Chebyshev, or Bessel response characteristics can be realized with most active filter circuit configurations by proper selection of certain component values. A general comparison of the three response characteristics for a low-pass filter response curve is shown in Figure 5–5. High-pass and band-pass filters can also be designed to have any one of the characteristics.

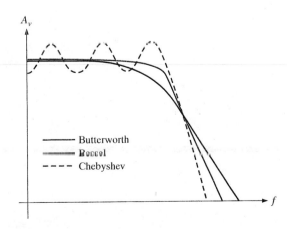

FIGURE 5–5
Comparative plots of three types of filter response characteristics.

The Butterworth Characteristic

The **Butterworth** characteristic provides a very flat amplitude response in the passband and a roll-off rate of -20 dB/decade/pole. The phase response is not linear, however, and the phase shift (thus, time delay) of signals passing through the filter varies nonlinearly with frequency. Therefore, a pulse applied to a filter with a Butterworth response will cause overshoots on the output because each frequency component of the pulse's rising and falling edges experiences a different time delay. Filters with the Butterworth response are normally used when all frequencies in the passband must have the same gain. The Butterworth response is often referred to as a *maximally flat response*.

The Chebyshev Characteristic

Filters with the **Chebyshev** response characteristic are useful when a rapid roll-off is required because it provides a roll-off rate greater than -20 dB/decade/pole. This is a greater rate than that of the Butterworth, so filters can be implemented with the Chebyshev response with fewer poles and less complex circuitry for a given roll-off rate. This type of filter response is characterized by overshoot or ripples in the passband (depending on the number of poles) and an even less linear phase response than the Butterworth.

The Bessel Characteristic

The **Bessel** response exhibits a linear phase characteristic, meaning that the phase shift increases linearly with frequency. The result is almost no overshoot on the output with a pulse input. For this reason, filters with the Bessel response are used for filtering pulse waveforms without distorting the shape of the waveform.

The Damping Factor

As mentioned, an active filter can be designed to have either a Butterworth, Chebyshev, or Bessel response characteristic regardless of whether it is a low-pass, high-pass, band-pass, or band-stop type. The **damping factor (DF)** of an active filter circuit determines which response characteristic the filter exhibits. To explain the basic concept, a generalized active filter is shown in Figure 5–6. It includes an amplifier, a negative feedback network, and a filter section. The amplifier and feedback are connected in a noninverting configuration. The damping factor is determined by the negative feedback network and is defined by the following equation:

$$DF = 2 - \frac{R_1}{R_2}$$

(5–5)

FIGURE 5–6
General diagram of an active filter.

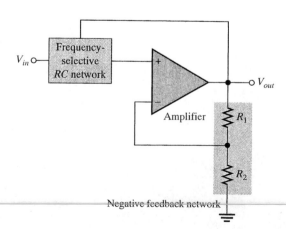

Basically, the damping factor affects the filter response by negative feedback action. Any attempted increase or decrease in the output voltage is offset by the opposing effect of the negative feedback. This tends to make the response curve flat in the passband of the filter if the value for the damping factor is precisely set. By advanced mathematics, which we will not cover, values for the damping factor have been derived for various orders of filters to achieve the maximally flat response of the Butterworth characteristic.

The value of the damping factor required to produce a desired response characteristic depends on the *order* (number of poles) of the filter. Recall that the more poles a filter has, the faster its roll-off rate is. To achieve a second-order Butterworth response, for example, the damping factor must be 1.414. To implement this damping factor, the feedback resistor ratio must be

$$\frac{R_1}{R_2} = 2 - DF = 2 - 1.414 = 0.586$$

This ratio gives the closed-loop gain of the noninverting filter amplifier, $A_{cl(NI)}$, a value of 1.586, derived as follows:

$$A_{cl(NI)} = \frac{1}{B} = \frac{1}{R_2/(R_1 + R_2)} = \frac{R_1 + R_2}{R_2} = \frac{R_1}{R_2} + 1 = 0.586 + 1 = 1.586$$

EXAMPLE 5–2

If resistor R_2 in the feedback network of an active two-pole filter of the type in Figure 5–6 is 10 kΩ, what value must R_1 be to obtain a maximally flat Butterworth response?

Solution

$$\frac{R_1}{R_2} = 0.586$$

$$R_1 = 0.586R_2 = 0.586(10 \text{ k}\Omega) = \textbf{5.86 k}\Omega$$

Using the nearest standard 5 percent value of 5600 Ω will get very close to the ideal Butterworth response.

Practice Exercise What is the damping factor for $R_2 = 10$ kΩ and $R_1 = 5.6$ kΩ?

Critical Frequency and Roll-Off Rate

The critical frequency is determined by the values of the resistor and capacitors in the RC network, as shown in Figure 5 6. For a single-pole (first-order) filter, as shown in Figure 5–7, the critical frequency is

$$f_c = \frac{1}{2\pi RC}$$

Although we show a low-pass configuration, the same formula is used for the f_c of a single-pole high-pass filter. The number of poles determines the roll-off rate of the filter. A Butterworth response produces −20 dB/decade/pole. So, a first-order (one-pole) filter has a roll-off of −20 dB/decade; a second-order (two-pole) filter has a roll-off rate of −40 dB/decade; a third-order (three-pole) filter has a roll-off rate of −60 dB/decade; and so on.

Generally, to obtain a filter with three poles or more, one-pole or two-pole filters are cascaded, as shown in Figure 5–8. To obtain a third-order filter, for example, cascade a

FIGURE 5–7
First-order (one-pole) low-pass filter.

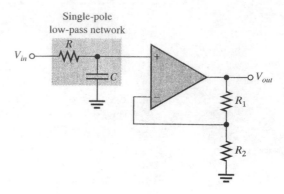

FIGURE 5–8
The number of filter poles can be increased by cascading.

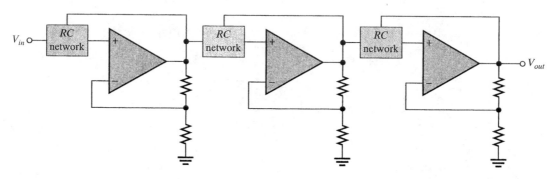

second-order and a first-order filter; to obtain a fourth-order filter, cascade two second-order filters; and so on. Each filter in a cascaded arrangement is called a *stage* or *section*.

Because of its maximally flat response, the Butterworth characteristic is the most widely used. Therefore, we will limit our coverage to the Butterworth response to illustrate basic filter concepts. Table 5–1 lists the roll-off rates, damping factors, and R_1/R_2 ratios for up to sixth-order Butterworth filters.

TABLE 5–1
Values for the Butterworth response.

Order	Roll-off dB/decade	1st stage			2nd stage			3rd stage		
		Poles	DF	R_1/R_2	Poles	DF	R_1/R_2	Poles	DF	R_1/R_2
1	−20	1	Optional							
2	−40	2	1.414	0.586						
3	−60	2	1.00	1	1	1.00	1			
4	−80	2	1.848	0.152	2	0.765	1.235			
5	−100	2	1.00	1	2	1.618	0.382	1	0.618	1.382
6	−120	2	1.932	0.068	2	1.414	0.586	2	0.518	1.482

5–2 REVIEW QUESTIONS

1. Explain how Butterworth, Chebyshev, and Bessel responses differ.
2. What determines the response characteristic of a filter?
3. Name the basic parts of an active filter.

5–3 ■ ACTIVE LOW-PASS FILTERS

Filters that use op-amps as the active element provide several advantages over passive filters (R, L, and C elements only). The op-amp provides gain, so that the signal is not attenuated as it passes through the filter. The high input impedance of the op-amp prevents excessive loading of the driving source, and the low output impedance of the op-amp prevents the filter from being affected by the load that it is driving. Active filters are also easy to adjust over a wide frequency range without altering the desired response.

After completing this section, you should be able to

❑ Understand active low-pass filters
    ❑ Identify a single-pole filter and determine its gain and critical frequency
    ❑ Identify a two-pole Sallen-Key filter and determine its gain and critical frequency
    ❑ Explain how a higher roll-off rate is achieved by cascading low-pass filters

A Single-Pole Filter

Figure 5–9(a) shows an active filter with a single low-pass RC network that provides a roll-off of -20 dB/decade above the critical frequency, as indicated by the response curve in Figure 5–9(b). The critical frequency of the single-pole filter is $f_c = 1/2\pi RC$. The op-amp

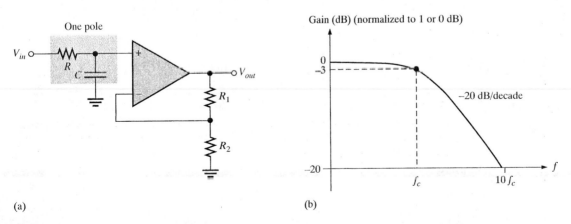

(a)   (b)

FIGURE 5–9
Single-pole active low-pass filter and response curve.

in this filter is connected as a noninverting amplifier with the closed-loop voltage gain in the passband set by the values of R_1 and R_2.

$$A_{cl} = \frac{R_1}{R_2} + 1$$

The Sallen-Key Low-Pass Filter

The Sallen-Key is one of the most common configurations for a second-order (two-pole) filter. It is also known as a VCVS (voltage-controlled voltage source) filter. A low-pass version of the Sallen-Key filter is shown in Figure 5–10. Notice that there are two low-pass RC networks that provide a roll-off of -40 dB/decade above the critical frequency (assuming a Butterworth characteristic). One RC network consists of R_A and C_A, and the second network consists of R_B and C_B. A unique feature is the capacitor C_A that provides feedback for shaping the response near the edge of the passband. The critical frequency for the second-order Sallen-Key filter is

$$f_c = \frac{1}{2\pi\sqrt{R_A R_B C_A C_B}} \tag{5–6}$$

For simplicity, the component values can be made equal so that $R_A = R_B = R$ and $C_A = C_B = C$. In this case, the expression for the critical frequency simplifies to $f_c = 1/2\pi RC$.

FIGURE 5–10
Basic Sallen-Key second-order low-pass filter.

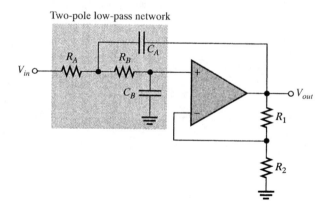

Two-pole low-pass network

As in the single-pole filter, the op-amp in the second-order Sallen-Key filter acts as a noninverting amplifier with the negative feedback provided by the R_1/R_2 network. As you have learned, the damping factor is set by the values of R_1 and R_2, thus making the filter response either Butterworth, Chebyshev, or Bessel. For example, from Table 5–1, the R_1/R_2 ratio must be 0.586 to produce the damping factor of 1.414 required for a second-order Butterworth response.

EXAMPLE 5–3

Determine the critical frequency of the low-pass filter in Figure 5–11, and set the value of R_1 for an approximate Butterworth response.

FIGURE 5–11

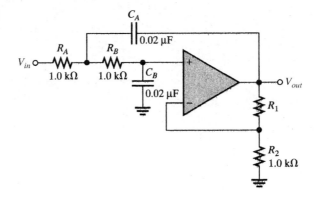

Solution   Since $R_A = R_B = 1.0$ kΩ and $C_A = C_B = 0.02$ μF,

$$f_c = \frac{1}{2\pi RC} = \frac{1}{2\pi(1.0 \text{ k}\Omega)(0.02 \text{ }\mu\text{F})} = \textbf{7.96 kHz}$$

For a Butterworth response, $R_1/R_2 = 0.586$.

$$R_1 = 0.586R_2 = 0.586(1.0 \text{ k}\Omega) = \textbf{586 }\boldsymbol{\Omega}$$

Select a standard value as near as possible to this calculated value.

Practice Exercise   Determine f_c for Figure 5–11 if $R_A = R_B = R_2 = 2.2$ kΩ and $C_A = C_B = 0.01$ μF. Also determine the value of R_1 for a Butterworth response.

Cascaded Low-Pass Filters Achieve a Higher Roll-Off Rate

A three-pole filter is required to get a third-order low-pass response (-60 dB/decade). This is done by cascading a two-pole low-pass filter and a single-pole low-pass filter, as shown in Figure 5–12(a). Figure 5–12(b) on the next page shows a four-pole configuration obtained by cascading two two-pole filters.

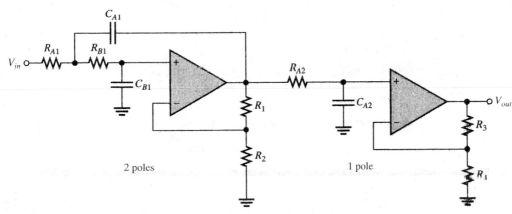

(a) Third-order configuration

FIGURE 5–12
Cascaded low-pass filters.

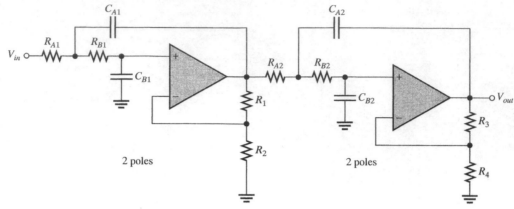

(b) Fourth-order configuration

FIGURE 5–12 (continued)

EXAMPLE 5–4

For the four-pole filter in Figure 5–12(b), determine the capacitance values required to produce a critical frequency of 2680 Hz if all the resistors in the RC low-pass networks are 1.8 kΩ. Also select values for the feedback resistors to get a Butterworth response.

Solution Both stages must have the same f_c. Assuming equal-value capacitors,

$$f_c = \frac{1}{2\pi RC}$$

$$C = \frac{1}{2\pi Rf_c} = \frac{1}{2\pi(1.8 \text{ k}\Omega)(2680 \text{ Hz})} = 0.033 \text{ }\mu\text{F}$$

$$C_{A1} = C_{B1} = C_{A2} = C_{B2} = \textbf{0.033 }\mu\text{F}$$

Also select $R_2 = R_4 = 1.8$ kΩ for simplicity. Refer to Table 5–1. For a Butterworth response in the first stage, $DF = 1.848$ and $R_1/R_2 = 0.152$. Therefore,

$$R_1 = 0.152R_2 = 0.152(1800 \text{ }\Omega) = \textbf{274 }\Omega$$

Choose $R_1 = 270$ Ω.

In the second stage, $DF = 0.765$ and $R_3/R_4 = 1.235$. Therefore,

$$R_3 = 1.235R_4 = 1.235(1800 \text{ }\Omega) = \textbf{2.22 k}\Omega$$

Choose $R_3 = 2.2$ kΩ.

Practice Exercise For the filter in Figure 5–12(b), determine the capacitance values for $f_c = 1$ kHz if all the filter resistors are 680 Ω. Also specify the values for the feedback resistors to produce a Butterworth response.

5–3 REVIEW QUESTIONS

1. How many poles does a second-order low-pass filter have? How many resistors and how many capacitors are used in the frequency-selective network?
2. Why is the damping factor of a filter important?
3. What is the primary purpose of cascading low-pass filters?

5–4 ■ ACTIVE HIGH-PASS FILTERS

In high-pass filters, the roles of the capacitor and resistor are reversed in the RC networks. Otherwise, the basic parameters are the same as for the low-pass filters.

After completing this section, you should be able to

❑ Understand active high-pass filters
   ❑ Identify a single-pole filter and determine its gain and critical frequency
   ❑ Identify a two-pole Sallen-Key filter and determine its gain and critical frequency
   ❑ Explain how a higher roll-off rate is achieved by cascading high-pass filters

A Single-Pole Filter

A high-pass active filter with a −20 dB/decade roll-off is shown in Figure 5–13(a). Notice that the input circuit is a single high-pass *RC* network. The negative feedback network is the same as for the low-pass filters previously discussed. The high-pass response curve is shown in Figure 5–13(b).

Ideally, a high-pass filter passes all frequencies above f_c without limit, as indicated in Figure 5–14(a), although in practice, this is not the case. As you have learned, all op-amps inherently have internal *RC* networks that limit the amplifier's response at high frequen-

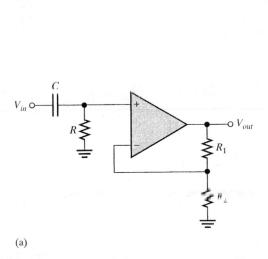

(a)

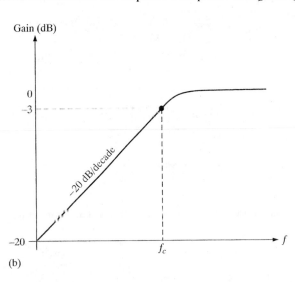

(b)

FIGURE 5–13
Single-pole active high-pass filter and response curve.

FIGURE 5–14
High-pass filter response.

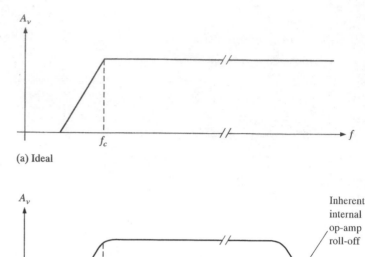

(a) Ideal

(b) Nonideal

cies. Therefore, there is an upper-frequency limit on the high-pass filter's response which, in effect, makes it a band-pass filter with a very wide bandwidth. In the majority of applications, the internal high-frequency limitation is so much greater than that of the filter's f_c that the limitation can be neglected. In some applications, special current-feedback op-amps or discrete transistors are used for the gain element to increase the high-frequency limitation beyond that realizable with standard op-amps.

The Sallen-Key High-Pass Filter

A high-pass second-order Sallen-Key configuration is shown in Figure 5–15. The components R_A, C_A, R_B, and C_B form the two-pole frequency-selective network. Notice that the positions of the resistors and capacitors in the frequency-selective network are opposite to those in the low-pass configuration. As with the other filters, the response characteristic can be optimized by proper selection of the feedback resistors, R_1 and R_2.

FIGURE 5–15
Basic Sallen-Key second-order high-pass filter.

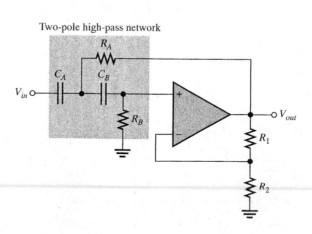

EXAMPLE 5–5

Choose values for the Sallen-Key high-pass filter in Figure 5–15 to implement an equal-value second-order Butterworth response with a critical frequency of approximately 10 kHz.

Solution   Start by selecting a value for R_A and R_B (R_1 or R_2 can also be the same value as R_A and R_B for simplicity).

$$R = R_A = R_B = R_2 = \textbf{3.3 k}\boldsymbol{\Omega} \qquad \text{(an arbitrary selection)}$$

Next, calculate the capacitance value from $f_c = 1/2\pi RC$.

$$C = C_A = C_B = \frac{1}{2\pi Rf_c} = \frac{1}{2\pi(3.3\text{ k}\Omega)(10\text{ kHz})} = \textbf{0.0048 }\boldsymbol{\mu}\textbf{F}$$

For a Butterworth response, the damping factor must be 1.414 and $R_1/R_2 = 0.586$.

$$R_1 = 0.586R_2 = 0.586(3.3\text{ k}\Omega) = \textbf{1.93 k}\boldsymbol{\Omega}$$

If you had chosen $R_1 = 3.3\text{ k}\Omega$, then

$$R_2 = \frac{R_1}{0.586} = \frac{3.3\text{ k}\Omega}{0.586} = 5.63\text{ k}\Omega$$

Either way, an approximate Butterworth response is realized by choosing the nearest standard values.

Practice Exercise   Select values for all the components in the high-pass filter of Figure 5–15 to obtain an $f_c = 300$ Hz. Use equal-value components and optimize for a Butterworth response.

Cascading High-Pass Filters

As with the low-pass configuration, first- and second-order high-pass filters can be cascaded to provide three or more poles and thereby create faster roll-off rates. Figure 5–16 shows a six-pole high-pass filter consisting of three two-pole stages. With this configuration optimized for a Butterworth response, a roll-off of -120 dB/decade is achieved.

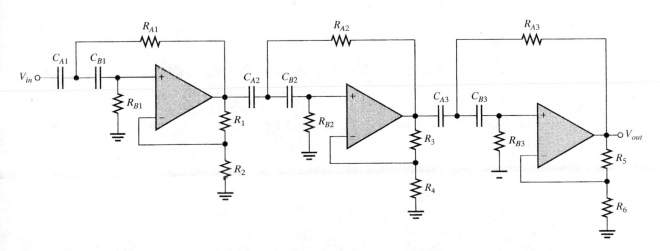

FIGURE 5–16
Sixth-order high-pass filter.

5–4 REVIEW QUESTIONS

1. How does a high-pass Sallen-Key filter differ from the low-pass configuration?
2. To increase the critical frequency of a high-pass filter, would you increase or decrease the resistor values?
3. If three two-pole high-pass filters and one single-pole high-pass filter are cascaded, what is the resulting roll-off?

5–5 ■ ACTIVE BAND-PASS FILTERS

As mentioned, band-pass filters pass all frequencies bounded by a lower-frequency limit and an upper-frequency limit and reject all others lying outside this specified band. A band-pass response can be thought of as the overlapping of a low-frequency response curve and a high-frequency response curve.

After completing this section, you should be able to

❑ Understand active band-pass filters
    ❑ Describe a band-pass filter composed of a low-pass and a high-pass filter
    ❑ Determine the critical frequencies and center frequency of a cascaded band-pass filter
    ❑ Determine center frequency, bandwidth, and gain of multiple-feedback band-pass filters
    ❑ Explain the operation of a state-variable band-pass filter

Cascaded Low-Pass and High-Pass Filters Achieve a Band-Pass Response

One way to implement a band-pass filter is a cascaded arrangement of a high-pass filter and a low-pass filter, as shown in Figure 5–17(a), as long as the critical frequencies are sufficiently separated. Each of the filters shown is a two-pole Sallen-Key Butterworth configuration so that the roll-off rates are −40 dB/decade, indicated in the composite response curve of Figure 5–17(b). The critical frequency of each filter is chosen so that the response curves overlap sufficiently, as indicated. The critical frequency of the high-pass filter must be sufficiently lower than that of the low-pass stage.

    The lower frequency, f_{c1}, of the passband is the critical frequency of the high-pass filter. The upper frequency, f_{c2}, is the critical frequency of the low-pass filter. Ideally, as discussed earlier, the center frequency, f_0, of the passband is the geometric mean of f_{c1} and f_{c2}. The following formulas express the three frequencies of the band-pass filter in Figure 5–17.

$$f_{c1} = \frac{1}{2\pi\sqrt{R_{A1}R_{B1}C_{A1}C_{B1}}}$$

$$f_{c2} = \frac{1}{2\pi\sqrt{R_{A2}R_{B2}C_{A2}C_{B2}}}$$

$$f_0 = \sqrt{f_{c1}f_{c2}}$$

Of course, if equal-value components are used in implementing each filter, the critical frequency equations simplify to the form $f_c = 1/2\pi RC$.

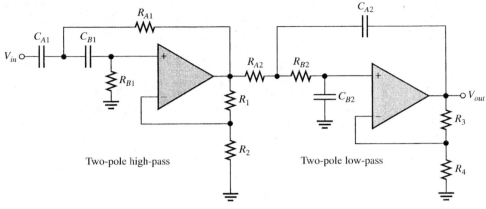

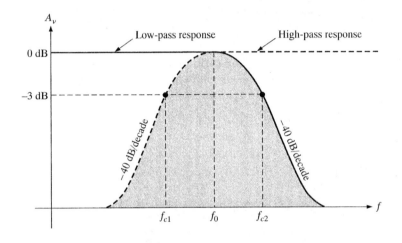

(b)

FIGURE 5–17

Band-pass filter formed by cascading a two-pole high-pass and a two-pole low-pass filter (it does not matter in which order the filters are cascaded).

Multiple-Feedback Band-Pass Filter

Another type of filter configuration, shown in Figure 5 18, is a multiple-feedback band-pass filter. The two feedback paths are through R_2 and C_1. Components R_1 and C_1 provide the low-pass response, and R_2 and C_2 provide the high-pass response. The maximum gain, A_0, occurs at the center frequency. Q values of less than 10 are typical in this type of filter. An expression for the center frequency follows, recognizing that R_1 and R_3 appear in parallel as viewed from the C_1 feedback path (with the V_{in} source replaced by a short).

$$f_0 = \frac{1}{2\pi\sqrt{(R_1\|R_3)R_2C_1C_2}}$$

Making $C_1 = C_2 = C$ gives the following formula (derived in Appendix B):

$$f_0 = \frac{1}{2\pi C}\sqrt{\frac{R_1 + R_3}{R_1R_2R_3}} \tag{5–7}$$

FIGURE 5–18
Multiple-feedback band-pass filter.

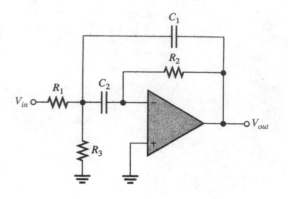

A convenient value for the capacitors is chosen; then the three resistor values are calculated based on the desired values for f_0, BW, and A_0. As you know, the Q can be determined from the relation $Q = f_0/BW$, and the resistors are found using the following formulas (stated without derivation).

$$R_1 = \frac{Q}{2\pi f_0 C A_0}$$

$$R_2 = \frac{Q}{\pi f_0 C}$$

$$R_3 = \frac{Q}{2\pi f_0 C(2Q^2 - A_0)}$$

To develop a gain expression, we solve for Q in the first two equations above.

$$Q = 2\pi f_0 A_0 C R_1$$
$$Q = \pi f_0 C R_2$$

Then,

$$2\pi f_0 A_0 C R_1 = \pi f_0 C R_2$$

An expression for the maximum gain at the center frequency is

$$A_0 = \frac{R_2}{2R_1} \tag{5–8}$$

In order for the denominator of the equation $R_3 = Q/[2\pi f_0 C(2Q^2 - A_0)]$ to be positive, $A_0 < 2Q^2$, which imposes a limitation on the gain.

EXAMPLE 5–6   Determine the center frequency, maximum gain, and bandwidth for the filter in Figure 5–19.

FIGURE 5–19

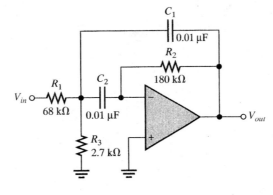

Solution

$$f_0 = \frac{1}{2\pi C} \sqrt{\frac{R_1 + R_3}{R_1 R_2 R_3}} = \frac{1}{2\pi(0.01\ \mu F)} \sqrt{\frac{68\ k\Omega + 2.7\ k\Omega}{(68\ k\Omega)(180\ k\Omega)(2.7\ k\Omega)}} = \textbf{736 Hz}$$

$$A_0 = \frac{R_2}{2R_1} = \frac{180\ k\Omega}{2(68\ k\Omega)} = \textbf{1.32}$$

$$Q = \pi f_0 C R_2 = \pi(736\ Hz)(0.01\ \mu F)(180\ k\Omega) = 4.16$$

$$BW = \frac{f_0}{Q} = \frac{736\ Hz}{4.16} = \textbf{177 Hz}$$

Practice Exercise   If R_2 in Figure 5–19 is increased to 330 kΩ, how does this affect the gain, center frequency, and bandwidth of the filter?

State-Variable Band-Pass Filter

The state-variable or universal active filter is widely used for band-pass applications. As shown in Figure 5–20, it consists of a summing amplifier and two op-amp integrators (which act as single-pole low-pass filters) that are combined in a cascaded arrangement to form a second-order filter. Although used primarily as a band-pass (BP) filter, the state-variable configuration also provides low-pass (LP) and high-pass (HP) outputs. The center frequency is set by the *RC* networks in both integrators. When used as a band-pass filter, the critical frequencies of the integrators are usually made equal, thus setting the center frequency of the passband.

Basic Operation   At input frequencies below f_c, the input signal passes through the summing amplifier and integrators and is fed back 180° out-of-phase. Thus, the feedback signal and input signal cancel for all frequencies below approximately f_c. As the low-pass response of the integrators rolls off, the feedback signal diminishes, thus allowing the input to pass through to the band-pass output. Above f_c the low-pass response disappears, thus preventing the input signal from passing through the integrators. As a result, the band-

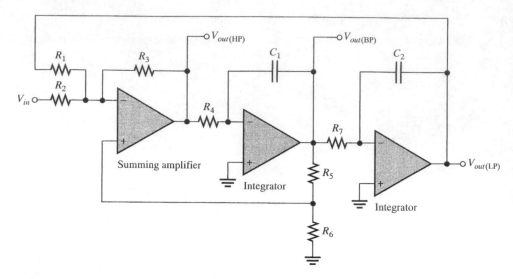

FIGURE 5–20
State-variable band-pass filter.

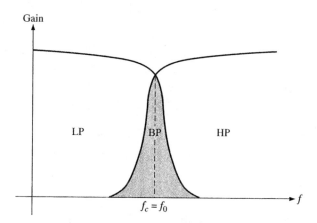

FIGURE 5–21
General state-variable response curves.

pass output peaks sharply at f_c, as indicated in Figure 5–21. Stable Qs up to 100 can be obtained with this type of filter. The Q is set by the feedback resistors R_5 and R_6 according to the following equation:

$$Q = \frac{1}{3}\left(\frac{R_5}{R_6} + 1\right)$$

The state-variable filter cannot be optimized for low-pass, high-pass, and band-pass performance simultaneously for this reason: To optimize for a low-pass or a high-pass Butterworth response, DF must equal 1.414. Since $Q = 1/DF$, a Q of 0.707 will result. Such a low Q provides a very poor band-pass response (large BW and poor selectivity). For optimization as a band-pass filter, the Q must be set high.

EXAMPLE 5–7   Determine the center frequency, Q, and BW for the band-pass output of the state-variable filter in Figure 5–22.

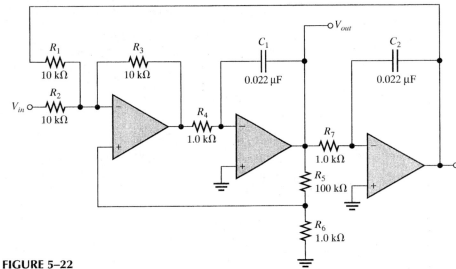

FIGURE 5–22

Solution   For each integrator,

$$f_c = \frac{1}{2\pi R_4 C_1} = \frac{1}{2\pi R_7 C_2} = \frac{1}{2\pi(1.0 \text{ k}\Omega)(0.022 \text{ }\mu\text{F})} = 7.23 \text{ kHz}$$

The center frequency is approximately equal to the critical frequencies of the integrators.

$$f_0 = f_c = \textbf{7.23 kHz}$$

$$Q = \frac{1}{3}\left(\frac{R_5}{R_6} + 1\right) = \frac{1}{3}\left(\frac{100 \text{ k}\Omega}{1.0 \text{ k}\Omega} + 1\right) = \textbf{33.7}$$

$$BW = \frac{f_0}{Q} = \frac{7.23 \text{ kHz}}{33.7} = \textbf{215 Hz}$$

Practice Exercise   Determine f_0, Q, and BW for the filter in Figure 5–22 if $R_4 = R_6 = R_7 = 330 \text{ }\Omega$ with all other component values the same as shown on the schematic.

5–5 REVIEW QUESTIONS

1. What determines selectivity in a band-pass filter?
2. One filter has a $Q = 5$ and another has a $Q = 25$. Which has the narrower bandwidth?
3. List the elements that make up a state-variable filter.

5–6 ■ ACTIVE BAND-STOP FILTERS

Band-stop filters reject a specified band of frequencies and pass all others. The response is opposite to that of a band-pass filter.

After completing this section, you should be able to

❑ Understand active band-stop filters
   ❑ Identify a multiple-feedback band-stop filter
   ❑ Explain the operation of a state-variable band-stop filter

Multiple-Feedback Band-Stop Filter

Figure 5–23 shows a multiple-feedback band-stop filter. Notice that this configuration is similar to the band-pass version except that R_3 has been moved and R_4 has been added.

FIGURE 5–23
Multiple-feedback band-stop filter.

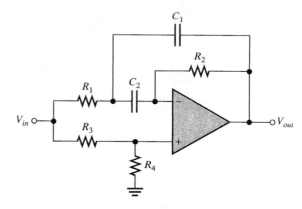

State-Variable Band-Stop Filter

Summing the low-pass and the high-pass responses of the state-variable filter covered in Section 5–5 creates a band-stop response as shown in Figure 5–24. One important application of this filter is minimizing the 60 Hz "hum" in audio systems by setting the center frequency to 60 Hz.

FIGURE 5–24
State-variable band-stop filter.

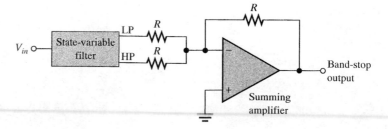

EXAMPLE 5–8   Verify that the band-stop filter in Figure 5–25 has a center frequency of 60 Hz, and optimize the filter for a Q of 30.

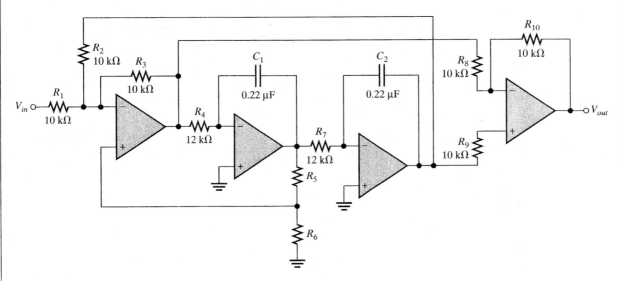

FIGURE 5–25

Solution f_0 equals the f_c of the integrator stages.

$$f_0 = \frac{1}{2\pi R_4 C_1} = \frac{1}{2\pi R_7 C_2} = \frac{1}{2\pi (12\ \text{k}\Omega)(0.22\ \mu\text{F})} = \textbf{60.3 Hz}$$

You can obtain a $Q = 30$ by choosing R_6 and then calculating R_5.

$$Q = \frac{1}{3}\left(\frac{R_5}{R_6} + 1\right)$$

$$R_5 = (3Q - 1)R_6$$

Choose $R_6 = \textbf{1.0 k}\Omega$. Then

$$R_5 = [3(30) - 1]1.0\ \text{k}\Omega = \textbf{89 k}\Omega$$

Practice Exercise   How would you change the center frequency to 120 Hz in Figure 5–25?

5–6 REVIEW QUESTIONS

1. How does a band-stop response differ from a band-pass response?
2. How is a state-variable band-pass filter converted to a band-stop filter?

5–7 ■ FILTER RESPONSE MEASUREMENTS

In this section, we discuss two methods of determining a filter's response by measurement—discrete point measurement and swept frequency measurement.

After completing this section, you should be able to

❑ Discuss two methods for measuring frequency response
   ❑ Explain the discrete point measurement method
   ❑ Explain the swept frequency measurement method

Discrete Point Measurement

Figure 5–26 shows an arrangement for taking filter output voltage measurements at discrete values of input frequency using common laboratory instruments. The general procedure is as follows:

1. Set the amplitude of the sine wave generator to a desired voltage level.
2. Set the frequency of the sine wave generator to a value well below the expected critical frequency of the filter under test. For a low-pass filter, set the frequency as near as possible to 0 Hz. For a band-pass filter, set the frequency well below the expected lower critical frequency.
3. Increase the frequency in predetermined steps sufficient to allow enough data points for an accurate response curve.
4. Maintain a constant input voltage amplitude while varying the frequency.
5. Record the output voltage at each value of frequency.
6. After recording a sufficient number of points, plot a graph of output voltage versus frequency.

If the frequencies to be measured exceed the response of the DMM, an oscilloscope may have to be used instead.

FIGURE 5–26
Test setup for discrete point measurement of the filter response. (Readings are arbitrary and for display only.)

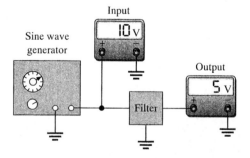

Swept Frequency Measurement

The swept frequency method requires more elaborate test equipment than does the discrete point method, but it is much more efficient and can result in a more accurate response curve. A general test setup is shown in Figure 5–27(a) using a swept frequency generator

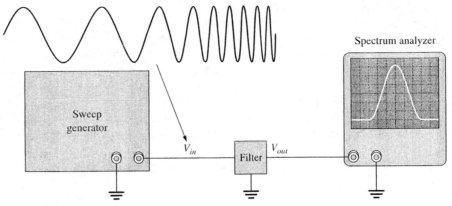

(a) Test setup for a filter response using a spectrum analyzer

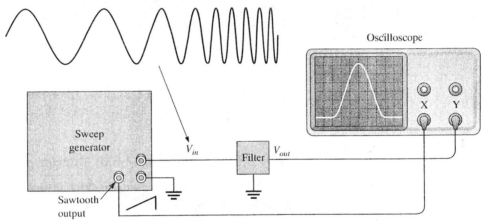

(b) Test setup for a filter response using an oscilloscope. The scope is put in X-Y mode. The sawtooth waveform from the sweep generator drives the X-channel of the oscilloscope.

FIGURE 5–27
Test setup for swept frequency measurement of the filter response.

and a spectrum analyzer. Figure 5–27(b) shows how the test can be made with an oscilloscope instead of a spectrum analyzer.

The swept frequency generator produces a constant amplitude output signal whose frequency increases linearly between two preset limits, as indicated in Figure 5–27. In part (a), the spectrum analyzer is an instrument that can be calibrated for a desired *frequency span/division* rather than for the usual *time/division* setting. Therefore, as the input frequency to the filter sweeps through a preselected range, the response curve is traced out on the screen of the spectrum analyzer. The test setup for using an oscilloscope to display the response curve is shown in part (b).

5–7 REVIEW QUESTIONS

1. What is the purpose of the two tests discussed in this section?
2. Name one disadvantage and one advantage of each test method.

5–8 ■ A SYSTEM APPLICATION

In this system application, the focus is on the filter board, which is part of the channel separation circuits in the FM stereo receiver. In addition to the active filters, the left and right channel separation circuit includes a demodulator, a frequency doubler, and a stereo matrix. Except for mentioning their purpose, we will not deal specifically with the demodulator, doubler, or matrix. However, the matrix is an interesting application of summing amplifiers, which were studied in Chapter 4 and these will be shown in detail on the schematic although we will not concentrate on them.

After completing this section, you should be able to

❑ Apply what you have learned in this chapter to a system application
    ❑ See how low-pass and band-pass active filters are used
    ❑ Use the schematic to locate and identify the components on the PC board
    ❑ Determine the operation of the filters
    ❑ Troubleshoot some common amplifier failures

A Brief Description of the System

Stereo FM (**frequency modulation**) signals are transmitted on a **carrier** frequency of 88 MHz to 108 MHz. The standard transmitted stereo signal consists of three modulating signals. These are the sum of the left and right channel audio (L + R), the difference of the left and right channel audio (L − R), and a 19 kHz pilot subcarrier.

The L + R audio extends from 30 Hz to 15 kHz and the L − R signal is contained in two sidebands extending from 23 kHz to 53 kHz as indicated in Figure 5–28. These frequencies come from the FM detector and go into the filter circuits where they are separated.

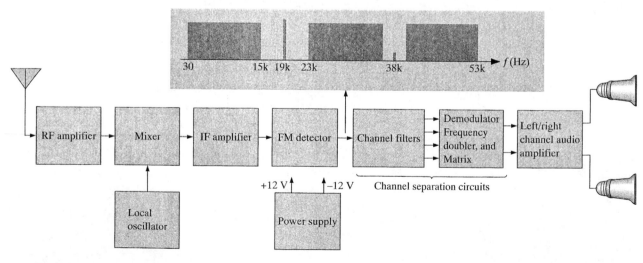

FIGURE 5–28
FM stereo receiver system.

The frequency doubler and demodulator are used to extract the audio signal from the 23 kHz to 53 kHz sidebands after which the 30 Hz to 15 kHz L − R signal is passed through a filter.

The L + R and L − R audio signals are then sent to the matrix where they are applied to the summing circuits to produce the left and right channel audio (−2L and −2R). Our focus in this application is on the filters.

Now, so that you can take a closer look at the filter board, let's take it out of the system and put it on the test bench.

ON THE TEST BENCH

■ ACTIVITY 1   Relate the PC Board to the Schematic

Locate and identify all components on the PC board in Figure 5–29, using the schematic in Figure 5–30. Label the PC board components to correspond with the schematic. Identify all inputs and outputs. Trace out the PC board to verify that it corresponds with the schematic. In this chapter, the backside of the PC board is shown instead of the x-ray view. The blue-shaded areas on the schematic are the filter networks contained on the board. The other blocks and circuits are on the demodulator, frequency doubler, and matrix board located elsewhere.

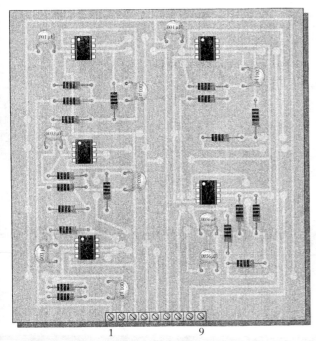

(a) Component side of board

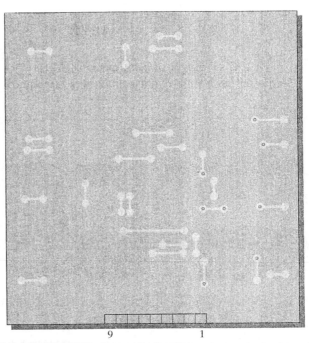

(b) Backside of board. The pads with small circles in the centers are component feedthrough pads.

FIGURE 5–29

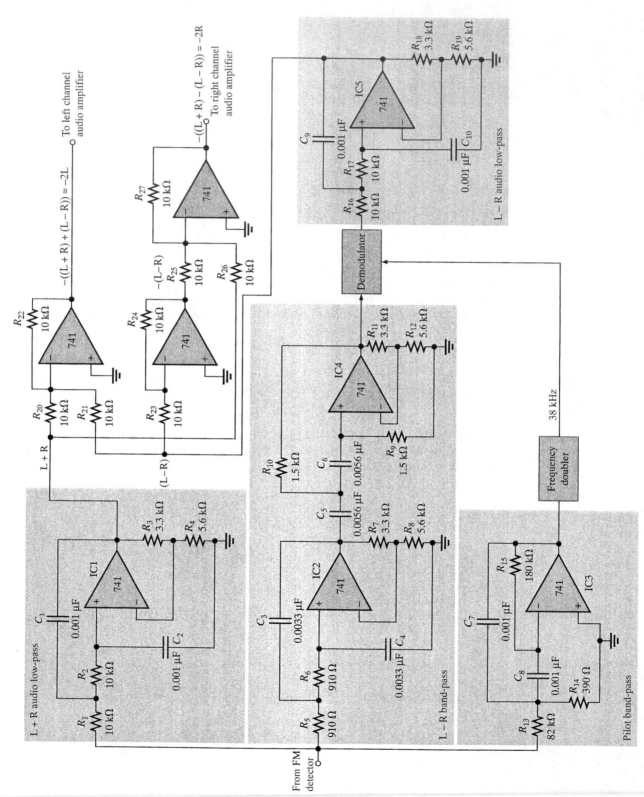

FIGURE 5-30
Left and right channel separation circuits.

■ ACTIVITY 2   Analyze the Filter Networks

Step 1:   Using the component values, determine the critical frequencies of each Sallen-Key-type filter.

Step 2:   Using the component values, determine the center frequency of the multiple-feedback filter.

Step 3:   Determine the bandwidth of each filter.

Step 4:   Determine the voltage gain of each filter.

Step 5:   Verify that the Sallen-Key filters have an approximate Butterworth response characteristic.

■ ACTIVITY 3   Write a Technical Report

Describe each filter in detail specifying the type of filter, the frequency responses, and the function of the filter within the overall circuitry. Also, describe the overall operation of the complete channel separation circuitry.

■ ACTIVITY 4   Troubleshoot the Filter Board

The filter board is plugged into a test fixture that permits access to each input and output, as shown in Figure 5–31, where the socket numbers correspond to the board pin numbers. The test instruments to be used are a sweep generator, a spectrum analyzer, and a dual power supply. For the sweep generator, a minimum and a maximum frequency is selected and the instrument produces an output that repetitively sweeps through all frequencies between the minimum and maximum setting. The spectrum analyzer will plot out a frequency response curve.

Develop a basic test procedure for completely testing the board in the fixture, using general references to instrument inputs, outputs, and settings. Include a diagram of a complete test setup.

FIGURE 5–31
Filter board in a test fixture.

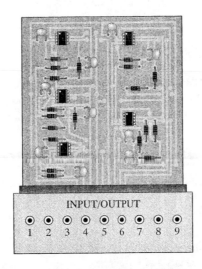

5-8 REVIEW QUESTIONS

1. What is the purpose of the filter board in this system?
2. What is the bandwidth of the L + R low-pass filter?
3. What is the bandwidth of the L − R low-pass filter?
4. Which filters on the board have approximate Butterworth responses?
5. What is the purpose of the stereo matrix circuitry?

■ SUMMARY

- The bandwidth in a low-pass filter equals the critical frequency because the response extends to 0 Hz.
- The bandwidth in a high-pass filter extends above the critical frequency and is limited only by the inherent frequency limitation of the active circuit.
- A band-pass filter passes all frequencies within a band between a lower and an upper critical frequency and rejects all others outside this band.
- The bandwidth of a band-pass filter is the difference between the upper critical frequency and the lower critical frequency.
- A band-stop filter rejects all frequencies within a specified band and passes all those outside this band.
- Filters with the Butterworth response characteristic have a very flat response in the passband, exhibit a roll-off of −20 dB/decade/pole, and are used when all the frequencies in the passband must have the same gain.
- Filters with the Chebyshev characteristic have ripples or overshoot in the passband and exhibit a faster roll-off per pole than filters with the Butterworth characteristic.
- Filters with the Bessel characteristic are used for filtering pulse waveforms. Their linear phase characteristic results in minimal waveshape distortion. The roll-off rate per pole is slower than for the Butterworth.
- In filter terminology, a single RC network is called a *pole*.
- Each pole in a Butterworth filter causes the output to roll off at a rate of −20 dB/decade.
- The quality factor Q of a band-pass filter determines the filter's selectivity. The higher the Q, the narrower the bandwidth and the better the selectivity.
- The damping factor determines the filter response characteristic (Butterworth, Chebyshev, or Bessel).

■ GLOSSARY

These terms are included in the end-of-book glossary.

Active filter   A frequency-selective circuit consisting of active devices such as transistors or op-amps coupled with reactive components.

Band-pass filter   A type of filter that passes a range of frequencies lying between a certain lower frequency and a certain higher frequency.

Band-stop filter   A type of filter that blocks or rejects a range of frequencies lying between a certain lower frequency and a certain higher frequency.

Bessel   A type of filter response having a linear phase characteristic and less than −20 dB/decade/pole roll-off.

Butterworth   A type of filter response characterized by flatness in the passband and a −20 dB/decade/pole roll-off.

Carrier   The high radio frequency (RF) signal that carries modulated information in AM, FM, or other communications systems.

Chebyshev   A type of filter response characterized by ripples in the passband and a greater than -20 dB/decade/pole roll-off.

Critical frequency (f_c)   The frequency that defines the end of the passband of a filter; also called *cutoff frequency*.

Damping factor (DF)   A filter characteristic that determines the type of response.

Filter   A type of electrical circuit that passes certain frequencies and rejects all others.

Frequency modulation (FM)   A communication method in which a lower frequency intelligence-carrying signal modulates (varies) the frequency of a higher frequency signal.

High-pass filter   A type of filter that passes frequencies above a certain frequency while rejecting lower frequencies.

Low-pass filter   A type of filter that passes frequencies below a certain frequency while rejecting higher frequencies.

Passband   The region of frequencies that are allowed to pass through a filter with minimum attenuation.

Pole   A network containing one resistor and one capacitor that contributes -20 dB/decade to a filter's roll-off rate.

Quality factor (Q)   The ratio of a band-pass filter's center frequency to its bandwidth.

■ KEY FORMULAS

(5–1)	$BW = f_c$		Low-pass bandwidth
(5–2)	$BW = f_{c2} - f_{c1}$		Filter bandwidth of a band-pass filter
(5–3)	$f_0 = \sqrt{f_{c1}f_{c2}}$		Center frequency of a band-pass filter
(5–4)	$Q = \dfrac{f_0}{BW}$		Quality factor of a band-pass filter
(5–5)	$DF = 2 - \dfrac{R_1}{R_2}$		Damping factor
(5–6)	$f_c = \dfrac{1}{2\pi\sqrt{R_A R_B C_A C_B}}$		Critical frequency for a second-order Sallen-Key filter
(5–7)	$f_0 = \dfrac{1}{2\pi C}\sqrt{\dfrac{R_1 + R_3}{R_1 R_2 R_3}}$		Center frequency of a multiple-feedback filter
(5–8)	$A_0 = \dfrac{R_2}{2R_1}$		Gain of a multiple-feedback filter

---

■ SELF-TEST

1. The term *pole* in filter terminology refers to
   (a) a high gain op amp   (b) one complete active filter
   (c) a single RC network   (d) the feedback circuit

2. An RC circuit produces a roll-off rate of
   (a) -20 dB/decade   (b) -40 dB/decade
   (c) -6 dB/octave   (d) answers (a) and (c)

3. A band-pass response has
   (a) two critical frequencies   (b) one critical frequency
   (c) a flat curve in the passband   (d) a wide bandwidth

4. The lowest frequency passed by a low-pass filter is
   (a) 1 Hz   (b) 0 Hz   (c) 10 Hz   (d) dependent on the critical frequency

5. The Q of a band-pass filter depends on
   (a) the critical frequencies   (b) only the bandwidth
   (c) the center frequency and the bandwidth   (d) only the center frequency

6. The damping factor of an active filter determines
   (a) the voltage gain   (b) the critical frequency
   (c) the response characteristic   (d) the roll-off rate

7. A maximally flat frequency response is known as
   (a) Chebyshev   (b) Butterworth
   (c) Bessel   (d) Colpitts

8. The damping factor of a filter is set by
   (a) the negative feedback circuit   (b) the positive feedback circuit
   (c) the frequency-selective circuit   (d) the gain of the op-amp

9. The number of poles in a filter affect the
   (a) voltage gain   (b) bandwidth
   (c) center frequency   (d) roll-off rate

10. Sallen-Key filters are
    (a) single-pole filters   (b) second-order filters
    (c) Butterworth filters   (d) band-pass filters

11. When filters are cascaded, the roll-off rate
    (a) increases   (b) decreases   (c) does not change

12. When a low-pass and a high-pass filter are cascaded to get a band-pass filter, the critical frequency of the low-pass filter must be
    (a) equal to the critical frequency of the high-pass filter
    (b) less than the critical frequency of the high-pass filter
    (c) greater than the critical frequency of the high-pass filter

13. A state-variable filter consists of
    (a) one op-amp with multiple-feedback paths
    (b) a summing amplifier and two integrators
    (c) a summing amplifier and two differentiators
    (d) three Butterworth stages

14. When the gain of a filter is minimum at its center frequency, it is a
    (a) band-pass filter   (b) a band-stop filter
    (c) a notch filter   (d) answers (b) and (c)

■ PROBLEMS

SECTION 5–1 Basic Filter Responses

1. Identify each type of filter response (low-pass, high-pass, band-pass, or band-stop) in Figure 5–32.

2. A certain low-pass filter has a critical frequency of 800 Hz. What is its bandwidth?

3. A single-pole high-pass filter has a frequency-selective network with $R = 2.2$ kΩ and $C = 0.0015$ μF. What is the critical frequency? Can you determine the bandwidth from the available information?

4. What is the roll-off rate of the filter described in Problem 3?

5. What is the bandwidth of a band-pass filter whose critical frequencies are 3.2 kHz and 3.9 kHz? What is the Q of this filter?

6. What is the center frequency of a filter with a Q of 15 and a bandwidth of 1.0 kHz?

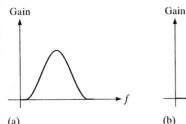

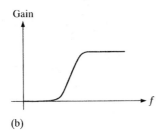

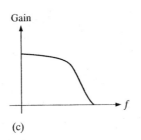

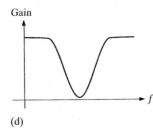

(a) (b) (c) (d)

FIGURE 5–32

SECTION 5–2 Filter Response Characteristics

7. What is the damping factor in each active filter shown in Figure 5–33? Which filters are approximately optimized for a Butterworth response characteristic?

8. For the filters in Figure 5–33 that do not have a Butterworth response, specify the changes necessary to convert them to Butterworth responses. (Use nearest standard values.)

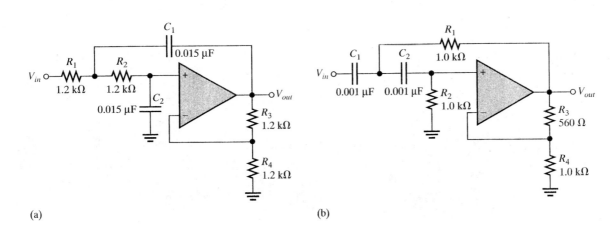

(a) (b)

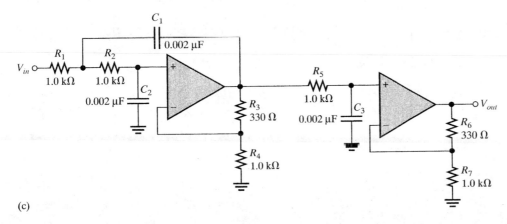

(c)

FIGURE 5–33

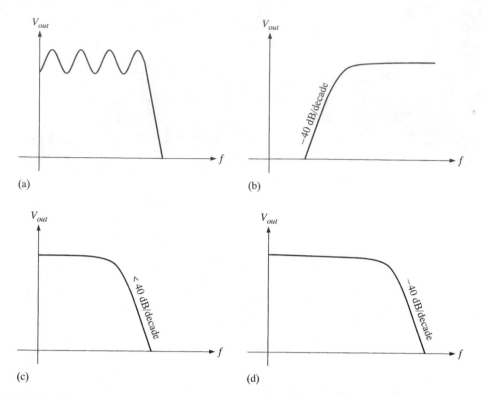

FIGURE 5–34

9. Response curves for second-order filters are shown in Figure 5–34. Identify each as Butterworth, Chebyshev, or Bessel.

SECTION 5–3   Active Low-Pass Filters

10. Is the four-pole filter in Figure 5–35 approximately optimized for a Butterworth response? What is the roll-off rate?

11. Determine the critical frequency in Figure 5–35.

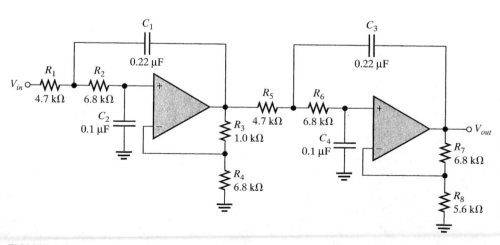

FIGURE 5–35

12. Without changing the response curve, adjust the component values in the filter of Figure 5–35 to make it an equal-value filter.

13. Modify the filter in Figure 5–35 to increase the roll-off rate to −120 dB/decade while maintaining an approximate Butterworth response.

14. Using a block diagram format, show how to implement the following roll-off rates using single-pole and two-pole low-pass filters with Butterworth responses.
    (a) −40 dB/decade   **(b)** −20 dB/decade   **(c)** −60 dB/decade
    (d) −100 dB/decade   **(e)** −120 dB/decade

SECTION 5–4   Active High-Pass Filters

15. Convert the equal-value filter from Problem 12 to a high-pass with the same critical frequency and response characteristic.

16. Make the necessary circuit modification to reduce by half the critical frequency in Problem 15.

17. For the filter in Figure 5–36,
    (a) How would you increase the critical frequency?
    (b) How would you increase the gain?

FIGURE 5–36

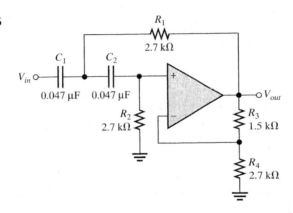

SECTION 5–5   Active Band-Pass Filters

18. Identify each band-pass filter configuration in Figure 5–37.

19. Determine the center frequency and bandwidth for each filter in Figure 5–37.

20. Optimize the state-variable filter in Figure 5–38 for $Q = 50$. What bandwidth is achieved?

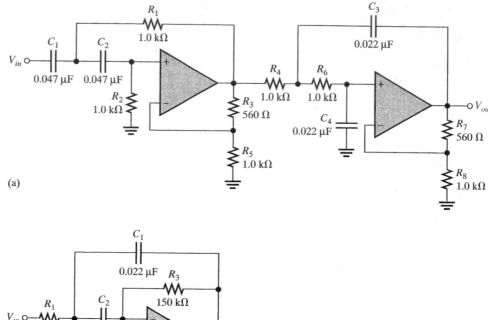

(a)

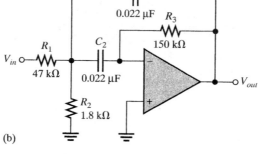

(b)

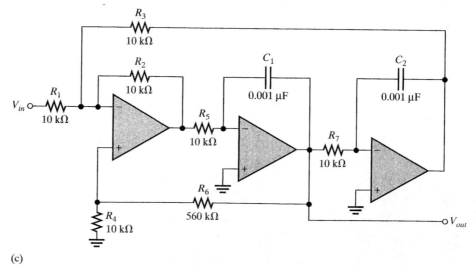

(c)

FIGURE 5–37

SECTION 5–6   Active Band-Stop Filters

21. Show how to make a notch (band-stop) filter using the basic circuit in Figure 5–38.

22. Modify the band-stop filter in Problem 21 for a center frequency of 120 Hz.

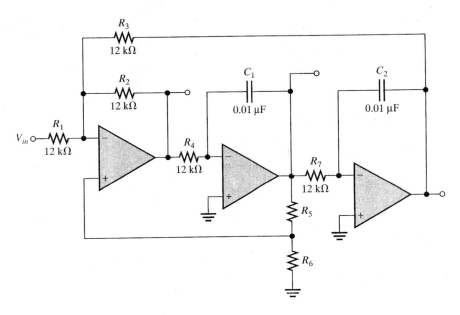

FIGURE 5–38

Section 5–1

1. The critical frequency determines the bandwidth.

2. The inherent frequency limitation of the op-amp limits the bandwidth.

3. Q and BW are inversely related. The higher the Q, the better the selectivity, and vice versa.

Section 5–2

1. Butterworth is very flat in the passband and has a -20 dB/decade/pole roll-off.
Chebyshev has ripples in the passband and has greater than -20 dB/decade/pole roll-off.
Bessel has a linear phase characteristic and less than -20 dB/decade/pole roll-off.

2. The damping factor determines the response characteristic.

3. Frequency-selection network, gain element, and negative feedback network are the parts of an active filter.

Section 5–3

1. A second-order filter has two poles. Two resistors and two capacitors make up the frequency-selective network.

2. The damping factor sets the response characteristic.

3. Cascading increases the roll-off rate.

Section 5–4

1. The positions of the Rs and Cs in the frequency-selection network are opposite for low-pass and high-pass configurations.
2. Decrease the R values to increase f_c.
3. -140 dB/decade

Section 5–5

1. Q determines selectivity.
2. $Q = 25$. Higher Q gives narrower BW.
3. A summing amplifier and two integrators make up a state-variable filter.

Section 5–6

1. A band-stop rejects frequencies within the stopband. A band-pass passes frequencies within the passband.
2. The low-pass and high-pass outputs are summed.

Section 5–7

1. To check the frequency response of a filter
2. Discrete point measurement—tedious and less complete; simpler equipment.
   Swept frequency measurement—uses more expensive equipment; more efficient, can be more accurate and complete.

Section 5–8

1. The filter board takes the detected FM signal and separates the L + R and L − R audio signals.
2. $BW = 15.9$ kHz
3. $BW = 15.9$ kHz
4. The L + R low-pass, the L − R low-pass, and the L − R band-pass
5. The stereo matrix combines the L + R and L − R signals and produces the separate left and right channel audio signals.

■ ANSWERS TO PRACTICE EXERCISES FOR EXAMPLES

5–1 500 Hz

5–2 1.44

5–3 7.23 kHz, 1.25 kΩ

5–4 $C_{A1} = C_{A2} = C_{B1} = C_{B2} = 0.234$ μF; $R_2 = R_4 = 680$ Ω, $R_1 = 103$ Ω, $R_3 = 840$ Ω

5–5 $R_A = R_B = R_2 = 10$ kΩ, $C_A = C_B = 0.053$ μF, $R_1 = 5.86$ kΩ

5–6 Gain increases to 2.43, frequency decreases to 544 Hz, and bandwidth decreases to 96.5 Hz.

5–7 $f_0 = 21.9$ kHz, $Q = 101$, $BW = 216$ Hz

5–8 Decrease the input resistors or the feedback capacitors of the two integrator stages by half.

6

OSCILLATORS AND TIMERS

Courtesy Hewlett-Packard Company

■ CHAPTER OBJECTIVES

☐ Describe the basic operating principles for all oscillators
☐ Explain the operation of feedback oscillators
☐ Describe and analyze the operation of basic *RC* sinsuoidal oscillators
☐ Describe and analyze the operation of basic relaxation oscillators
☐ Use a 555 timer in an oscillation application
☐ Use a 555 timer as a one-shot device
☐ Apply what you have learned in this chapter to a system application

Oscillators are circuits that generate a periodic waveform to perform timing, control, or communication functions. They are found in nearly all electronic systems, including analog and digital systems, and in most test instruments such as oscilloscopes and function generators.

Oscillators require a form of positive feedback, where a portion of the output signal is fed back to the input in a way that causes it to reinforce itself and thus sustain a continuous output signal. Although an external input is not strictly necessary, many oscillators use an external signal to control the frequency or to synchronize it with another source. Oscillators are designed to produce a controlled oscillation with one of two basic methods: the unity-gain method used with feedback oscillators and the timing method used with relaxation oscillators. Both will be discussed in this chapter.

Different types of oscillators produce various types of outputs including sine waves, square waves, triangular waves, and sawtooth waves. In this chapter, several types of basic oscillator circuits using an op-amp as the gain element are introduced. Also, a very popular integrated circuit, called the 555 timer, is discussed.

The function generator shown in Figure 6–37 is a good illustration of a system application for oscillators. The oscillator is a major part of this particular system. No doubt, you are already familiar with the use of the signal or function generator in your lab. As with most types of systems, a function generator can be implemented in more than one way. The system in this chapter uses circuits with which you are already familiar without some of the refinements and features found in many commercial instruments. The system reinforces what you have studied and lets you see these circuits "at work" in a specific application.

For the system application in Section 6–7, in addition to the other topics, be sure you understand

☐ How *RC* oscillators work
☐ How a zero-level detector works
☐ How an integrator works

6–1 ■ THE OSCILLATOR

An oscillator is a circuit that produces a periodic waveform on its output with only the dc supply voltage as a required input. A repetitive input signal is not required but sometimes used to synchronize oscillations. The output voltage can be either sinusoidal or non-sinusoidal, depending on the type of oscillator. Two major classifications for oscillators are feedback oscillators and relaxation oscillators.

After completing this section, you should be able to

❑ Describe the basic operating principles for all oscillators
   ❑ Explain the purpose of an oscillator
   ❑ Discuss two important classifications for oscillators
   ❑ List the basic elements of a feedback oscillator

Types of Oscillators

Essentially, all **oscillators** convert electrical energy from the dc power supply to periodic waveforms that can be used for various timing, control, or signal-generating applications. A basic oscillator is illustrated in Figure 6–1. Oscillators are classified according to the technique for generating a signal.

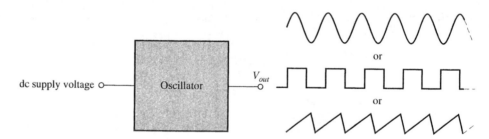

FIGURE 6–1
The basic oscillator concept showing three common types of output waveforms.

Feedback Oscillators   **Feedback oscillators** return a fraction of the output signal to the input with no net phase shift, resulting in a reinforcement of the output signal. After oscillations are started, the loop gain is maintained at 1.0 to maintain oscillations. A feedback oscillator consists of an amplifier for gain (either a discrete transistor or an op-amp) and a positive feedback network that produces phase shift and provides attenuation, as shown in Figure 6–2.

Relaxation Oscillators   A second type of oscillator is the **relaxation oscillator.** A relaxation oscillator uses an *RC* timing circuit to generate a waveform that is generally a square wave or other nonsinusoidal waveform. Typically, a relaxation oscillator uses a Schmitt trigger or other device that changes states to alternately charge and discharge a capacitor through a resistor. Relaxation oscillators are discussed in Section 6–4.

FIGURE 6–2
Basic elements of a feedback oscillator.

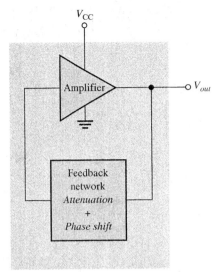

Feedback oscillator

6–1 REVIEW QUESTIONS

1. What is an oscillator?
2. What type of feedback does a feedback oscillator require?
3. What is the purpose of the feedback network?

6–2 ■ FEEDBACK OSCILLATOR PRINCIPLES

Feedback oscillator operation is based on the principle of positive feedback. In this section, we will examine this concept and look at the general conditions required for oscillation to occur. Feedback oscillators are widely used to generate sinusoidal waveforms.

After completing this section, you should be able to

❑ Explain the operation of feedback oscillators
   ❑ Explain positive feedback
   ❑ Describe the conditions for oscillation
   ❑ Discuss the start-up conditions

Positive Feedback

Positive feedback is characterized by the condition wherein a portion of the output voltage of an amplifier is fed back to the input. This basic idea is illustrated with the sinusoidal oscillator shown in Figure 6–3. As you can see, the in-phase feedback voltage is amplified to produce the output voltage, which in turn produces the feedback voltage. That is, a loop is created in which the signal sustains itself and a continuous sinusoidal output is produced. This phenomenon is called *oscillation*.

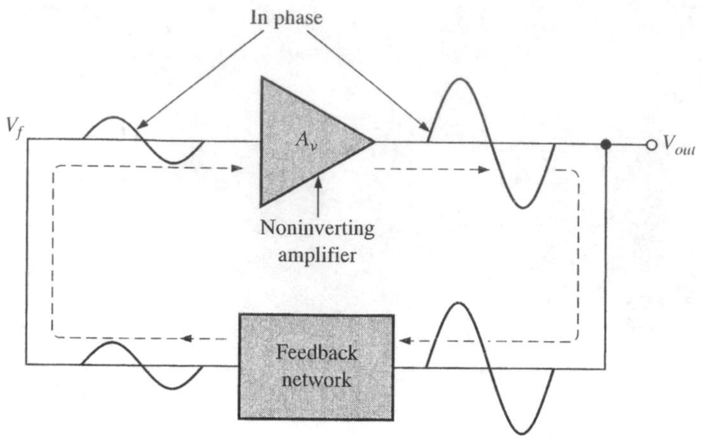

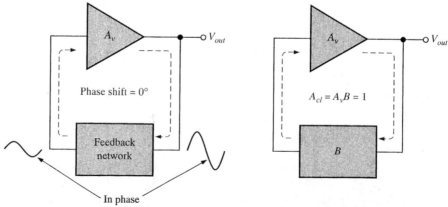

FIGURE 6–3
Positive feedback produces oscillation.

Conditions for Oscillation

Two conditions, illustrated in Figure 6–4, are required to sustain oscillations:

1. The phase shift around the feedback loop must be 0° (360°).

2. The loop gain (voltage gain around the closed feedback loop) must be at least 1 (unity).

The loop gain, A_{cl} is the product of the amplifier gain, A_v, and the attenuation, B, of the feedback circuit.

$$A_{cl} = A_v B$$

If a sinusoidal wave is the desired output, a loop gain greater than 1 will rapidly cause the output to saturate at both peaks of the waveform, producing unacceptable distortion. To avoid this, some form of gain control must be used to keep the loop gain at exactly 1, once oscillations have started. For example, if the attenuation of the feedback network is 0.01, the amplifier must have a gain of exactly 100 to overcome this attenuation and not

(a) The phase shift around the loop is 0°.

(b) The closed loop gain is 1.

FIGURE 6–4
Conditions for oscillation.

create unacceptable distortion ($0.01 \times 100 = 1.0$). An amplifier gain of greater than 100 will cause the oscillator to limit both peaks of the waveform.

Start-Up Conditions

So far, you have seen what it takes for an oscillator to produce a continuous sinusoidal output. Now let's examine the requirements for the oscillation to start when the dc supply voltage is turned on. As you know, the unity-gain condition must be met for oscillation to be sustained. For oscillation to *begin*, the voltage gain around the positive feedback loop must be greater than 1 so that the amplitude of the output can build up to a desired level. The gain must then decrease to 1 so that the output stays at the desired level and oscillation is sustained. (Several ways to achieve this reduction in gain after start-up are discussed in the next section.) The conditions for both starting and sustaining oscillation are illustrated in Figure 6–5.

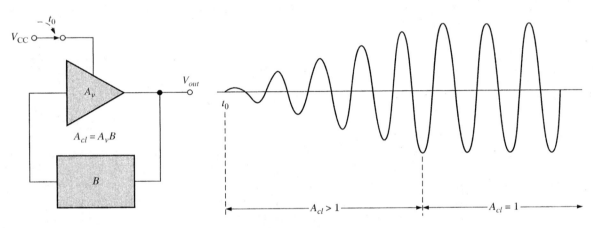

FIGURE 6–5

When oscillation starts at t_0, the condition $A_{cl} > 1$ causes the sinusoidal output voltage amplitude to build up to a desired level, where A_{cl} decreases to 1 and maintains the desired amplitude.

A question that normally arises is this: If the oscillator is off (no dc voltage) and there is no output voltage, how does a feedback signal originate to start the positive feedback build-up process? Initially, a small positive feedback voltage develops from thermally produced broad-band noise in the resistors or other components or from power supply turn-on transients. The feedback circuit permits only a voltage with a frequency equal to the selected oscillation frequency to appear in phase on the amplifier's input. This initial feedback voltage is amplified and continually reinforced, resulting in a buildup of the output voltage as previously discussed.

6–2 REVIEW QUESTIONS

1. What are the conditions required for a circuit to oscillate?
2. Define positive feedback.
3. What are the start-up conditions for an oscillator?

6–3 ■ SINUSOIDAL OSCILLATORS

In this section, you will learn about three types of RC oscillator circuits that produce sinusoidal outputs: the Wien-bridge oscillator, the phase-shift oscillator, and the twin-T oscillator. Generally, RC oscillators are used for frequencies up to about 1 MHz. The Wien-bridge is by far the most widely used type of RC oscillator for this range of frequencies.

After completing this section, you should be able to

❑ Describe and analyze the operation of basic *RC* sinusoidal oscillators
    ❑ Identify a Wien-bridge oscillator
    ❑ Determine the resonant frequency of a Wien-bridge oscillator
    ❑ Analyze oscillator feedback conditions
    ❑ Analyze oscillator start-up conditions
    ❑ Describe a self-starting Wien-bridge oscillator
    ❑ Identify a phase-shift oscillator
    ❑ Calculate the resonant frequency and analyze the feedback conditions for a phase-shift oscillator
    ❑ Identify a twin-T oscillator and describe its operation

The Wien-Bridge Oscillator

One type of sinusoidal oscillator is the *Wien-bridge* oscillator. A fundamental part of the Wien-bridge oscillator is a lead-lag network like that shown in Figure 6–6(a). R_1 and C_1 together form the lag portion of the network; R_2 and C_2 form the lead portion. The operation of this circuit is as follows. At lower frequencies, the lead network dominates due to the high reactance of C_2. As the frequency increases, X_{C2} decreases, thus allowing the output voltage to increase. At some specified frequency, the response of the lag network takes over, and the decreasing value of X_{C1} causes the output voltage to decrease.

The response curve for the lead-lag network shown in Figure 6–6(b) indicates that the output voltage peaks at a frequency called the resonant frequency, f_r. At this point, the attenuation (V_{out}/V_{in}) of the network is ⅓ if $R_1 = R_2$ and $X_{C1} = X_{C2}$ as stated by the following equation, which is derived in Appendix B:

$$\frac{V_{out}}{V_{in}} = \frac{1}{3}$$

(6–1)

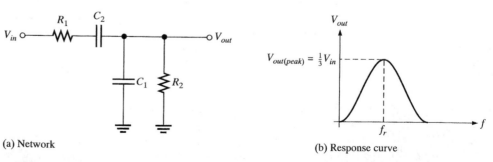

(a) Network

(b) Response curve

FIGURE 6–6

A lead-lag network and its response curve.

The formula for the resonant frequency is also derived in Appendix B and is

$$f_r = \frac{1}{2\pi RC} \tag{6-2}$$

To summarize, the lead-lag network in the Wien-bridge oscillator has a resonant frequency, f_r, at which the phase shift through the network is 0° and the attenuation is ⅓. Below f_r, the lead network dominates and the output leads the input. Above f_r, the lag network dominates and the output lags the input.

The Basic Circuit The lead-lag network is used in the positive feedback loop of an op-amp, as shown in Figure 6–7(a). A voltage divider is used in the negative feedback loop. The Wien-bridge oscillator circuit can be viewed as a noninverting amplifier configuration with the input signal fed back from the output through the lead-lag network. Recall that the closed-loop gain of the amplifier is determined by the voltage divider.

$$A_{cl} = \frac{1}{B} = \frac{1}{R_2/(R_1 + R_2)} = \frac{R_1 + R_2}{R_2}$$

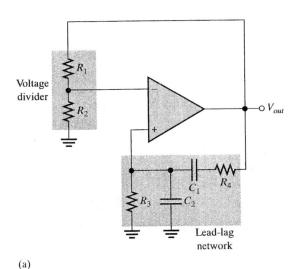

(a)

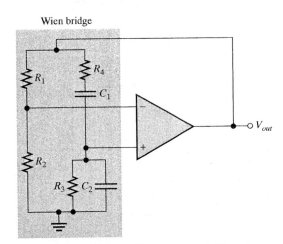

Wien bridge

(b) Wien bridge circuit combines a voltage divider and a lead-lag network.

FIGURE 6–7
Two ways to draw the schematic of a Wien-bridge oscillator.

The circuit is redrawn in Figure 6–7(b) to show that the op-amp is connected across the bridge circuit. One leg of the bridge is the lead-lag network, and the other is the voltage divider.

Positive Feedback Conditions for Oscillation As you know, for the circuit to produce a sustained sinusoidal output (oscillate), the phase shift around the positive feedback loop must be 0° and the gain around the loop must be at least unity (1). The 0° phase-shift condition is met when the frequency is f_r because the phase shift through the lead-lag network is 0° and there is no inversion from the noninverting (+) input of the op-amp to the output. This is shown in Figure 6–8(a).

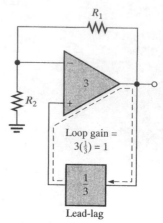

(a) The phase shift around the loop is 0°.

(b) The voltage gain around the loop is 1.

FIGURE 6–8
Conditions for oscillation.

The unity-gain condition in the feedback loop is met when

$$A_{cl} = 3$$

This offsets the ⅓ attenuation of the lead-lag network, thus making the total gain around the positive feedback loop equal to 1, as depicted in Figure 6–8(b). To achieve a closed-loop gain of 3,

$$R_1 = 2R_2$$

Then

$$A_{cl} = \frac{R_1 + R_2}{R_2} = \frac{2R_2 + R_2}{R_2} = \frac{3R_2}{R_2} = 3$$

Start-Up Conditions Initially, the closed-loop gain of the amplifier itself must be more than 3 ($A_{cl} > 3$) until the output signal builds up to a desired level. The gain of the amplifier must then decrease to 3 so that the total gain around the loop is 1 and the output signal stays at the desired level, thus sustaining oscillation. This is illustrated in Figure 6–9.

The circuit in Figure 6–10 illustrates a basic method for achieving the condition just described. Notice that the voltage-divider network has been modified to include an additional resistor R_3 in parallel with a back-to-back zener diode arrangement. When dc power is first applied, both zener diodes appear as opens. This places R_3 in series with R_1, thus increasing the closed-loop gain of the amplifier as follows ($R_1 = 2R_2$):

$$A_{cl} = \frac{R_1 + R_2 + R_3}{R_2} = \frac{3R_2 + R_3}{R_2} = 3 + \frac{R_3}{R_2}$$

Initially, a small positive feedback signal develops from noise or turn-on transients. The lead-lag network permits only a signal with a frequency equal to f_r to appear in phase on the noninverting input. This feedback signal is amplified and continually reinforced, resulting in a buildup of the output voltage. When the output signal reaches the zener breakdown voltage, the zeners conduct and effectively short out R_3. This lowers the amplifier's closed-loop gain to 3. At this point the total loop gain is 1 and the output signal levels off

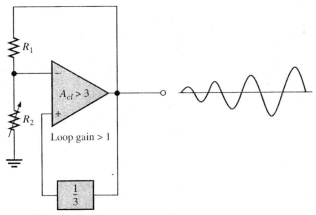

(a) Loop gain greater than 1 causes output to build up.

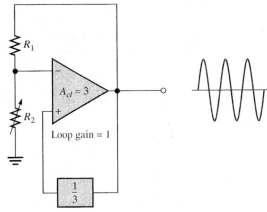

(b) Loop gain of 1 causes a sustained constant output.

FIGURE 6–9
Oscillator start-up conditions.

FIGURE 6–10
Self-starting Wien-bridge oscillator using back-to-back zener diodes.

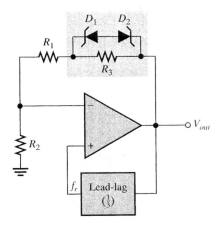

and the oscillation is sustained. All practical methods to achieve stability for feedback oscillators require the gain to be self-adjusting. This requirement is a form of **automatic gain control (AGC).** The zener diodes in this example limit the gain at the onset of a nonlinearity, in this case, zener conduction.

Another method to control the gain uses a JFET as a voltage-controlled resistor in a negative feedback path. A JFET operating with a small or zero V_{DS} is operating in the ohmic region. As the gate voltage increases, the drain-source resistance increases. If the JFET is placed in the negative feedback path, automatic gain control can be achieved because of this voltage-controlled resistance.

A JFET stabilized Wien-bridge oscillator is shown in Figure 6–11. The gain of the op-amp is controlled by the components shown in the shaded box, which include the JFET. The JFET's drain-source resistance depends on the gate voltage. With no output signal, the gate is at zero volts, causing the drain-source resistance to be at the minimum. With this condition, the loop gain is greater than 1. Oscillations begin and rapidly build to a large output signal. Negative excursions of the output signal forward-bias D_1, causing capacitor C_3 to charge to a negative voltage. This voltage increases the drain-source resistance of the

FIGURE 6–11
Self-starting Wien-bridge oscillator using a JFET in the negative feedback loop.

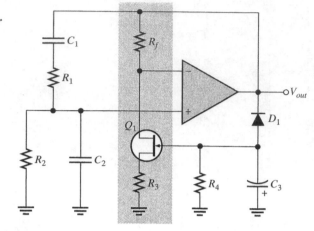

JFET and reduces the gain (and hence the output). This is classic negative feedback at work. With the proper selection of components, the gain can be stabilized at the required level.

EXAMPLE 6–1

Determine the frequency of oscillation for the Wien-bridge oscillator in Figure 6–12. Also, verify that oscillations will start and then continue when the output signal reaches 5.4 V.

FIGURE 6–12

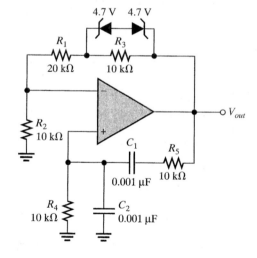

Solution For the lead-lag network, $R_4 = R_5 = R = 10 \text{ k}\Omega$ and $C_1 = C_2 = C = 0.001 \text{ } \mu\text{F}$. The frequency is

$$f_r = \frac{1}{2\pi RC} = \frac{1}{2\pi(10 \text{ k}\Omega)(0.001 \text{ } \mu\text{F})} = \textbf{15.9 kHz}$$

Initially, the closed-loop gain is

$$A_{cl} = \frac{R_1 + R_2 + R_3}{R_2} = \frac{40 \text{ k}\Omega}{10 \text{ k}\Omega} = 4$$

Since $A_{cl} > 3$, the start-up condition is met.

When the output reaches 5.4 V (4.7 V + 0.7 V), the zeners conduct (their forward resistance is assumed small, compared to 10 kΩ), and the closed-loop gain is reached. Thus, oscillation is sustained.

$$A_{cl} = \frac{R_1 + R_2}{R_2} = \frac{30 \text{ k}\Omega}{10 \text{ k}\Omega} = 3$$

Practice Exercise What change is required in the oscillator in Figure 6–12 to produce an output with an amplitude of 6.8 V?

The Phase-Shift Oscillator

Figure 6–13 shows a type of sinusoidal oscillator called the *phase-shift oscillator*. Each of the three *RC* networks in the feedback loop can provide a maximum phase shift approaching 90°. Oscillation occurs at the frequency where the total phase shift through the three *RC* networks is 180°. The inversion of the op-amp itself provides the additional 180° to meet the requirement for oscillation of a 360° (or 0°) phase shift around the feedback loop.

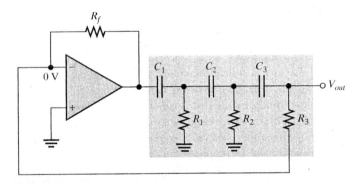

FIGURE 6–13
Op-amp phase-shift oscillator.

The attenuation *B* of the three-section *RC* feedback network is

$$B = \frac{1}{29} \qquad (6\text{–}3)$$

where $B = R_3/R_f$. The derivation of this unusual result is given in Appendix B. To meet the greater-than-unity loop gain requirement, the closed-loop voltage gain of the op-amp must be greater than 29 (set by R_f and R_3). The frequency of oscillation is also derived in Appendix B and stated in the following equation, where $R_1 = R_2 = R_3 = R$ and $C_1 = C_2 = C_3 = C$.

$$f_r = \frac{1}{2\pi\sqrt{6}RC} \qquad (6\text{–}4)$$

EXAMPLE 6–2

(a) Determine the value of R_f necessary for the circuit in Figure 6–14 to operate as an oscillator.

(b) Determine the frequency of oscillation.

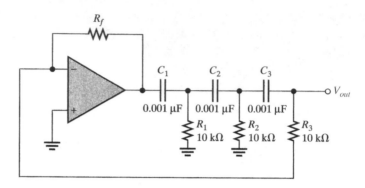

FIGURE 6–14

Solution

(a) $A_{cl} = 29$, and $B = \dfrac{1}{29} = \dfrac{R_3}{R_f}$. Therefore,

$$\frac{R_f}{R_3} = 29$$

$$R_f = 29R_3 = 29(10 \text{ k}\Omega) = \mathbf{290 \text{ k}\Omega}$$

(b) $R_1 = R_2 = R_3 = R$ and $C_1 = C_2 = C_3 = C$. Therefore,

$$f_r = \frac{1}{2\pi\sqrt{6}RC} = \frac{1}{2\pi\sqrt{6}(10 \text{ k}\Omega)(0.001 \text{ }\mu\text{F})} \cong \mathbf{6.5 \text{ kHz}}$$

Practice Exercise

(a) If R_1, R_2, and R_3 in Figure 6–14 are changed to 8.2 kΩ, what value must R_f be for oscillation?

(b) What is the value of f_r?

Twin-T Oscillator

Another type of *RC* oscillator is called the *twin-T* because of the two T-type *RC* filters used in the negative feedback loop, as shown in Figure 6–15(a). One of the twin-T filters has a low-pass response, and the other has a high-pass response. The combined parallel filters produce a band-stop or notch response with a center frequency equal to the desired frequency of oscillation, f_r, as shown in Figure 6–15(b).

Oscillation cannot occur at frequencies above or below f_r because of the negative feedback through the filters. At f_r, however, there is negligible negative feedback; thus, the positive feedback through the voltage divider (R_1 and R_2) allows the circuit to oscillate.

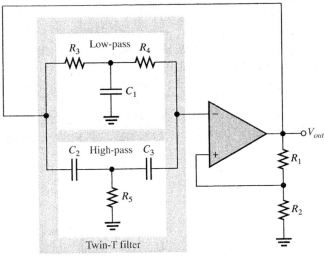

(a) Oscillator circuit

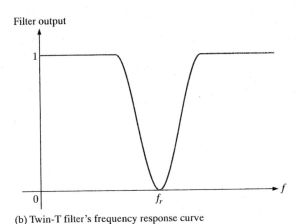

(b) Twin-T filter's frequency response curve

FIGURE 6–15
Twin-T oscillator and twin-T filter response.

6–3 REVIEW QUESTIONS

1. There are two feedback loops in the Wien-bridge oscillator. What is the purpose of each?

2. A certain lead-lag network has $R_1 = R_2$ and $C_1 = C_2$. An input voltage of 5 V rms is applied. The input frequency equals the resonant frequency of the network. What is the rms output voltage?

3. Why is the phase shift through the RC feedback network in a phase-shift oscillator equal to 180°.

6–4 ■ RELAXATION OSCILLATOR PRINCIPLES

The second major category of oscillators is the relaxation oscillator. Relaxation oscillators use an RC timing circuit and a device that changes states to generate a periodic waveform. In this section, you will learn about several circuits that are used to produce nonsinusoidal waveforms.

After completing this section, you should be able to

❑ Describe and analyze the operation of basic relaxation oscillators
  ❑ Discuss the operation of basic triangular-wave oscillators
  ❑ Discuss the operation of a voltage-controlled oscillator (VCO)
  ❑ Discuss the operation of a square-wave relaxation oscillator

A Triangular-Wave Oscillator

The op-amp integrator covered in Chapter 4 can be used as the basis for a triangular-wave generator. The basic idea is illustrated in Figure 6–16(a) where a dual-polarity, switched input is used. We use the switch only to introduce the concept; it is not a practical way to implement this circuit. When the switch is in position 1, the negative voltage is applied, and the output is a positive-going ramp. When the switch is thrown into position 2, a negative-going ramp is produced. If the switch is thrown back and forth at fixed intervals, the output is a triangular wave consisting of alternating positive-going and negative-going ramps, as shown in Figure 6–16(b).

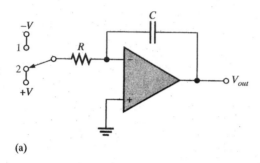

(a)

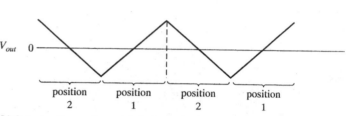

(b) Output voltage as the switch is thrown back and forth at regular intervals

FIGURE 6–16
Basic triangular-wave generator.

A Practical Triangular-Wave Circuit One practical implementation of a triangular-wave generator utilizes an op-amp comparator to perform the switching function, as shown in Figure 6–17. The operation is as follows. To begin, assume that the output voltage of the comparator is at its maximum negative level. This output is connected to the inverting input of the integrator through R_1, producing a positive-going ramp on the output of the integrator. When the ramp voltage reaches the upper trigger point (UTP), the comparator switches to its maximum positive level. This positive level causes the integrator ramp to change to a negative-going direction. The ramp continues in this direction until the lower trigger point (LTP) of the comparator is reached. At this point, the comparator output switches back to the maximum negative level and the cycle repeats. This action is illustrated in Figure 6–18.

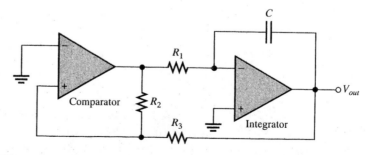

FIGURE 6–17
A triangular-wave generator using two op-amps.

FIGURE 6–18
Waveforms for the circuit in Figure 6–17.

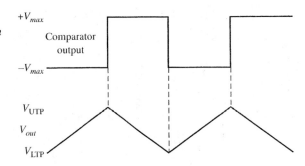

Since the comparator produces a square-wave output, the circuit in Figure 6–17 can be used as both a triangular-wave generator and a square-wave generator. Devices of this type are commonly known as *function generators* because they produce more than one output function. The output amplitude of the square wave is set by the output swing of the comparator, and resistors R_2 and R_3 set the amplitude of the triangular output by establishing the UTP and LTP voltages according to the following formulas:

$$V_{\text{UTP}} = +V_{max}\left(\frac{R_3}{R_2}\right)$$

$$V_{\text{LTP}} = -V_{max}\left(\frac{R_3}{R_2}\right)$$

where the comparator output levels, $+V_{max}$ and $-V_{max}$, are equal. The frequency of both waveforms depends on the R_1C time constant as well as the amplitude-setting resistors, R_2 and R_3. By varying R_1, the frequency of oscillation can be adjusted without changing the output amplitude.

$$f = \frac{1}{4R_1C}\left(\frac{R_2}{R_3}\right) \tag{6–5}$$

EXAMPLE 6–3 Determine the frequency of the circuit in Figure 6–19. To what value must R_1 be changed to make the frequency 20 kHz?

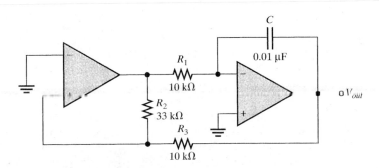

FIGURE 6–19

Solution

$$f = \frac{1}{4R_1C}\left(\frac{R_2}{R_3}\right) = \left(\frac{1}{4(10\ k\Omega)(0.01\ \mu F)}\right)\left(\frac{33\ k\Omega}{10\ k\Omega}\right) = \textbf{8.25 kHz}$$

To make $f = 20$ kHz,

$$R_1 = \frac{1}{4fC}\left(\frac{R_2}{R_3}\right) = \left(\frac{1}{4(20\ kHz)(0.01\ \mu F)}\right)\left(\frac{33\ k\Omega}{10\ k\Omega}\right) = \textbf{4.13 k}\Omega$$

Practice Exercise What is the amplitude of the triangular wave in Figure 6–19 if the comparator output is ± 10 V?

A Voltage-Controlled Sawtooth Oscillator (VCO)

The voltage-controlled oscillator (VCO) is an oscillator whose frequency can be changed by a variable dc control voltage. VCOs can be either sinusoidal or nonsinusoidal. One way to build a voltage-controlled sawtooth oscillator is with an op-amp integrator that uses a switching device (PUT) in parallel with the feedback capacitor to terminate each ramp at a prescribed level and effectively "reset" the circuit. Figure 6–20(a) shows the implementation.

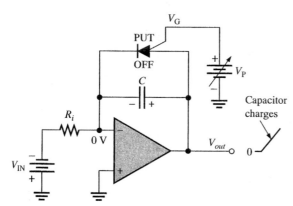

(a) Initially, the capacitor charges, the output ramp begins, and the PUT is off.

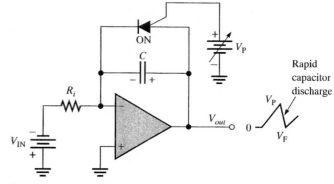

(b) The capacitor rapidly discharges when the PUT momentarily turns on.

FIGURE 6–20
Voltage-controlled sawtooth oscillator operation.

The PUT is a programmable unijunction transistor with an anode, a cathode, and a gate terminal. The gate is always biased positively with respect to the cathode. When the anode voltage exceeds the gate voltage by approximately 0.7 V, the PUT turns on and acts as a forward-biased diode. When the anode voltage falls below this level, the PUT turns off. Also, the current must be above the holding value to maintain conduction.

The operation of the sawtooth generator begins when the negative dc input voltage, $-V_{IN}$, produces a positive-going ramp on the output. During the time that the ramp is increasing, the circuit acts as a regular integrator. The PUT triggers on when the output ramp (at the anode) exceeds the gate voltage by 0.7 V. The gate is set to the approximate desired sawtooth peak voltage. When the PUT turns on, the capacitor rapidly discharges, as shown in Figure 6–20(b). The capacitor does not discharge completely to zero because of the PUT's forward voltage, V_F. Discharge continues until the PUT current falls below the holding value. At this point, the PUT turns off and the capacitor begins to charge again, thus generating a new output ramp. The cycle continually repeats, and the resulting output is a repetitive sawtooth waveform, as shown. The sawtooth amplitude and period can be adjusted by varying the PUT gate voltage.

The frequency is determined by the R_iC time constant of the integrator and the peak voltage set by the PUT. Recall that the charging rate of the capacitor is V_{IN}/R_iC. The time it takes the capacitor to charge from V_F to V_P is the period, T, of the sawtooth (neglecting the rapid discharge time).

$$T = \frac{V_P - V_F}{|V_{IN}|/R_iC}$$

From $f = 1/T$,

$$f = \frac{|V_{IN}|}{R_iC}\left(\frac{1}{V_P - V_F}\right) \tag{6–6}$$

EXAMPLE 6–4

(a) Find the amplitude and frequency of the sawtooth output in Figure 6–21. Assume that the forward PUT voltage, V_F, is approximately 1 V.

(b) Sketch the output waveform.

FIGURE 6–21

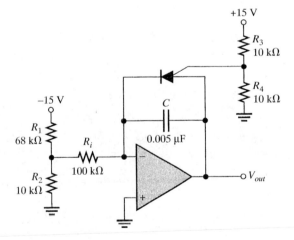

Solution

(a) First, find the gate voltage in order to establish the approximate voltage at which the PUT turns on.

$$V_G = \frac{R_4}{R_3 + R_4}(+V) = \frac{10\ k\Omega}{20\ k\Omega}(+15\ V) = 7.5\ V$$

This voltage sets the approximate maximum peak value of the sawtooth output (neglecting the 0.7 V).

$$V_P \cong 7.5\ V$$

The minimum peak value (low point) is

$$V_F \cong 1 \text{ V}$$

So the peak-to-peak amplitude is

$$V_{pp} = V_P - V_F = 7.5 \text{ V} - 1 \text{ V} = \mathbf{6.5 \text{ V}}$$

The period is determined as follows:

$$V_{IN} = \frac{R_2}{R_1 + R_2}(-V) = \frac{10 \text{ k}\Omega}{78 \text{ k}\Omega}(-15 \text{ V}) = -1.92 \text{ V}$$

$$f = \frac{|V_{IN}|}{R_i C}\left(\frac{1}{V_P - V_F}\right) = \left(\frac{1.92 \text{ V}}{(100 \text{ k}\Omega)(0.005 \text{ }\mu\text{F})}\right)\left(\frac{1}{7.5 \text{ V} - 1 \text{ V}}\right) \cong \mathbf{591 \text{ Hz}}$$

(b) The output waveform is shown in Figure 6–22. The period is

$$T = \frac{1}{f} = \frac{1}{591 \text{ Hz}} = 1.69 \text{ ms}$$

FIGURE 6–22
Output of the circuit in Figure 6–21.

Practice Exercise If R_i is changed to 56 kΩ in Figure 6–21, what is the frequency?

A Square-Wave Oscillator

The basic square-wave generator shown in Figure 6–23 is a type of relaxation oscillator because its operation is based on the charging and discharging of a capacitor. Notice that the op-amp's inverting ($-$) input is the capacitor voltage and the noninverting ($+$) input is a portion of the output fed back through resistors R_2 and R_3. When the circuit is first turned on, the capacitor is uncharged, and thus the inverting input is at 0 V. This makes the output

FIGURE 6–23
A square-wave relaxation oscillator.

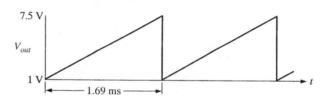

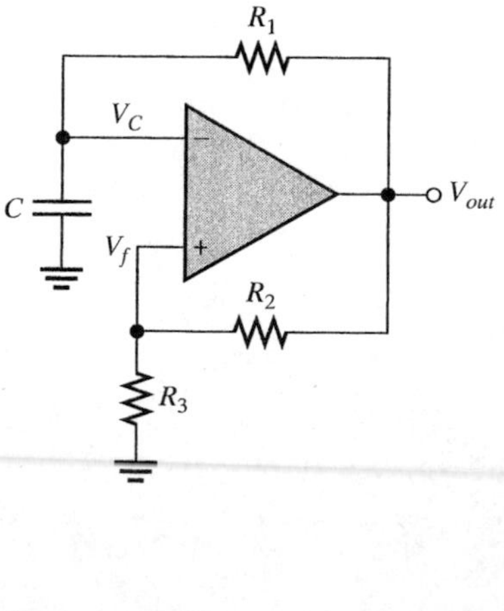

a positive maximum, and the capacitor begins to charge toward V_{out} through R_1. When the capacitor voltage (V_C) reaches a value equal to the feedback voltage (V_f) on the noninverting input, the op-amp switches to the maximum negative state. At this point, the capacitor begins to discharge from $+V_f$ toward $-V_f$. When the capacitor voltage reaches $-V_f$, the op-amp switches back to the maximum positive state. This action continues to repeat, as shown in Figure 6–24, and a square-wave output voltage is obtained.

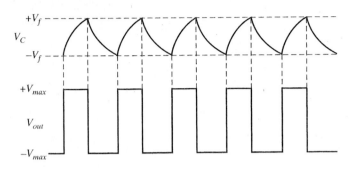

FIGURE 6–24
Waveforms for the square-wave relaxation oscillator.

6–4 REVIEW QUESTIONS

1. What is a VCO, and basically, what does it do?
2. Upon what principle does a relaxation oscillator operate?

6–5 ■ THE 555 TIMER AS AN OSCILLATOR

The 555 timer is a versatile integrated circuit with many applications. In this section, you will see how the 555 is configured as an astable or free-running multivibrator, which is essentially a square-wave oscillator. The use of the 555 timer as a voltage-controlled oscillator (VCO) is also discussed.

After completing this section, you should be able to

❑ Use a 555 timer in an oscillator application
   ❑ Discuss astable operation of the 555 timer
   ❑ Explain how to use the 555 timer as a VCO

Astable Operation

A 555 timer connected to operate as an astable **multivibrator,** which is a free-running non-sinusoidal oscillator that produces a pulse waveform on its output, is shown in Figure 6–25. Notice that the threshold input (THRESH) is now connected to the trigger input (TRIG). The external components R_1, R_2, and C_{ext} form the timing network that sets the frequency of oscillation. The 0.01 μF capacitor connected to the control (CONT) input is strictly for decoupling and has no effect on the operation.

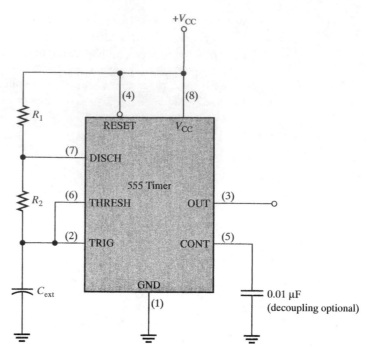

FIGURE 6–25
The 555 timer connected as an astable multivibrator.

The frequency of oscillation is given by Equation (6–7), or it can be found using the graph in Figure 6–26.

$$f = \frac{1.44}{(R_1 + 2R_2)C_{ext}} \tag{6–7}$$

By selecting R_1 and R_2, the duty cycle of the output can be adjusted. Since C_{ext} charges through $R_1 + R_2$ and discharges only through R_2, duty cycles approaching a mini-

FIGURE 6–26
Frequency of oscillation (free-running frequency) of a 555 timer in the astable mode as a function of C_{ext} and $R_1 + 2R_2$. The sloped lines are values of $R_1 + 2R_2$.

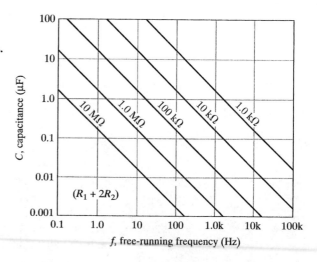

mum of 50 percent can be achieved if $R_2 >> R_1$ so that the charging and discharging times are approximately equal.

A formula to calculate the duty cycle is developed as follows. The time that the output is high (t_H) is expressed as

$$t_H = 0.693(R_1 + R_2)C_{ext}$$

The time that the output is low (t_L) is expressed as

$$t_L = 0.693R_2C_{ext}$$

The period, T, of the output waveform is the sum of t_H and t_L.

$$T = t_H + t_L = 0.693(R_1 + 2R_2)C_{ext}$$

This is the reciprocal of f in Equation (6–7). Finally, the duty cycle is

$$\text{Duty cycle} = \frac{t_H}{T} = \frac{t_H}{t_H + t_L}$$

$$\text{Duty cycle} = \left(\frac{R_1 + R_2}{R_1 + 2R_2}\right)100\% \tag{6–8}$$

To achieve duty cycles of less than 50 percent, the circuit in Figure 6–25 can be modified so that C_{ext} charges through only R_1 and discharges through R_2. This is achieved with a diode, D_1, placed as shown in Figure 6–27. The duty cycle can be made less than 50 percent by making R_1 less than R_2. Under this condition, the formula for the duty cycle is

$$\text{Duty cycle} = \left(\frac{R_1}{R_1 + R_2}\right)100\% \tag{6–9}$$

FIGURE 6–27

The addition of diode D_1 allows the duty cycle of the output to be adjusted to less than 50 percent by making $R_1 < R_2$.

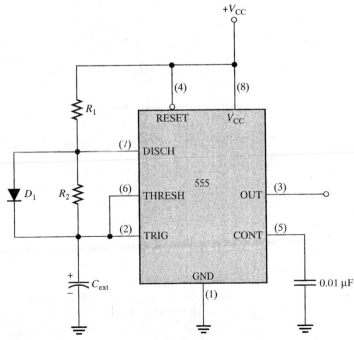

EXAMPLE 6–5   A 555 timer configured to run in the astable mode (oscillator) is shown in Figure 6–28. Determine the frequency of the output and the duty cycle.

FIGURE 6–28

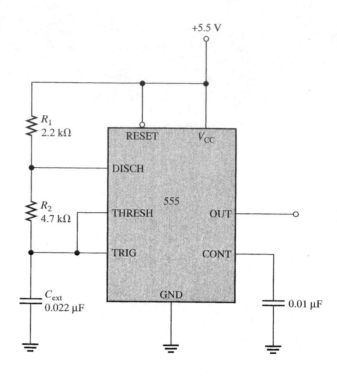

Solution

$$f = \frac{1.44}{(R_1 + 2R_2)C_{ext}} = \frac{1.44}{(2.2 \text{ k}\Omega + 9.4 \text{ k}\Omega)0.022 \text{ }\mu\text{F}} = \textbf{5.64 kHz}$$

$$\text{Duty cycle} = \left(\frac{R_1 + R_2}{R_1 + 2R_2}\right)100\% = \left(\frac{2.2 \text{ k}\Omega + 4.7 \text{ k}\Omega}{2.2 \text{ k}\Omega + 9.4 \text{ k}\Omega}\right)100\% = \textbf{59.5\%}$$

Practice Exercise   Determine the duty cycle in Figure 6–28 if a diode is connected across R_2 as indicated in Figure 6–27.

Operation as a Voltage-Controlled Oscillator (VCO)

A 555 timer can be set up to operate as a VCO by using the same external connections as for astable operation, with the exception that a variable control voltage is applied to the CONT input (pin 5), as indicated in Figure 6–29.

For the capacitor voltage, as shown in Figure 6–30, the upper value is V_{CONT} and the lower value is $\frac{1}{2}V_{CONT}$. When the control voltage is varied, the output frequency also varies. An increase in V_{CONT} increases the charging and discharging time of the external capacitor and causes the frequency to decrease. A decrease in V_{CONT} decreases the charging and discharging time of the capacitor and causes the frequency to increase.

An interesting application of the VCO is in phase-locked loops, which are used in various types of communications receivers to track variations in the frequency of incoming signals. You will learn about the basic operation of a phase-locked loop in Chapter 9.

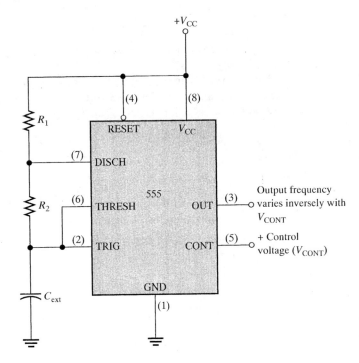

FIGURE 6–29

The 555 timer connected as a voltage-controlled oscillator (VCO). Note the variable control voltage input on pin 5.

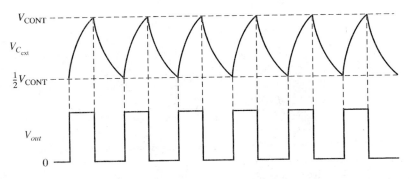

FIGURE 6–30

The VCO output frequency varies inversely with V_{CONT} because the charging and discharging time of C_{ext} is directly dependent on the control voltage.

6–5 REVIEW QUESTIONS

1. When the 555 timer is configured as an astable multivibrator, how is the duty cycle determined?

2. When the 555 timer is used as a VCO, how is the frequency varied?

6–6 ■ THE 555 TIMER AS A ONE-SHOT

A one-shot is a monostable multivibrator that produces a single output pulse for each input trigger pulse. The term monostable means that the device has only one stable state. When a one-shot is triggered, it temporarily goes to its unstable state but it always returns to its stable state. The time that it remains in its unstable state establishes the width of the output pulse and is set by the values of an external resistor and capacitor.

After completing this section, you should be able to

❑ Use a 555 timer as a one-shot device
   ❑ Discuss monostable operation
   ❑ Explain how to set the output pulse width

A 555 timer connected for **monostable** operation is shown in Figure 6–31. Compare this configuration to the one used for **astable** operation in Figure 6–25 and note the difference in the external circuit.

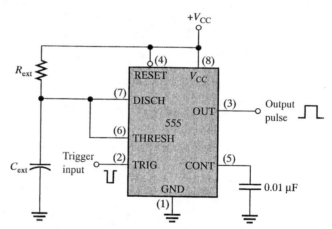

FIGURE 6–31
The 555 timer connected as a monostable multivibrator (one-shot).

Monostable Operation

A single input trigger pulse produces a single output pulse with a predetermined width. Once triggered, the one-shot cannot be retriggered until it completely times out; that is, it completes a full output pulse. Once it times out, the one-shot can then be triggered again to produce another output pulse. A low level on the reset input can be used to prematurely terminate the output pulse. The width of the output pulse is determined by the following formula:

$$t_W = 1.1R_{ext}C_{ext} \qquad\qquad (6\text{--}10)$$

The graph in Figure 6–32 shows various combinations of R_{ext} and C_{ext} and the associated output pulse widths. This graph can be used to select component values for a desired pulse width.

FIGURE 6–32
555 one-shot timing.

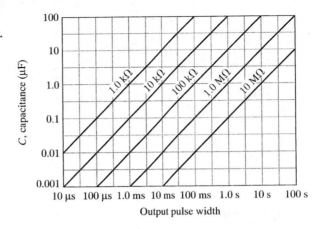

EXAMPLE 6–6

A 555 timer is connected as a one-shot with $R_{ext} = 10$ kΩ and $C_{ext} = 0.1$ μF. What is the pulse width of the output?

Solution   You can determine the pulse width in two ways. You can use either Equation (6–10) or the graph in Figure 6–32. Using the formula,

$$t_W = 1.1R_{ext}C_{ext} = 1.1(10 \text{ kΩ})(0.1 \text{ μF}) = \textbf{1.1 ms}$$

To use the graph, move along the $C = 0.1$ μF line until it intersects with the sloped line corresponding to $R = 10$ kΩ. At that point, project down to the horizontal axis and you get a pulse width of 1.1 ms as illustrated in Figure 6–33.

FIGURE 6–33

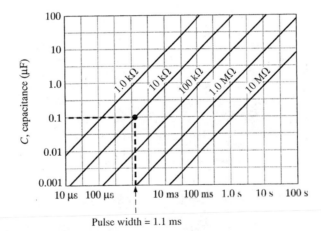

Pulse width = 1.1 ms

Practice Exercise   To what value must R_{ext} be changed to increase the one-shot's output pulse width to 5 ms?

Using One-Shots for Time Delay

In many applications, it is necessary to have a fixed time delay between certain events. Figure 6–34(a) shows two 555 timers connected as one-shots. The output of the first goes to the input of the second. When the first one-shot is triggered, it produces an output pulse

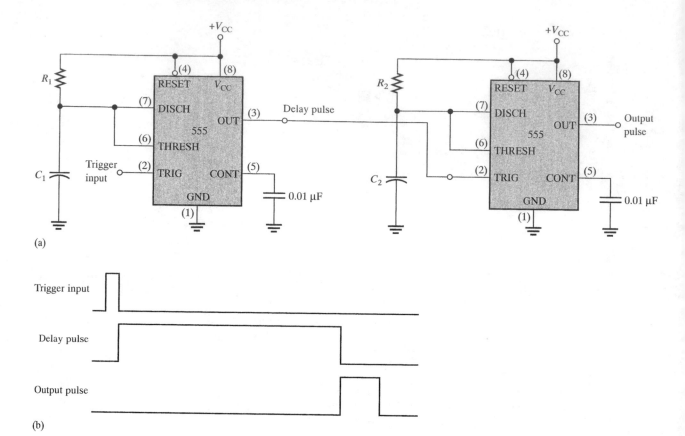

FIGURE 6–34
Two one-shots produce a delayed output pulse.

whose width establishes a time delay. At the end of this pulse, the second one-shot is triggered. Therefore, we have an output pulse from the second one-shot that is delayed from the input trigger to the first one-shot by a time equal to the pulse width of the first one-shot, as indicated in the timing diagram in Figure 6–34(b).

EXAMPLE 6–7   Determine the pulse widths and show the timing diagram (relationships of the input and output pulses) for the circuit in Figure 6–35.

Solution   The time relationship of the inputs and outputs are shown in Figure 6–36. The pulse widths for the two one-shots are

$$t_{W1} = 1.1R_1C_1 = 1.1(100 \text{ k}\Omega)(1.0 \text{ }\mu\text{F}) = \textbf{110 ms}$$
$$t_{W2} = 1.1R_2C_2 = 1.1(2.2 \text{ k}\Omega)(0.47 \text{ }\mu\text{F}) = \textbf{1.14 ms}$$

Practice Exercise   Suggest a way that the circuit in Figure 6–35 can be modified so that the delay can be made adjustable from 10 ms to 200 ms.

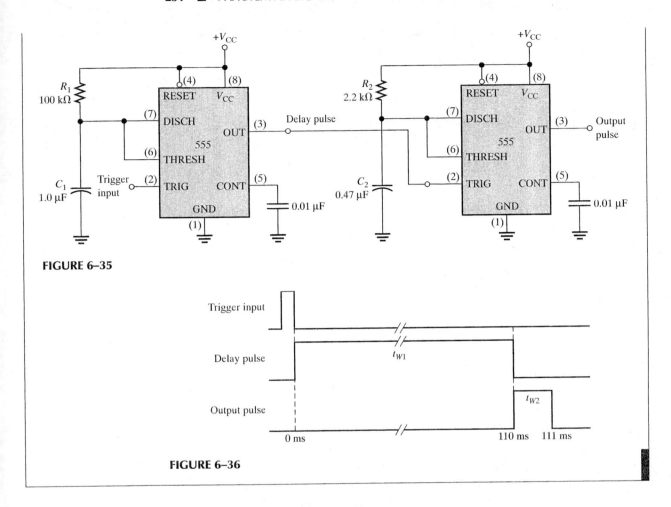

FIGURE 6–35

FIGURE 6–36

6–6 REVIEW QUESTIONS

1. How many stable states does a one-shot have?
2. A certain 555 one-shot circuit has a time constant of 5 ms. What is the output pulse width?
3. How can you decrease the pulse width of a one-shot?

6–7 ■ A SYSTEM APPLICATION

The function generator presented at the beginning of the chapter is a laboratory instrument used as a source for sine waves, square waves, and triangular waves.

After completing this section, you should be able to

❑ Apply what you have learned in this chapter to a system application
  ❑ Describe how an oscillator is used as a signal source
  ❑ State how the frequency and amplitude of the generated signal are varied
  ❑ Translate between printed circuit boards and a schematic

❑ Interconnect the front panel controls and two PC boards
❑ Troubleshoot some common system problems

A Brief Description of the System

The function generator in this system application produces either a sinusoidal wave, a square wave, or a triangular wave depending on the function selected by the front panel switches. The frequency of the selected waveform can be varied from less than 1 Hz to greater than 80 kHz using the range switches and the frequency dial. The amplitude of the output waveform can be adjusted up to approximately +10 V. Also, any dc offset voltage can be nulled out with the front panel dc offset control.

The system block diagram is shown in Figure 6–37. The concept of this particular function generator is very simple. The oscillator produces a sinusoidal output voltage that drives a zero-level detector (comparator) to produce a square wave of the same frequency as the oscillator output. The level detector output goes to an integrator, which generates a triangular output voltage also with the same frequency as the oscillator output.

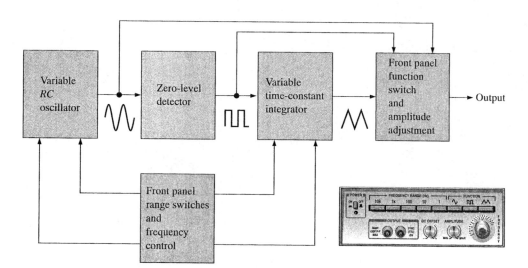

FIGURE 6–37
Function generator block diagram with front panel inset.

The schematic of the function generator is shown in Figure 6–38, where the portions in blue are the front panel components. The frequency of the sinusoidal oscillator is controlled by the selection of any two of ten capacitors (C_1 through C_{10}) in the oscillator feedback circuit. These capacitors produce the five frequency ranges indicated on the front panel switches, which are multiplication factors for the setting of the frequency dial. The adjustment of the frequency within each range is accomplished by varying resistors R_8 and R_9 in the feedback circuit of the oscillator.

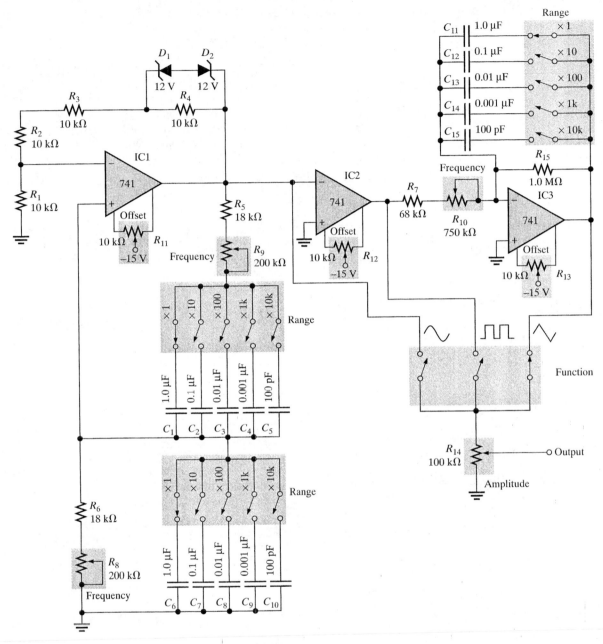

FIGURE 6–38

The integrator time constant is adjusted in step with the frequency by selection of the appropriate capacitor (C_{11} through C_{15}) and adjustment of resistor R_{10}. Resistors R_8, R_9, and R_{10} are potentiometers that are ganged together so that they change resistance together as the frequency dial is turned. For example, if the 1k switch is selected and the frequency dial is set at 5, then the resulting output frequency for any of the three types of waveforms is 1 kHz × 5 = 5 kHz.

Now, so that you can take a closer look at the oscillator boards, let's take them out of the system and put them on the test bench.

ON THE TEST BENCH

■ ACTIVITY 1   Relate the PC Boards to the Schematic

Locate and identify each component on the PC boards, shown in Figure 6–39, using the system schematic in Figure 6–38. The blue-shaded portions are front panel components and are not on the boards. The range switches are mechanically linked and the frequency rheostats are mechanically linked.

Develop a board-to-board wiring list specifying which pins on the two PC boards connect to each other and also indicate which pins go to the front panel.

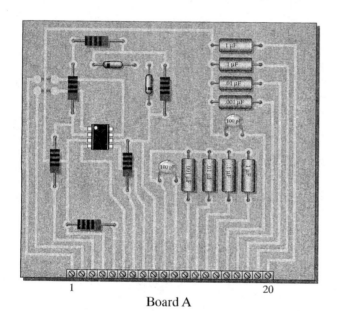

Board A

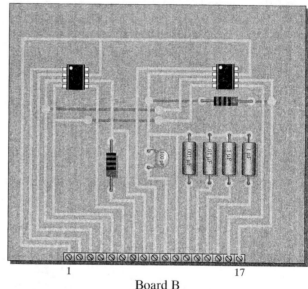

Board B

FIGURE 6–39

■ ACTIVITY 2   Analyze the System

Step 1:   Determine the maximum frequency of the oscillator for each range switch position ($\times 1$, $\times 10$, and so on). Only one set of three switches corresponding to a given range setting can be closed at a time. There is a set of three switches for $\times 1$, a set of three switches for $\times 10$, and so on.

Step 2:   Determine the minimum frequency of the oscillator for each range switch.

Step 3:   Determine the approximate maximum peak-to-peak output voltages for each function. The dc supply voltages are $+15$ V and -15 V.

■ ACTIVITY 3   Write a Technical Report

Describe the overall operation of the function generator. Specify how each circuit works and what its purpose is. Identify the type of oscillator circuit used. Explain how the function, frequency, and amplitude are selected. Use the results of Activity 2 where appropriate.

■ ACTIVITY 4 Troubleshoot the System for Each of the Following Problems by Stating the Probable Cause or Causes

1. There is a square wave output when a triangular wave output is selected and only when the ×1k range is selected.

2. There is no output on any function setting.

3. There is no output when the square or triangular function is selected, but the sinusoidal output is OK.

4. Both the sinusoidal and the square wave outputs are OK, but there is no triangular wave output.

6–7 REVIEW QUESTIONS

1. What type of oscillator is used in this function generator?
2. How many frequency ranges are available?
3. List the components that determine the output frequency.
4. What is the purpose of the zener diodes in the oscillator circuit?

■ SUMMARY

- Feedback oscillators operate with positive feedback.
- The two conditions for positive feedback are the phase shift around the feedback loop must be 0° and the voltage gain around the feedback loop must equal 1.
- For initial start-up, the loop gain must be greater than 1.
- Sinusoidal *RC* oscillators include the Wien-bridge, phase-shift, and twin-T.
- A relaxation oscillator uses an *RC* timing circuit and a device that changes states to generate a periodic waveform.
- The frequency in a voltage-controlled oscillator (VCO) can be varied with a dc control voltage.
- The 555 timer is an integrated circuit that can be used as an oscillator or as a one-shot by proper connection of external components.

■ GLOSSARY

These terms are included in the end-of-book glossary.

Astable Characterized by having no stable states; a type of oscillator.

Automatic gain control (AGC) A feedback system that reduces the gain for larger signals and increases the gain for smaller signals.

Feedback oscillator A type of oscillator that returns a fraction of output signal to the input with no net phase shift resulting in a reinforcement of the output signal.

Monostable Characterized by having one stable state.

Multivibrator A type of circuit that can operate as an oscillator or as a one-shot.

One-shot A monostable multivibrator.

Oscillator An electronic circuit that generates a periodic waveform to perform timing, control, or communications functions.

Relaxation oscillator A type of oscillator that uses an *RC* timing circuit to generate a nonsinusoidal waveform.

■ KEY FORMULAS

(6–1) $\dfrac{V_{out}}{V_{in}} = \dfrac{1}{3}$ Wien-bridge positive feedback attenuation

(6–2) $f_r = \dfrac{1}{2\pi RC}$ Wien-bridge frequency

(6–3) $B = \dfrac{1}{29}$ Phase-shift feedback attenuation

(6–4) $f_r = \dfrac{1}{2\pi\sqrt{6}RC}$ Phase-shift oscillator frequency

(6–5) $f = \dfrac{1}{4R_1 C}\left(\dfrac{R_2}{R_3}\right)$ Triangular wave generator frequency

(6–6) $f = \dfrac{|V_{IN}|}{R_i C}\left(\dfrac{1}{V_P - V_F}\right)$ Sawtooth VCO frequency

(6–7) $f = \dfrac{1.44}{(R_1 + 2R_2)C_{ext}}$ 555 astable frequency

(6–8)   Duty cycle $= \left(\dfrac{R_1 + R_2}{R_1 + 2R_2}\right)100\%$ 555 astable (duty cycle $\geq$ 50%)

(6–9)   Duty cycle $= \left(\dfrac{R_1}{R_1 + R_2}\right)100\%$ 555 astable (duty cycle $<$ 50%)

(6–10) $t_W = 1.1 R_{ext} C_{ext}$ 555 one-shot pulse width

■ SELF-TEST

1. An oscillator differs from an amplifier because
   (a) it has more gain
   (b) it requires no input signal
   (c) it requires no dc supply
   (d) it always has the same output

2. All oscillators are based on
   (a) positive feedback
   (b) negative feedback
   (c) the piezoelectric effect
   (d) high gain

3. One condition for oscillation is
   (a) a phase shift around the feedback loop of 180°
   (b) a gain around the feedback loop of one-third
   (c) a phase shift around the feedback loop of 0°
   (d) a gain around the feedback loop of less than one

4. A second condition for oscillation is
   (a) no gain around the feedback loop
   (b) a gain of one around the feedback loop
   (c) the attenuation of the feedback network must be one-third
   (d) the feedback network must be capacitive

5. In a certain oscillator, $A_v = 50$. The attenuation of the feedback network must be
   (a) 1   (b) 0.01   (c) 10   (d) 0.02

6. For an oscillator to properly start, the gain around the feedback loop must initially be
   (a) 1   (b) less than 1   (c) greater than 1   (d) equal to B

7. In a Wien-bridge oscillator, if the resistances in the feedback circuit are decreased, the frequency
   (a) decreases   (b) increases   (c) remains the same

8. The Wien-bridge oscillator's positive feedback circuit is
   (a) an RL network   (b) an LC network
   (c) a voltage divider   (d) a lead-lag network

9. A phase-shift oscillator has
   (a) three RC networks   (b) three LC networks
   (c) a T-type network   (d) a π-type network

10. An oscillator whose frequency is changed by a variable dc voltage is known as
    (a) a Wien-bridge oscillator   (b) a VCO
    (c) a phase-shift oscillator   (d) an astable multivibrator

11. Which one of the following is not an input or output of the 555 timer?
    (a) Threshold   (b) Control voltage   (c) Clock
    (d) Trigger   (e) Discharge   (f) Reset

12. An astable multivibrator is
    (a) an oscillator   (b) a one-shot
    (c) a time-delay circuit   (d) characterized by having no stable states
    (e) answers (a) and (d)

13. The output frequency of a 555 timer connected as an oscillator is determined by
    (a) the supply voltage   (b) the frequency of the trigger pulses
    (c) the external RC time constant (d) the internal RC time constant
    (e) answers (a) and (d)

14. The term *monostable* means
    (a) one output   (b) one frequency
    (c) one time constant   (d) one stable state

15. A 555 timer connected as a one-shot has $R_{ext} = 2.0$ kΩ and $C_{ext} = 2.0$ μF. The output pulse has a width of
    (a) 1.1 ms   (b) 4 ms   (c) 4 μs   (d) 4.4 ms

■ PROBLEMS

SECTION 6–1   The Oscillator

1. What type of input is required for an oscillator?

2. What are the basic components of an oscillator circuit?

SECTION 6–2   Feedback Oscillator Principles

3. If the voltage gain of the amplifier portion of an oscillator is 75, what must be the attenuation of the feedback circuit to sustain the oscillation?

4. Generally describe the change required in the oscillator of Problem 3 in order for oscillation to begin when the power is initially turned on.

SECTION 6–3   Sinusoidal Oscillators

5. A certain lead-lag network has a resonant frequency of 3.5 kHz. What is the rms output voltage if an input signal with a frequency equal to f_r and with an rms value of 2.2 V is applied to the input?

6. Calculate the resonant frequency of a lead-lag network with the following values: $R_1 = R_2 = 6.2$ kΩ, and $C_1 = C_2 = 0.02$ μF.

7. Determine the necessary value of R_2 in Figure 6–40 so that the circuit will oscillate. Neglect the forward resistance of the zener diodes.

8. Explain the purpose of R_3 in Figure 6–40.

9. What is the initial closed-loop gain in Figure 6–40? At what value of output voltage does A_{cl} change and to what value does it change? (The value of R_2 was found in Problem 7.)

10. Find the frequency of oscillation for the Wien-bridge oscillator in Figure 6–40.

11. What value of R_f is required in Figure 6–41? What is f_r?

FIGURE 6–40

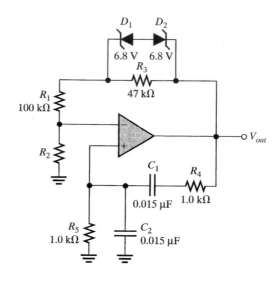

FIGURE 6–41

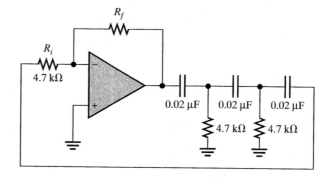

SECTION 6–4   Relaxation Oscillator Principles

12. What type of signal does the circuit in Figure 6–42 produce? Determine the frequency of the output.

13. Show how to change the frequency of oscillation in Figure 6–42 to 10 kHz.

FIGURE 6–42

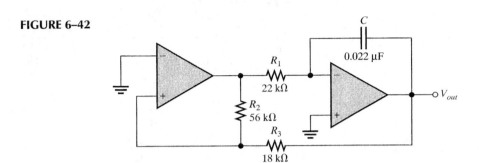

14. Determine the amplitude and frequency of the output voltage in Figure 6–43. Use 1 V as the forward PUT voltage.

15. Modify the sawtooth generator in Figure 6–43 so that its peak-to-peak output is 4 V.

16. A certain sawtooth generator has the following parameter values: $V_{IN} = 3$ V, $R = 4.7$ kΩ, $C = 0.001$ μF, and V_F for the PUT is 1.2 V. Determine its peak-to-peak output voltage if the period is 10 μs.

FIGURE 6–43

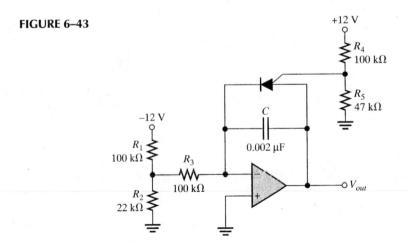

SECTION 6–5   The 555 Timer as an Oscillator

17. What are the two comparator reference voltages in a 555 timer when $V_{CC} = 10$ V?

18. Determine the frequency of oscillation for the 555 astable oscillator in Figure 6–44.

19. To what value must C_{ext} be changed in Figure 6–44 to achieve a frequency of 25 kHz?

20. In an astable 555 configuration, the external resistor $R_1 = 3.3$ kΩ. What must R_2 equal to produce a duty cycle of 75 percent?

FIGURE 6–44

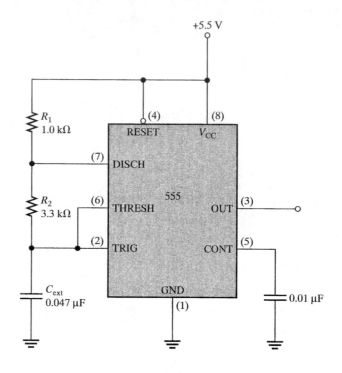

SECTION 6–6   The 555 Timer as a One-Shot

21. A 555 timer connected in the monostable configuration has a 56 kΩ external resistor and a 0.22 μF external capacitor. What is the pulse width of the output?

22. The output pulse width of a certain 555 one-shot is 12 ms. If $C_{ext} = 2.2$ μF, what is R_{ext}?

23. Suppose that you need to hook up a 555 timer as a one-shot in the lab to produce an output pulse with a width of 100 μs. Select the appropriate values for the external components.

24. Devise a circuit to produce two sequential 50 μs pulses. The first pulse must occur 100 ms after an initial trigger and the second pulse must occur 300 ms after the first pulse.

■ ANSWERS TO REVIEW QUESTIONS

Section 6–1

1. An oscillator is a circuit that produces a repetitive output waveform with only the dc supply voltage as an input.

2. Positive feedback

3. The feedback network provides attenuation and phase shift.

Section 6–2

1. Zero phase shift and unity voltage gain around the closed feedback

2. Positive feedback is when a portion of the output signal is fed back to the input of the amplifier such that it reinforces itself.

3. Loop gain greater than 1; zero phase shift and unity voltage gain

Section 6–3

1. The negative feedback loop sets the closed-loop gain; the positive feedback loop sets the frequency of oscillation.

2. 1.67 V

3. The three RC networks each contribute 60°.

Section 6–4

1. A voltage-controlled oscillator exhibits a frequency that can be varied with a dc control voltage.

2. The basis of a relaxation oscillator is the charging and discharging of a capacitor.

Section 6–5

1. The duty cycle is set by the external resistors and the external capacitor.

2. The frequency of a VCO is varied by changing V_{CONT}.

Section 6–6

1. A one-shot has one stable state.

2. $t_W = 5.5$ ms

3. The pulse width can be decreased by decreasing the external resistance or capacitance.

Section 6–7

1. A Wien-bridge oscillator

2. There are five frequency ranges.

3. R_5, R_6, R_8, R_9, C_1–C_5, C_6–C_{10}

4. To limit the oscillator input amplitude and to help ensure start-up

---

■ ANSWERS TO PRACTICE EXERCISES FOR EXAMPLES

6–1	Change the zener diodes to 6.1 V devices.
6–2	(a) 238 kΩ (b) 7.92 kHz
6–3	6.06 V peak-to-peak
6–4	1055 Hz
6–5	31.9%
6–6	45.5 kΩ
6–7	Replace R_1 with a potentiometer with a maximum resistance of at least 182 kΩ.

7

VOLTAGE REGULATORS

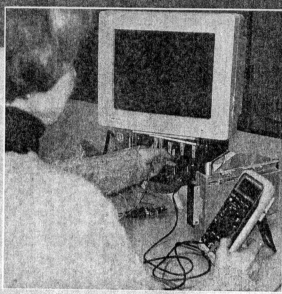

Courtesy Sunsweet Growers

■ CHAPTER OBJECTIVES

☐ Describe line and load regulation
☐ Discuss the principles of series voltage regulators
☐ Discuss the principles of shunt voltage regulators
☐ Discuss the principles of switching regulators
☐ Discuss integrated circuit voltage regulators
☐ Discuss applications of IC voltage regulators
☐ Apply what you have learned in this chapter to a system application

A voltage **regulator** provides a constant dc output voltage that is practically independent of the input voltage, output load current, and temperature. The voltage regulator is one part of a power supply. Its input voltage comes from the filtered output of a rectifier derived from an ac voltage or from a battery in the case of portable systems.

Most voltage regulators fall into two broad categories—linear regulators and switching regulators. In the linear regulator category, two general types are the linear series regulator and the linear shunt regulator. These are normally available for either positive or negative output voltages. A dual regulator provides both positive and negative outputs. In the switching regulator category, three general configurations are step-down, step-up, and inverting.

Many types of integrated circuit (IC) regulators are available. The most popular types of linear regulator are the three-terminal fixed voltage regulator and the three-terminal adjustable voltage regulator. Switching regulators are also widely used. In this chapter, specific IC devices are introduced as representative of the wide range of available devices.

A dual-polarity regulated power supply is used for the FM stereo system that you worked with in Chapter 5. Two regulators, one positive and the other negative, provide the positive voltage required for the receiver circuits and the dual polarity voltages for the op-amp circuits. The regulator input voltages come from a full-wave rectifier with filtered outputs.

For the system application in Section 7–7, in addition to the other topics, be sure you understand

☐ How three-terminal fixed-voltage regulators are used

☐ The basic operation of a power supply rectifier and filter

☐ How to set the current limit of a regulator

☐ How to determine power dissipation in a pass transistor

7–1 ■ VOLTAGE REGULATION

The requirement for a reliable source of constant voltage in virtually all electronic systems has led to many advances in power supply design. Designers have used feedback and operational amplifiers, as well as pulse circuit techniques to develop reliable constant-voltage (and constant-current) power supplies. The heart of any regulated supply is the ability to establish a constant-voltage reference. In this section, you will learn about line and load regulation.

After completing this section, you should be able to

❑ Describe line and load voltage regulation
    ❑ Express line regulation as either a percentage or as a percentage per volt
    ❑ Calculate line regulation
    ❑ Express load regulation as either a percentage or as a percentage per milliamp
    ❑ Calculate load regulation from either voltage data or resistance data

Line Regulation

Line regulation is a measure of the ability of a power supply to maintain a constant output for changes in the input voltage. It is typically defined as a ratio of a change in output for a corresponding change in the input and expressed as a percentage.

$$\text{Line regulation} = \left(\frac{\Delta V_{\text{OUT}}}{\Delta V_{\text{IN}}}\right)100\% \tag{7–1}$$

Some specification sheets show line regulation differently. It can be specified as a percentage change in the output voltage per volt divided by change in the input voltage. In this case, line regulation is defined and expressed as a percentage as

$$\text{Line regulation} = \left(\frac{\Delta V_{\text{OUT}}/V_{\text{OUT}}}{\Delta V_{\text{IN}}}\right)100\% \tag{7–2}$$

Because this definition is different, you need to be sure which definition is used when reading specifications. The key in a specification sheet is to look at the units. If the specification is a ratio of mV/V or other pure number, then Equation (7–1) is the defining equation. If the units are shown as %/mV or %/V, then Equation (7–2) is the defining equation.

EXAMPLE 7–1

When the input to a particular voltage regulator decreases by 5 V, the output decreases by 0.25 V. The nominal output is 15 V. Determine the line regulation expressed as a percentage and in units of %/V.

Solution From Equation (7–1), the percent line regulation is

$$\text{Line regulation} = \left(\frac{\Delta V_{\text{OUT}}}{\Delta V_{\text{IN}}}\right)100\% = \left(\frac{0.25\text{ V}}{5\text{ V}}\right)100\% = \mathbf{5\%}$$

From Equation (7–2), the percent line regulation is

$$\text{Line regulation} = \left(\frac{\Delta V_{\text{OUT}}/V_{\text{OUT}}}{\Delta V_{\text{IN}}}\right)100\% = \left(\frac{0.25\text{ V}/15\text{ V}}{5\text{ V}}\right)100\% = \mathbf{0.33\%/V}$$

Practice Exercise The input of a certain regulator increases by 3.5 V. As a result, the output voltage increases by 0.42 V. The nominal output is 20 V. Determine the regulation expressed as a percentage and in units of %/V.

Load Regulation

Load regulation is a measure of the change in output voltage for a given change in load current. When the amount of current through a load changes due to a varying load resistance, the voltage regulator must maintain a nearly constant output voltage across the load. The percent load regulation specifies how much change occurs in the output voltage over a certain range of load current values, usually from minimum current (no load, NL) to maximum current (full load, FL). Ideally, the percent load regulation is 0%. It can be calculated and expressed as a percentage with the following formula:

$$\text{Load regulation} = \left(\frac{V_{\text{NL}} - V_{\text{FL}}}{V_{\text{FL}}}\right)100\% \tag{7–3}$$

where V_{NL} is the output voltage with no load, and V_{FL} is the output voltage with full (maximum) load. Equation (7–3) is expressed as a change due only to changes in load conditions; all other factors (such as input voltage and operating temperature) must remain constant. Normally, the operating temperature is specified as 25°C.

Sometimes power supply manufacturers specify the equivalent output resistance of a power supply (R_{OUT}) instead of its load regulation. Recall (Section 1–3) that an equivalent Thevenin circuit can be drawn for any two-terminal linear circuit. Figure 7–1 shows the equivalent Thevenin circuit for a power supply with a load resistor. The Thevenin voltage is the voltage from the supply with no load (V_{NL}), and the Thevenin resistance is the specified output resistance, R_{OUT}. Ideally, R_{OUT} is zero, corresponding to 0% load regulation, but in practical power supplies R_{OUT} is a small value. With the load resistor in place, the output voltage is found by applying the voltage-divider rule:

$$V_{\text{OUT}} = V_{\text{NL}}\left(\frac{R_L}{R_{\text{OUT}} + R_L}\right)$$

FIGURE 7–1
Thevenin equivalent circuit for a power supply with a load resistor.

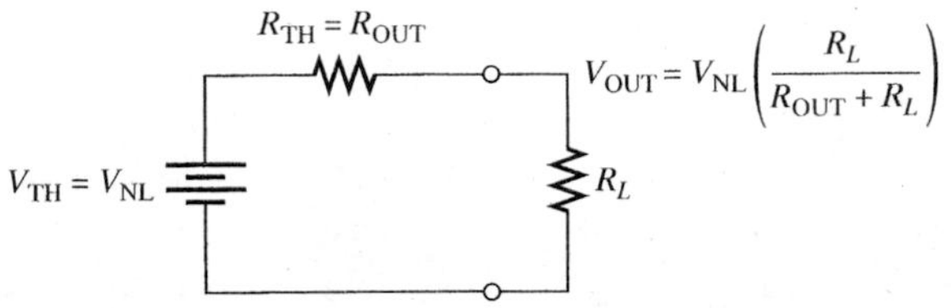

If we let R_{FL} equal the smallest-rated load resistance (largest-rated current), then the full-load output voltage (V_{FL}) is

$$V_{FL} = V_{NL}\left(\frac{R_{FL}}{R_{OUT} + R_{FL}}\right)$$

By rearranging and substituting into Equation (7–3),

$$V_{NL} = V_{FL}\left(\frac{R_{OUT} + R_{FL}}{R_{FL}}\right)$$

$$\text{Load regulation} = \frac{V_{FL}\left(\dfrac{R_{OUT} + R_{FL}}{R_{FL}}\right) - V_{FL}}{V_{FL}} \times 100\% = \left(\frac{R_{OUT} + R_{FL}}{R_{FL}} - 1\right)100\%$$

$$\text{Load regulation} = \left(\frac{R_{OUT}}{R_{FL}}\right)100\% \qquad (7\text{--}4)$$

Equation (7–4) is a useful way of finding the percent load regulation when the output resistance and minimum load resistance are specified.

Alternately, the load regulation can be expressed as a percentage change in output voltage for each mA change in load current. For example, a load regulation of 0.01%/mA means that the output voltage changes 0.01 percent when the load current increases or decreases by 1 mA.

EXAMPLE 7–2

A certain voltage regulator has a +12.1 V output when there is no load ($I_L = 0$) and has a rated output current of 200 mA. With maximum current, the output voltage drops to +12.0 V. Determine the percentage load regulation and find the percent load regulation per mA change in load current.

Solution The no-load output voltage is

$$V_{NL} = 12.1 \text{ V}$$

The full-load output voltage is

$$V_{FL} = 12.0 \text{ V}$$

The percent load regulation is

$$\text{Load regulation} = \left(\frac{V_{NL} - V_{FL}}{V_{FL}}\right)100\% = \left(\frac{12.1 \text{ V} - 12.0 \text{ V}}{12.0 \text{ V}}\right)100\% = \mathbf{0.83\%}$$

The load regulation can also be expressed as

$$\text{Load regulation} = \frac{0.83\%}{200 \text{ mA}} = \mathbf{0.0042\%/mA}$$

Practice Exercise Prove that the results of this example are consistent with a specified output resistance of 0.5 Ω.

7–1 REVIEW QUESTIONS

1. Define *line regulation.*
2. Define *load regulation.*
3. The input of a certain regulator increases by 3.5 V. As a result, the output voltage increases by 0.042 V. The nominal output is 20 V. Determine the line regulation in both % and in %/V.
4. If a 5.0 V power supply has an output resistance of 80 mΩ and a specified maximum output current of 1.0 A, what is the load regulation? Give the result as a % and as a %/mA.

7–2 ■ BASIC SERIES REGULATORS

The fundamental classes of voltage regulators are linear regulators and switching regulators. Both of these are available in integrated circuit form. There are two basic types of linear regulator. One is the series regulator and the other is the shunt regulator. In this section, we will look at the series regulator. The shunt and switching regulators are covered in the next two sections.

After completing this section, you should be able to

❑ Discuss the principles of series voltage regulators
   ❑ Explain regulating action
   ❑ Calculate output voltage of an op-amp series regulator
   ❑ Discuss overload protection and explain how to use current limiting
   ❑ Describe a regulator with fold-back current limiting

A simple representation of a series type of linear regulator is shown in Figure 7–2(a), and the basic components are shown in the block diagram in Figure 7–2(b). Notice that the control element is in series with the load between input and output. The output sample cir-

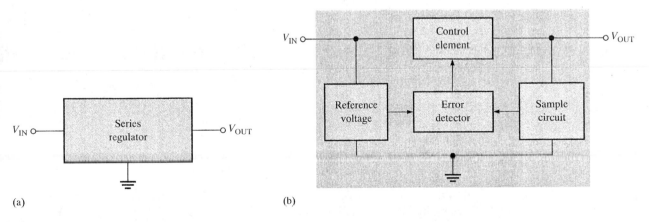

(a)

(b)

FIGURE 7–2
Simple series voltage regulator block diagram.

cuit senses a change in the output voltage. The error detector compares the sample voltage with a reference voltage and causes the control element to compensate in order to maintain a constant output voltage.

Voltage References

The ability of a voltage regulator to provide a constant output is dependent on the stability of a voltage reference to maintain a constant voltage for any change in temperature or other condition. Traditionally zener diodes were used as references and are shown in many of the circuits in this chapter. Zeners are designed to break down at a specific voltage and maintain a fairly constant voltage if the current in the zener is constant and the temperature does not change. The drawback to zener diodes is they tend to be noisy and the zener voltage may change slightly as the zener ages (this is called *drift*). An even more serious effect is that the zener voltage is sensitive to temperature changes; the zener voltage can change hundreds of parts per million (ppm) for a change of just 1°C in temperature. This temperature effect varies widely among different types of zeners.

Special zener diode ICs have been designed to serve as references with very low temperature drift (less than 10 ppm/°C). For low-voltage applications, zener-diode references are available that look and behave as diodes (but actually contain circuits to enhance their specifications). In the 8 V to 12 V range, two-terminal devices such as the LM329, LM399, and LM369 provide high stability and low noise, and have excellent temperature stability. The circuit symbol and internal construction of a representative IC reference is shown in Figure 7–3. The reference shown is called a *bandgap reference*. It is designed so that positive and negative temperature coefficients cancel, producing a reference with almost no temperature coefficient. It uses a current mirror (Q_1) to set a particular current in Q_2. The output of the reference is the sum of V_{BE} (from Q_3) and the voltage drop across R_2 (V_{R2}).

FIGURE 7–3
An IC reference. The reference shown is a bandgap type that has a very small temperature coefficient.

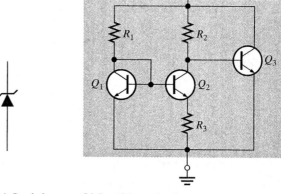

$V_{REF} = V_{BE} + V_{R2}$

(a) Symbol   (b) Internal construction

A more complicated voltage reference is the REF10 precision-voltage reference (from Burr-Brown). It is laser trimmed to 1 ppm/°C over its full range. It is a 10.00 V reference that is within 5 mV of this value. It uses a zener diode and op-amp in an 8-pin package.

Regulating Action

A basic op-amp series regulator circuit is shown in Figure 7–4. The operation of the series regulator is illustrated in Figure 7–5. The resistive voltage divider formed by R_2 and R_3 senses any change in the output voltage.

FIGURE 7–4
Basic op-amp series regulator.

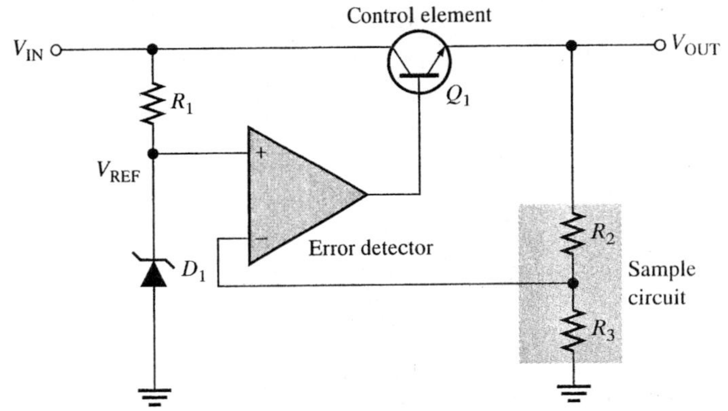

Figure 7–5(a) illustrates what happens when the output tries to decrease because of a decrease in V_{IN} or because of a change in load current. A proportional voltage decrease is applied to the op-amp's inverting input by the voltage divider. Since the zener diode (D_1) holds the other op-amp input at a nearly fixed reference voltage, V_{REF}, a small difference voltage (error voltage) is developed across the op-amp's inputs. This difference voltage is amplified, and the op-amp's output voltage increases. For highest accuracy, D_1 is replaced with an IC reference. This increase is applied to the base of Q_1, causing the emitter voltage V_{OUT} to increase until the voltage to the inverting input again equals the reference (zener) voltage. This action offsets the attempted decrease in output voltage, thus keeping it nearly constant, as shown in part (b). The power transistor, Q_1, is used with a heat sink because it must handle all of the load current.

The opposite action occurs when the output tries to increase, as indicated in Figure 7–5(c) and (d). The op-amp in the series regulator is actually connected as a noninverting amplifier where the reference voltage V_{REF} is the input at the noninverting terminal, and the R_2/R_3 voltage divider forms the negative feedback network. The closed-loop voltage gain is

$$A_{cl} = 1 + \frac{R_2}{R_3}$$

Therefore, the regulated output voltage of the series regulator is

$$V_{OUT} \simeq \left(1 + \frac{R_2}{R_3}\right)V_{REF} \tag{7–5}$$

From this analysis, you can see that the output voltage is determined by the zener voltage (V_{REF}) and the resistors R_2 and R_3. It is relatively independent of the input voltage, and therefore, regulation is achieved (as long as the input voltage and load current are within specified limits).

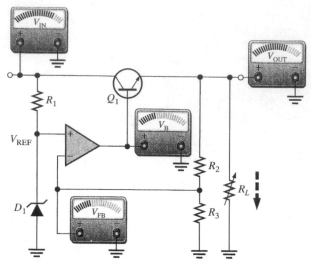

(a) When V_{IN} or R_L decreases, V_{OUT} attempts to decrease. The feedback voltage, V_{FB}, also attempts to decrease, and as a result, the op-amp's output voltage V_B attempts to increase, thus compensating for the attempted decrease in V_{OUT} by increasing the Q_1 emitter voltage. Changes in V_{OUT} are exaggerated for illustration.

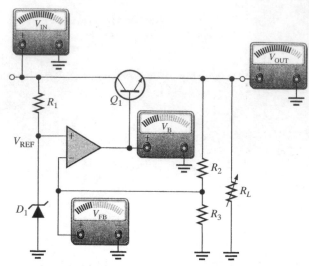

(b) When V_{IN} (or R_L) stabilizes at its new lower value, the voltages return to their original values, thus keeping V_{OUT} constant as a result of the negative feedback.

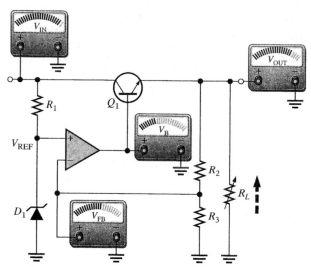

(c) When V_{IN} or R_L increases, V_{OUT} attempts to increase. The feedback voltage, V_{FB}, also attempts to increase, and as a result, V_B, applied to the base of the control transistor, attempts to decrease, thus compensating for the attempted increase in V_{OUT} by decreasing the Q_1 emitter voltage.

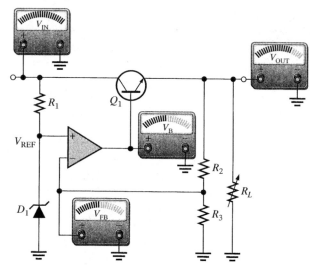

(d) When V_{IN} (or R_L) stabilizes at its new higher value, the voltages return to their original values, thus keeping V_{OUT} constant as a result of the negative feedback.

FIGURE 7–5

Illustration of series regulator action that keeps V_{OUT} constant when V_{IN} or R_L changes.

EXAMPLE 7–3

Determine the output voltage for the regulator in Figure 7–6 and the base voltage of Q_1.

FIGURE 7–6

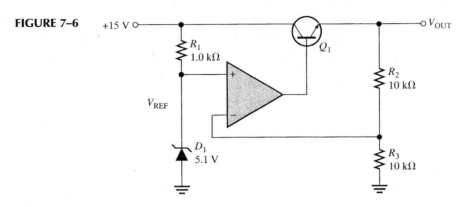

Solution $V_{REF} = 5.1$ V, the zener voltage. The regulated output voltage is therefore

$$V_{OUT} = \left(1 + \frac{R_2}{R_3}\right)V_{REF} = \left(1 + \frac{10\ k\Omega}{10\ k\Omega}\right)5.1\ V = (2)5.1\ V = \mathbf{10.2\ V}$$

The base voltage of Q_1 is

$$V_B = 10.2\ V + V_{BE} = 10.2\ V + 0.7\ V = \mathbf{10.9\ V}$$

Practice Exercise   The following changes are made in the circuit in Figure 7–6: A 3.3 V zener replaces the 5.1 V zener, $R_1 = 1.8$ kΩ, $R_2 = 22$ kΩ, and $R_3 = 18$ kΩ. What is the output voltage?

Short-Circuit or Overload Protection

If an excessive amount of load current is drawn, the series-pass transistor can be quickly damaged or destroyed. Most regulators use some type of current-limiting mechanism. Figure 7–7 shows one method of current limiting to prevent overloads called *constant-current limiting*. The current-limiting circuit consists of transistor Q_2 and resistor R_4.

FIGURE 7–7
Series regulator with constant-current limiting.

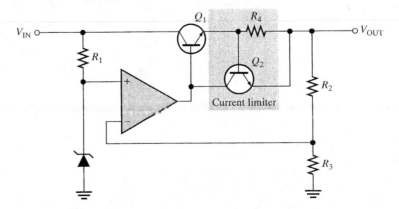

The load current through R_4 creates a voltage from base to emitter of Q_2. When I_L reaches a predetermined maximum value, the voltage drop across R_4 is sufficient to forward-bias the base-emitter junction of Q_2, thus causing it to conduct. Enough Q_1 base current is diverted into the collector of Q_2 so that I_L is limited to its maximum value $I_{L(max)}$. Since the base-to-emitter voltage of Q_2 cannot exceed about 0.7 V, the voltage across R_4 is held to this value, and the load current is limited to

$$I_{L(max)} = \frac{0.7 \text{ V}}{R_4} \tag{7-6}$$

EXAMPLE 7–4

Determine the maximum current that the regulator in Figure 7–8 can provide to a load.

FIGURE 7–8

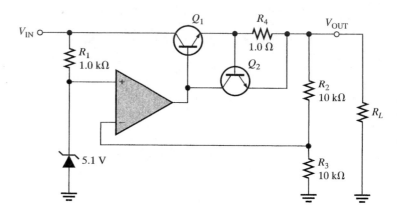

Solution

$$I_{L(max)} = \frac{0.7 \text{ V}}{R_4} = \frac{0.7 \text{ V}}{1.0 \ \Omega} = \textbf{0.7 A}$$

Practice Exercise If the output of the regulator in Figure 7–8 is shorted to ground, what is the current?

Regulator with Fold-Back Current Limiting

In the previous current-limiting technique, the current is restricted to a maximum constant value. **Fold-back current limiting** is a method used particularly in high-current regulators whereby the output current under overload conditions drops to a value well below the peak load current capability to prevent excessive power dissipation.

Basic Idea The basic concept of fold-back current limiting is as follows, with reference to Figure 7–9. The circuit is similar to the constant current-limiting arrangement in Figure 7–7, with the exception of resistors R_5 and R_6. The voltage drop developed across R_4 by the load current must not only overcome the base-emitter voltage required to turn on Q_2, but it must also overcome the voltage across R_5. That is, the voltage across R_4 must be

$$V_{R4} = V_{R5} + V_{BE}$$

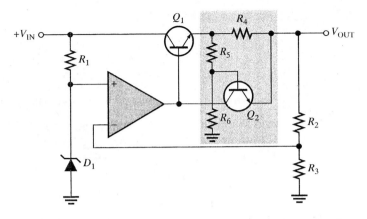

FIGURE 7–9
Series regulator with fold-back current limiting.

In an overload or short-circuit condition, the load current increases to a value, $I_{L(max)}$, that is sufficient to cause Q_2 to conduct. At this point the current can increase no further. The decrease in output voltage results in a proportional decrease in the voltage across R_5; thus less current through R_4 is required to maintain the forward-biased condition of Q_1. So, as V_{OUT} decreases, I_L decreases, as shown in the graph of Figure 7–10.

The advantage of this technique is that the regulator is allowed to operate with peak load current up to $I_{L(max)}$; but when the output becomes shorted, the current drops to a lower value to prevent overheating of the device.

FIGURE 7–10
Fold-back current limiting (output voltage versus load current).

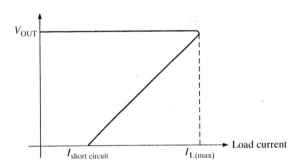

7-3 ■ BASIC SHUNT REGULATORS

The second basic type of linear voltage regulator is the shunt regulator. As you have learned, the control element in the series regulator is the series-pass transistor. In the shunt regulator, the control element is a transistor in parallel (shunt) with the load.

After completing this section, you should be able to

❑ Discuss the principles of shunt voltage regulators
   ❑ Describe the operation of a basic op-amp shunt regulator
   ❑ Compare series and shunt regulators

A simple representation of a shunt type of linear regulator is shown in Figure 7–11(a), and the basic components are shown in the block diagram in part (b) of the figure.

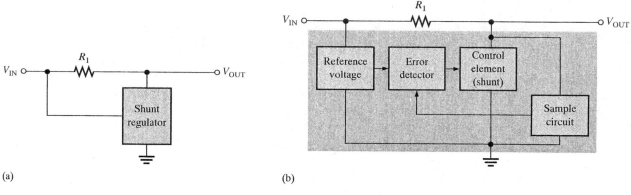

(a)   (b)

FIGURE 7–11
Simple shunt regulator and block diagram.

In the basic shunt regulator, the control element is a transistor, Q_1, in parallel with the load, as shown in Figure 7–12. A resistor, R_1, is in series with the load. The operation of the circuit is similar to that of the series regulator, except that regulation is achieved by controlling the current through the parallel transistor Q_1.

When the output voltage tries to decrease due to a change in input voltage or load current caused by a change in load resistance, as shown in Figure 7–13(a), the attempted

FIGURE 7–12
Basic op-amp shunt regulator.

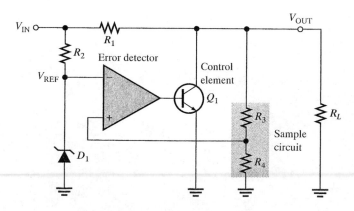

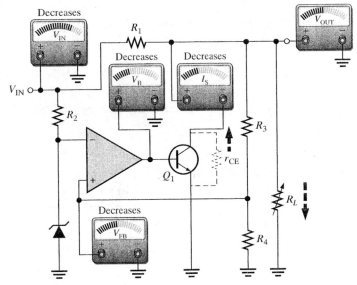

(a) Response to a decrease in V_{IN} or R_L

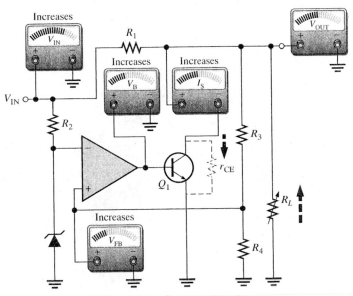

(b) Response to an increase in V_{IN} or R_L

FIGURE 7–13

Sequence of responses when V_{OUT} tries to decrease as a result of a decrease in R_L or V_{IN} (opposite responses for an attempted increase).

decrease is sensed by R_3 and R_4 and applied to the op-amp's noninverting input. The resulting difference voltage reduces the op-amp's output (V_B), driving Q_1 less, thus reducing its collector current (shunt current) and increasing its internal collector-to-emitter resistance r_{CE}. Since r_{CE} acts as a voltage divider with R_1, this action offsets the attempted decrease in V_{OUT} and maintains it at an almost constant level.

The opposite action occurs when the output tries to increase, as indicated in Figure 7–13(b). With I_L and V_{OUT} constant, a change in the input voltage produces a change in shunt current (I_S) as follows:

$$\Delta I_S = \frac{\Delta V_{IN}}{R_1}$$

With a constant V_{IN} and V_{OUT}, a change in load current causes an opposite change in shunt current.

$$\Delta I_S = -\Delta I_L$$

This formula says that if I_L increases, I_S decreases, and vice versa. The shunt regulator is less efficient than the series type but offers inherent short-circuit protection. If the output is shorted ($V_{OUT} = 0$), the load current is limited by the series resistor R_1 to a maximum value as follows ($I_S = 0$).

$$I_{L(max)} = \frac{V_{IN}}{R_1} \tag{7–7}$$

EXAMPLE 7–5

In Figure 7–14, what power rating must R_1 have if the maximum input voltage is 12.5 V?

FIGURE 7–14

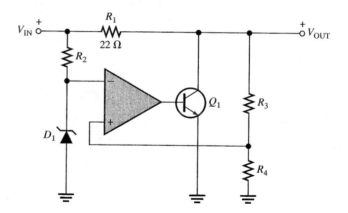

Solution The worst-case power dissipation in R_1 occurs when the output is short-circuited. $V_{OUT} = 0$, and when $V_{IN} = 12.5$ V, the voltage dropped across R_1 is $V_{IN} - V_{OUT} = 12.5$ V. The power dissipation in R_1 is

$$P_{R1} = \frac{V_{R1}^2}{R_1} = \frac{(12.5 \text{ V})^2}{22 \text{ }\Omega} = 7.1 \text{ W}$$

Therefore, a resistor with at least a **10 W** rating should be used.

Practice Exercise In Figure 7–14, R_1 is changed to 33 Ω. What must be the power rating of R_1 if the maximum input voltage is 24 V?

7-3 REVIEW QUESTIONS

1. How does the control element in a shunt regulator differ from that in a series regulator?
2. What is one advantage of a shunt regulator over a series type? What is a disadvantage?

7-4 ■ BASIC SWITCHING REGULATORS

The two types of linear regulators—series and shunt—have control elements (transistors) that are conducting all the time, with the amount of conduction varied as demanded by changes in the output voltage or current. The switching regulator is different; the control element operates as a switch. A greater efficiency can be realized with this type of voltage regulator than with the linear types because the transistor is not always conducting. Therefore, switching regulators can provide greater load currents at low voltage than linear regulators because the control transistor doesn't dissipate as much power. Three basic configurations of switching regulators are step-down, step-up, and inverting.

After completing this section, you should be able to

❑ Discuss the principles of switching regulators
    ❑ Describe the step-down configuration of a switching regulator
    ❑ Determine the output voltage of the step-down configuration
    ❑ Describe the step-up configuration of a switching regulator
    ❑ Determine the output voltage of the step-up configuration
    ❑ Describe the voltage-inverter configuration

Step-Down Configuration

In the step-down configuration, the output voltage is always less than the input voltage. A basic step-down switching regulator is shown in Figure 7–15(a), and its simplified equivalent is shown in Figure 7–15(b). Transistor Q_1 is used to switch the input voltage at a duty cycle that is based on the regulator's load requirement. The *LC* filter is then used to average the switched voltage. Since Q_1 is either *on* (saturated) or *off*, the power lost in the control element is relatively small. Therefore, the switching regulator is useful primarily in higher power applications or in applications such as computers where efficiency is of utmost concern.

    The on and off intervals of Q_1 are shown in the waveform of Figure 7–16(a). The capacitor charges during the on-time (t_{on}) and discharges during the off-time (t_{off}). When the on-time is increased relative to the off-time, the capacitor charges more, thus increasing the output voltage, as indicated in Figure 7–16(b). When the on-time is decreased relative to the off-time, the capacitor discharges more, thus decreasing the output voltage, as in Figure 7–16(c). Therefore, by adjusting the duty cycle, $t_{on}/(t_{on} + t_{off})$, of Q_1, the output voltage can be varied. The inductor further smooths the fluctuations of the output voltage caused by the charging and discharging action.

FIGURE 7–15

Basic step-down switching regulator.

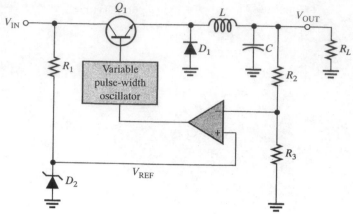

(a) Typical circuit

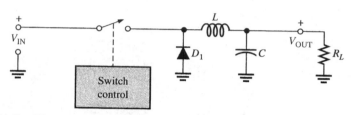

(b) Simplified equivalent circuit

FIGURE 7–16

Switching regulator waveforms. The V_C waveform is for no inductive filtering to illustrate the charge and discharge action. L and C smooth V_C to a nearly constant level, as indicated by the dashed line for V_{OUT}.

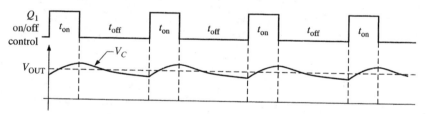

(a) V_{OUT} depends on the duty cycle.

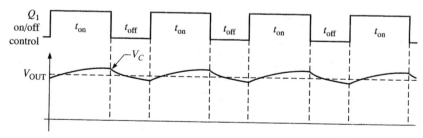

(b) Increase the duty cycle and V_{OUT} increases.

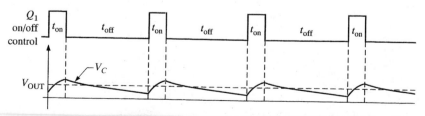

(c) Decrease the duty cycle and V_{OUT} decreases.

278

The output voltage is expressed as

$$V_{OUT} = \left(\frac{t_{on}}{T}\right)V_{IN} \qquad (7\text{--}8)$$

T is the period of the on-off cycle of Q_1 and is related to the frequency by $T = 1/f$. The period is the sum of the on-time and the off-time.

$$T = t_{on} + t_{off}$$

The ratio t_{on}/T is called the *duty cycle*.

The regulating action is as follows and is illustrated in Figure 7–17. When V_{OUT} tries to decrease, the on-time of Q_1 is increased, causing an additional charge on the

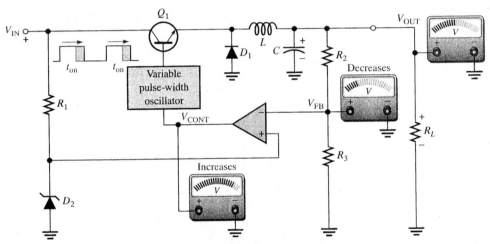

(a) When V_{OUT} attempts to decrease, the on-time of Q_1 increases.

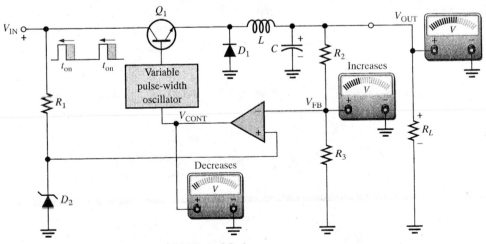

(b) When V_{OUT} attempts to increase, the on-time of Q_1 decreases.

FIGURE 7–17
Regulating action of the basic step-down switching regulator.

capacitor, C, or, C, to offset the attempted decrease. When V_{OUT} tries to increase, the on-time of Q_1 is decreased, causing C to discharge enough to offset the attempted increase.

Step-Up Configuration

A basic step-up type of switching regulator is shown in Figure 7–18.

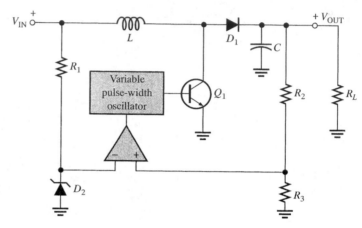

FIGURE 7–18
Basic step-up switching regulator.

The switching action is illustrated in Figure 7–19. When Q_1 turns on, voltage across L increases instantaneously to $V_{IN} - V_{CE(sat)}$, and the inductor's magnetic field expands quickly, as indicated in Figure 7–19(a). During the on-time (t_{on}) of Q_1, V_L decreases from its initial maximum, as shown. The longer Q_1 is on, the smaller V_L becomes. When Q_1 turns off, the inductor's magnetic field collapses; and its polarity reverses so that its voltage adds to V_{IN}, thus producing an output voltage greater than the input, as indicated in Figure 7–19(b). During the off-time (t_{off}) of Q_1, the diode is forward-biased, allowing the capacitor to charge. The variations in the output voltage due to the charging and discharging action are sufficiently smoothed by the filtering action of L and C.

The regulating action is illustrated in Figure 7–20. The shorter the on-time of Q_1, the greater the inductor voltage is, and thus the greater the output voltage is (greater V_L adds to V_{IN}). The longer the on-time of Q_1, the smaller are the inductor voltage and the output voltage (small V_L adds to V_{IN}). When V_{OUT} tries to decrease because of increasing load or decreasing input voltage, t_{on} decreases and the attempted decrease in V_{OUT} is offset. When V_{OUT} tries to increase, t_{on} increases and the attempted increase in V_{OUT} is offset. As you can see, the output voltage is inversely related to the duty cycle of Q_1 and can be expressed as follows:

$$V_{OUT} = \left(\frac{T}{t_{on}}\right)V_{IN} \tag{7–9}$$

where $T = t_{on} + t_{off}$.

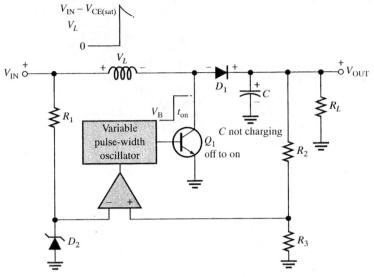

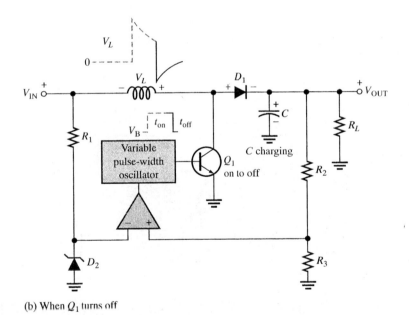

FIGURE 7–19
Switching action of the basic step-up regulator.

FIGURE 7–20

Regulating action of the basic step-up switching regulator.

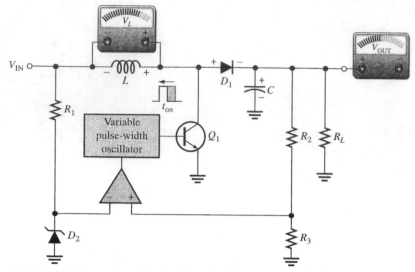

(a) When V_{OUT} tries to decrease, t_{on} decreases, causing V_L to increase. This compensates for the attempted decrease in V_{OUT}.

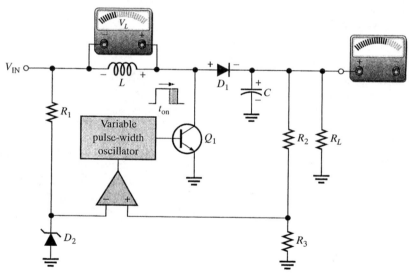

(b) When V_{OUT} tries to increase, t_{on} increases, causing V_L to decrease. This compensates for the attempted increase in V_{OUT}.

Voltage-Inverter Configuration

A third type of switching regulator produces an output voltage that is opposite in polarity to the input. A basic diagram is shown in Figure 7–21.

When Q_1 turns on, the inductor voltage jumps to $V_{IN} - V_{CE(sat)}$ and the magnetic field rapidly expands, as shown in Figure 7–22(a). While Q_1 is on, the diode is reverse-biased and the inductor voltage decreases from its initial maximum. When Q_1 turns off, the magnetic field collapses and the inductor's polarity reverses, as shown in Figure 7–22(b). This forward-biases the diode, charges C, and produces a negative output voltage, as indicated. The repetitive on-off action of Q_1 produces a repetitive charging and discharging that is smoothed by the LC filter action.

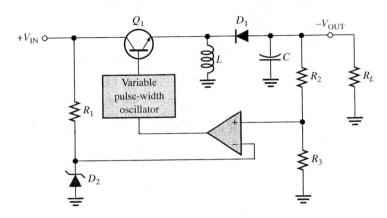

FIGURE 7–21
Basic inverting switching regulator.

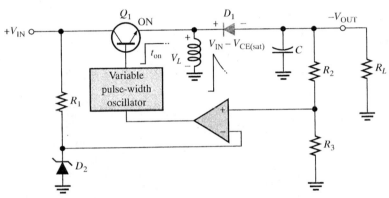

(a) When Q_1 is on, D_1 is reverse-biased.

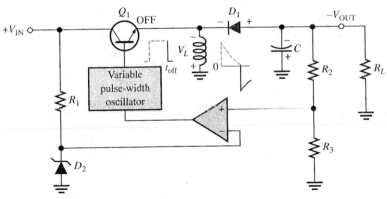

(b) When Q_1 turns off, D_1 is forward biased.

FIGURE 7–22
Inverting action of the basic inverting switching regulator.

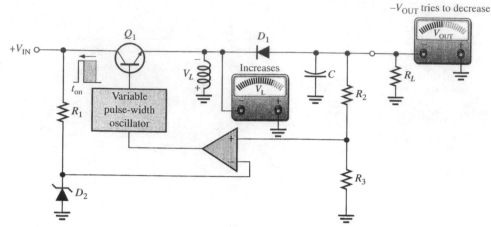

(a) When $-V_{OUT}$ tries to decrease, t_{on} decreases, causing V_L to increase. This compensates for the attempted decrease in $-V_{OUT}$.

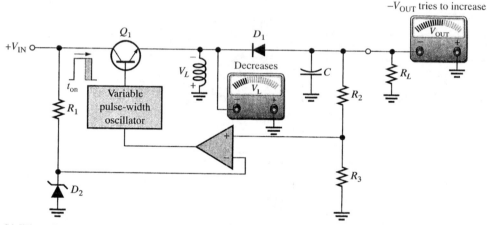

(b) When $-V_{OUT}$ tries to increase, t_{on} increases, causing V_L to decrease. This compensates for the attempted increase in $-V_{OUT}$.

FIGURE 7–23
Regulating action of the basic inverting switching regulator.

As with the step-up regulator, the less time Q_1 is on, the greater the output voltage is, and vice versa. This regulating action is illustrated in Figure 7–23. Switching regulator efficiencies can be greater than 90 percent.

7–4 REVIEW QUESTIONS

1. What are three types of switching regulators?
2. What is the primary advantage of switching regulators over linear regulators?
3. How are changes in output voltage compensated for in the switching regulator?

7–5 ■ INTEGRATED CIRCUIT VOLTAGE REGULATORS

In the previous sections, the basic voltage regulator configurations were presented. Several types of both linear and switching regulators are available in integrated circuit (IC) form. Generally, the linear regulators are three-terminal devices that provide either positive or negative output voltages that can be either fixed or adjustable. In this section, typical linear and switching IC regulators are covered in more detail.

After completing this section, you should be able to

❑ Discuss integrated circuit voltage regulators
   ❑ Describe the 7800 series of positive regulators
   ❑ Describe the 7900 series of negative regulators
   ❑ Describe the LM317 adjustable positive regulator
   ❑ Describe the LM337 adjustable negative regulator
   ❑ Describe IC switching regulators

Fixed Positive Linear Voltage Regulators

Although many types of IC regulators are available, the 7800 series of IC regulators is representative of three-terminal devices that provide a fixed positive output voltage. The three terminals are input, output, and ground as indicated in the standard fixed voltage configuration in Figure 7–24(a). The last two digits in the part number designate the output voltage. For example, the 7805 is a +5.0 V regulator. Other available output voltages are given in Figure 7–24(b). (Common IC regulator packages are shown in Appendix A.)

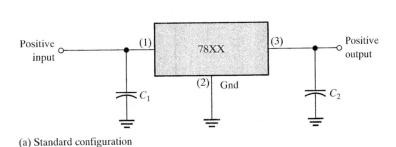

(a) Standard configuration

Type number	Output voltage
7805	+5.0 V
7806	+6.0 V
7808	+8.0 V
7809	+9.0 V
7812	+12.0 V
7815	+15.0 V
7818	+18.0 V
7824	+24.0 V

(b) The 7800 series

FIGURE 7–24
The 7800 series three-terminal fixed positive voltage regulators.

Capacitors are used on the input and output as indicated. The output capacitor acts basically as a line filter to improve transient response. The input capacitor is used to prevent unwanted oscillations when the regulator is some distance from the power supply filter such that the line has a significant inductance.

The 7800 series can produce output current in excess of 1 A when used with an adequate heat sink. The 78L00 series can provide up to 100 mA, the 78M00 series can provide up to 500 mA, and the 78T00 series can provide in excess of 3 A. These devices are available with either a 2% or 4% output voltage tolerance.

The input voltage must be at least 2 V above the output voltage in order to maintain regulation. The circuits have internal thermal overload protection and short-circuit current-limiting features.

Thermal overload occurs when the internal power dissipation becomes excessive and the temperature of the device exceeds a certain value. Thermal overload is a problem if inadequate heat sinking is provided or the regulator is not properly secured to the heat sink. Almost all applications of regulators require a heat sink. Heat generated by the regulator must move to the heat sink and then to the surrounding air. A good heat sink is massive and has fins (to increase area). A regulator that is too hot may show symptoms of drift, excess ripple, or the output may fall out of regulation.

Fixed Negative Linear Voltage Regulators

The 7900 series is typical of three-terminal IC regulators that provide a fixed negative output voltage. This series is the negative-voltage counterpart of the 7800 series and shares most of the same features and characteristics. Figure 7–25 indicates the standard configuration and part numbers with corresponding output voltages that are available. Be aware that the pins on the 7900 series regulators do not have the same function as the 7800 series pins.

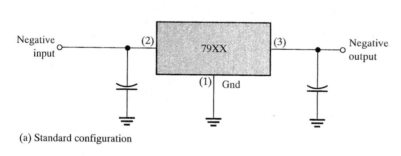

(a) Standard configuration

Type number	Output voltage
7905	−5.0 V
7905.2	−5.2 V
7906	−6.0 V
7908	−8.0 V
7912	−12.0 V
7915	−15.0 V
7918	−18.0 V
7924	−24.0 V

(b) The 7900 series

FIGURE 7–25
The 7900 series three-terminal fixed negative voltage regulators.

Adjustable Positive Linear Voltage Regulators

The LM317 is an excellent example of a three-terminal positive regulator with an adjustable output voltage. A data sheet for this device is given in Appendix A. The standard configuration is shown in Figure 7–26. Input and output capacitors, C_1 and C_3 respectively, are used for the reasons discussed previously. The capacitor, C_2, at the adjustment terminal also acts as a filter to improve transient response. Notice that there is an input, an output, and an adjustment terminal. The external fixed resistor, R_1, and the external variable resistor, R_2, provide the output voltage adjustment. V_{OUT} can be varied from 1.2 V to 37 V depending on the resistor values. The LM317 can provide over 1.5 A of output current to a load.

The LM317 is operated as a "floating" regulator because the adjustment terminal is not connected to dc ground, but floats to whatever voltage is across R_2. This allows the output voltage to be much higher than that of a fixed-voltage regulator.

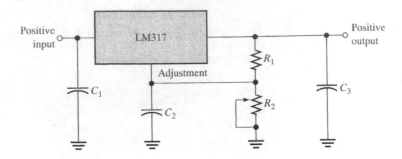

FIGURE 7–26
The LM317 three-terminal adjustable positive voltage regulator.

Basic Operation   As indicated in Figure 7–27, a constant 1.25 V reference voltage (V_{REF}) is maintained by the regulator between the output terminal and the adjustment terminal. This constant reference voltage produces a constant current (I_{REF}) through R_1, regardless of the value of R_2. I_{REF} is also through R_2.

$$I_{REF} = \frac{V_{REF}}{R_1} = \frac{1.25 \text{ V}}{R_1}$$

In addition, there is a very small constant current into the adjustment terminal of approximately 50 μA called I_{ADJ} through R_2. A formula for the output voltage is as follows:

$$V_{OUT} = V_{REF}\left(1 + \frac{R_2}{R_1}\right) + I_{ADJ}R_2 \qquad (7\text{–}10)$$

As you can see, the output voltage is a function of both R_1 and R_2. Once the value of R_1 is set, the output voltage is adjusted by varying R_2.

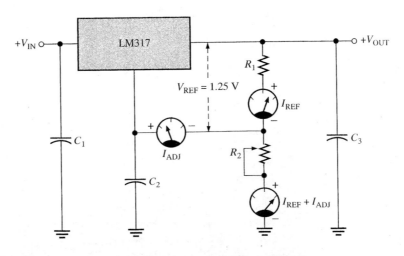

FIGURE 7–27
Operation of the LM317 adjustable voltage regulator.

EXAMPLE 7–6

Determine the minimum and maximum output voltages for the voltage regulator in Figure 7–28. Assume $I_{ADJ} = 50\ \mu A$.

FIGURE 7–28

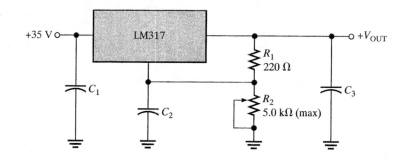

Solution
$$V_{R1} = V_{REF} = 1.25\ V$$

When R_2 is set at its minimum of $0\ \Omega$,

$$V_{OUT(min)} = V_{REF}\left(1 + \frac{R_2}{R_1}\right) + I_{ADJ}R_2 = 1.25\ V(1) = \mathbf{1.25\ V}$$

When R_2 is set at its maximum of $5.0\ k\Omega$,

$$V_{OUT(max)} = V_{REF}\left(1 + \frac{R_2}{R_1}\right) + I_{ADJ}R_2 = 1.25\ V\left(1 + \frac{5.0\ k\Omega}{220\ \Omega}\right) + (50\ \mu A)5.0\ k\Omega$$

$$= 29.66\ V + 0.25\ V = \mathbf{29.9\ V}$$

Practice Exercise What is the output voltage of the regulator if R_2 is set at $2.0\ k\Omega$?

Adjustable Negative Linear Voltage Regulators

The LM337 is the negative output counterpart of the LM317 and is a good example of this type of IC regulator. Like the LM317, the LM337 requires two external resistors for output voltage adjustment as shown in Figure 7–29. The output voltage can be adjusted from $-1.2\ V$ to $-37\ V$, depending on the external resistor values.

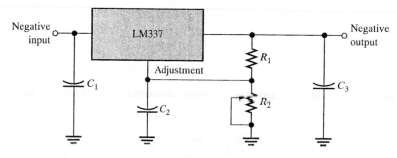

FIGURE 7–29
The LM337 three-terminal adjustable negative voltage regulator.

Troubleshooting Three-Terminal Regulators

Three-terminal regulators are very reliable devices. When problems occur, the indication is usually an incorrect voltage, high ripple, noisy or oscillating output, or drift. Troubleshooting a regulator circuit is best done with an oscilloscope as problems such as excessive ripple or noise won't show up using a DMM. Before starting, it is useful to review the possible causes of a failure (analysis) and plan measurements that will point to the failure.

If the output voltage is too low, the input voltage should be checked; the problem may be in the circuit preceding the regulator. Also check the load resistor: Does the problem go away when the load is removed? If so, it may be that the load draws too much current. A high output can occur with adjustable regulators if the feedback resistors are the wrong value or open. If there is ripple or noise on the output, check the capacitors for an open, a wrong value, or that they are installed with the proper polarity. A useful quick check of a capacitor is to place another capacitor of the same or larger size in parallel with the capacitor to be tested. If the output is oscillating, has high ripple, or drifting, check that the regulator is not too hot or supplying more than its rated current. If heat is a problem, make sure the regulator is firmly secured to the heat sink.

Switching Voltage Regulators

As an example of an IC switching voltage regulator, let's look at the 78S40. This is a universal device that can be used with external components to provide step-up, step-down, and inverting operation.

The internal circuitry of the 78S40 is shown in Figure 7–30. This circuit can be compared to the basic switching regulators that were covered in Section 7–4. For example, look back at Figure 7–15(a). The oscillator and comparator functions are directly comparable. The gate and flip-flop which are digital devices were not included in the basic circuit of Figure 7–15(a), but they provide additional regulating action. Transistors Q_1 and Q_2 effectively perform the same function as Q_1 in the basic circuit. The 1.25 V reference block

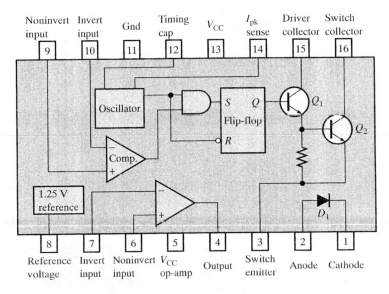

FIGURE 7–30
The 78S40 switching regulator.

in the 78S40 has the same purpose as the zener diode in the basic circuit, and diode D_1 in the 78S40 corresponds to D_1 in the basic circuit.

The 78S40 also has an "uncommitted" op-amp thrown in for good measure. It is not used in any of the regulator configurations. External circuitry is required to make this device operate as a regulator, as you will see in Section 7–6.

7–5 REVIEW QUESTIONS

1. What are the three terminals of a fixed-voltage regulator?
2. What is the output voltage of a 7809? Of a 7915?
3. What are the three terminals of an adjustable-voltage regulator?
4. What external components are required for a basic LM317 configuration?

7–6 ■ APPLICATIONS OF IC VOLTAGE REGULATORS

In the last section, you saw several devices that are representative of the general types of IC voltage regulators. Now, several different ways these devices can be modified with external circuitry to improve or alter their performance are examined.

After completing this section, you should be able to

❑ Discuss applications of IC voltage regulators
   ❑ Explain the use of an external pass transistor
   ❑ Explain the use of current limiting
   ❑ Explain how to use a voltage regulator as a constant-current source
   ❑ Discuss some application considerations for switching regulators

The External Pass Transistor

As you know, an IC voltage regulator is capable of delivering only a certain amount of output current to a load. For example, the 7800 series regulators can handle a maximum output current of at least 1.3 A and 2.5 A under certain conditions. If the load current exceeds the maximum allowable value, there will be thermal overload and the regulator will shut down. A thermal overload condition means that there is excessive power dissipation inside the device.

If an application requires more than the maximum current that the regulator can deliver, an external pass transistor can be used. Figure 7–31 illustrates a three-terminal regulator with an external pass transistor for handling currents in excess of the output current capability of the basic regulator.

The value of the external current-sensing resistor R_{ext} determines the value of current at which Q_{ext} begins to conduct because it sets the base-to-emitter voltage of the transistor. As long as the current is less than the value set by R_{ext}, the transistor Q_{ext} is off, and the regulator operates normally as shown in Figure 7–32(a). This is because the voltage drop across R_{ext} is less than the 0.7 V base-to-emitter voltage required to turn Q_{ext} on. R_{ext} is de-

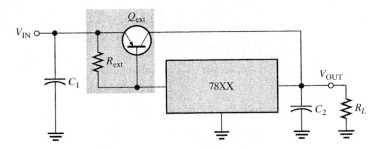

FIGURE 7–31
A 7800-series three-terminal regulator with an external pass transistor.

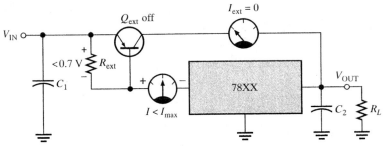

(a) When the regulator current is less than I_{max}, the external pass transistor is off and the regulator is handling all of the current.

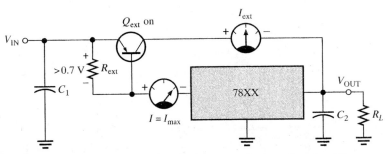

(b) When the load current exceeds I_{max}, the drop across R_{ext} turns Q_{ext} on and the transistor conducts the excess current.

FIGURE 7–32
Operation of the regulator with an external pass transistor.

termined by the following formula, where I_{max} is the highest current that the voltage regulator is to handle internally.

$$R_{ext} = \frac{0.7\ V}{I_{max}}$$

When the current is sufficient to produce at least a 0.7 V drop across R_{ext}, the external pass transistor Q_{ext} turns on and conducts any current in excess of I_{max}, as indicated in Figure 7–32(b). Q_{ext} will conduct more or less, depending on the load requirements. For example, if the total load current is 3 A and I_{max} was selected to be 1 A, the external pass transistor will conduct 2 A, which is the excess over the internal regulator current I_{max}.

EXAMPLE 7–7

What value is R_{ext} if the maximum current to be handled internally by the voltage regulator in Figure 7–31 is set at 700 mA?

Solution

$$R_{ext} = \frac{0.7 \text{ V}}{I_{max}} = \frac{0.7 \text{ V}}{0.7 \text{ A}} = 1 \text{ }\Omega$$

Practice Exercise If R_{ext} is changed to 1.5 Ω, at what current value will Q_{ext} turn on?

The external pass transistor is typically a power transistor with heat sink that must be capable of handling a maximum power of

$$P_{ext} = I_{ext}(V_{IN} - V_{OUT})$$

EXAMPLE 7–8

What must be the minimum power rating for the external pass transistor used with a 7824 regulator in a circuit such as that shown in Figure 7–31? The input voltage is 30 V and the load resistance is 10 Ω. The maximum internal current is to be 700 mA. Assume that there is no heat sink for this calculation. Keep in mind that the use of a heat sink increases the effective power rating of the transistor and you can use a lower-rated transistor.

Solution The load current is

$$I_L = \frac{V_{OUT}}{R_L} = \frac{24 \text{ V}}{10 \text{ }\Omega} = 2.4 \text{ A}$$

The current through Q_{ext} is

$$I_{ext} = I_L - I_{max} = 2.4 \text{ A} - 0.7 \text{ A} = 1.7 \text{ A}$$

The power dissipated by Q_{ext} is

$$P_{ext(min)} = I_{ext}(V_{IN} - V_{OUT}) = 1.7 \text{ A}(30 \text{ V} - 24 \text{ V}) = 1.7 \text{ A}(6 \text{ V}) = \textbf{10.2 W}$$

For a safety margin, choose a power transistor with a rating greater than 10.2 W, say at least 15 W.

Practice Exercise Rework this example using a 7815 regulator.

Current Limiting

A drawback of the circuit in Figure 7–31 is that the external transistor is not protected from excessive current, such as would result from a shorted output. An additional current-limiting network (Q_{lim} and R_{lim}) can be added as shown in Figure 7–33 to protect Q_{ext} from excessive current and possible burn out.

The following describes the way the current-limiting network works. The current-sensing resistor R_{lim} sets the V_{BE} of transistor Q_{lim}. The base-to-emitter voltage of Q_{ext} is now determined by $V_{R_{ext}} - V_{R_{lim}}$ because they have opposite polarities. So, for normal operation, the drop across R_{ext} must be sufficient to overcome the opposing drop across R_{lim}. If the current through Q_{ext} exceeds a certain maximum ($I_{ext(max)}$) because of a shorted out-

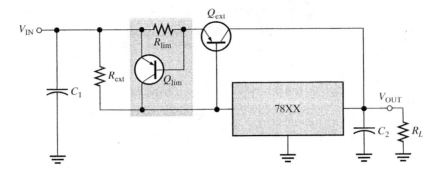

FIGURE 7–33
Regulator with current limiting.

put or a faulty load, the voltage across R_{lim} reaches 0.7 V and turns Q_{lim} on. Q_{lim} now conducts current away from Q_{ext} and through the regulator, forcing a thermal overload to occur and shut down the regulator. Remember, the IC regulator is internally protected from thermal overload as part of its design.

This action is shown in Figure 7–34. In part (a), the circuit is operating normally with Q_{ext} conducting less than the maximum current that it can handle with Q_{lim} off. Part (b)

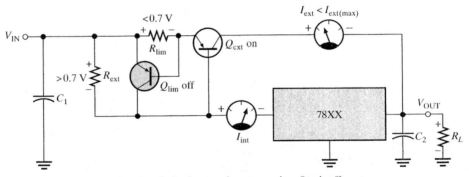

(a) During normal operation, when the load current is not excessive, Q_{lim} is off.

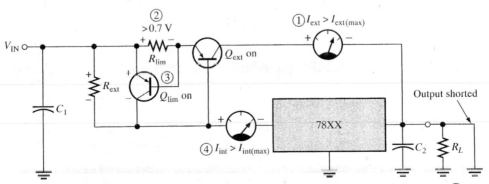

(b) When short occurs ①, the external current becomes excessive and the voltage across R_{lim} increases ② and turns on Q_{lim} ③, which then conducts current away from Q_{ext} and routes it through the regulator, causing the internal regulator current to become excessive ④ and to force the regulator into thermal shut down.

FIGURE 7–34
The current-limiting action of the regulator circuit.

shows what happens when there is a short across the load. The current through Q_{ext} suddenly increases and causes the voltage drop across R_{lim} to increase, which turns Q_{lim} on. The current is now diverted through the regulator, which causes it to shut down due to thermal overload.

A Current Regulator

The three-terminal regulator can be used as a current source when an application requires that a constant current be supplied to a variable load. The basic circuit is shown in Figure 7–35 where R_1 is the current-setting resistor. The regulator provides a fixed constant voltage, V_{OUT}, between the ground terminal (not connected to ground in this case) and the output terminal. This determines the constant current supplied to the load.

$$I_L = \frac{V_{OUT}}{R_1} + I_G$$

The current, I_G, from the ground terminal is very small compared to the output current and can often be neglected.

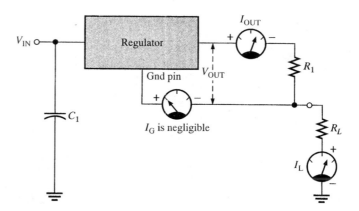

FIGURE 7–35
The three-terminal regulator as a current source.

EXAMPLE 7–9

What value of R_1 is necessary in a 7805 regulator to provide a constant current of 1 A to a variable load? The input must be at least 2 V greater than the output and $I_G = 1.5$ mA.

Solution   First, 1 A is within the limits of the 7805's capability (remember, it can handle at least 1.3 A without an external pass transistor).

The 7805 produces 5 V between its ground terminal and its output terminal. Therefore, if you want 1 A of current, the current-setting resistor must be (neglecting I_G)

$$R_1 = \frac{V_{OUT}}{I_L} = \frac{5\ V}{1\ A} = \textbf{5.0 } \boldsymbol{\Omega}$$

The circuit is shown in Figure 7–36.

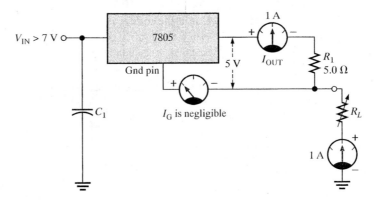

FIGURE 7–36
A 1 A constant-current source.

Practice Exercise If a 7812 regulator is used instead of the 7805, to what value would you change R_1 to maintain a constant current of 1 A?

Switching Regulator Configurations

In Section 7–5, the 78S40 was introduced as an example of an IC switching voltage regulator. Figure 7–37 shows the external connections for a step-down configuration where the

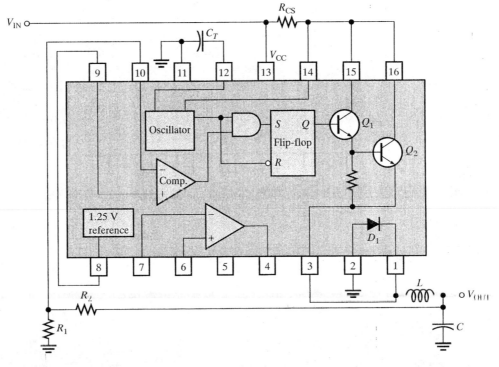

FIGURE 7–37
The step-down configuration of the 78S40 switching regulator.

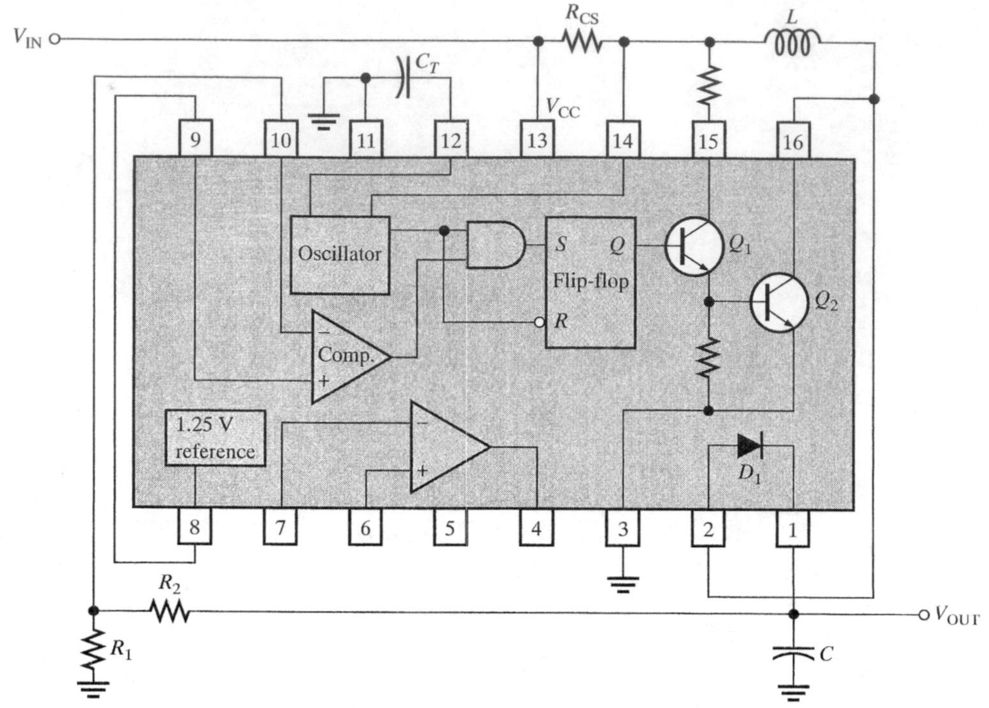

FIGURE 7–38
The step-up configuration of the 78S40 switching regulator.

output voltage is less than the input voltage, and Figure 7–38 shows a step-up configuration in which the output voltage is greater than the input voltage. An inverting configuration is also possible, but it is not shown here.

The timing capacitor, C_T, controls the pulse width and frequency of the oscillator and thus establishes the on-time of transistor Q_2. The voltage across the current-sensing resistor R_{CS} is used internally by the oscillator to vary the duty cycle based on the desired peak load current. The voltage divider, made up of R_1 and R_2, reduces the output voltage to a nominal value equal to the reference voltage. If V_{OUT} exceeds its set value, the output of the comparator switches to its low state, disabling the gate to turn Q_2 off until the output decreases. This regulating action is in addition to that produced by the duty cycle variation of the oscillator as described in Section 7–4 in relation to the basic switching regulator.

7–6 REVIEW QUESTIONS

1. What is the purpose of using an external pass transistor with an IC voltage regulator?
2. What is the advantage of current limiting in a voltage regulator?
3. How can you configure a three-terminal regulator as a current source?

7–7 ■ A SYSTEM APPLICATION

In this system application, the focus is on the regulated power supply which provides the FM stereo receiver with dual polarity dc voltages. Recall from previous system applications that the op-amps in the channel separation circuits and the audio amplifiers operate from ±12 V. Both positive and negative voltage regulators are used to regulate the rectified and filtered voltages from a bridge rectifier.

After completing this section, you should be able to

❑ Apply what you have learned in this chapter to a system application
    ❑ Discuss how dual supply voltages are produced by a rectifier
    ❑ Explain how positive and negative three-terminal IC regulators are used in a power supply
    ❑ Relate a schematic to a PC board
    ❑ Analyze the operation of the power supply circuit
    ❑ Troubleshoot some common power supply failures

About the Power Supply

This power supply utilizes a full-wave bridge **rectifier** with both the positive and negative rectified voltages taken off the bridge at the appropriate points and filtered by electrolytic capacitors. A 7812 and a 7912 provide regulation.

Now, so that you can take a closer look at the dual power supply, let's take it out of the system and put it on the test bench.

ON THE TEST BENCH

■ ACTIVITY 1 Relate the PC Board to the Schematic

Develop a schematic for the power supply in Figure 7–39. Add any missing labels and include the IC pin numbers by referring to the voltage regulator data sheets in Appendix A. The rectifier diodes are type 1N4001, the filter capacitors C1 and C2 are 1000 μF, and the transformer has a turns ratio of 5:1.

■ ACTIVITY 2 Analyze the Power Supply Circuits

Step 1: Determine the approximate voltage at each of the four "corners" of the bridge with respect to ground.

Step 2: Calculate the peak inverse voltage of the rectifier diodes.

Step 3: Determine the voltage at the inputs of the voltage regulators.

Step 4: In this stereo system, assume that op-amps are used only in the channel separation circuits and the channel audio amplifiers. If all of the other circuits in the receiver use +12 V and draw an average dc current of 500 mA, determine how much total current each regulator must supply. Refer to the system applications in Chapters 3 and 5. Use the appropriate data sheets.

Step 5: Based on the results in Step 4, do the IC regulators have to be attached to the heat sink or is this just for a safety margin?

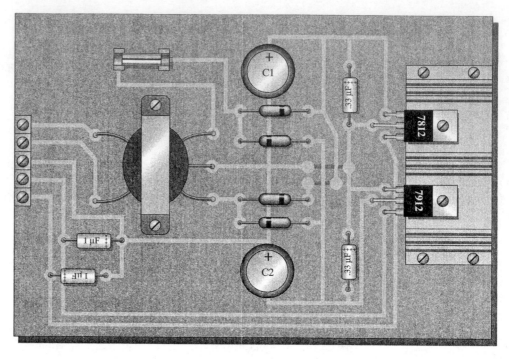

FIGURE 7–39

■ ACTIVITY 3   Write a Technical Report

Describe the operation of the power supply with an emphasis on how both positive and negative voltages are obtained. State the purpose of each component. Use the results of Activity 2 where appropriate.

■ ACTIVITY 4   Troubleshoot the Power Supply by Stating the Probable Cause or Causes in Each Case

1. Both positive and negative output voltages are zero.
2. Positive output voltage is zero and the negative output voltage is -12 V.
3. Negative output voltage is zero and the positive output voltage is $+12$ V.
4. Radical voltage fluctuations on output of positive regulator.

7–7 REVIEW QUESTIONS

1. What should be the rating of the power supply fuse?
2. What purpose do the 0.33 μF capacitors serve?
3. Which regulator provides the negative voltage?
4. Would you recommend that an external pass transistor be used with the regulators in this power supply? Why?

■ **SUMMARY**

- Voltage regulators keep a constant dc output voltage when the input or load varies within limits.
- A basic voltage regulator consists of a reference voltage source, an error detector, a sampling element, and a control device. Protection circuitry is also found in most regulators.
- Two basic categories of voltage regulators are linear and switching.
- Two basic types of linear regulators are series and shunt.
- In a series linear regulator, the control element is a transistor in series with the load.
- In a shunt linear regulator, the control element is a transistor in parallel with the load.
- Three configurations for switching regulators are step-down, step-up, and inverting.
- Switching regulators are more efficient than linear regulators and are particularly useful in low-voltage, high-current applications.
- Three-terminal linear IC regulators are available for either fixed output or variable output voltages of positive or negative polarities.
- An external pass transistor increases the current capability of a regulator.
- The 7800 series are three-terminal IC regulators with fixed positive output voltage.
- The 7900 series are three-terminal IC regulators with fixed negative output voltage.
- The LM317 is a three-terminal IC regulator with a positive variable output voltage.
- The LM337 is a three-terminal IC regulator with a negative variable output voltage.
- The 78S40 is a switching voltage regulator.

■ **GLOSSARY**

These terms are included in the end-of-book glossary.

Fold-back current limiting   A method of current limiting in voltage regulators.

Line regulation   The percentage change in output voltage for a given change in line (input) voltage.

Load regulation   The percentage change in output voltage for a given change in load current.

Rectifier   An electronic circuit that converts ac into pulsating dc.

Regulator   An electronic circuit that maintains an essentially constant output voltage with a changing input voltage or load current.

Thermal overload   A condition in a rectifier where the internal power dissipation of the circuit exceeds a certain maximum due to excessive current.

■ **KEY FORMULAS**

(7–1) $\text{Line regulation} = \left(\dfrac{\Delta V_{\text{OUT}}}{\Delta V_{\text{IN}}}\right)100\%$ Percent line regulation

(7–2) $\text{Line regulation} = \left(\dfrac{\Delta V_{\text{OUT}}/V_{\text{OUT}}}{\Delta V_{\text{IN}}}\right)100\%$ Percent line regulation per volt

(7 3) $\text{Load regulation} = \left(\dfrac{V_{\text{NL}} - V_{\text{FL}}}{V_{\text{FL}}}\right)100\%$ Percent load regulation

(7–4) $\text{Load regulation} = \left(\dfrac{R_{\text{OUT}}}{R_{\text{FL}}}\right)100\%$ Percent load regulation given output resistance and minimum load resistance

(7–5) $V_{\text{OUT}} \cong \left(1 + \dfrac{R_2}{R_3}\right)V_{\text{REF}}$ Series regulator output

(7–6) $I_{L(max)} = \dfrac{0.7 \text{ V}}{R_4}$ Constant current limiting

(7–7) $I_{L(max)} = \dfrac{V_{IN}}{R_1}$ Maximum load current for a shunt regulator

(7–8) $V_{OUT} = \left(\dfrac{t_{on}}{T}\right)V_{IN}$ Output voltage for step-down switching regulator

(7–9) $V_{OUT} = \left(\dfrac{T}{t_{on}}\right)V_{IN}$ Output voltage for step-up switching regulator

(7–10) $V_{OUT} = V_{REF}\left(1 + \dfrac{R_2}{R_1}\right) + I_{ADJ}R_2$ Output voltage for IC voltage regulator

■ SELF-TEST

1. In the case of line regulation,
   (a) when the temperature varies, the output voltage stays constant
   (b) when the output voltage changes, the load current stays constant
   (c) when the input voltage changes, the output voltage stays constant
   (d) when the load changes, the output voltage stays constant

2. In the case of load regulation,
   (a) when the temperature varies, the output voltage stays constant
   (b) when the input voltage changes, the load current stays constant
   (c) when the load changes, the load current stays constant
   (d) when the load changes, the output voltage stays constant

3. All of the following are parts of a basic voltage regulator *except*
   (a) control element   (b) sampling circuit   (c) voltage follower
   (d) error detector   (e) reference voltage

4. The basic difference between a series regulator and a shunt regulator is
   (a) the amount of current that can be handled
   (b) the position of the control element
   (c) the type of sample circuit
   (d) the type of error detector

5. In a basic series regulator, V_{OUT} is determined by
   (a) the control element   (b) the sample circuit
   (c) the reference voltage   (d) answers (b) and (c)

6. The main purpose of current limiting in a regulator is
   (a) protection of the regulator from excessive current
   (b) protection of the load from excessive current
   (c) to keep the power supply transformer from burning up
   (d) to maintain a constant output voltage

7. In a linear regulator, the control transistor is conducting
   (a) a small part of the time   (b) half the time
   (c) all of the time   (d) only when the load current is excessive

8. In a switching regulator, the control transistor is conducting
   (a) part of the time
   (b) all of the time
   (c) only when the input voltage exceeds a set limit
   (d) only when there is an overload

9. The LM317 is an example of an IC
   (a) three-terminal negative voltage regulator   (b) fixed positive voltage regulator
   (c) switching regulator   (d) linear regulator
   (e) variable positive voltage regulator   (f) answers (b) and (d) only
   (g) answers (d) and (e) only

10. An external pass transistor is used for
    (a) increasing the output voltage
    (b) improving the regulation
    (c) increasing the current that the regulator can handle
    (d) short-circuit protection

PROBLEMS

SECTION 7–1 Voltage Regulation

1. The nominal output voltage of a certain regulator is 8 V. The output changes 2 mV when the input voltage goes from 12 V to 18 V. Determine the line regulation and express it as a percentage change.

2. Express the line regulation found in Problem 1 in units of %/V.

3. A certain regulator has a no-load output voltage of 10 V and a full-load output voltage of 9.90 V. What is the percent load regulation?

4. In Problem 3, if the full-load current is 250 mA, express the load regulation in %/mA.

SECTION 7–2 Basic Series Regulators

5. Label the functional blocks for the voltage regulator in Figure 7–40.

6. Determine the output voltage for the regulator in Figure 7–41.

7. Determine the output voltage for the series regulator in Figure 7–42.

8. If R_3 in Figure 7–42 is increased to 4.7 kΩ, what happens to the output voltage?

FIGURE 7–40

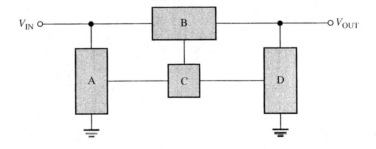

FIGURE 7–41

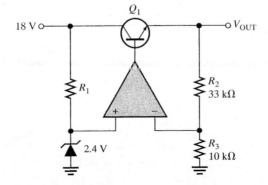

FIGURE 7–42

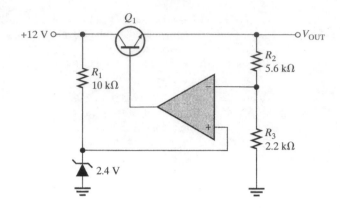

FIGURE 7–43

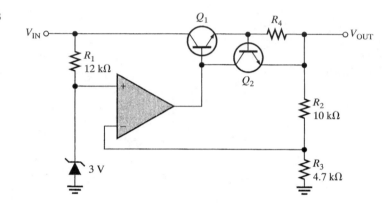

9. If the zener voltage is 2.7 V instead of 2.4 V in Figure 7–42, what is the output voltage?

10. A series voltage regulator with constant current limiting is shown in Figure 7–43. Determine the value of R_4 if the load current is to be limited to a maximum value of 250 mA. What power rating must R_4 have?

11. If the R_4 determined in Problem 10 is halved, what is the maximum load current?

SECTION 7–3   Basic Shunt Regulators

12. In the shunt regulator of Figure 7–44, when the current through R_L increases, does Q_1 conduct more or less? Why?

FIGURE 7–44

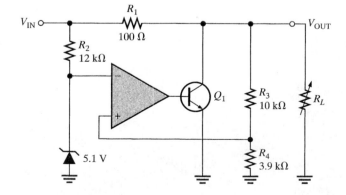

13. Assume the current through R_L remains constant and V_{IN} changes by 1 V in Figure 7–44. What is the change in the collector current of Q_1?

14. With a constant input voltage of 17 V, the load resistance in Figure 7–44 is varied from 1.0 kΩ to 1.2 kΩ. Neglecting any change in output voltage, how much does the shunt current through Q_1 change?

15. If the maximum allowable input voltage in Figure 7–44 is 25 V, what is the maximum possible output current when the output is short-circuited? What power rating should R_1 have?

SECTION 7–4 Basic Switching Regulators

16. A basic switching regulator is shown in Figure 7–45. If the switching frequency of the transistor is 100 Hz with an off-time of 6 ms, what is the output voltage?

17. What is the duty cycle of the transistor in Problem 16?

18. Determine the output voltage for the switching regulator in Figure 7–46 when the duty cycle is 40 percent.

19. If the on-time of Q_1 in Figure 7–46 is decreased, does the output voltage increase or decrease?

FIGURE 7–45

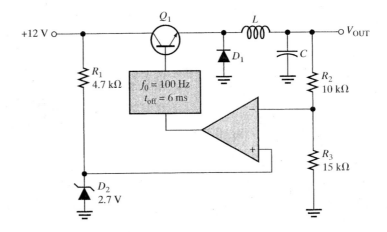

FIGURE 7–46

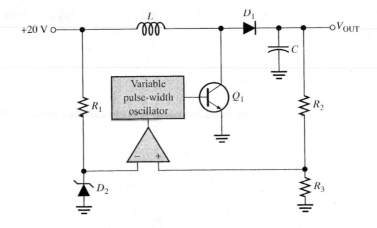

SECTION 7–5   Integrated Circuit Voltage Regulators

20. What is the output voltage of each of the following IC regulators?
   (a) 7806   **(b)** 7905.2   **(c)** 7818   **(d)** 7924

21. Determine the output voltage of the regulator in Figure 7–47. $I_{ADJ} = 50\ \mu A$.

22. Determine the minimum and maximum output voltages for the circuit in Figure 7–48. $I_{ADJ} = 50\ \mu A$.

23. With no load connected, how much current is there through the regulator in Figure 7–47? Neglect the adjustment terminal current.

24. Select the values for the external resistors to be used in an LM317 circuit that is required to produce an output voltage of 12 V with an input of 18 V. The maximum regulator current with no load is to be 2 mA. There is no external pass transistor.

FIGURE 7–47

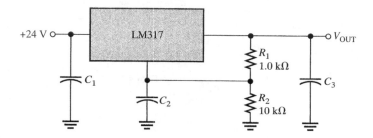

FIGURE 7–48

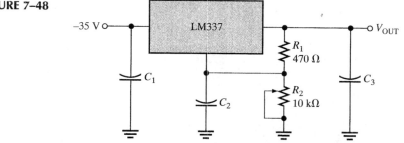

SECTION 7–6   Applications of IC Voltage Regulators

25. In the regulator circuit of Figure 7–49, determine R_{ext} if the maximum internal regulator current is to be 250 mA.

26. Using a 7812 voltage regulator and a 10 Ω load in Figure 7–49, how much power will the external pass transistor have to dissipate? The maximum internal regulator current is set at 500 mA by R_{ext}.

FIGURE 7–49

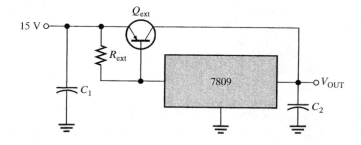

27. Show how to include current limiting in the circuit of Figure 7–49. What should the value of the limiting resistor be if the external current is to be limited to 2 A?

28. Using an LM317, design a circuit that will provide a constant current of 500 mA to a load.

29. Repeat Problem 28 using a 7909.

30. If a 78S40 switching regulator is to be used to regulate a 12V input down to a 6 V output, calculate the values of the external voltage-divider resistors.

■ ANSWERS TO REVIEW QUESTIONS

Section 7–1

1. The percentage change in the output voltage for a given change in input voltage.

2. The percentage change in output voltage for a given change in load current.

3. 1.2%; 0.06%/V

4. 1.6%; 0.0016%/mA

Section 7–2

1. Control element, error detector, sampling element, reference source

2. 2 V

Section 7–3

1. In a shunt regulator, the control element is in parallel with the load rather than in series.

2. A shunt regulator has inherent current limiting. A disadvantage is that a shunt regulator is less efficient than a series regulator.

Section 7–4

1. Step-down, step-up, inverting

2. Switching regulators operate at a higher efficiency.

3. The duty cycle varies to regulate the output.

Section 7–5

1. Input, output, and ground

2. A 7809 has a +9 V output; A 7915 has a −15 V output.

3. Input, output, adjustment

4. A two-resistor voltage divider

Section 7–6

1. A pass transistor increases the current that can be handled.

2. Current limiting prevents excessive current and prevents damage to the regulator.

3. See Figure 7–35.

Section 7–7

1. 1 A

2. Those optional capacitors on the regulator inputs prevent oscillations.

3. The 7909 is a negative-voltage regulator.

4. No. The current that either regulator must supply is less than 1 A.

■ **ANSWERS TO PRACTICE EXERCISES FOR EXAMPLES**

7–1 12%, 0.6%/V

7–2 The voltage drop of 12.1 V − 12.0 V = 0.1 V is across the specified output resistance. Since $I = 0.2$ A at 12.0 V, 0.1/0.2 = 0.5 Ω.

7–3 7.33 V

7–4 0.7 A

7–5 17.5 W

7–6 12.7 V

7–7 467 mA

7–8 12 W dissipated; choose a larger practical value (e.g., 20 W).

7–9 12 Ω

8

SPECIAL-PURPOSE AMPLIFIERS

Courtesy Hewlett-Packard Company

■ CHAPTER OBJECTIVES

☐ Understand and explain the operation of an instrumentation amplifier (IA)

☐ Understand and explain the operation of an isolation amplifier

☐ Understand and explain the operation of an operational transconductance amplifier (OTA)

☐ Understand and explain the operation of log and antilog amplifiers

☐ Apply what you have learned in this chapter to a system application

A general-purpose op-amp, such as the 741, is an extremely versatile and widely used device. However, many specialized IC amplifiers have been designed with certain types of applications in mind or with certain special features or characteristics. Most of these devices are actually derived from the basic op-amp. These special amplifiers include the instrumentation amplifier (IA) that is used in high-noise environments, the isolation amplifier that is used in high-voltage and medical applications, the operational transconductance amplifier (OTA) that is used as a voltage-to-current amplifier, and the logarithmic amplifiers that are used for linearizing certain types of inputs and for mathematical operations. In this chapter, you will learn about each of these devices and some of their basic applications.

Medical electronics is a very important application area for electronic devices and, without doubt, one of the most beneficial. The electrocardiograph (ECG), one of the most common and important instruments in use for medical purposes, is used to monitor the heart function of patients in order to detect any irregularities or abnormalities in the heartbeat. Sensors called electrodes are placed at points on the body to pick up the small electrical signal produced by the heart. This signal goes through an amplification process and is fed to a video monitor or chart recorder for viewing. Because of the safety hazards related to electrical equipment, it is important that the patient be protected from the possibility of unpleasant or even fatal electrical shock. For this reason, the isolation amplifier is used in medical equipment that comes in contact with the human body. The diagram in Figure 8–36 shows a basic block diagram for a simplified ECG system. Our focus in this system application is on the amplifier section.

For the system application in Section 8–5, in addition to the other topics, be sure you understand

☐ Basic op-amp operation
☐ Isolation amplifiers

8–1 ■ INSTRUMENTATION AMPLIFIERS

An instrumentation amplifier is a differential voltage-gain device that amplifies the difference between the voltages existing at its two input terminals. The main purpose of an instrumentation amplifier is to amplify small signals that are riding on large common-mode voltages. The key characteristics are high input impedance, high common-mode rejection, low output offset, and low output impedance. A basic instrumentation amplifier is made up of three op-amps and several resistors. The voltage gain is set with an external resistor. Instrumentation amplifiers (IAs) are commonly used in environments with high common-mode noise such as in data acquisition systems where remote sensing of input variables is required.

After completing this section, you should be able to

❑ Understand and explain the operation of an instrumentation amplifier (IA)
   ❑ Explain how op-amps are connected to form an IA
   ❑ Describe how the voltage gain is set
   ❑ Discuss an application
   ❑ Describe the features of the AD521 instrumentation amplifier

The Basic Instrumentation Amplifier

One of the most common problems in measuring systems is the contamination of the signal from a transducer with unwanted noise (such as 60 Hz power line interference). The transducer signal is typically a small differential signal carrying the desired information. Noise that is added to both signal conductors in the same amount is called a common-mode noise (discussed in Section 2–2). Ideally, the differential signal should be amplified and the common-mode noise should be rejected.

A second problem for measuring systems is that many transducers have a high output impedance and can easily be loaded down when connected to an amplifier. An amplifier for small transducer signals needs to have a very high input impedance to avoid this loading effect.

The solution to these measurement problems is the **instrumentation amplifier (IA),** a specially designed differential amplifier with ultra-high input impedance and extremely good common-mode rejection (up to 130 dB) as well as being able to achieve high, stable gains. Instrumentation amplifiers can faithfully amplify low-level signals in the presence of high common-mode noise. They are used in a variety of signal-processing applications where accuracy is important and where low drift, low bias currents, precise gain, and very high CMRR are required.

Figure 8–1 shows a basic instrumentation amplifier (IA) constructed from three op-amps. Op-amps 1 and 2 are modified voltage-followers, each containing a feedback resistor (R_1 and R_2). The feedback resistors have no effect in this circuit (and could be left out) but will be used when the circuit is modified in the next step. The voltage-followers provide high input impedance with a gain of 1. Op-amp 3 is a differential amplifier that amplifies the difference between V_{out1} and V_{out2}. Although this circuit has the advantage of high input impedance, it requires extremely high precision matching of the gain resistors to achieve a high CMRR (R_3 must match R_4 and R_5 must match R_6). Further, it still has two resistors that must be changed if variable gain is desired (typically R_3 and R_4), and they must track each other with high precision over the operating temperature range.

A clever alternate configuration which solves the difficulties of the circuit in Figure 8–1 and provides high gain is the three op-amp IA shown in Figure 8–2. The inputs are

FIGURE 8–1
The basic instrumentation amplifier.

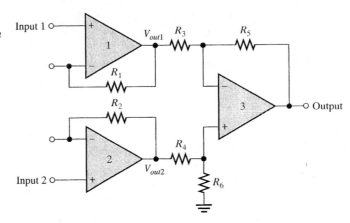

buffered by op-amp 1 and op-amp 2, providing a very high input impedance. Op-amps 1 and 2 can now provide gain. The entire assembly (except for R_G) is contained in a single IC. In this design, the common-mode gain still depends on very precisely matched resistors. However, these resistors can be critically matched during manufacture (by laser trimming) within the IC. Resistors R_3, R_4, R_5, and R_6 are generally set by the manufacturer for a gain of 1.0 for the differential amplifier. Resistors R_1 and R_2 are precision-matched resistors set equal to each other. This means the overall differential gain can be controlled by the size of just resistor R_G, supplied by the user. Appendix B derives the following equation for the output voltage:

$$V_{out} = \left(1 + \frac{2R}{R_G}\right)(V_{in2} - V_{in1}) \tag{8–1}$$

where the closed-loop gain is

$$A_{cl} = 1 + \frac{2R}{R_G}$$

and where $R_1 = R_2 = R$. The last equation shows that the gain of the instrumentation amplifier can be set by the value of the external resistor R_G when R_1 and R_2 have known fixed values.

FIGURE 8–2
The instrumentation amplifier with the external gain-setting resistor R_G. Differential and common-mode signals are indicated.

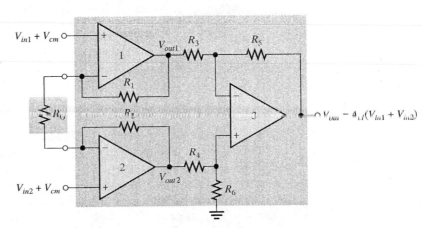

The external gain-setting resistor R_G can be calculated for a desired voltage gain by using the following formula:

$$R_G = \frac{2R}{A_{cl} - 1} \tag{8–2}$$

EXAMPLE 8–1   Determine the value of the external gain-setting resistor R_G for a certain IC instrumentation amplifier with $R_1 = R_2 = 25$ kΩ. The voltage gain is to be 500.

Solution
$$R_G = \frac{2R}{A_{cl} - 1} = \frac{50 \text{ k}\Omega}{500 - 1} \cong \mathbf{100\ \Omega}$$

Practice Exercise   What value of external gain-setting resistor is required for an instrumentation amplifier with $R_1 = R_2 = 39$ kΩ to produce a gain of 325?

Applications

As mentioned in the introduction to this section, the instrumentation amplifier is normally used to measure small differential signal voltages that are superimposed on a common-mode noise voltage often much larger than the signal voltage. Applications include situations where a quantity is sensed by a remote device, such as a temperature- or pressure-sensitive transducer, and the resulting small electrical signal is sent over a long line subject to electrical noise that produces common-mode voltages in the line. The instrumentation amplifier at the end of the line must amplify the small signal from the remote sensor and reject the large common-mode voltage. Figure 8–3 illustrates this.

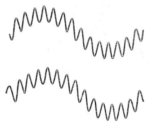
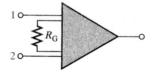
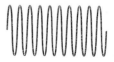

Small differential signals riding on larger common-mode signals

Instrumentation amplifier

Amplified differential signal No common-mode signal.

FIGURE 8–3

Illustration of the rejection of large common-mode voltages and the amplification of smaller signal voltages by an instrumentation amplifier.

A Specific Instrumentation Amplifier

Now that you have the basic idea of how an instrumentation amplifier works, let's look at a specific device. A representative device, the AD521, is shown in Figure 8–4 where IC pin numbers are given for reference. As you can see, there are some additional inputs and outputs that did not appear on the basic circuit. These provide additional features that are typical of many IAs on the market.

FIGURE 8–4
The AD521 instrumentation amplifier.

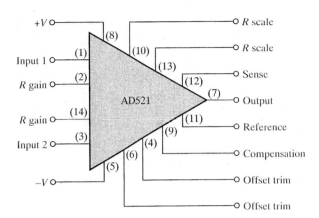

Some of the features of the AD521 are as follows. The voltage gain can be adjusted from 0.1 to 1000 with two external resistors. The input impedance is 3 GΩ. The common-mode rejection ratio (CMRR$'$) has a minimum value of 110 dB. Recall that a higher CMRR$'$ means better rejection of common-mode voltages. The AD521 has a gain-bandwidth product of 40 MHz. There is also an external provision for limiting the bandwidth, and the device is protected against excessive input voltages.

Setting the Voltage Gain   For the AD521, two external resistors must be used to set the voltage gain as indicated in Figure 8–5. Resistor R_G is connected between the R-gain terminals (pins 2 and 14). Resistor R_S is connected between the R-scale terminals (pins 10 and 13). R_S must be within ±15% of 100 kΩ, and R_G is selected for the desired gain based on the following formula:

$$A_v = \frac{R_S}{R_G} \tag{8–3}$$

Don't be concerned about the difference in this gain expression and the closed-loop gain for the basic circuit, which was given as $1 + 2R/R_G$. This difference is due to the subtle differences in design. If you were constructing an instrumentation amplifier from separate op-amps and discrete resistors, you would use the formulas discussed earlier.

FIGURE 8–5
The AD521 with gain-setting resistors and output offset adjustment.

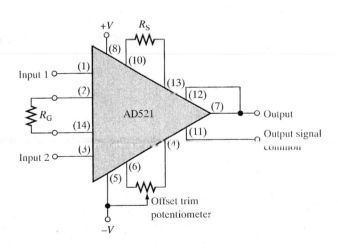

Offset Trim The offset **trim** adjustment (pins 4 and 6) is used to zero any output offset voltage caused by an input offset voltage multiplied by the gain. A potentiometer connected between pins 4 and 6, as shown in Figure 8–5, can be used to adjust the offset.

Bandwidth Control When you need to set the amplifier's bandwidth to a desired value, the compensation input (pin 9) can be used in conjunction with an external *RC* network. A recommended configuration is shown in Figure 8–6. The values of the two resistors and one of the capacitors are set as recommended by the manufacturer and then the value of the capacitor C_x is selected for the desired bandwidth according to the following formula:

$$C_x = \frac{1}{100\pi f_c} \tag{8–4}$$

where $BW = f_c$ is in kilohertz (kHz) and C_x is in microfarads (μF).

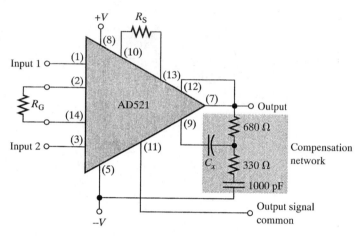

FIGURE 8–6
The AD521 with a compensation network for controlling the bandwidth.

EXAMPLE 8–2   Determine the gain and the bandwidth for the instrumentation amplifier in Figure 8–7.

FIGURE 8–7

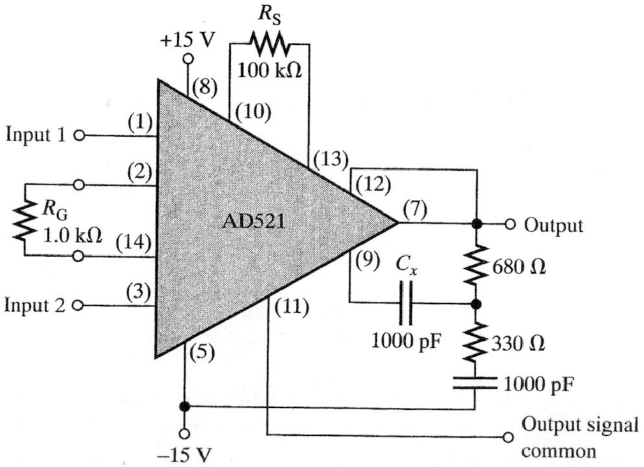

Solution   The voltage gain is determined by R_S and R_G as follows:

$$A_v = \frac{R_S}{R_G} = \frac{100 \text{ k}\Omega}{1.0 \text{ k}\Omega} = 100$$

The bandwidth is determined as follows:

$$C_x = \frac{1}{100\pi f_c}$$

$$BW = f_c = \frac{1}{100\pi C_x} = \frac{1}{100\pi(0.001)} = 3.18 \text{ kHz}$$

Notice that the number of microfarads (0.001) is substituted into the formula, not the number of farads (0.001×10^{-6}).

Practice Exercise   Modify the circuit in Figure 8–7 for a gain of approximately 45 and a bandwidth of approximately 10 kHz.

8–1 REVIEW QUESTIONS

1. What is the main purpose of an instrumentation amplifier and what are three of its key characteristics?
2. What components do you need to construct a basic instrumentation amplifier?
3. How is the gain determined in a basic instrumentation amplifier?
4. In a certain AD521 configuration, $R_S = 91$ kΩ and $R_G = 56$ kΩ. Is the voltage gain less than or greater than unity?

8–2 ■ ISOLATION AMPLIFIERS

An isolation amplifier provides dc isolation between input and output for the protection of human life or sensitive equipment in those applications where hazardous power-line leakage or high-voltage transients are possible. The principal areas of application for isolation amplifiers are in medical instrumentation, power plant instrumentation, industrial processing, and automated testing.

After completing this section, you should be able to

❑ Understand and explain the operation of an isolation amplifier
   ❑ Explain the basic configuration of an isolation amplifier
   ❑ Discuss an application in medical electronics
   ❑ Describe the features of the AD295 isolation amplifier

The Basic Isolation Amplifier

In some ways, the isolation amplifier can be viewed as an elaborate op-amp or instrumentation amplifier. The isolation amplifier has an input circuit that is electrically isolated from the output and power supply circuits using transformers or optical coupling. Transformer coupling is more common. The circuits are in IC form, but the transformers are not integrated. Although the packages that contain both the circuits and transformers are somewhat larger than standard IC packages, they are generally pin compatible for easy circuit board assembly.

The typical three-port isolation amplifier has three basic isolated sections that are transformer coupled. As shown in the block diagram of Figure 8–8, the sections are the input circuit, the output circuit, and the power source. The input section contains an instrumentation amplifier or an op-amp (in this case, it is an instrumentation amplifier), a power supply, and a modulator. The output section contains an op-amp, a power supply, and a demodulator. The power section contains an oscillator that produces an ac voltage from the dc input.

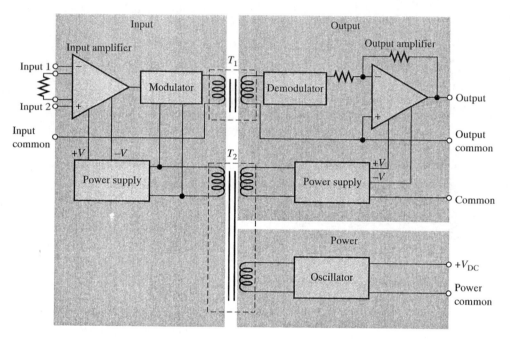

FIGURE 8–8
Basic isolation amplifier block diagram.

Operation An external dc supply voltage is applied to the power section where the internal oscillator is energized and converts the dc input power to ac power. The frequency of the oscillator is fairly high in order to keep the size of the transformers small; for example, it is 80 kHz in the AD295 isolation amplifier. The ac power signal from the oscillator is coupled to both the input and output sections through transformer T_2. In the input and output sections, the ac power signal is rectified and filtered by the power-supply circuits to provide dc power for the amplifiers.

The ac power signal is also sent to the modulator in the input section where it is modulated by the output signal from the input amplifier. The modulated signal is coupled through transformer T_1 to the output section where it is demodulated. The demodulation process recovers the original signal from the ac power signal. The output amplifier provides further gain before the signal goes to the final output.

Although the isolation amplifier is a fairly complex system in itself, in terms of its overall function, it is still simply an amplifier. You apply a dc voltage, put a signal in, and you get an amplified signal out. The isolation function is an unseen process.

Applications

As previously mentioned, the isolation amplifier is mainly used in medical equipment and for remote sensing in high-noise industrial environments where interfacing to sensitive equipment is required. In medical applications where body functions such as heart and blood pressure are monitored, the very small monitored signals are combined with large common-mode signals, such as 60 Hz power line pickup from the skin. In these situations, without isolation, dc leakage or equipment failure could be fatal. In chemical, nuclear, and metal-processing industries, for example, millivolt signals typically exist in the presence of common-mode voltages that can be in the kilovolt range. In this type of environment, the isolation amplifier can amplify small signals from very noisy equipment and provide a safe output to sensitive equipment such as computers.

Figure 8–9 shows a simplified diagram of an isolation amplifier in a cardiac monitoring application. In this situation, heart signals, which are very small, are combined with much larger common-mode signals caused by muscle noise, electrochemical noise, residual electrode voltages, and 60 Hz line pickup from the skin. The monitoring of fetal heartbeat, as illustrated, is the most demanding type of cardiac monitoring because in addition to the fetal heartbeat that typically generates 50 μV, there is also the mother's heartbeat that typically generates 1 mV. The common-mode voltages can run from about 1 mV to about 100 mV. The CMR (common-mode rejection) of the isolation amplifier separates the signal of the fetal heartbeat from that of the mother's heartbeat and from those signals of the common-mode voltages. So, the signal from the fetal heartbeat is essentially all that the amplifier sends to the monitoring equipment.

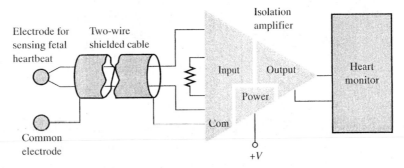

FIGURE 8–9
Fetal heartbeat monitoring using an isolation amplifier.

A Specific Isolation Amplifier

Now that you have learned basically what an isolation amplifier is and what it does, let's look at a representative device, the AD295, for a good introduction to practical IC isolation amplifiers. As you can see in Figure 8–10, the AD295 is similar to the basic isolation am-

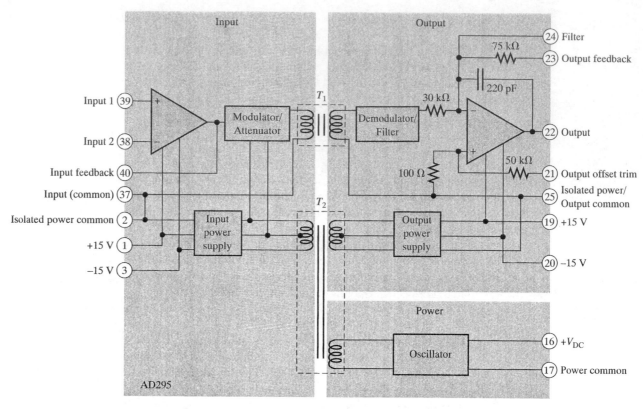

FIGURE 8–10
The AD295 isolation amplifier.

plifier in Figure 8–8 except that the input amplifier is an op-amp and, also, it has a few more inputs and outputs. These additional pins provide for gain adjustments, offset adjustments, isolated dc voltage outputs, and other functions.

Isolated Power Outputs The AD295 isolation amplifier provides ±15 V from both isolated power supplies. These voltages are available for powering associated external circuits such as preamplifiers, transducers, and the like.

Voltage Gain The gains of both the input and the output amplifiers can be set with external resistors. The overall gain of the device can be set at any value from 1 to 1000. Figure 8–11 shows the circuit connected for unity gain. In the input circuit, the amplifier output is connected directly back to the inverting input (pin 40 to pin 38) creating a voltage-follower configuration with a gain of 1. The attenuation network within the modulator/attenuator block has an inherent fixed attenuation of 0.4. To overcome this attenuation, the output amplifier has a gain of 2.5 ($A_v = 75\text{ k}\Omega/30\text{ k}\Omega = 2.5$) when the output is connected directly to the 75 kΩ feedback resistor (pin 22 to pin 23). This produces a combined gain of 1 ($0.4 \times 2.5 = 1$).

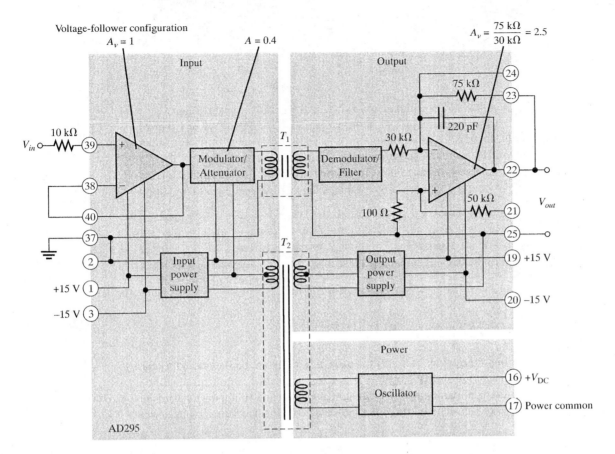

FIGURE 8–11
Unity-gain connections for the AD295 isolation amplifier.

Gains up to 1000 can be achieved by connecting external resistors as shown in Figure 8–12. Although the connection for the input amplifier is for a noninverting configuration, it can also be connected in an inverting configuration. The voltage gain of the input amplifier is

$$A_{v(input)} = \frac{R_f}{R_i} + 1 \tag{8–5}$$

For the AD295, there must also be a 10 kΩ resistor in series with the input as indicated in the figure. Also, $R_f + R_i$ must be equal to or greater than 10 kΩ. R_f is the feedback resistor and R_i is the input resistor for the op-amp configuration.

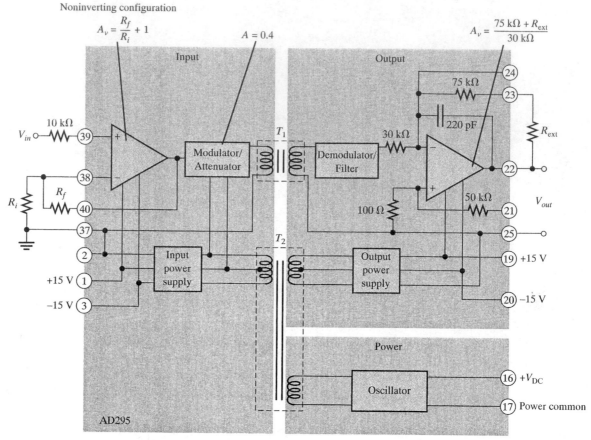

FIGURE 8–12
Nonunity gain connections for the AD295 isolation amplifier.

The voltage gain of the output amplifier can be increased above 2.5 by adding an external resistor in series with the internal 75 kΩ resistor as shown in Figure 8–12. For this case, the voltage gain of the output amplifier is

$$A_{v(output)} = \frac{75 \text{ k}\Omega + R_{ext}}{30 \text{ k}\Omega} \tag{8–6}$$

Offset Adjustments The external connections for adjustment of the input and output offset voltages are shown in Figure 8–13 in conjunction with a unity-gain configuration. The resistor values shown are recommended by the manufacturer for this particular device.

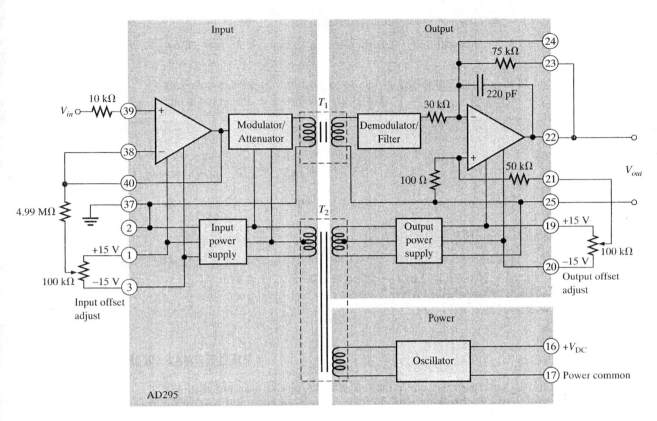

FIGURE 8–13
Offset voltage adjustments for the AD295 isolation amplifier.

EXAMPLE 8–3

Determine the overall voltage gain of the AD295 isolation amplifier in Figure 8–14.

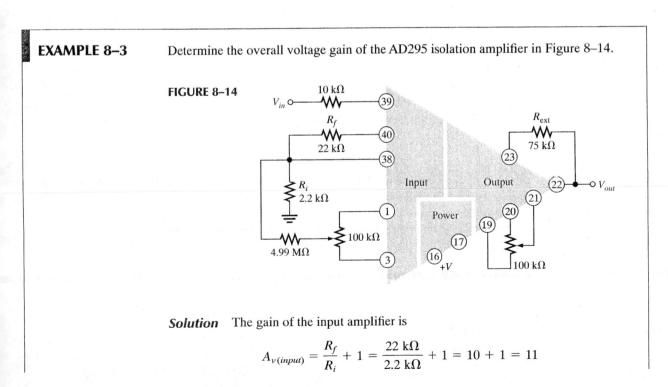

FIGURE 8–14

Solution The gain of the input amplifier is

$$A_{v(input)} = \frac{R_f}{R_i} + 1 = \frac{22 \text{ k}\Omega}{2.2 \text{ k}\Omega} + 1 = 10 + 1 = 11$$

The gain of the output amplifier is

$$A_{v(output)} = \frac{75 \text{ k}\Omega + R_{ext}}{30 \text{ k}\Omega} = \frac{75 \text{ k}\Omega + 75 \text{ k}\Omega}{30 \text{ k}\Omega} = 5$$

Since the fixed-internal attenuation is 0.4 for the AD295, the overall gain of the isolation amplifier is

$$A_v = A_{v(input)} \times A_{v(output)} \times \text{Atten} = (11)(5)(0.4) = \mathbf{22}$$

Practice Exercise Select resistor values and specify the connections in Figure 8–14 that will produce an overall gain of approximately 10.

8–2 REVIEW QUESTIONS

1. In what types of applications are isolation amplifiers used?
2. What are the three sections in a typical isolation amplifier?
3. How are the sections in an isolation amplifier connected?
4. What is the purpose of the oscillator in an isolation amplifier?

8–3 ■ OPERATIONAL TRANSCONDUCTANCE AMPLIFIERS (OTAs)

Conventional op-amps are, as you know, primarily voltage amplifiers in which the output voltage equals the gain times the input voltage. The OTA is primarily a voltage-to-current amplifier in which the output current equals the gain times the input voltage.

After completing this section, you should be able to

❑ Understand and explain the operation of an operational transconductance amplifier (OTA)
  ❑ Identify the OTA symbol
  ❑ Discuss the relationship between transconductance and bias current
  ❑ Describe the features of the CA3080 OTA
  ❑ Discuss OTA applications

Figure 8–15 shows the symbol for an operational transconductance amplifier (OTA). The double circle symbol at the output represents an output current source that is dependent on a bias current. Like the conventional op-amp, the OTA has two differential input terminals, a high input impedance, and a high CMRR. Unlike the conventional op-amp, the OTA has a bias-current input terminal, a high output impedance, and no fixed open-loop voltage gain.

FIGURE 8–15

Symbol for an operational transconductance amplifier (OTA).

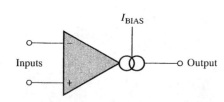

The Transconductance Is the Gain of an OTA

In general, the transconductance of an electronic device is the ratio of the output current to the input voltage. For an OTA, voltage is the input variable and current is the output variable; therefore, the ratio of output current to input voltage is its gain. Consequently, the voltage-to-current gain of an OTA is the transconductance, g_m.

$$A = g_m = \frac{I_{out}}{V_{in}}$$

In an OTA, the transconductance is dependent on a constant (K) times the bias current (I_{BIAS}) as indicated in Equation (8–7). The value of the constant is dependent on the internal circuit design.

$$g_m = KI_{BIAS} \tag{8–7}$$

The output current is controlled by the input voltage and the bias current as shown by the following formula:

$$I_{out} = g_m V_{in} = KI_{BIAS} V_{in}$$

The Transconductance Is a Function of Bias Current

The relationship of the transconductance and the bias current in an OTA is an important characteristic. The graph in Figure 8–16 illustrates a typical relationship. Notice that the transconductance increases linearly with the bias current. The constant of proportionality, K, is the slope of the line and has a value of approximately 16 μS/μA.

FIGURE 8–16
Graph of transconductance versus bias current for a typical OTA.

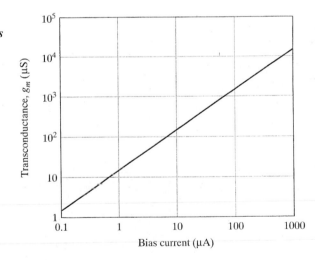

Bias current (μA)

EXAMPLE 8–4   If an OTA has a $g_m = 1000$ μS, what is the output current when the input voltage is 50 mV?

Solution $I_{out} = g_m V_{in} = (1000 \text{ μS})(50 \text{ mV}) = \textbf{50 μA}$

Practice Exercise   Based on the graph in Figure 8–16 where $K \cong 16$ μS/μA, determine the bias current required to produce $g_m = 1000$ μS.

Basic OTA Circuits

Figure 8–17 shows the OTA used as an inverting amplifier with fixed-voltage gain. The voltage gain is set by the transconductance and the load resistance as follows.

$$V_{out} = I_{out}R_L$$

Dividing both sides by V_{in},

$$\frac{V_{out}}{V_{in}} = \left(\frac{I_{out}}{V_{in}}\right)R_L$$

Since V_{out}/V_{in} is the voltage gain and $I_{out}/V_{in} = g_m$,

$$A_v = g_m R_L$$

FIGURE 8–17
An OTA as an inverting amplifier with a fixed-voltage gain.

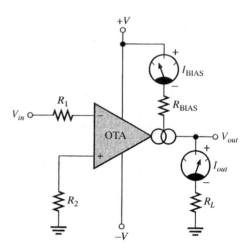

The transconductance of the amplifier in Figure 8–17 is determined by the amount of bias current, which is set by the dc supply voltages and the bias resistor R_{BIAS}.

One of the most useful features of an OTA is that the voltage gain can be controlled by the amount of bias current. This can be done manually, as shown in Figure 8–18(a), by using a variable resistor in series with R_{BIAS} in the circuit of Figure 8–17. By changing the resistance, you can produce a change in I_{BIAS}, which changes the transconductance. A change in the transconductance changes the voltage gain. The voltage gain can also be controlled with an externally applied variable voltage as shown in Figure 8–18(b). A variation in the applied bias voltage causes a change in the bias current.

A Specific OTA

The CA3080 is a typical OTA and serves as a representative device. Figure 8–19 shows its pin configuration for an eight-pin DIP. The maximum dc supply voltages are ±15 V, and its transconductance characteristic happens to be the same as indicated by the

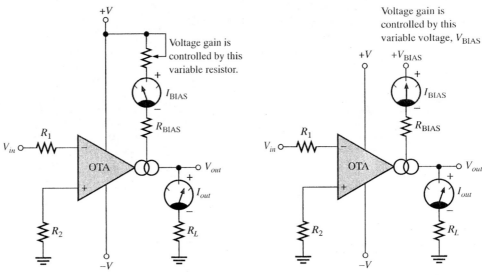

(a) Amplifier with resistance-controlled gain   (b) Amplifier with voltage-controlled gain

FIGURE 8–18
An OTA as an inverting amplifier with a variable-voltage gain.

FIGURE 8–19
The CA3080 OTA.

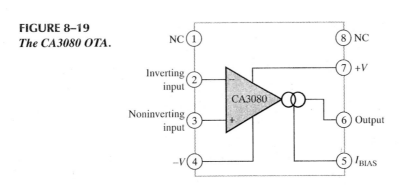

graph in Figure 8–16. For a CA3080, the bias current is determined by the following formula:

$$I_{BIAS} = \frac{+V_{BIAS} - (-V) - 0.7\ V}{R_{BIAS}} \qquad (8\text{–}8)$$

The 0.7 V is due to the internal circuit where a base-emitter junction connects the external R_{BIAS} with the negative supply voltage $(-V)$. The positive bias voltage may be obtained from the positive supply voltage.

Not only does the transconductance of an OTA vary with bias current, but so does the input and output resistances. Both the input and output resistances decrease as the bias current increases, as shown in Figure 8–20 for a CA3080.

FIGURE 8–20
Input and output resistances versus bias current.

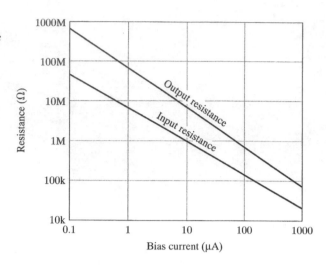

EXAMPLE 8–5

The OTA in Figure 8–21 is connected as an inverting fixed-gain amplifier. Determine the voltage gain.

FIGURE 8–21

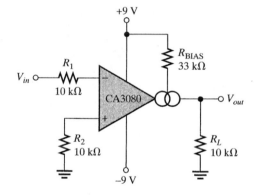

Solution   Calculate the bias current as follows:

$$I_{BIAS} = \frac{+V_{BIAS} - (-V) - 0.7\ V}{R_{BIAS}} = \frac{9\ V - (-9\ V) - 0.7\ V}{33\ k\Omega} = 524\ \mu A$$

Using $K \cong 16\ \mu S/\mu A$ from the graph in Figure 8–16, the value of transconductance corresponding to $I_{BIAS} = 524\ \mu A$ is approximately

$$g_m = KI_{BIAS} \cong (16\ \mu S/\mu A)(524\ \mu A) = 8.38 \times 10^3\ \mu S$$

Using this value of g_m, calculate the voltage gain.

$$A_v = g_m R_L \cong (8.38 \times 10^3\ \mu S)(10\ k\Omega) = \mathbf{83.8}$$

Practice Exercise   If the OTA in Figure 8–21 is operated with dc supply voltages of ±12 V, will this change the voltage gain and, if so, to what value?

Two OTA Applications

Amplitude Modulator   Figure 8–22 illustrates an OTA connected as an amplitude modulator. The voltage gain is varied by applying a modulation voltage to the bias input. When a constant-amplitude input signal is applied, the amplitude of the output signal will vary according to the modulation voltage on the bias input. The gain is dependent on bias current, and bias current is related to the modulation voltage by the following relationship:

$$I_{BIAS} = \frac{V_{mod} - (-V) - 0.7 \text{ V}}{R_{BIAS}}$$

This modulating action is shown in Figure 8–22 for a higher frequency sinusoidal input voltage and a lower frequency sinusoidal modulating voltage.

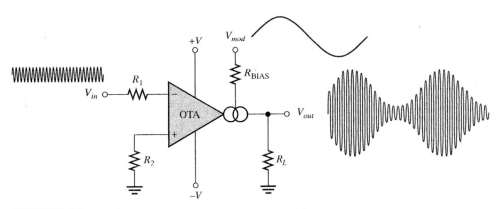

FIGURE 8–22
The OTA as an amplitude modulator.

EXAMPLE 8–6   The input to the OTA amplitude modulator in Figure 8–23 is a 50 mV peak-to-peak, 1 MHz sine wave. Determine the output signal, given the modulation voltage shown is applied to the bias input.

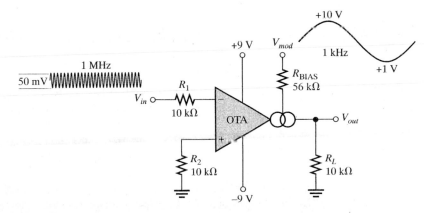

FIGURE 8–23

Solution The maximum voltage gain is when I_{BIAS}, and thus g_m, is maximum. This occurs at the maximum peak of the modulating voltage, V_{mod}.

$$I_{\text{BIAS(max)}} = \frac{V_{mod(max)} - (-V) - 0.7 \text{ V}}{R_{\text{BIAS}}} = \frac{10 \text{ V} - (-9 \text{ V}) - 0.7 \text{ V}}{56 \text{ k}\Omega} = 327 \text{ }\mu\text{A}$$

From the graph in Figure 8–16, the constant K is approximately 16 μS/μA.

$$g_m = KI_{\text{BIAS(max)}} = (16 \text{ }\mu\text{S/}\mu\text{A})(327 \text{ }\mu\text{A}) = 5.23 \text{ mS}$$
$$A_{v(max)} = g_m R_L = (5.23 \text{ mS})(10 \text{ k}\Omega) = 52.3$$
$$V_{out(max)} = A_{v(min)}V_{in} = (52.3)(50 \text{ mV}) = 2.62 \text{ V}$$

The minimum bias current is

$$I_{\text{BIAS(min)}} = \frac{V_{mod(min)} - (-V) - 0.7 \text{ V}}{R_{\text{BIAS}}} = \frac{1 \text{ V} - (-9 \text{ V}) - 0.7 \text{ V}}{56 \text{ k}\Omega} = 166 \text{ }\mu\text{A}$$

$$g_m = KI_{\text{BIAS(min)}} = (16 \text{ }\mu\text{S/}\mu\text{A})(166 \text{ }\mu\text{A}) = 2.66 \text{ mS}$$
$$A_{v(min)} = g_m R_L = (2.66 \text{ mS})(10 \text{ k}\Omega) = 26.6$$
$$V_{out(min)} = A_{v(min)}V_{in} = (26.6)(50 \text{ mV}) = 1.33 \text{ V}$$

The resulting output voltage is shown in Figure 8–24.

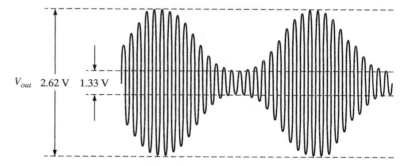

FIGURE 8–24

Practice Exercise Repeat this example with the sinusoidal modulating signal replaced by a square wave with the same maximum and minimum levels and a bias resistor of 39 kΩ.

Schmitt Trigger Figure 8–25 shows an OTA used in a Schmitt-trigger configuration. (Refer to Section 4–1.) Basically, a Schmitt trigger is a comparator with hysteresis where the input voltage drives the device into either positive or negative saturation. When the input voltage exceeds a certain threshold value or trigger point, the device switches to one of its saturated output states. When the input falls back below another threshold value, the device switches back to its other saturated output state.

In the case of the OTA Schmitt trigger, the threshold levels are set by the current through resistor R_1. The maximum output current in an OTA equals the bias current. Therefore, in the saturated output states, $I_{out} = I_{\text{BIAS}}$. The maximum positive output voltage is $I_{out}R_1$, and this voltage is the positive threshold value or upper trigger point. When the input voltage exceeds this value, the output switches to its maximum negative voltage, which

FIGURE 8–25
The OTA as a Schmitt trigger.

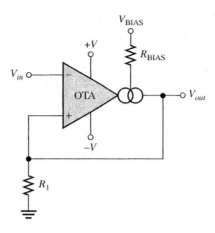

FIGURE 8–26
Basic operation of the OTA Schmitt trigger.

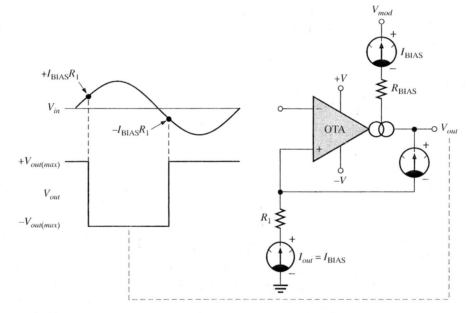

is $-I_{out}R_1$. Since $I_{out} = I_{BIAS}$, the trigger points can be controlled by the bias current. Figure 8–26 illustrates this operation.

8–3 REVIEW QUESTIONS

1. What does OTA stand for?
2. If the bias current in an OTA is increased, does the transconductance increase or decrease?
3. What happens to the voltage gain if the OTA is connected as a fixed-voltage amplifier and the supply voltages are increased?
4. What happens to the voltage gain if the OTA is connected as a variable-gain voltage amplifier and the voltage at the bias terminal is decreased?

8–4 ■ LOG AND ANTILOG AMPLIFIERS

A logarithmic (log) amplifier produces an output that is proportional to the logarithm of the input. Log amplifiers are used in applications that require compression of analog input data, linearization of transducers that have exponential outputs, optical density measurements and more. An antilogarithmic (antilog) amplifier takes the antilog or inverse log of the input. In this section, the principles of these amplifiers are discussed.

After completing this section, you should be able to

❑ Understand and explain the operation of log and antilog amplifiers
   ❑ Define *logarithm* and *natural logarithm*
   ❑ Describe the feedback configurations
   ❑ Discuss signal compression with logarithmic amplifiers

Logarithms

A **logarithm** (log) is basically a power. It is defined as the power to which a base, *b*, must be raised to yield a particular number, *N*. The defining formula for a logarithm is

$$b^x = N$$

In this formula, *x* represents the log of *N*. For example, you know that $10^2 = 100$. In this example, 2 is the power of ten that yields the number 100. In other words, 2 is the log of 100 (with base ten implied).

There are two practical bases used for logarithms. Because our counting system is base ten, base ten is used for what are called common logs. The abbreviation *log* in a mathematical expression or on your calculator implies base ten. Sometimes the subscript 10 is included with the abbreviation as $\log_{10}$. The second base is derived from an important mathematical series which gives the number 2.71828.[1] This number is represented by the letter *e* (mathematicians use ϵ). Base *e* is used because it is part of mathematical equations that describe natural phenomena such as the charging and discharging of a capacitor and the relationship between voltage and current in certain semiconductor devices. Logarithms that use base *e* are said to be **natural logarithms** and are shown with the abbreviation *ln* in mathematical expressions and on your calculator.

A useful conversion between the two bases is given by the equation

$$\ln x = 2.303 \log_{10} x$$

The Basic Logarithmic Amplifier

A log amp produces an output that is proportional to the logarithm of the input voltage. The key element in a basic log amplifier is a semiconductor *pn* junction in the form of either a diode or the base-emitter junction of a bipolar transistor. A *pn* junction exhibits a natural logarithmic current for many decades of input voltage. Figure 8–27(a) shows this characteristic for a typical small-signal diode, plotted as a linear plot; and Figure 8–27(b) shows the same characteristic plotted as a log plot (the *y*-axis is logarithmic). I_D is the forward diode current and V_D is the forward diode voltage. The logarithmic relationship between diode current and voltage is clearly seen in the plot in part (b). Although the plot only shows four decades of

---

[1] The series is $e = \lim_{n \to \inf} \left(1 + \dfrac{1}{n}\right)^n$.

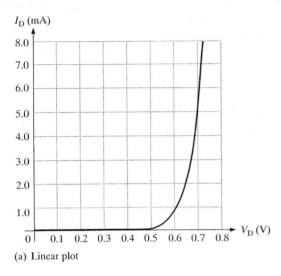

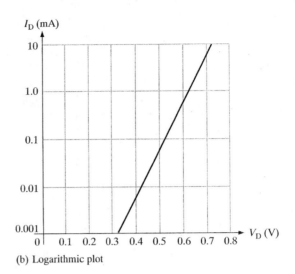

(a) Linear plot

(b) Logarithmic plot

FIGURE 8–27
The characteristic curve for a typical diode.

data, the logarithmic relationship extends over seven decades! The relationship between the current and voltage is expressed by the following general equation for a diode:

$$V_D = K \ln\left(\frac{I_D}{I_R}\right)$$

In this equation, K is a constant that is determined by several factors including the temperature and is approximately 0.025 V at 25°C. I_R, the reverse leakage current, is a constant for a given diode.

Log Amplifier with a Diode   When the feedback resistor in an inverting amplifier is replaced with a diode, the result is a basic log amp, as shown in Figure 8–28. The output voltage, V_{out}, is equal to $-V_D$. Because of the virtual ground, the input current can be expressed as V_{in}/R_1. By substituting these quantities into the diode equation, the output voltage is

$$V_{out} \cong -(0.025 \text{ V})\ln\left(\frac{V_{in}}{I_R R_1}\right) \qquad (8\text{–}9)$$

From Equation (8–9), you can see that the output voltage is the negative of a logarithmic function of the input voltage. The value of the output is controlled by the value of the input voltage and the value of the resistor R_1.

FIGURE 8–28
A basic log amplifier using a diode as the feedback element.

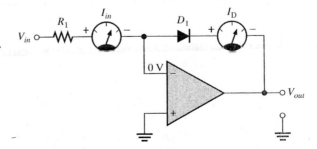

EXAMPLE 8–7   Determine the output voltage for the log amplifier in Figure 8–29. Assume $I_R = 50$ nA.

FIGURE 8–29

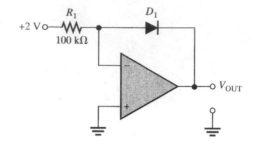

Solution   The input voltage and the resistor value are given in Figure 8–29.

$$V_{OUT} = -(0.025 \text{ V})\ln\left(\frac{V_{in}}{I_R R_1}\right) = -(0.025 \text{ V})\ln\left(\frac{2 \text{ V}}{(50 \text{ nA})(100 \text{ k}\Omega)}\right)$$
$$= -(0.025 \text{ V})\ln(400) = -(0.025 \text{ V})(5.99) = \mathbf{-0.150 \text{ V}}$$

Practice Exercise   Calculate the output voltage of the log amplifier with a +4 V input.

Log Amplifier with a BJT   The base-emitter junction of a bipolar transistor exhibits the same type of natural logarithmic characteristic as a diode because it is also a *pn* junction. A log amplifier with a BJT connected in a common-base form in the feedback loop is shown in Figure 8–30. Notice that V_{out} with respect to ground is equal to $-V_{BE}$.

FIGURE 8–30

A basic log amplifier using a transistor as the feedback element.

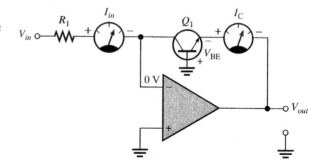

The analysis for this circuit is the same as for the diode log amplifier except that $-V_{BE}$ replaces V_D, I_C replaces I_D, and I_{EBO} replaces I_R. The emitter-to-base leakage current is I_{EBO}. The expression for the output voltage is

$$V_{out} = -(0.025 \text{ V})\ln\left(\frac{V_{in}}{I_{EBO} R_1}\right) \tag{8–10}$$

EXAMPLE 8–8

What is V_{out} for a transistor log amplifier with $V_{in} = 3$ V and $R_1 = 68$ kΩ? Assume $I_{EBO} = 40$ nA.

Solution

$$V_{out} = -(0.025 \text{ V})\ln\left(\frac{V_{in}}{I_{EBO}R_1}\right) = -(0.025 \text{ V})\ln\left(\frac{3 \text{ V}}{(40 \text{ nA})(68 \text{ kΩ})}\right)$$

$$= -(0.025 \text{ V})\ln(1103) = \mathbf{-0.175 \text{ V}}$$

Practice Exercise   Calculate V_{out} if R_1 is changed to 33 kΩ.

The Basic Antilog Amplifier

The antilog amplifier is the complement of a log amplifier. If you know the logarithm of a number, you know the power the base is raised to. To obtain the **antilog,** you must take the *exponential* of the logarithm.

$$x = e^{\ln x}$$

This is equivalent to saying the antilog$_e$ of ln x is just x. (Notice that the antilog is base e in this statement.) On many calculators the antilog of a base 10 logarithm is labelled $\boxed{10^x}$ and in some cases $\boxed{\text{INV}}\boxed{\text{LOG}}$. The antilog of a base e logarithm is labelled $\boxed{e^x}$ or $\boxed{\text{INV}}\boxed{\text{LN}}$.

The basic antilog amplifier is formed by reversing the position of the transistor (or diode) with the resistor in the log amp circuit. The antilog circuit is shown in Figure 8–31 using a transistor base-emitter junction as the input element and a resistor as the feedback element. The relationship between the current and voltage for a diode still applies.

$$V_D = K \ln\left(\frac{I_D}{I_R}\right)$$

FIGURE 8–31
A basic antilog amplifier.

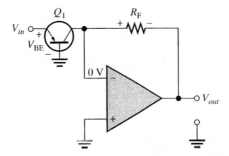

For the antilog amplifier, V_D is the negative input voltage and I_D represents the current in the feedback resistor, which by Ohm's law is V_{out}/R_F. Since a transistor is used, $I_R = I_{EBO}$. Making these substitutions in the diode equation,

$$V_{in} = -K \ln\left(\frac{V_{out}}{I_{EBO}R_F}\right)$$

Rearranging,

$$V_{out} = -I_{EBO}R_F e^{V_{in}/K}$$

Substituting $K \cong 0.025$ V and clearing the exponent,

$$V_{out} \cong -I_{EBO}R_F \text{antilog}_e\left(\frac{V_{in}}{25 \text{ mV}}\right)$$

(8–11)

EXAMPLE 8–9   For the antilog amplifier in Figure 8–32, find the output voltage. Assume $I_{EBO} = 40$ nA.

FIGURE 8–32

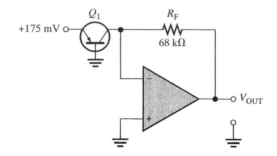

Solution   First of all, notice that the input voltage in Figure 8–32 is the inverted output voltage of the log amplifier in Example 8–8. In this case, the antilog amplifier reverses the process and produces an output that is proportional to the antilog of the input. So, the output voltage of the antilog amplifier in Figure 8–32 should have the same magnitude as the input voltage of the log amplifier in Example 8–8 because all the constants are the same. Let's see if it does.

$$V_{OUT} \cong -I_{EBO}R_F \text{antilog}_e\left(\frac{V_{in}}{25 \text{ mV}}\right) = -(40 \text{ nA})(68 \text{ k}\Omega)\text{antilog}_e\left(\frac{0.175 \text{ V}}{25 \text{ mV}}\right)$$

$$= -(40 \text{ nA})(68 \text{ k}\Omega)(1100) = -3 \text{ V}$$

Practice Exercise   Determine V_{OUT} for the amplifier in Figure 8–32 if the feedback resistor is changed to 100 kΩ.

IC Log, Log-Ratio, and Antilog Amplifiers

Several factors make the basic log amp and the basic antilog amp circuit with a diode and op-amp unsatisfactory for many applications. The basic circuit is temperature sensitive and tends to have error at very low diode currents; also, components need to be precisely matched, and the output level is not a convenient value. These problems are difficult to address with off-the-shelf components; however, manufacturers have designed precision integrated circuit logarithmic and log-ratio amplifiers with temperature compensation, low bias currents, and high accuracy that require no user adjustments. Log-ratio measurements produce an output that is proportional to the log ratio of *two* inputs.

The Burr-Brown LOG100[2] is an example of a log, log-ratio, and antilog amplifier in one 14 pin IC. It has a maximum accuracy specification of 0.37% at full-scale output (FSO) and a six-decade range of input current (1 nA to 1 mA). With a few external resistors, the user can connect the amplifier as either a log, a log-ratio, or an antilog amplifier. Scaling of the output voltage can be done by simply selecting an appropriate output pin. As with most log amps, the input will function with only one polarity of input current.

Another interesting device from Burr-Brown is the 24 pin 4127 log amp that features the ability to accept input voltages or currents of *either* polarity, the first log amp to do so. It maintains high accuracy over a six-decade range of input current (1 nA to 1 mA) or a four-decade range of input voltage. Only a few external resistors are required to complete a log, log-ratio, or an antilog amplifier. Within this IC is an uncommitted op-amp that can be used for any purpose such as a gain stage, buffer, filter, or inverter.

Signal Compression with Logarithmic Amplifiers

In certain applications, a signal may be too large in magnitude for a particular system to handle. The term *dynamic range* is often used to describe the range of voltages contained in a signal. In these cases, the signal voltage must be scaled down by a process called **signal compression** so that it can be properly handled by the system. If a linear circuit is used to scale a signal down in amplitude, the lower voltages are reduced by the same percentage as the higher voltages. Linear signal compression often results in the lower voltages becoming obscured by noise and difficult to accurately distinguish, as illustrated in Figure 8–33(a). To overcome this problem, a signal with a large dynamic range can be com-

---

[2] The data sheet is given in Appendix A.

FIGURE 8–33

The basic concept of signal compression with a logarithmic amplifier.

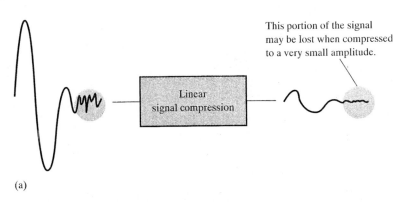

This portion of the signal may be lost when compressed to a very small amplitude.

Linear signal compression

(a)

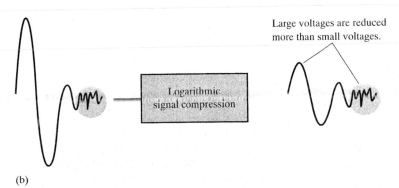

Large voltages are reduced more than small voltages.

Logarithmic signal compression

(b)

pressed using a logarithmic response, as shown in Figure 8–33(b). In logarithmic signal compression, the higher voltages are reduced more than the lower voltages, thus keeping the lower voltage signals from being lost in noise. A log amplifier preceding an 8-bit ADC can replace a more expensive 20-bit ADC because of signal compression.

A Basic Multiplier with Log and Antilog Amps

Multipliers are based on the fundamental logarithmic relationship that states that the product of two terms equals the sum of the logarithms of each term. This relationship is shown in the following formula:

$$\ln(a \times b) = \ln a + \ln b$$

This formula shows that two signal voltages are effectively multiplied if the logarithms of the signal voltages are added.

You know how to get the logarithm of a signal voltage by using a log amplifier. By summing the outputs of two log amplifiers, you get the logarithm of the product of the two original input voltages. Then, by taking the antilogarithm, you get the product of the two input voltages as indicated in the following equations:

$$\ln V_1 + \ln V_2 = \ln(V_1 V_2)$$
$$\text{antilog}_e[\ln(V_1 V_2)] = V_1 V_2$$

The block diagram in Figure 8–34 shows how the functions are connected to multiply two input voltages. Constant terms are omitted for simplicity.

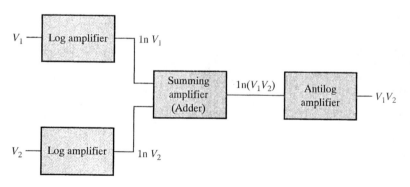

FIGURE 8–34
Basic block diagram of an analog multiplier.

Figure 8–35 shows the basic multiplier circuitry. The outputs of the log amplifiers are stated as follows:

$$V_{out(log1)} = -K_1 \ln\left(\frac{V_{in1}}{K_2}\right)$$

$$V_{out(log2)} = -K_1 \ln\left(\frac{V_{in2}}{K_2}\right)$$

FIGURE 8–35
A basic multiplier.

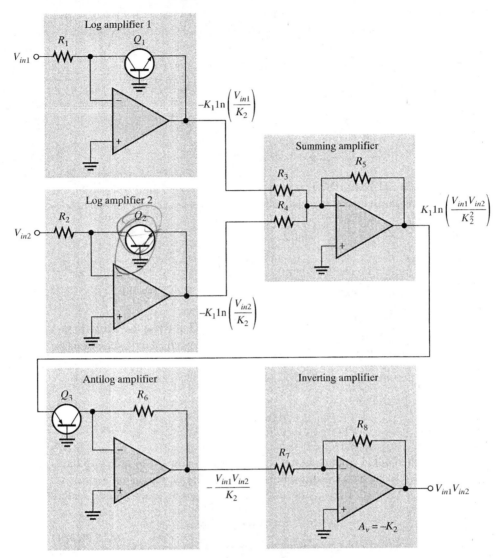

where $K_1 = 0.025$ V, $K_2 = RI_{EBO}$, and $R = R_1 = R_2 = R_6$. The two output voltages from the log amplifiers are added and inverted by the unity-gain summing amplifier to produce the following result:

$$V_{out(sum)} = K_1 \ln\left[\left(\frac{V_{in1}}{K_2}\right) + \ln\left(\frac{V_{in2}}{K_2}\right)\right] = K_1 \ln\left(\frac{V_{in1}V_{in2}}{K_2^2}\right)$$

This expression is then applied to the antilog amplifier; the expression for the multiplier output voltage is as follows:

$$V_{out(antilog)} = -K_2 \text{antilog}_e\left(\frac{V_{out(sum)}}{K_1}\right) = -K_2 \text{antilog}_e\left[\frac{K_1 \ln\left(\frac{V_{in1}V_{in2}}{K_2^2}\right)}{K_1}\right]$$

$$= -K_2\left(\frac{V_{in1}V_{in2}}{K_2^2}\right) = -\frac{V_{in1}V_{in2}}{K_2}$$

As you can see, the output of the antilog amplifier is a constant $(1/K_2)$ times the *product* of the input voltages. The final output is developed by an inverting amplifier with a voltage gain of $-K_2$.

$$V_{out} = -K_2\left(-\frac{V_{in1}V_{in2}}{K_2}\right)$$

$$V_{out} = V_{in1}V_{in2} \tag{8–12}$$

As in the case of log amps, analog multipliers are available in IC form. These are covered in Chapter 9.

8–4 REVIEW QUESTIONS

1. What purpose does the diode or transistor perform in the feedback loop of a log amplifier?
2. Why is the output of a basic log amplifier limited to about 0.7 V?
3. What are the factors that determine the output voltage of a basic log amplifier?
4. In terms of implementation, how does a basic antilog amplifier differ from a basic log amplifier?
5. What circuits make up a basic analog multiplier?

8–5 ■ A SYSTEM APPLICATION

The electrocardiograph, described at the beginning of the chapter, is a medical instrument used for monitoring heart signals of patients. From the output waveform of an ECG, the doctor can detect abnormalities in the heartbeat.

After completing this section, you should be able to

❑ Apply what you have learned in this chapter to a system application
   ❑ Discuss how an isolation amplifier is used in medical instrumentation
   ❑ Describe how other op-amp circuits are also used as part of the ECG system
   ❑ Translate between a printed circuit board and a schematic
   ❑ Analyze the amplifier board
   ❑ Troubleshoot some common problems

A Brief Description of the System

The human heart produces an electrical signal that can be picked up by electrodes in contact with the skin. When the heart signal is displayed on a chart recorder or on a video monitor, it is called an electrocardiograph or ECG. Typically, the heart signal picked up by the electrode is about 1 mV and has significant frequency components from less than 1 Hz to about 100 Hz.

As indicated in the block diagram in Figure 8–36, an ECG system has at least three electrodes. There is a right-arm (RA) electrode, a left-arm (LA) electrode, and a right-leg

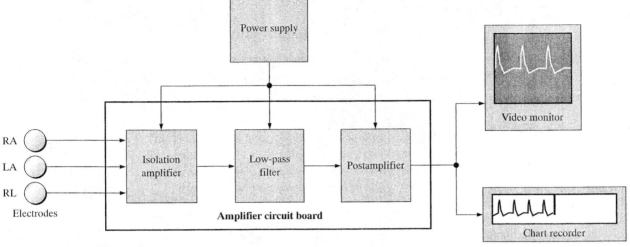

FIGURE 8–36
ECG block diagram.

(RL) electrode that is used as the common. The isolation amplifier provides for differential inputs from the electrodes, provides a high CMR to eliminate the relatively high common-mode noise voltages associated with heart signals, and provides electrical isolation for protection of the patient. The low-pass active filter rejects frequencies above those contained in the heart signal. The postamplifier provides most of the amplification in the system and drives a video monitor and/or a chart recorder. The three op-amp circuits—the isolation amplifier, the low-pass filter, and the postamplifier—are on a single PC board called the amplifier board.

Now, so that you can take a closer look at the amplifier board, let's take it out of the system and put it on the test bench.

ON THE TEST BENCH

A Brief Description of the Circuits

The inputs from the electrode sensors come into the amplifier board shown in Figure 8–37 with a shielded cable to prevent noise pickup. The schematic for the amplifier board is shown in Figure 8–38. The shielded cable is basically a twisted pair of wires surrounded by a braided metal sheathing that is covered by an insulated sheathing. The braided metal shield serves as the conduit for the common connection. The incoming differential signal is amplified by the fixed gain of the AD295 isolation amplifier. The AD295 is housed in a package that is significantly larger than standard IC DIPs or SOPs. The larger configuration is required to accommodate the coupling transformers used in the isolation circuitry. The particular package in this system is a 40-pin package beginning with pin 1 at the dot. Pin 40 is directly across from pin 1.

The low-pass filter is a Sallen-Key two-pole filter, and the postamplifier is an inverting amplifier with adjustable gain. The inverting input of the postamplifier also serves as a summing point for the signal voltage and a dc voltage used for adding a dc level to the output for purposes of adjusting the vertical position of the display.

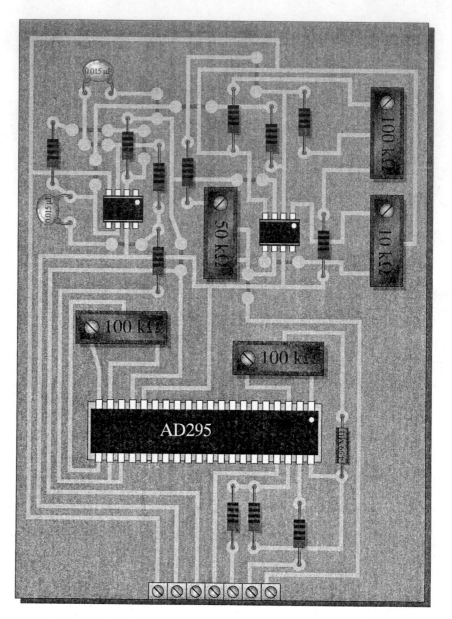

FIGURE 8–37

■ ACTIVITY 1 Relate the PC Board to the Schematic

Locate and identify each component and each input/output pin on the PC board in Figure 8–37 using the schematic in Figure 8–38. Verify that the board and the schematic agree.

■ ACTIVITY 2 Analyze the System

Step 1: Determine the voltage gain of the isolation amplifier.

Step 2: Determine the bandwidth of the active filter.

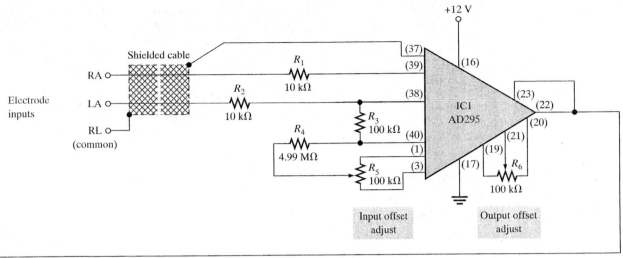

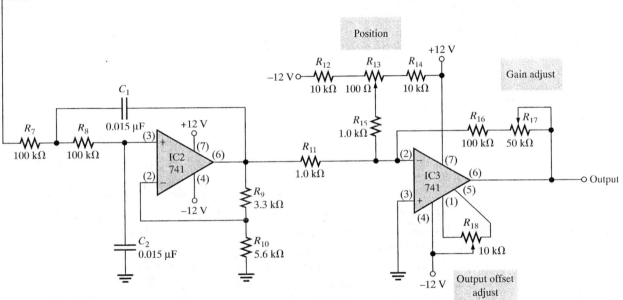

FIGURE 8-38

Step 3: Determine the minimum and maximum voltage gain of the postamplifier.

Step 4: Determine the overall gain range of the amplifier board.

Step 5: Determine the voltage range at the wiper of the position adjustment potentiometer.

■ ACTIVITY 3   Write a Technical Report

Describe the overall operation of the amplifier board. Specify how each circuit works and what its purpose is. Explain how the gain is adjusted and how the dc level of the output can be changed. Use the results of Activity 2 as appropriate.

■ **ACTIVITY 4 Troubleshoot the System for Each of the Following Problems by Stating the Probable Cause or Causes**

1. There is no final output voltage when there is a verified 1 mV input signal.
2. There is a 10 mV signal at the output of IC1, but no signal at pin 3 of IC2.
3. There is a 15 mV signal at the output of IC2, but no signal at pin 2 of IC3.
4. With a valid input signal, IC3 is being driven into its saturated states and is basically acting as a comparator.

8–5 REVIEW QUESTIONS

1. Which resistors determine the voltage gain of the isolation amplifier?
2. What is the voltage gain of the output section of IC1?
3. What are the lower and upper critical frequencies of the active filter?
4. What is the gain of the postamplifier if R_{17} is set at its midpoint resistance?

■ SUMMARY

- A basic instrumentation amplifier is formed by three op-amps and seven resistors, including the gain-setting resistor, R_G.
- An instrumentation amplifier has high input impedance, high CMRR, low output offset, and low output impedance.
- The voltage gain of a basic instrumentation amplifier is set by a single external resistor.
- An instrumentation amplifier is useful in applications where small signals are embedded in large common-mode noise.
- A basic isolation amplifier has three electrically isolated sections: input, output, and power.
- Most isolation amplifiers use transformer coupling for isolation.
- Isolation amplifiers are used to interface sensitive equipment with high-voltage environments and to provide protection from electrical shock in certain medical applications.
- The operational transconductance amplifier (OTA) is a voltage-to-current amplifier.
- The output current of an OTA is the input voltage times the transconductance.
- In an OTA, transconductance varies with bias current; therefore, the gain of an OTA can be varied with a bias voltage.
- The operation of log and antilog amplifiers is based on the nonlinear (logarithmic) characteristic of a *pn* junction.
- A log amplifier has a BJT in the feedback loop.
- An antilog amplifier has a BJT in series with the input.
- Logarithmic amplifiers are used for signal compression, analog multiplication, and log-ratio measurements.
- An analog multiplier is based on the mathematical principle that states the logarithm of the product of two variables equals the sum of the logarithms of the variables.

■ GLOSSARY

These terms are included in the end-of-book glossary.

Antilog   The number corresponding to a given logarithm.

Instrumentation amplifier   A differential voltage-gain device that amplifies the differences between the voltages existing at its two input terminals.

Logarithm   An exponent; the logarithm of a quantity is the exponent or power to which a given number called the base must be raised in order to equal the quantity.

Natural logarithm   The exponent to which the base e ($e = 2.71828$) must be raised in order to equal a given quantity.

Signal compression   The process of scaling down the amplitude of a signal voltage.

Trim   To precisely adjust or fine tune a value.

■ **KEY FORMULAS**

Instrumentation Amplifier

$$(8\text{–}1) \qquad V_{out} = \left(\frac{1 + 2R}{R_G}\right)(V_{in2} - V_{in1})$$

$$(8\text{–}2) \qquad R_G = \frac{2R}{A_{cl} - 1}$$

$$(8\text{–}3) \qquad A_v = \frac{R_S}{R_G}$$

$$(8\text{–}4) \qquad C_x = \frac{1}{100\pi f_c}$$

Isolation Amplifier

$$(8\text{–}5) \qquad A_{v(input)} = \frac{R_f}{R_i} + 1$$

$$(8\text{–}6) \qquad A_{v(output)} = \frac{75\text{ k}\Omega + R_{ext}}{30\text{ k}\Omega}$$

Operational Transconductance Amplifier (OTA)

$$(8\text{–}7) \qquad g_m = KI_{BIAS}$$

$$(8\text{–}8) \qquad I_{BIAS} = \frac{+V_{BIAS} - (-V) - 0.7\text{ V}}{R_{BIAS}}$$

Logarithmic Amplifier

$$(8\text{–}9) \qquad V_{out} \cong -(0.025\text{ V})\ln\left(\frac{V_{in}}{I_R R_1}\right)$$

$$(8\text{–}10) \qquad V_{out} = -(0.025\text{ V})\ln\left(\frac{V_{in}}{I_{EBO} R_1}\right)$$

$$(8\text{–}11) \qquad V_{out} = -I_{EBO}R_F \text{antilog}_e\left(\frac{V_{in}}{25\text{ mV}}\right)$$

$$(8\text{–}12) \qquad V_{out} = V_{in1}V_{in2}$$

■ **SELF-TEST**

1. To make a basic instrumentation amplifier, it takes
   - (a) one op-amp with a certain feedback arrangement
   - (b) two op-amps and seven resistors
   - (c) three op-amps and seven capacitors
   - (d) three op-amps and seven resistors

2. Typically, an instrumentation amplifier has an external resistor used for
   (a) establishing the input impedance   (b) setting the voltage gain
   (c) setting the current gain   (d) for interfacing with an instrument

3. Instrumentation amplifiers are used primarily in
   (a) high-noise environments   (b) medical equipment
   (c) test instruments   (d) filter circuits

4. Isolation amplifiers are used primarily in
   (a) remote, isolated locations
   (b) systems that isolate a single signal from many different signals
   (c) applications where there are high voltages and sensitive equipment
   (d) applications where human safety is a concern
   (e) answers (c) and (d)

5. The three sections of a basic isolation amplifier are
   (a) amplifier, filter, and power   (b) input, output, and coupling
   (c) input, output, and power   (d) gain, attenuation, and offset

6. The sections of most isolation amplifiers are connected by
   (a) copper strips   (b) transformers
   (c) microwave links   (d) current loops

7. The characteristic that allows an isolation amplifier to amplify small signal voltages in the presence of much greater noise voltages is its
   (a) CMRR   (b) high gain
   (c) high input impedance   (d) magnetic coupling between input and output

8. The term *OTA* means
   (a) operational transistor amplifier
   (b) operational transformer amplifier
   (c) operational transconductance amplifier
   (d) output transducer amplifier

9. In an OTA, the transconductance is controlled by
   (a) the dc supply voltage   (b) the input signal voltage
   (c) the manufacturing process (d) a bias current

10. The voltage gain of an OTA circuit is set by
    (a) a feedback resistor
    (b) the transconductance only
    (c) the transconductance and the load resistor
    (d) the bias current and supply voltage

11. An OTA is basically a
    (a) voltage-to-current amplifier   (b) current-to-voltage amplifier
    (c) current-to-current amplifier   (d) voltage-to-voltage amplifier

12. The operation of a logarithmic amplifier is based on
    (a) the nonlinear operation of an op-amp
    (b) the logarithmic characteristic of a *pn* junction
    (c) the reverse breakdown characteristic of a *pn* junction
    (d) the logarithmic charge and discharge of an *RC* circuit

13. If the input to a log amplifier is *x*, the output is proportional to
    (a) e^x (b) $\ln x$ (c) $\log_{10}x$ (d) $2.3 \log_{10}x$
    (e) answers (a) and (c)   (f) answers (b) and (d)

14. If the input to an antilog amplifier is *x*, the output is proportional to
    (a) $e^{\ln x}$ (b) e^x (c) $\ln x$ (d) e^{-x}

15. The logarithm of the product of two numbers is equal to the
   (a) sum of the two numbers
   (b) sum of the logarithms of each of the numbers
   (c) difference of the logarithms of each of the numbers
   (d) ratio of the logarithms of the numbers

■ PROBLEMS

SECTION 8–1   Instrumentation Amplifiers

1. Determine the voltage gains of op-amps 1 and 2 for the instrumentation amplifier configuration in Figure 8–39.

2. Find the overall voltage gain of the instrumentation amplifier in Figure 8–39.

3. The following voltages are applied to the instrumentation amplifier in Figure 8–39.
   $V_{in1} = 5$ mV, $V_{in2} = 10$ mV, and $V_{cm} = 225$ mV. Determine the final output voltage.

4. What value of R_G must be used to change the gain of the instrumentation amplifier in Figure 8–39 to 1000?

FIGURE 8–39

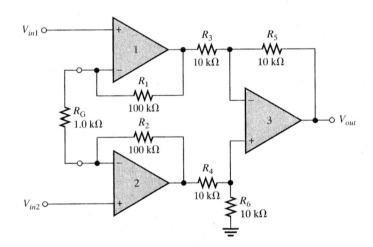

5. What is the voltage gain of the AD521 instrumentation amplifier in Figure 8–40?

6. Determine the bandwidth of the amplifier in Figure 8–40.

FIGURE 8–40

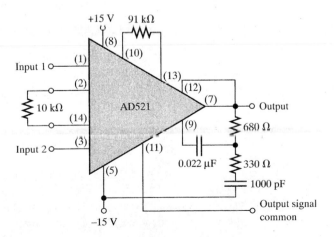

7. Specify what you must do to change the gain of the amplifier in Figure 8–40 to approximately 50.

8. Specify what you must do to change the bandwidth of the amplifier in Figure 8–40 to approximately 15 kHz.

SECTION 8–2 Isolation Amplifiers

9. The op-amp in the input section of a certain isolation amplifier has a voltage gain of 30 and the attenuation network has an attenuation of 0.75. The output section is set for a gain of 10. What is the overall voltage gain of this device?

10. Determine the overall voltage gain of each AD295 in Figure 8–41.

11. Specify how you would change the overall gain of the amplifier in Figure 8–41(a) to approximately 100 by changing only the gain of the input section.

12. Specify how you would change the overall gain in Figure 8–41(b) to approximately 450 by changing only the gain of the output section.

13. Specify how you would connect each amplifier in Figure 8–41 for unity gain.

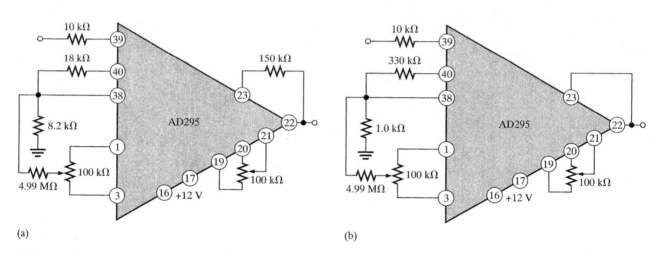

(a)   (b)

FIGURE 8–41

SECTION 8–3 Operational Transconductance Amplifiers (OTAs)

14. A certain OTA has an input voltage of 10 mV and an output current of 10 μA. What is the transconductance?

15. A certain OTA with a transconductance of 5000 μS has a load resistance of 10 kΩ. If the input voltage is 100 mV, what is the output current? What is the output voltage?

16. The output voltage of a certain OTA with a load resistance is determined to be 3.5 V. If its transconductance is 4000 μS and the input voltage is 100 mV, what is the value of the load resistance?

17. Determine the voltage gain of the OTA in Figure 8–42. Assume $K = 16$ μS/μA for the graph in Figure 8–43.

18. If a 10 kΩ rheostat is added in series with the bias resistor in Figure 8–42, what are the minimum and maximum voltage gains?

19. The OTA in Figure 8–44 functions as an amplitude modulation circuit. Determine the output voltage waveform for the given input waveforms assuming $K = 16$ μS/μA.

FIGURE 8–42

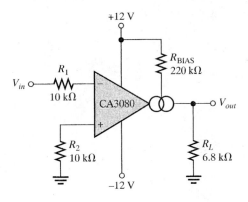

FIGURE 8–43

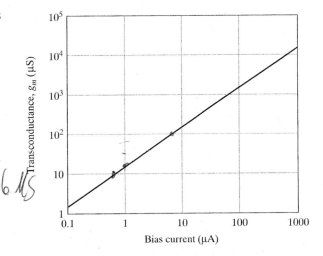

FIGURE 8–44

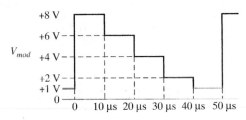

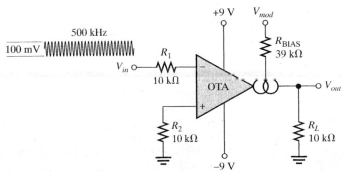

FIGURE 8–45

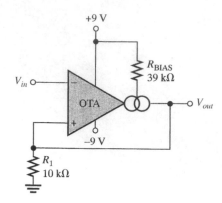

20. Determine the trigger points for the Schmitt-trigger circuit in Figure 8–45.

21. Determine the output voltage waveform for the Schmitt trigger in Figure 8–45 in relation to a 1 kHz sine wave with peak values of ±10 V.

SECTION 8–4   Log and Antilog Amplifiers

22. Using your calculator, find the natural logarithm (ln) of each of the following numbers:
    (a) 0.5   (b) 2   (c) 50   (d) 130

23. Repeat Problem 22 for $\log_{10}$.

24. What is the antilog of 1.6?

25. Explain why the output of a log amplifier is limited to approximately 0.7 V.

26. What is the output voltage of a certain log amplifier with a diode in the feedback path when the input voltage is 3 V? The input resistor is 82 kΩ and the reverse leakage current is 100 nA.

27. Determine the output voltage for the amplifier in Figure 8–46. Assume $I_{EBO} = 60$ nA.

28. Determine the output voltage for the amplifier in Figure 8–47. Assume $I_{EBO} = 60$ nA.

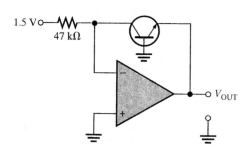

FIGURE 8–46

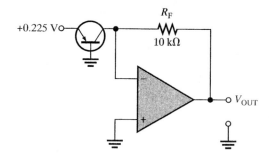

FIGURE 8–47

29. Signal compression is one application of logarithmic amplifiers. Suppose an audio signal with a maximum voltage of 1 V and a minimum voltage of 100 mV is applied to the log amplifier in Figure 8–46. What will be the maximum and minimum output voltages? What conclusion can you draw from this result?

SECTION 8–5 A System Application

30. With a 1 mV, 50 Hz signal applied to the ECG amplifier board in Figure 8–48, what voltage would you expect to see at each of the probed points? All voltages are measured with respect to ground (not shown). Assume that all offset voltages are nulled out and the position control is adjusted for zero deflection. The gain adjustment potentiometer is set at midrange. Refer to the schematic in Figure 8–38.

31. Repeat Problem 30 for a 2 mV, 1 kHz input signal.

FIGURE 8–48

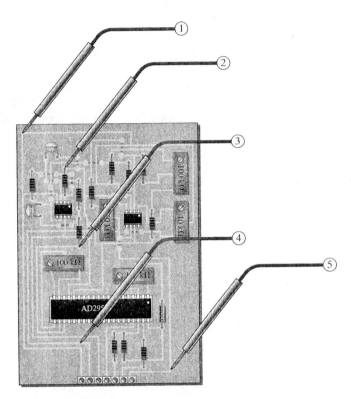

ANSWERS TO REVIEW QUESTIONS

Section 8–1

1. The main purpose of an instrumentation amplifier is to amplify small signals that occur on large common-mode voltages. The key characteristics are high input impedance, high CMRR, low output impedance, and low output offset.

2. Three op-amps and seven resistors are required to construct a basic instrumentation amplifier.

3. The gain is set by the internal feedback resistors and an external resistor.

4. The gain is greater than unity.

Section 8–2

1. Isolation amplifiers are used in medical equipment, power plant instrumentation, industrial processing, and automated testing.

2. The three sections of an isolation amplifier are input, output, and power.

3. The sections are connected by transformer coupling and in some devices by optical coupling.

4. The oscillator acts as a dc to ac converter so that the dc power can be ac coupled to the input and output sections.

Section 8–3

1. OTA stands for Operational Transconductance Amplifier.

2. Transconductance increases with bias current.

3. Assuming that the bias input is connected to the supply voltage, the voltage gain increases when the supply voltage is increased because this increases the bias current.

4. The gain decreases as the bias voltage decreases.

Section 8–4

1. A diode or transistor in the feedback loop provides the exponential (nonlinear) characteristic.

2. The output of a basic log amplifier is limited to the barrier potential of the *pn* junction (about 0.7 V).

3. The output voltage is determined by the input voltage, the input resistor, and the emitter-to-base leakage current.

4. The transistor in an antilog amplifier is in series with the input rather than in the feedback loop.

5. A multiplier is made of two log amplifiers, a summing amplifier, an antilog amplifier, and an inverting amplifier.

Section 8–5

1. The gain is set by R_2 and R_3.

2. 75 kΩ/30 kΩ = 2.5

3. Lower: 0 Hz; Upper: $1/2\pi RC = 1/2\pi(0.015\ \mu F)(100\ k\Omega) = 106$ Hz

4. $-(125\ k\Omega/1\ k\Omega) = -125$

■ ANSWERS TO PRACTICE EXERCISES FOR EXAMPLES

8–1 240 Ω

8–2 Leave R_S = 100 kΩ and make R_G = 2.2 kΩ, C_x = 320 pF.

8–3 Many combinations are possible. Here is one: Connect the input for unity gain (pin 38 to 40, no R_i or R_f). Connect a 680 kΩ resistor from pin 22 to pin 23.

8–4 62.5 μA. Note the scale is logarithmic.

8–5 Yes. Approximately 113.

8–6 $V_{out(max)}$ = 3.75 V; $V_{out(min)}$ = 1.91 V

8–7 -0.167 V

8–8 -0.193 V

8–9 -4.4 V

9

COMMUNICATIONS CIRCUITS

Courtesy Hewlett-Packard Company

■ CHAPTER OBJECTIVES

☐ Describe basic superheterodyne receivers
☐ Discuss the function of a linear multiplier
☐ Discuss the fundamentals of amplitude modulation
☐ Discuss the basic function of a mixer
☐ Describe the basic principles of AM demodulation
☐ Describe the basic aspects of IF and audio amplifiers
☐ Describe the basic principles of frequency modulation
☐ Describe the basic principles of the phase-locked loop (PLL)
☐ Apply what you have learned in this chapter to a system application

Communications electronics encompasses a wide range of systems, including both analog (linear) and digital. Any system that sends information from one point to another over relatively long distances can be classified as a communications system. Some of the categories of communications systems are radio (broadcast, ham, CB, marine), television, telephony, radar, navigation, satellite, data (digital), and telemetry.

Many communications systems use either amplitude modulation (AM) or frequency modulation (FM) to send information. Other modulation methods include pulse modulation, phase modulation, and frequency shift keying (FSK) as well as more specialized techniques. By necessity, the scope of this chapter is limited and is intended to introduce you to basic AM and FM communications systems and circuits. You will cover communications electronics more thoroughly in another course.

Digital data consisting of a series of binary digits (1s and 0s) are commonly sent from one computer to another over the telephone lines. Two voltage levels are used to represent the two types of bits, a high-voltage level and a low-voltage level. The data stream is made up of time intervals when the voltage has a constant high value or a constant low value with very fast transitions from one level to the other. In other words, the data stream contains very low frequencies (constant-voltage intervals) and very high frequencies (transitions). Since the telephone system has a bandwidth of approximately 300 Hz to 3000 Hz, it cannot handle the very low and the very high frequencies that make up a typical data stream without losing most of the information. Because of the bandwidth limitation of the telephone system, it is necessary to modify digital data before they are sent out; and one method of doing this is with frequency shift keying (FSK), which is a form of frequency modulation.

A simplified block diagram of a digital communications equipment (DCE) system for interfacing digital terminal equipment (DTE), such as a computer, to the telephone network is shown in Figure 9–56. The system FSK-modulates digital data before they are transmitted over the phone line and demodulates FSK signals received from another computer. Because the DTE's basic function is to *mod*ulate and *demod*ulate, it is called a *modem*. Although the modem performs many associated functions, as indicated by the different blocks, in this system application our focus will be on the modulation and demodulation circuits.

For the system application in Section 9–9, in addition to the other topics, be sure you understand

☐ The basic operation of a VCO
☐ The basic operation of a PLL
☐ How to use an NE565 PLL

9–1 ■ BASIC RECEIVERS

Receivers based on the superheterodyne principle are standard in one form or another in most types of communications systems and are found in familiar systems such as standard broadcast radio, stereo, and television. In several of the system applications in previous chapters, we presented the superheterodyne receiver in order to focus on a given circuit; now we cover it from a system viewpoint. This section provides a basic introduction to amplitude and frequency modulation and an overview of the complete AM and FM receiver.

After completing this section, you should be able to

❑ Describe basic superheterodyne receivers
    ❑ Define *AM* and *FM*
    ❑ Discuss the major functional blocks of an AM receiver
    ❑ Discuss the major functional blocks of an FM receiver

Amplitude Modulation (AM)

Amplitude modulation (AM) is a method for sending audible information, such as voice and music, by electromagnetic waves that are broadcast through the atmosphere. In AM, the amplitude of a signal with a specific frequency (f_c), called the *carrier,* is varied according to a modulating signal which can be an audio signal (such as voice or music), as shown in Figure 9–1. The carrier frequency permits the receiver to be tuned to a specific known frequency. The resulting AM waveform contains the carrier frequency, an upper-side frequency equal to the carrier frequency plus the modulation frequency ($f_c + f_m$), and a lower-side frequency equal to the carrier frequency minus the modulation frequency ($f_c - f_m$). Harmonics of these frequencies are also present. For example, if a 1 MHz carrier is amplitude modulated with a 5 kHz audio signal, the frequency components in the AM waveform are 1 MHz (carrier), 1 MHz + 5 kHz = 1,005,000 Hz (upper side), and 1 MHz − 5 kHz = 995,000 Hz (lower side).

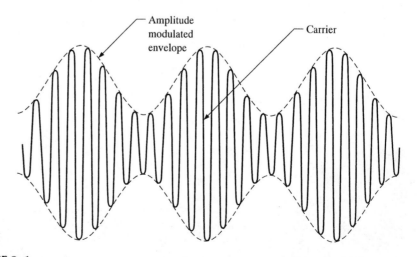

Amplitude modulated envelope

Carrier

FIGURE 9–1

An example of an amplitude modulated signal. In this case, the higher-frequency carrier is modulated by a lower-frequency sinusoidal signal.

The frequency band for AM broadcast receivers is 540 kHz to 1640 kHz. This means that an AM receiver can be tuned to pick up a specific carrier frequency that lies in the broadcast band. Each AM radio station transmits at a specific carrier frequency that is different from any other station in the area, so you can tune the receiver to pick up any desired station.

The Superheterodyne AM Receiver

A block diagram of a superheterodyne AM receiver is shown in Figure 9–2. The receiver shown consists of an antenna, an RF (radio frequency) amplifier, a mixer, a local oscillator (LO), an IF (intermediate frequency) amplifier, a detector, an audio amplifier and a power amplifier, and a speaker.

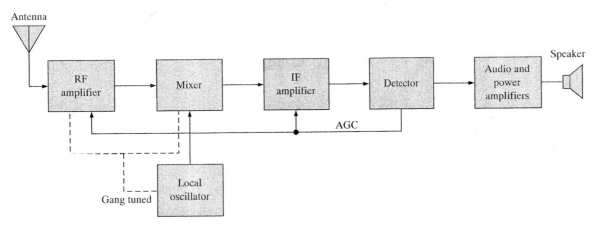

FIGURE 9–2
Superheterodyne AM receiver block diagram.

Antenna The antenna picks up all radiated signals and feeds them into the RF amplifier. These signals are very small (usually only a few microvolts).

RF Amplifier This circuit can be adjusted (tuned) to select and amplify any frequency within the AM broadcast band. Only the selected frequency and its two side bands pass through the amplifier. (Some AM receivers do not have a separate RF amplifier stage.)

Local Oscillator This circuit generates a steady sine wave at a frequency 455 kHz above the selected RF frequency.

Mixer This circuit accepts two inputs, the amplitude modulated RF signal from the output of the RF amplifier (or the antenna when there is no RF amplifier) and the sinusoidal output of the local oscillator. These two signals are then "mixed" by a nonlinear process called *heterodyning* to produce sum and difference frequencies. For example, if the RF carrier has a frequency of 1000 kHz, the LO frequency is 1455 kHz and the sum and difference frequencies out of the mixer are 2455 kHz and 455 kHz, respectively. The difference frequency is always 455 kHz no matter what the RF carrier frequency.

IF Amplifier The input to the IF amplifier is the 455 kHz AM signal, a replica of the original AM carrier signal except that the frequency has been lowered to 455 kHz. The IF amplifier significantly increases the level of this signal.

Detector   This circuit recovers the modulating signal (audio signal) from the 455 kHz IF. At this point the IF is no longer needed, so the output of the detector consists of only the audio signal.

Audio and Power Amplifiers   This circuit amplifies the detected audio signal and drives the speaker to produce sound.

AGC   The automatic gain control (AGC) provides a dc level out of the detector that is proportional to the strength of the received signal. This level is fed back to the IF amplifier, and sometimes to the mixer and RF amplifier, to adjust the gains so as to maintain constant signal levels throughout the system over a wide range of incoming carrier signal strengths.

     Figure 9–3 shows the signal flow through an AM superheterodyne receiver. The receiver can be tuned to accept any frequency in the AM band. The RF amplifier, mixer, and local oscillator are tuned simultaneously so that the LO frequency is always 455 kHz above the incoming RF signal frequency. This is called *gang tuning.*

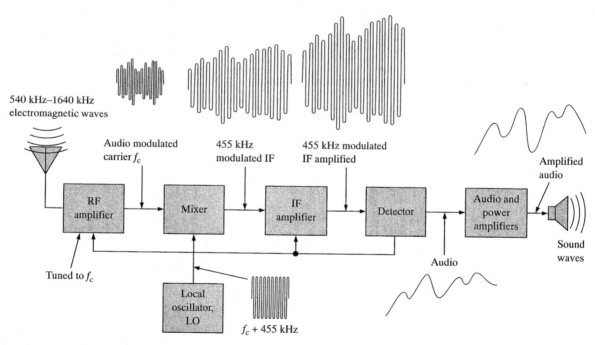

FIGURE 9–3
Illustration of signal flow through an AM receiver.

Frequency Modulation (FM)

In this method of **modulation,** the modulating signal (audio) varies the frequency of a carrier as opposed to the amplitude, as in the case of AM. Figure 9–4 illustrates basic **frequency modulation (FM).** The standard FM broadcast band consists of carrier frequencies from 88 MHz to 108 MHz, which is significantly higher than AM. The FM receiver is similar to the AM receiver in many ways, but there are several differences.

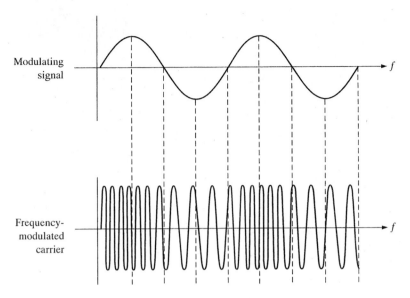

FIGURE 9–4
An example of frequency modulation.

The Superheterodyne FM Receiver

A block diagram of a superheterodyne FM receiver is shown in Figure 9–5. Notice that it includes an RF amplifier, mixer, local oscillator, and IF amplifier just as in the AM receiver. These circuits must, however, operate at higher frequencies than in the AM system. A significant difference in FM is the way the audio signal must be recovered from the modulated IF. This is accomplished by the limiter, discriminator, and de-emphasis network. Figure 9–6 depicts the signal flow through an FM receiver.

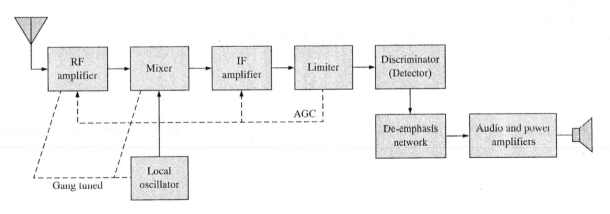

FIGURE 9–5
Superheterodyne FM receiver block diagram.

RF Amplifier This circuit must be capable of amplifying any frequency between 88 MHz and 108 MHz. It is highly selective so that it passes only the selected carrier frequency and significant side-band frequencies that contain the **audio.**

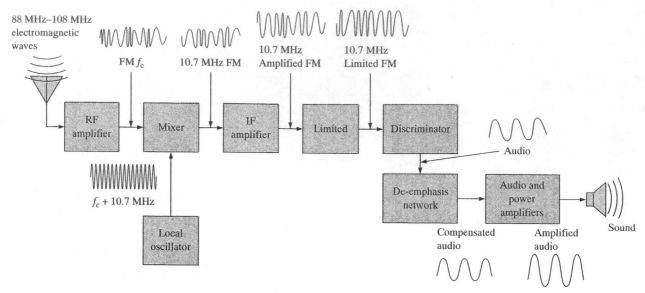

FIGURE 9–6
Example of signal flow through an FM receiver.

Local Oscillator   This circuit produces a sine wave at a frequency 10.7 MHz above the selected RF frequency.

Mixer   This circuit performs the same function as in the AM receiver, except that its output is a 10.7 MHz FM signal regardless of the RF carrier frequency.

IF Amplifier   This circuit amplifies the 10.7 MHz FM signal.

Limiter   The limiter removes any unwanted variations in the amplitude of the FM signal as it comes out of the IF amplifier and produces a constant amplitude FM output at the 10.7 MHz intermediate frequency.

Discriminator   This circuit performs the equivalent function of the detector in an AM system and is often called a detector rather than a discriminator. The **discriminator** recovers the audio from the FM signal.

De-emphasis Network   For certain reasons, the higher modulating frequencies are amplified more than the lower frequencies at the transmitting end of an FM system by a process called *preemphasis*. The de-emphasis circuit in the FM receiver brings the high-frequency audio signals back to the proper amplitude relationship with the lower frequencies.

Audio and Power Amplifiers   This circuit is the same as in the AM system and can be shared when there is a dual AM/FM configuration.

9–1 REVIEW QUESTIONS

1. What do *AM* and *FM* mean?
2. How do AM and FM differ?
3. What are the standard broadcast frequency bands for AM and FM?

9–2 ■ THE LINEAR MULTIPLIER

The linear multiplier is a key circuit in many types of communications systems. In this section, you will examine the basic principles of IC linear multipliers and look at a few applications that are found in communications as well as other areas. In the following sections, we will concentrate on multiplier applications in AM and FM systems.

After completing this section, you should be able to

❑ Discuss the function of a linear multiplier
    ❑ Describe multiplier quadrants and transfer characteristic
    ❑ Discuss scale factor
    ❑ Show how to use a multiplier circuit as a multiplier, squaring circuit, divide circuit, square root circuit, and mean square circuit

Multiplier Quadrants

There are one-quadrant, two-quadrant, and four-quadrant **multipliers.** The quadrant classification indicates the number of input polarity combinations that the multiplier can handle. A graphical representation of the quadrants is shown in Figure 9–7. A four-quadrant multiplier can accept any of the four possible input polarity combinations and produce an output with the corresponding polarity.

FIGURE 9–7
Four-quadrant polarities and their products.

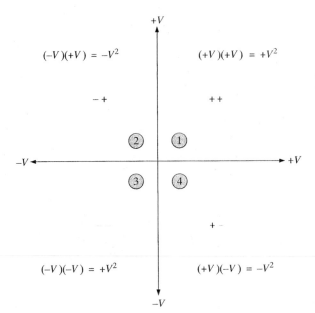

The Multiplier Transfer Characteristic

Figure 9–8 shows the transfer characteristic for a typical IC linear multiplier. To find the output voltage from the transfer characteristic graph, you find the intersection of the two input voltages V_X and V_Y. Values of V_X run along the horizontal axis and values of V_Y are the sloped lines. The output voltage is found by projecting the point of intersection over to the vertical axis. An example will illustrate this.

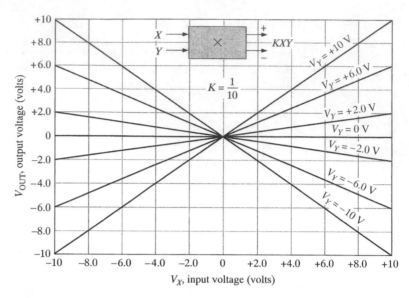

FIGURE 9–8
A four-quadrant multiplier transfer characteristic.

EXAMPLE 9–1

Determine the output voltage for a four-quadrant linear multiplier whose transfer characteristic is given in Figure 9–8. The input voltages are $V_X = -4$ V and $V_Y = +10$ V.

Solution The output voltage is **−4 V** as illustrated in Figure 9–9. For this transfer characteristic, the output voltage is a factor of ten smaller than the actual product of the two input voltages. This is due to the *scale factor* of the multiplier, which is discussed next.

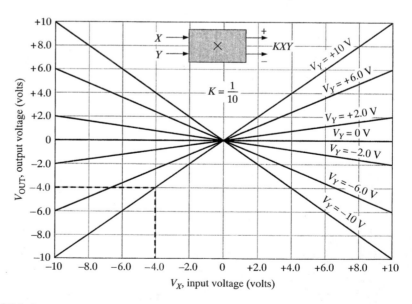

FIGURE 9–9

Practice Exercise Find V_{OUT} if $V_X = -6$ V and $V_Y = +6$ V.

The Scale Factor, *K*

The scale factor, *K*, is basically an internal attenuation that reduces the output by a fixed amount. The scale factor on most IC multipliers is adjustable and has a typical value of 0.1. Figure 9–10 shows an MC1595 configured as a basic multiplier. The scale factor is determined by external resistors, which include two equal load resistors, according to the following formula.

$$K = \frac{2R_L}{R_X R_Y I_{R2}}$$

The current I_{R2} is set by internal and external parameters according to this formula:

$$I_{R2} = \frac{|-V| - 0.7 \text{ V}}{R_2 + 500 \text{ }\Omega}$$

where R_2 is the combination of the fixed resistor and the potentiometer. The potentiometer provides for fine adjustment by controlling I_{R2}.

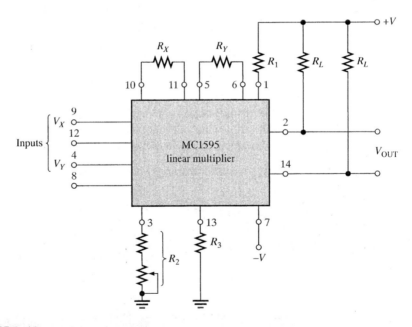

FIGURE 9–10

Basic MC1595 linear multiplier with external circuitry for setting the scale factor.

The expression for the output voltage of the IC linear multiplier includes the scale factor, *K*, as indicated in Equation (9–1)

$$V_{OUT} = K V_X V_Y \tag{9–1}$$

EXAMPLE 9–2

Determine the scale factor for the basic MC1595 multiplier in Figure 9–11. Assume the 5 kΩ potentiometer portion of R_2 is set to 2.5 kΩ. Also, determine the output voltage for the given inputs.

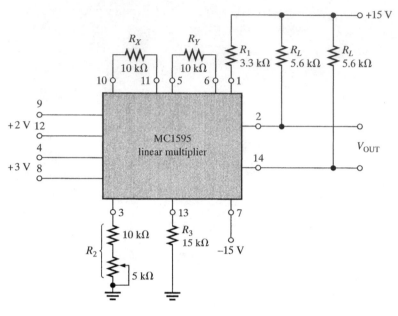

FIGURE 9–11

Solution Calculate I_{R2} as follows:

$$I_{R2} = \frac{|-V| - 0.7\ V}{R_2 + 500\ \Omega} = \frac{15\ V - 0.7\ V}{12.5\ k\Omega + 500\ \Omega} = \frac{14.3\ V}{13\ k\Omega} = 1.1\ mA$$

The scale factor is

$$K = \frac{2R_L}{R_X R_Y I_{R2}} = \frac{2(5.6\ k\Omega)}{(10\ k\Omega)(10\ k\Omega)(1.1\ mA)} = \mathbf{0.102}$$

The output voltage is

$$V_{OUT} = K V_X V_Y = 0.102(+2\ V)(+3\ V) = \mathbf{0.611\ V}$$

Practice Exercise What is the output voltage in Figure 9–11 if the 5 kΩ potentiometer is set to its maximum resistance?

Offset Adjustment

Due to internal mismatches, generally small offset voltages are at the inputs and the output of an IC linear multiplier. External circuits to null out the offset voltages are shown in Figure 9–12. The resistive voltage dividers on the inputs allow the actual input voltages to be greater than the recommended maximum for the device. For example, the MC1595 has a maximum input voltage of 5 V. The voltage dividers allow a maximum of 10 V to be applied if the resistors are of equal value. The zener diodes in the input offset adjust circuit keep the inputs on pins 8 and 12 from exceeding the maximum of 5 V.

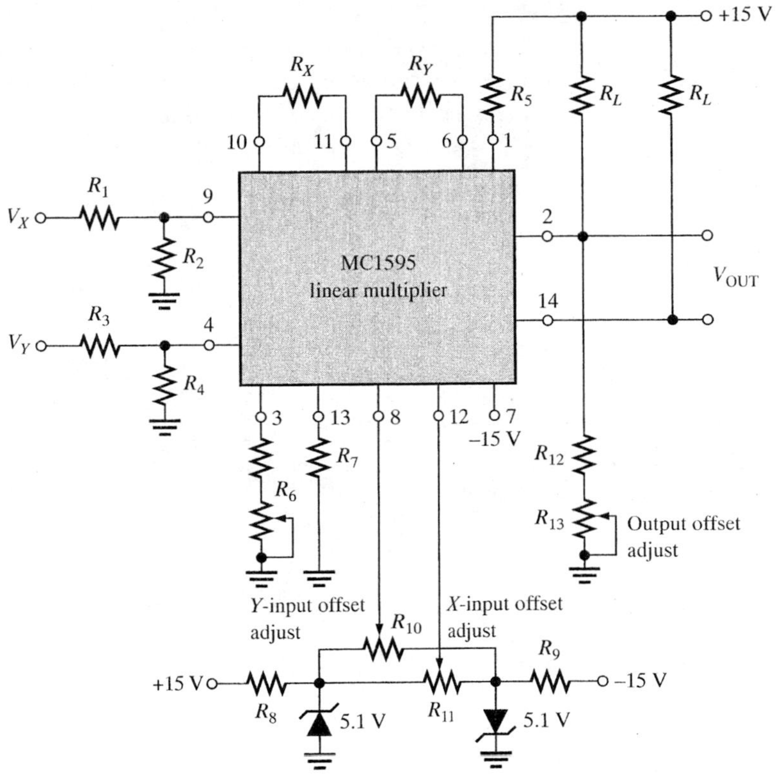

FIGURE 9–12
Basic MC1595 multiplier with both scale factor and offset circuitry.

Basic Applications of the Multiplier

Applications of linear multipliers are numerous. Some basic applications are now presented.

Multiplier   The most obvious application of a linear multiplier is, of course, to multiply two voltages as indicated in Figure 9–13.

FIGURE 9–13
Multiplier.

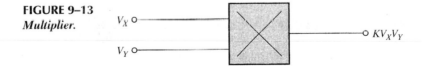

V_X o———
V_Y o———
o KV_XV_Y

Squaring Circuit   A special case of the multiplier is a squaring circuit that is realized by simply applying the same variable to both inputs by connecting the inputs together as shown in Figure 9–14.

FIGURE 9–14
Squaring circuit.

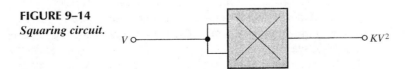

V o———
o KV^2

Divide Circuit   The circuit in Figure 9–15 shows the multiplier placed in the feedback loop of an op-amp. The basic operation is as follows. There is a virtual ground at the inverting ($-$) input of the op-amp and therefore the current at the inverting input is negligible. Therefore, I_1 and I_2 are equal. Since the inverting input voltage is 0 V, the voltage across R_1 is KV_YV_{OUT} and the current through R_1 is

$$I_1 = \frac{KV_YV_{OUT}}{R_1}$$

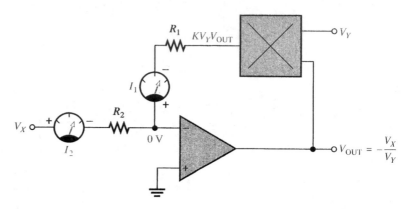

FIGURE 9–15
Divide circuit.

The voltage across R_2 is V_X, so the current through R_2 is

$$I_2 = \frac{V_X}{R_2}$$

Since $I_1 = -I_2$,

$$\frac{KV_YV_{OUT}}{R_1} = -\frac{V_X}{R_2}$$

Solving for V_{OUT},

$$V_{OUT} = -\frac{V_XR_1}{KV_YR_2}$$

If $R_1 = KR_2$,

$$V_{OUT} = -\frac{V_X}{V_Y}$$

Square Root Circuit   The square root circuit is a special case of the divide circuit where V_{OUT} is applied to both inputs of the multiplier as shown in Figure 9–16.

Mean Square Circuit   In this application, the multiplier is used as a squaring circuit with its output connected to an op-amp integrator as shown in Figure 9–17. The integrator produces the average or mean value of the squared input over time, as indicated by the integration sign ($\int$).

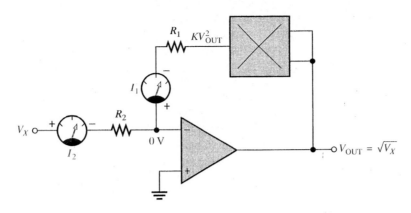

FIGURE 9–16
Square root circuit.

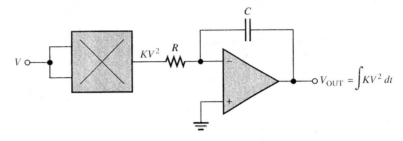

FIGURE 9–17
Mean square circuit.

9–2 REVIEW QUESTIONS

1. Compare a four-quadrant multiplier to a one-quadrant multiplier in terms of the inputs that can be handled.

2. If 5 V and 1 V are applied to the inputs of a multiplier and its output is 0.5 V, what is the scale factor? What must the scale factor be for an output of 5 V?

3. How do you convert a basic multiplier to a squaring circuit?

9–3 ■ AMPLITUDE MODULATION

Amplitude modulation (AM) is an important method for transmitting information. Of course, the AM superheterodyne receiver is designed to receive transmitted AM signals. In this section, we further define amplitude modulation and show how the linear multiplier can be used as an amplitude-modulated device.

After completing this section, you should be able to

❑ Discuss the fundamentals of amplitude modulation
  ❑ Explain how AM is basically a multiplication process
  ❑ Describe sum and difference frequencies

❑ Discuss balanced modulation
❑ Describe the frequency spectra
❑ Explain standard AM

As you learned in Section 9–1, amplitude modulation is the process of varying the amplitude of a signal of a given frequency (carrier) with another signal of much lower frequency (modulating signal). One reason that the higher-frequency carrier signal is necessary is because audio or other signals with relatively low frequencies cannot be transmitted with antennas of a practical size. The basic concept of standard amplitude modulation is illustrated in Figure 9–18.

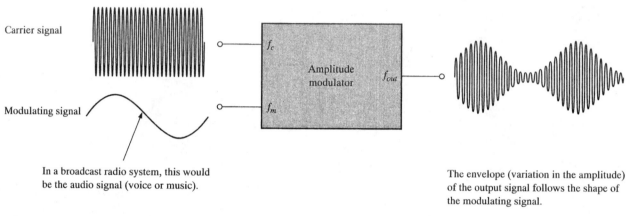

FIGURE 9–18
Basic concept of amplitude modulation.

A Multiplication Process

If a signal is applied to the input of a variable-gain device, the resulting output is an amplitude-modulated signal because $V_{out} = A_v V_{in}$. The output voltage is the input voltage multiplied by the voltage gain. For example, if the gain of an amplifier is made to vary sinusoidally at a certain frequency and an input signal is applied at a higher frequency, the output signal will have the higher frequency. However, its amplitude will vary according to the variation in gain as illustrated in Figure 9–19. Amplitude modulation is basically a multiplication process (input voltage multiplied by a variable gain).

Sum and Difference Frequencies

If the expressions for two sinusoidal signals of different frequencies are multiplied mathematically, a term containing both the difference and the sum of the two frequencies is produced. Recall from ac circuit theory that a sinusoidal voltage can be expressed as

$$v = V_p \sin 2\pi f t$$

where V_p is the peak voltage and f is the frequency. Two different sinusoidal signals can be expressed as follows:

$$v_1 = V_{1(p)} \sin 2\pi f_1 t$$
$$v_2 = V_{2(p)} \sin 2\pi f_2 t$$

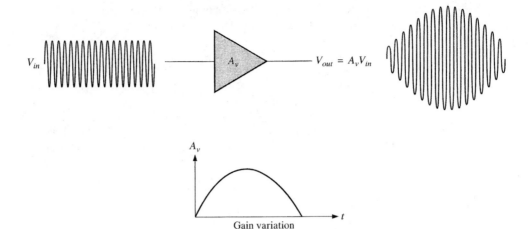

FIGURE 9–19
The amplitude of the output voltage varies according to the gain and is the product of voltage gain and input voltage.

Multiplying these two sinusoidal wave terms,

$$v_1 v_2 = (V_{1(p)}\sin 2\pi f_1 t)(V_{2(p)}\sin 2\pi f_2 t)$$
$$= V_{1(p)}V_{2(p)}(\sin 2\pi f_1 t)(\sin 2\pi f_2 t)$$

The basic trigonometric identity for the product of two sinusoidal functions is

$$(\sin A)(\sin B) = \frac{1}{2}[\cos(A - B) - \cos(A + B)]$$

Applying this identity to the previous formula for $v_1 v_2$,

$$v_1 v_2 = \frac{V_{1(p)}V_{2(p)}}{2}[(\cos 2\pi f_1 t - 2\pi f_2 t) - (\cos 2\pi f_1 t + 2\pi f_2 t)]$$

$$= \frac{V_{1(p)}V_{2(p)}}{2}[(\cos 2\pi(f_1 - f_2)t) - (\cos 2\pi(f_1 + f_2)t)]$$

$$v_1 v_2 = \frac{V_{1(p)}V_{2(p)}}{2}\cos 2\pi(f_1 - f_2)t - \frac{V_{1(p)}V_{2(p)}}{2}\cos 2\pi(f_1 + f_2)t \qquad (9\text{–}2)$$

You can see in Equation (9–2) that the product of the two sinusoidal voltages V_1 and V_2 contains a difference frequency $(f_1 - f_2)$ and a sum frequency $(f_1 + f_2)$. The fact that the product terms are cosine simply indicates a 90° phase shift in the multiplication process.

Analysis of Balanced Modulation

Since amplitude modulation is simply a multiplication process, the preceding analysis is now applied to carrier and modulating signals. The expression for the sinusoidal carrier signal can be written as

$$v_c = V_{c(p)}\sin 2\pi f_c t$$

Assuming a sinusoidal modulating signal, it can be expressed as

$$v_m = V_{m(p)}\sin 2\pi f_m t$$

Substituting these two signals in Equation (9–2),

$$v_c v_m = \frac{V_{c(p)}V_{m(p)}}{2}\cos 2\pi(f_c - f_m)t - \frac{V_{c(p)}V_{m(p)}}{2}\cos 2\pi(f_c + f_m)t$$

An outout signal described by this expression for the product of two sinusoidal signals is produced by a linear multiplier. Notice that there is a difference frequency term $(f_c - f_m)$ and a sum frequency term $(f_c + f_m)$, but the original frequencies, f_c and f_m, do not appear alone in the expression. Thus, the product of two sinusoidal signals contains no signal with the carrier frequency, f_c, or with the modulating frequency, f_m. This is called **balanced modulation** because there is no carrier frequency in the output. The carrier frequency is "balanced out."

The Frequency Spectra of a Balanced Modulator

A graphical picture of the frequency content of a signal is called its frequency spectrum (see Sec. 1–2). A frequency spectrum shows voltage on a frequency base rather than on a time base as a waveform diagram does. The frequency spectra of the product of two sinusoidal signals is shown in Figure 9–20. Part (a) shows the two input frequencies and part (b) shows the output frequencies. In communications terminology, the sum frequency is called the *upper-side frequency* and the difference frequency is called the *lower-side frequency* because the frequencies appear on each side of the missing carrier frequency.

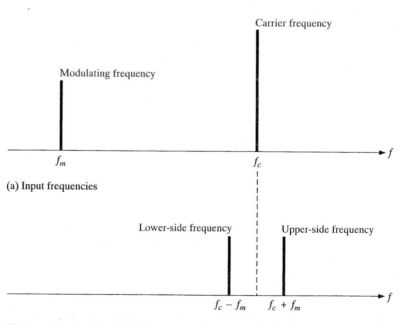

(a) Input frequencies

(b) Output frequencies

FIGURE 9–20
Illustration of the input and output frequency spectra for a linear multiplier.

The Linear Multiplier as a Balanced Modulator

As mentioned, the linear multiplier acts as a balanced modulator when a carrier signal and a modulating signal are applied to its inputs, as illustrated in Figure 9–21. A balanced modulator produces an upper-side frequency and a lower-side frequency, but it does not produce a carrier frequency. Since there is no carrier signal, balanced modulation is sometimes known as *suppressed-carrier modulation*. Balanced modulation is used in certain types of communications such as single side-band systems, but it is not used in standard AM broadcast systems.

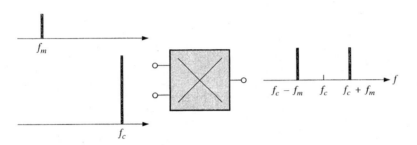

FIGURE 9–21
The linear multiplier as a balanced modulator.

EXAMPLE 9–3 Determine the frequencies contained in the output signal of the balanced modulator in Figure 9–22.

FIGURE 9–22

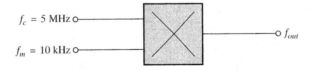

Solution The upper-side frequency is

$$f_c + f_m = 5 \text{ MHz} + 10 \text{ kHz} = \textbf{5.01 MHz}$$

The lower-side frequency is

$$f_c - f_m = 5 \text{ MHz} - 10 \text{ kHz} = \textbf{4.99 MHz}$$

Practice Exercise Explain how the separation between the side frequencies can be increased using the same carrier frequency.

Standard Amplitude Modulation (AM)

In standard AM systems, the output signal contains the carrier frequency as well as the sum and difference side frequencies. Standard amplitude modulation is illustrated by the frequency spectrum in Figure 9–23.

FIGURE 9–23
The output frequency spectrum of a standard amplitude modulator.

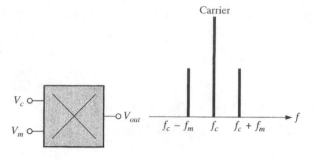

The expression for a standard amplitude-modulated signal is

$$V_{out} = V_{c(p)}^2 \sin 2\pi f_c t + \frac{V_{c(p)}V_{m(p)}}{2}\cos 2\pi(f_c - f_m)t - \frac{V_{c(p)}V_{m(p)}}{2}\cos 2\pi(f_c + f_m)t \qquad (9\text{–}3)$$

Notice in Equation (9–3) that the first term is for the carrier frequency and the other two terms are for the side frequencies. Let's see how the carrier-frequency term gets into the equation.

If a dc voltage equal to the peak of the carrier voltage is added to the modulating signal before the modulating signal is multiplied by the carrier signal, a carrier-signal term appears in the final result as shown in the following steps. Add the peak carrier voltage to the modulating signal, and you get the following expression:

$$V_{c(p)} + V_{m(p)}\sin 2\pi f_m t$$

Multiply by the carrier signal.

$$V_{out} = (V_{c(p)}\sin 2\pi f_c t)(V_{c(p)} + V_{m(p)}\sin 2\pi f_m t)$$
$$= \underbrace{V_{c(p)}^2 \sin 2\pi f_c t}_{\text{carrier term}} + \underbrace{V_{c(p)}V_{m(p)}(\sin 2\pi f_c t)(\sin 2\pi f_m t)}_{\text{product term}}$$

Apply the basic trigonometric identity to the product term.

$$V_{out} = V_{c(p)}^2 \sin 2\pi f_c t + \frac{V_{c(p)}V_{m(p)}}{2}\cos 2\pi(f_c - f_m)t - \frac{V_{c(p)}V_{m(p)}}{2}\cos 2\pi(f_c + f_m)t$$

This result shows that the output of the multiplier contains a carrier term and two side-frequency terms. Figure 9–24 illustrates how a standard amplitude modulator can be implemented by a summing circuit followed by a linear multiplier. Figure 9–25 shows a possible implementation of the summing circuit.

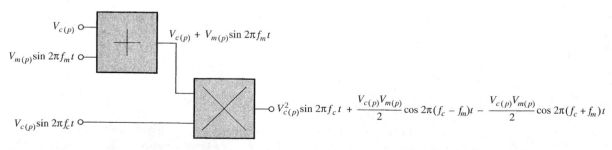

FIGURE 9–24
Basic block diagram of an amplitude modulator.

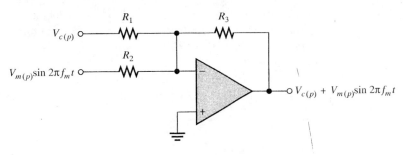

FIGURE 9–25
Implementation of the summing circuit in the amplitude modulator.

EXAMPLE 9–4

A carrier frequency of 1200 kHz is modulated by a sinusoidal wave with a frequency of 25 kHz by a standard amplitude modulator. Determine the output frequency spectrum.

Solution   The lower-side frequency is

$$f_c - f_m = 1200 \text{ kHz} - 25 \text{ kHz} = \textbf{1175 kHz}$$

The upper-side frequency is

$$f_c + f_m = 1200 \text{ kHz} + 25 \text{ kHz} = \textbf{1225 kHz}$$

The output contains the carrier frequency and the two side frequencies as shown in Figure 9–26.

FIGURE 9–26

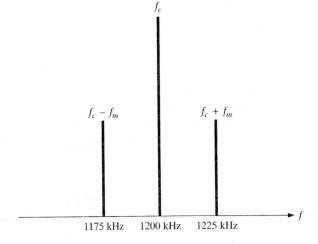

Practice Exercise   Compare the output frequency spectrum in this example to that of a balanced modulator having the same inputs.

Amplitude Modulation with Voice or Music

To this point in our discussion, we have considered the modulating signal to be a pure sinusoidal signal just to keep things fairly simple. If you receive an AM signal modulated by a pure sinusoidal signal in the audio frequency range, you will hear a single tone from the receiver's speaker.

A voice or music signal consists of many sinusoidal components within a range of frequencies from about 20 Hz to 20 kHz. For example, if a carrier frequency is amplitude modulated with voice or music with frequencies from 100 Hz to 10 kHz, the frequency spectrum is as shown in Figure 9–27. Instead of one lower-side and one upper-side frequency as in the case of a single-frequency modulating signal, a band of lower-side frequencies and a band of upper-side frequencies correspond to the sum and difference frequencies of each sinusoidal component of the voice or music signal.

FIGURE 9–27
Example of a frequency spectrum for a voice or music signal.

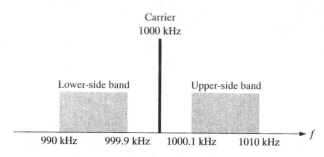

9–3 REVIEW QUESTIONS

1. What is amplitude modulation?
2. What is the difference between balanced modulation and standard AM?
3. What two input signals are used in amplitude modulation? Explain the purpose of each signal.
4. What are the upper-side frequency and the lower-side frequency?
5. How can a balanced modulator be changed to a standard amplitude modulator?

9–4 ■ THE MIXER

The mixer in the receiver system discussed in Section 9–1 can be implemented with a linear multiplier. Principles of linear multiplication of sinusoidal signals are covered next, and you will see how sum and difference frequencies are produced. The difference frequency is a critical part of the operation of many types of receiver systems.

After completing this section, you should be able to

❑ Discuss the basic function of a mixer
   ❑ Explain why a mixer is a linear multiplier
   ❑ Describe the frequencies in the mixer and IF portion of a receiver

The **mixer** is a nonlinear circuit that combines two signals and produces the sum and difference frequencies. Basically a mixer is a frequency converter because it changes the frequency of a signal to another value. The mixer in a receiver system takes the incoming modulated RF signal (which is sometimes amplified by an RF amplifier and sometimes not) along with the signal from the local oscillator and produces a modulated signal with a frequency equal to the difference of its two input frequencies (RF and LO). The mixer also produces a frequency equal to the sum of the input frequencies. The mixer function is illustrated in Figure 9–28.

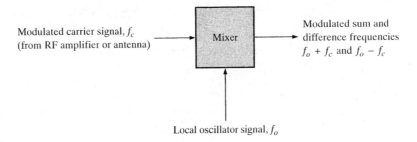

FIGURE 9–28
The mixer function.

The Mixer Is a Linear Multiplier

In the case of receiver applications, the mixer must produce an output that has a frequency component equal to the difference of its input frequencies. From the mathematical analysis in Section 9–3, you can see that if two sinusoidal signals are multiplied, the product contains the difference frequency and the sum frequency. Thus, the mixer is actually a linear multiplier as indicated in Figure 9–29.

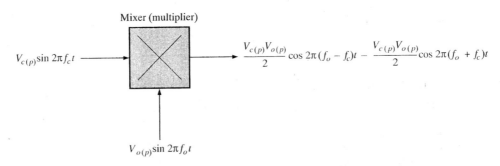

FIGURE 9–29
The mixer as a linear multiplier.

EXAMPLE 9–5

Determine the output expression for a multiplier with one sinusoidal input having a peak voltage of 5 mV and a frequency of 1200 kHz and the other input having a peak voltage of 10 mV and a frequency of 1655 kHz.

Solution   The two input expressions are

$$v_1 = (5 \text{ mV})\sin 2\pi(1200 \text{ kHz})t$$
$$v_2 = (10 \text{ mV})\sin 2\pi(1655 \text{ kHz})t$$

Multiplying,

$$v_1 v_2 = (5 \text{ mV})(10 \text{ mV})[\sin 2\pi(1200 \text{ kHz})t][\sin 2\pi(1655 \text{ kHz})t]$$

Applying the trigonometric identity, $(\sin A)(\sin B) = \frac{1}{2}[\cos(A - B) - \cos(A + B)]$,

$$V_{out} = \frac{(5 \text{ mV})(10 \text{ mV})}{2}\cos 2\pi(1655 \text{ kHz} - 1200 \text{ kHz})t$$

$$- \frac{(5 \text{ mV})(10 \text{ mV})}{2}\cos 2\pi(1655 \text{ kHz} + 1200 \text{ kHz})t$$

$$V_{out} = (25 \; \mu V)\cos 2\pi (455 \; \text{kHz})t - (25 \; \mu V)\cos 2\pi (2855 \; \text{kHz})t$$

Practice Exercise What is the value of the peak amplitude and frequency of the difference frequency component in this example?

In the receiver system, both the sum and difference frequencies from the mixer are applied to the IF (intermediate frequency) amplifier. The IF amplifier is actually a tuned amplifier that is designed to respond to the difference frequency while rejecting the sum frequency. You can think of the IF amplifier section of a receiver as a band-pass filter plus an amplifier because it uses resonant circuits to provide the frequency selectivity. This is illustrated in Figure 9–30.

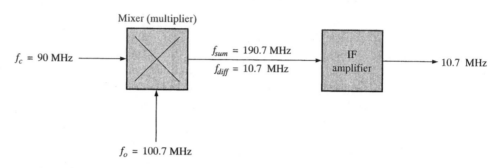

FIGURE 9–30
Example of frequencies in the mixer and IF portion of a receiver.

EXAMPLE 9–6   Determine the output frequency of the IF amplifier for the conditions shown in Figure 9–31.

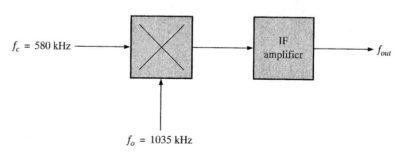

FIGURE 9–31

Solution The IF amplifier produces only the difference frequency signal on its output.

$$f_{out} = f_{diff} = f_o - f_c = 1035 \; \text{kHz} - 580 \; \text{kHz} = \textbf{455 kHz}$$

Practice Exercise Based on your basic knowledge of the superheterodyne receiver from Section 9–1, determine the IF output frequency when the incoming RF signal changes to 1550 kHz.

9–4 REVIEW QUESTIONS

1. What is the purpose of the mixer in a superheterodyne receiver?
2. How does the mixer produce its output?
3. If a mixer has 1000 kHz on one input and 350 kHz on the other, what frequencies appear on the output?

9–5 ■ AM DEMODULATION

The linear multiplier can be used to demodulate or detect an AM signal as well as to perform the modulation process that was discussed in Section 9–3. Demodulation *can be thought of as reverse modulation. The purpose is to get back the original modulating signal (voice or music in the case of standard AM receivers). The detector in the AM receiver can be implemented using a multiplier, although another method using peak envelope detection is common.*

After completing this section, you should be able to

❑ Describe the basic principle of AM demodulation
   ❑ Discuss a basic AM demodulator
   ❑ Discuss the frequency spectra

The Basic AM Demodulator

An AM demodulator can be implemented with a linear multiplier followed by a low-pass filter, as shown in Figure 9–32. The critical frequency of the filter is the highest audio frequency that is required for a given application (15 kHz, for example).

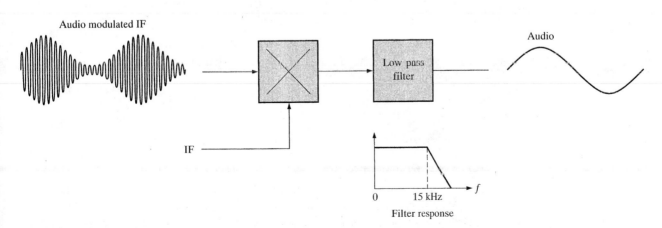

FIGURE 9–32
Basic AM demodulator.

Operation in Terms of the Frequency Spectra

Let's assume a carrier modulated by a single tone with a frequency of 10 kHz is received and converted to a modulated intermediate frequency of 455 kHz, as indicated by the frequency spectra in Figure 9–33. Notice that the upper-side and lower-side frequencies are separated from both the carrier and the IF by 10 kHz.

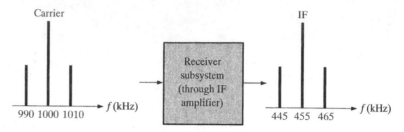

FIGURE 9–33
An AM signal converted to IF.

When the modulated output of the IF amplifier is applied to the demodulator along with the IF, sum and difference frequencies for each input frequency are produced as shown in Figure 9–34. Only the 10 kHz audio frequency is passed by the filter. A drawback to this type of AM detection is that a pure IF must be produced to mix with the modulated IF.

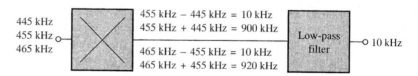

FIGURE 9–34
Example of demodulation.

9–5 REVIEW QUESTIONS

1. What is the purpose of the filter in the linear multiplier demodulator?
2. If a 455 kHz IF modulated by a 1 kHz audio frequency is demodulated, what frequency or frequencies appear on the output of the demodulator?

9–6 ■ IF AND AUDIO AMPLIFIERS

In this section, IC amplifiers for intermediate and audio frequencies are introduced. A typical IF amplifier is discussed and audio preamplifiers and power amplifiers are covered. As you know, the IF amplifier in a communications receiver provides amplification of the modulated IF signal out of the mixer before it is applied to the detector. After the audio signal is recovered by the detector, it goes to the audio preamp where it is amplified and applied to the power amplifier that drives the speaker.

After completing this section, you should be able to

❑ Describe the basic principles of IF and audio amplifiers
    ❑ Discuss the function of an IF amplifier
    ❑ Explain how the local oscillator and mixer operate with the IF amplifier
    ❑ Discuss the MC1350 IF amplifier
    ❑ State the purpose of the audio amplifier
    ❑ Discuss the LM386 audio power amplifier

The Basic Function of the IF Amplifier

The IF amplifier in a receiver is a tuned amplifier with a specified bandwidth operating at a center frequency of 455 kHz for AM and 10.7 kHz for FM. The IF amplifier is one of the key features of a superheterodyne receiver because it is set to operate at a single resonant frequency that remains the same over the entire band of carrier frequencies that can be received. Figure 9–35 illustrates the basic function of an IF amplifier in terms of the frequency spectra.

Assume, for example, that the received carrier frequency of $f_c = 1$ MHz is modulated by an audio signal with a maximum frequency of $f_m = 5$ kHz, indicated in Figure

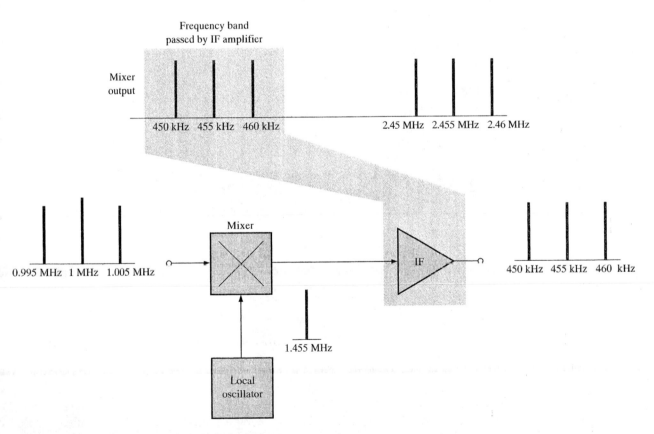

FIGURE 9–35

An illustration of the basic function of the IF amplifier in an AM receiver.

9–35 by the frequency spectrum on the input to the mixer. For this frequency, the local oscillator is at a frequency of

$$f_o = 1 \text{ MHz} + 455 \text{ kHz} = 1.455 \text{ MHz}$$

The mixer produces the following sum and difference frequencies as indicated in Figure 9–35.

$$f_o + f_c = 1.455 \text{ MHz} + 1 \text{ MHz} = 2.455 \text{ MHz}$$
$$f_o - f_c = 1.455 \text{ MHz} - 1 \text{ MHz} = 455 \text{ kHz}$$
$$f_o + (f_c + f_m) = 1.455 \text{ MHz} + 1.005 \text{ MHz} = 2.46 \text{ MHz}$$
$$f_o + (f_c - f_m) = 1.455 \text{ MHz} + 0.995 \text{ MHz} = 2.45 \text{ MHz}$$
$$f_o - (f_c + f_m) = 1.455 \text{ MHz} - 1.005 \text{ MHz} = 450 \text{ kHz}$$
$$f_o - (f_c - f_m) = 1.455 \text{ MHz} - 0.995 \text{ MHz} = 460 \text{ kHz}$$

Since the IF amplifier is a frequency-selective circuit, it responds only to 455 kHz and any side frequencies lying in the 10 kHz band centered at 455 kHz. So, all of the frequencies out of the mixer are rejected except the 455 kHz IF, all lower-side frequencies down to 450 kHz, and all upper-side frequencies up to 460 kHz. This frequency spectrum is the audio modulated IF.

A Basic IF Amplifier

Although the detailed circuitry of the IF amplifier may differ from one system to another, it always has a tuned (resonant) circuit on the input or on the output or on both. Figure 9–36(a) shows a basic IF amplifier with tuned transformer coupling at the input and output. The general frequency response curve is shown in Figure 9–36(b).

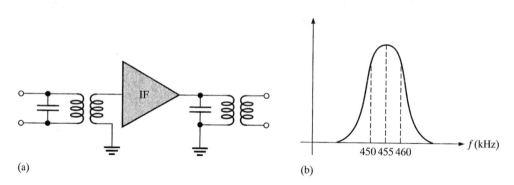

(a)   (b)

FIGURE 9–36
A basic IF amplifier with a tuned circuit on the input and output.

The MC1350   This device is representative of integrated circuit IF amplifiers. It can be used in either AM or FM systems and has a typical power gain of 62 dB at 455 kHz. Figure 9–37 shows packaging and a typical circuit diagram for application in an AM receiver. This configuration has a single-tuned transformer-coupled output. The AGC input is normally fed back from the detector in an AM receiver and is used to keep the IF gain at a constant level so that variations in the strength of the incoming RF signal does not cause the audio output to vary significantly. When the AGC voltage increases, the IF gain decreases and when the AGC voltage decreases, the IF gain increases.

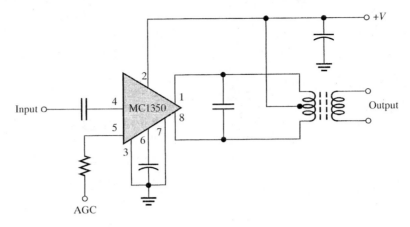

FIGURE 9–37
A typical circuit configuration using the MC1350 IF amplifier.

Audio Amplifiers

Audio amplifiers are used in a receiver system following the detector to provide amplification of the recovered audio signal and audio power to drive the speaker(s), as indicated in Figure 9–38. Audio amplifiers typically have bandwidths of 3 kHz to 15 kHz depending on the requirements of the system. IC audio amplifiers are available with a range of capabilities. To complete the radio, the versatile LM386 is selected here.

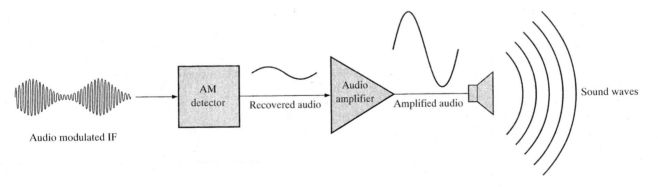

FIGURE 9–38
The audio amplifier in a receiver system.

The LM386 Audio-Power Amplifier This device is an example of a low-power audio amplifier that is capable of providing several hundred milliwatts to an 8 Ω speaker. It operates from any dc supply voltage in the 4 V to 12 V range, making it a good choice for battery operation. The pin configuration of the LM386 is shown in Figure 9–39(a). The voltage gain of the LM386 is 20 without external connections to the gain terminals, as shown in Figure 9–39(b). A voltage gain of 200 is achieved by connecting a 10 μF capacitor from pin 1 to pin 8, as shown in Figure 9–39(c). Voltage gains between 20 and 200 can be realized by a resistor (R_G) and capacitor (C_G) connected in series from pin 1 to pin 8 as shown in Figure 9–39(d). These external components are effectively placed in parallel with an internal gain-setting resistor.

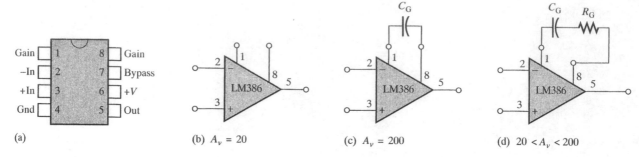

FIGURE 9–39
Pin configuration and gain connections for the LM386 audio amplifier.

A typical application of the LM386 as a power amplifier in a radio receiver is shown in Figure 9–40. Here the detected AM signal is fed to the inverting input through the volume control potentiometer, R_1, and resistor R_2. C_1 is the input coupling capacitor and C_2 is the power supply decoupling capacitor. R_2 and C_3 filter out any residual RF or IF signal that may be on the output of the detector. R_3 and C_6 provide additional filtering before the audio signal is applied to the speaker through the coupling capacitor C_7.

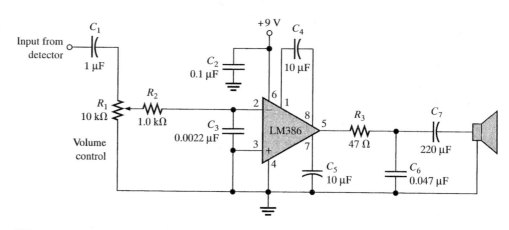

FIGURE 9–40
The LM386 as an AM audio power amplifier.

9–6 REVIEW QUESTIONS

1. What is the purpose of the IF amplifier in an AM receiver?
2. What is the center frequency of an AM IF amplifier?
3. Why is the bandwidth of an AM receiver IF amplifier 10 kHz?
4. Why must the audio amplifier follow the detector in a receiver system?
5. Compare the frequency response of the IF amplifier to that of the audio amplifier.

9–7 ■ FREQUENCY MODULATION

As you have seen, modulation is the process of varying a parameter of a carrier signal with an information signal. Recall that in amplitude modulation the parameter of amplitude is varied. In frequency modulation (FM), the frequency of a carrier is varied above and below its normal or at-rest value by a modulating signal. This section provides a basic introduction to FM and discusses the differences between an AM and an FM receiver.

After completing this section, you should be able to

❑ Describe the basic principles of frequency modulation
   ❑ Discuss the voltage-controlled oscillator
   ❑ Describe the MC2833 FM modulator
   ❑ Explain frequency demodulation

In a frequency-modulated (FM) signal, the carrier frequency is increased or decreased according to the modulating signal. The amount of deviation above or below the carrier frequency depends on the amplitude of the modulating signal. The rate at which the frequency deviation occurs depends on the frequency of the modulating signal.

Figure 9–41 illustrates both a square wave and a sine wave modulating the frequency of a carrier. The carrier frequency is highest when the modulating signal is at its maximum positive amplitude and is lowest when the modulating signal is at its maximum negative amplitude.

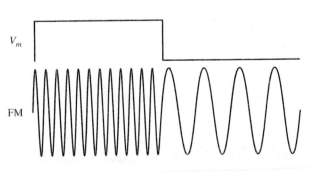

(a) Frequency modulation with a square wave

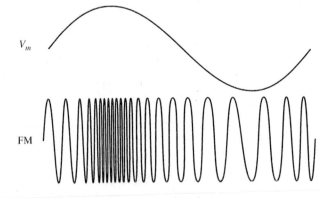

(b) Frequency modulation with a sine wave

FIGURE 9–41
Examples of frequency modulation.

A Basic Frequency Modulator

Frequency modulation is achieved by varying the frequency of an oscillator with the modulating signal. A **voltage-controlled oscillator (VCO)** is typically used for this purpose, as illustrated in Figure 9–42.

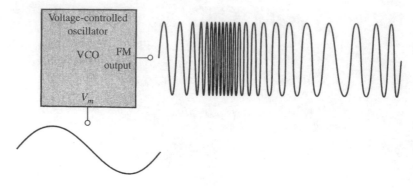

FIGURE 9–42

Frequency modulation with a voltage-controlled oscillator.

Generally, a variable-reactance type of voltage-controlled oscillator is used in FM applications. The variable-reactance VCO uses the varactor diode as a voltage-variable capacitance, as illustrated in Figure 9–43, where the capacitance is varied with the modulating voltage, V_m.

FIGURE 9–43

Basic variable-reactance VCO.

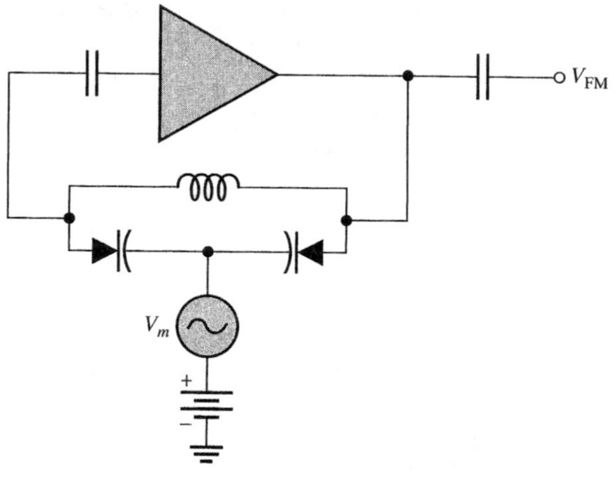

An Integrated Circuit FM Transmitter

An example of a single-chip FM transmitter is the MC2833, which is designed for cordless telephone and other FM communications equipment. The 16-pin package configuration showing the basic functional blocks is illustrated in Figure 9–44. The microphone amplifier (mic amp) amplifies the low-level input from a microphone and feeds its output to the variable-reactance circuit which uses a varactor diode tuning circuit to control the RF oscillator. The reference voltage circuit (V_{REF}) provides stable bias to the reactance circuit. The two individual transistors can be connected as tuned amplifiers to boost the power output.

A typical VHF narrow-band FM transmitter using the MC2833 appropriate external circuitry is shown in Figure 9–45. This particular implementation has an output frequency of 49.7 MHz. The frequency of the oscillator is set by the external 16.5667 MHz crystal. The reactance circuit deviates this frequency with the amplified audio input to produce an

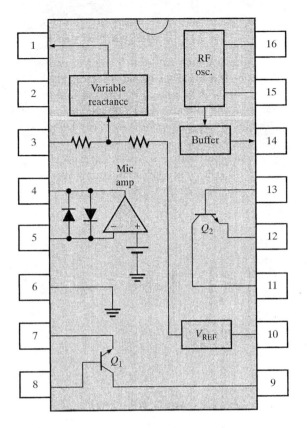

FIGURE 9–44
The MC2833 FM modulator.

FM signal. The 16.5667 MHz output of the oscillator goes to the buffer and is then applied to the input of Q_2, which is operated as a frequency tripler (16.5667 MHz $\times$ 3 = 49.7 MHz). A frequency tripler is basically a class C amplifier with the output tuned to a resonant frequency equal to three times the input. The signal from the Q_2 frequency tripler goes to the input of Q_1, which functions as a linear amplifier that drives the transmitting antenna from its resonant output circuit.

FM Demodulation

Except for the higher frequencies, the standard broadcast FM receiver is basically the same as the AM receiver up through the IF amplifier. The main difference between an FM receiver and an AM receiver is the method used to recover the audio signal from the modulated IF.

There are several methods for demodulating an FM signal. These include slope detection, phase-shift discrimination, ratio detection, quadrature detection, and phase-locked loop demodulation. Most of these methods are covered in detail in communications courses. However, because of its importance in many types of applications, we will cover the phase-locked loop (PLL) demodulation in the next section.

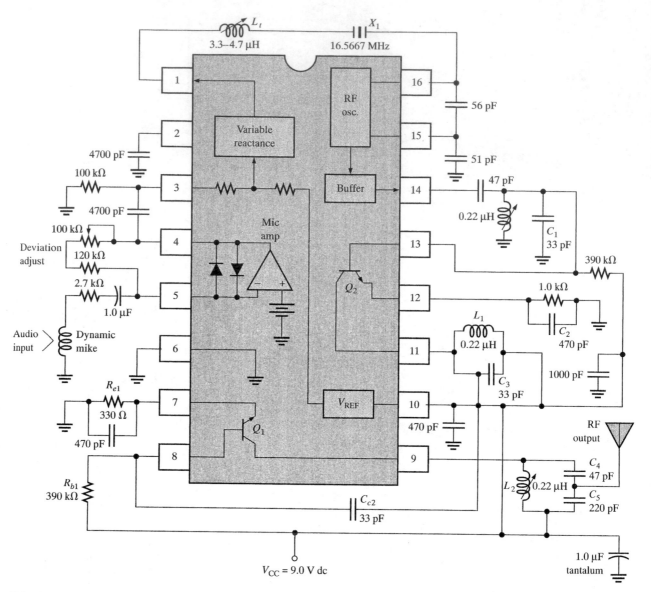

FIGURE 9–45
The MC2833 connected as a 49.7 MHz FM transmitter.

9–7 REVIEW QUESTIONS

1. How does an FM signal carry information?
2. What does VCO stand for?
3. On what principle are most VCOs used in FM based?

9–8 ■ THE PHASE-LOCKED LOOP (PLL)

In the last section, the PLL was mentioned as a way to demodulate an FM signal. In addition to FM demodulation, PLLs are used in a wide variety of communications applications, which include TV receivers, tone decoders, telemetry receivers, modems, and data synchronizers, to name a few. Many of these applications are covered in an electronic communications course. In fact, entire books have been written on the finer points of PLL operation, analysis, and applications. The approach in this section is intended only to present the basic concept and give you an intuitive idea of how PLLs work and how they are used in FM demodulation. A specific PLL integrated circuit is also introduced.

After completing this section, you should be able to

❑ Describe the basic principles of the phase-locked loop (PLL)
    ❑ Draw a basic block diagram for the PLL
    ❑ Discuss the phase detector and state its purpose
    ❑ State the purpose of the VCO
    ❑ State the purpose of the low-pass filter
    ❑ Explain lock range and capture range
    ❑ Discuss the NE565 PLL and explain how it can be used as an FM demodulator

The Basic PLL Concept

The **phase-locked loop (PLL)** is a feedback circuit consisting of a phase detector, a low-pass filter, and a voltage-controlled oscillator (VCO). Some PLLs also include an amplifier in the loop, and in some applications the filter is not used.

    The PLL is capable of locking onto or synchronizing with an incoming signal. When the phase of the incoming signal changes, indicating a change in frequency, the phase detector's output increases or decreases just enough to keep the VCO frequency the same as the frequency of the incoming signal. A basic PLL block diagram is shown in Figure 9–46.

    The general operation of the PLL is as follows. The phase detector compares the phase difference between the incoming signal, V_i, and the VCO signal, V_o. When the fre-

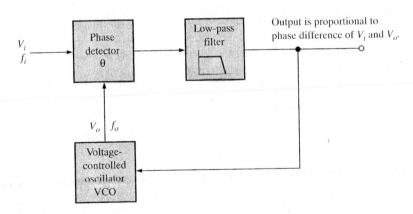

FIGURE 9–46
Basic PLL block diagram.

quency of the incoming signal, f_i, is different from that of the VCO frequency, f_o, the phase angle between the two signals is also different. The output of the phase detector and the filter is proportional to the phase difference of the two signals. This proportional voltage is fed to the VCO, forcing its frequency to move toward the frequency of the incoming signal until the two frequencies are equal. At this point, the PLL is locked onto the incoming frequency. If f_i changes, the phase difference also changes, forcing the VCO to track the incoming frequency.

The Phase Detector

The phase-detector circuit in a PLL is basically a linear multiplier. The following analysis illustrates how it works in a PLL application. The incoming signal, V_i, and the VCO signal, V_o, applied to the phase detector can be expressed as

$$v_i = V_i \sin(2\pi f_i t + \theta_i)$$
$$v_o = V_o \sin(2\pi f_o t + \theta_o)$$

where θ_i and θ_o are the relative phase angles of the two signals. The phase detector multiplies these two signals and produces a sum and difference frequency output, V_d, as follows:

$$
\begin{aligned}
V_d &= V_i \sin(2\pi f_i t + \theta_i) \times V_o \sin(2\pi f_o t + \theta_o) \\
&= \frac{V_i V_o}{2} \cos[(2\pi f_i t + \theta_i) - (2\pi f_o t + \theta_o)] - \frac{V_i V_o}{2} \cos[(2\pi f_i t + \theta_i) + (2\pi f_o t + \theta_o)]
\end{aligned}
$$

When the PLL is locked,

$$f_i = f_o$$

and

$$2\pi f_i t = 2\pi f_o t$$

Therefore, the detector output voltage is

$$V_d = \frac{V_i V_o}{2}[\cos(\theta_i - \theta_o) - \cos(4\pi f_i t + \theta_i + \theta_o)]$$

The second cosine term in the above equation is a second harmonic term $(2 \times 2\pi f_i t)$ and is filtered out by the low-pass filter. The control voltage on the output of the filter is expressed as

$$V_c = \frac{V_i V_o}{2}\cos\theta_e \qquad\qquad (9\text{--}4)$$

where $\theta_e = \theta_i - \theta_o$, where θ_e is the *phase error*. The filter output voltage is proportional to the phase difference between the incoming signal and the VCO signal and is used as the control voltage for the VCO. This operation is illustrated in Figure 9–47.

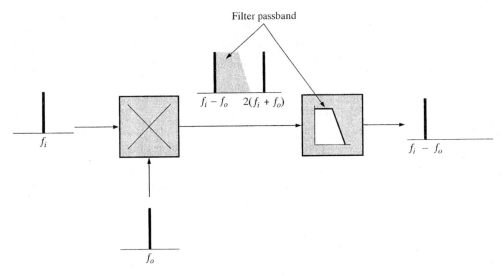

FIGURE 9–47
Basic phase detector/filter operation.

EXAMPLE 9–7

A PLL is locked onto an incoming signal with a frequency of 1 MHz at a phase angle of 50°. The VCO signal is at a phase angle of 20°. The peak amplitude of the incoming signal is 0.5 V and that of the VCO output signal is 0.7 V.

(a) What is the VCO frequency?
(b) What is the value of the control voltage being fed back to the VCO at this point?

Solution

(a) Since the PLL is in lock, $f_i = f_o = $ **1 MHz.**

(b) $\theta_e = \theta_i - \theta_o = 50° - 20° = 30°$

$$V_c = \frac{V_i V_o}{2} \cos \theta_e = \frac{(0.5 \text{ V})(0.7 \text{ V})}{2} \cos 30° = (0.175 \text{ V}) \cos 30° = \textbf{0.152 V}$$

Practice Exercise If the phase angle of the incoming signal changes instantaneously to 30°, indicating a change in frequency, what is the instantaneous VCO control voltage?

The Voltage-Controlled Oscillator (VCO)

Voltage-controlled oscillators can take many forms. A VCO can be some type of *LC* or crystal oscillator as was shown in Section 9–7 or it can be some type of *RC* oscillator or multivibrator. No matter the exact type, most VCOs employed in PLLs operate on the principle of *variable reactance* using the varactor diode as a voltage-variable capacitor.

The capacitance of a varactor diode varies inversely with reverse-bias voltage. The capacitance decreases as reverse voltage increases and vice versa.

In a PLL, the control voltage fed back to the VCO is applied as a reverse-bias voltage to the varactor diode within the VCO. The frequency of oscillation is inversely related to capacitance by the formula

$$f_o = \frac{1}{2\pi RC}$$

for an RC type oscillator and

$$f_o = \frac{1}{2\pi\sqrt{LC}}$$

for an LC type oscillator. These formulas show that frequency increases as capacitance decreases and vice versa.

Capacitance decreases as reverse voltage (control voltage) increases. Therefore, an increase in control voltage to the VCO causes an increase in frequency and vice versa. Basic VCO operation is illustrated in Figure 9–48. The graph in part (b) shows that at the nominal control voltage, $V_{c(nom)}$, the oscillator is running at its nominal or free-running frequency, f_o. An increase in V_c above the nominal value forces the oscillator frequency to increase, and a decrease in V_c below the nominal value forces the oscillator frequency to decrease. There are, of course, limits on the operation as indicated by the minimum and maximum points. The transfer function or conversion gain, K, of the VCO is normally expressed as a certain frequency deviation per unit change in control voltage.

$$K = \frac{\Delta f_o}{\Delta V_c}$$

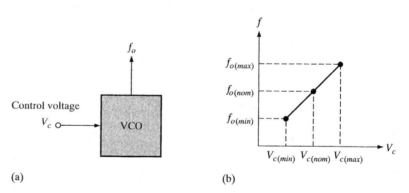

(a)   (b)

FIGURE 9–48
Basic VCO operation.

EXAMPLE 9–8

The output frequency of a certain VCO changes from 50 kHz to 65 kHz when the control voltage increases from 0.5 V to 1 V. What is the conversion gain, K?

Solution

$$K = \frac{\Delta f_o}{\Delta V_c} = \frac{65\ \text{kHz} - 50\ \text{kHz}}{1\ \text{V} - 0.5\ \text{V}} = \frac{15\ \text{kHz}}{0.5\ \text{V}} = \textbf{30 kHz/V}$$

Practice Exercise   If the conversion gain of a certain VCO is 20 kHz/V, how much frequency deviation does a change in control voltage from 0.8 V to 0.5 V produce? If the VCO frequency is 250 kHz at 0.8 V, what is the frequency at 0.5 V?

Basic PLL Operation

When the PLL is locked, the incoming frequency, f_i, and the VCO frequency, f_o, are equal. However, there is always a phase difference between them called the *static phase error*. The phase error, θ_e, is the parameter that keeps the PLL locked in. As you have seen, the filtered voltage from the phase detector is proportional to θ_e (Equation (9–4)). This voltage controls the VCO frequency and is always just enough to keep $f_o = f_i$.

Figure 9–49 shows the PLL and two sinusoidal signals of the same frequency but with a phase difference, θ_e. For this condition the PLL is in lock and the VCO control voltage is constant. If f_i decreases, θ_e increases to θ_{e1} as illustrated in Figure 9–50. This increase in θ_e is sensed by the phase detector causing the VCO control voltage to decrease, thus decreasing f_o until $f_o = f_i$ and keeping the PLL in lock. If f_i increases, θ_e decreases to θ_{e1} as illustrated in Figure 9–51. This decrease in θ_e causes the VCO control voltage to increase, thus increasing f_o until $f_o = f_i$ and keeping the PLL in lock.

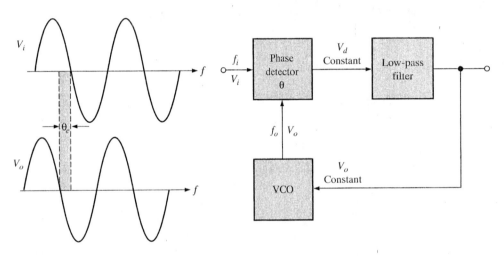

FIGURE 9–49
PLL in lock under static condition ($f_o = f_i$ and constant θ_e).

Lock Range   Once the PLL is locked, it will track frequency changes in the incoming signal. The range of frequencies over which the PLL can maintain lock is called the *lock* or *tracking range*. Limitations on the hold-in range are the maximum frequency deviations of the VCO and the output limits of the phase detector. The hold-in range is independent of the bandwidth of the low-pass filter because, when the PLL is in lock, the difference frequency ($f_i - f_o$) is zero or a very low instantaneous value that falls well within the bandwidth. The hold-in range is usually expressed as a percentage of the VCO frequency.

Capture Range   Assuming the PLL is not in lock, the range of frequencies over which it can acquire lock with an incoming signal is called the *capture range*. Two basic conditions are required for a PLL to acquire lock. First, the difference frequency ($f_o - f_i$) must be low enough to fall within the filter's bandwidth. This means that the incoming frequency must not be separated from the nominal or free-running frequency of the VCO by more than the bandwidth of the low-pass filter. Second, the maximum deviation, Δf_{max}, of the VCO frequency must be sufficient to allow f_o to increase or decrease to a value equal to

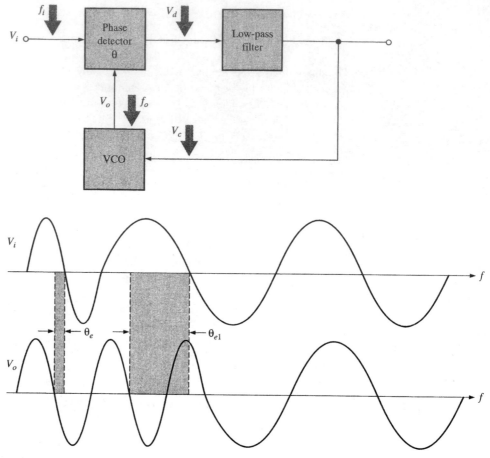

FIGURE 9–50
PLL action when f_i decreases.

f_i. These conditions are illustrated in Figure 9–52; and when they exist, the PLL will "pull" the VCO frequency toward the incoming frequency until $f_o = f_i$.

The NE565 Phase-Locked Loop

The NE565 is a good example of an integrated circuit PLL. The circuit consists of a VCO, phase detector, a low-pass filter formed by an internal resistor and an external capacitor, and an amplifier. The free-running VCO frequency can be set with external components. A block diagram is shown in Figure 9–53. The NE565 can be used for the frequency range from 0.001 Hz to 500 kHz.

The free-running frequency of the VCO is set by the values of R_1 and C_1 in Figure 9–53 according to the following formula. The frequency is in hertz when the resistance is in ohms and the capacitance is in farads.

$$f_o \cong \frac{1.2}{4R_1C_1} \qquad \qquad (9\text{–}5)$$

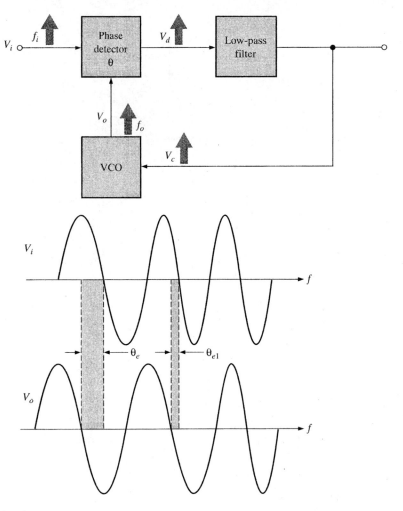

FIGURE 9–51
PLL action when f_i increases.

The lock range is given by

$$f_{lock} = \pm \frac{8f_o}{V_{CC}} \qquad (9\text{–}6)$$

where V_{CC} is the total voltage between the positive and negative dc supply voltage terminals.

The capture range is given by

$$f_{cap} \cong \pm \frac{1}{2\pi} \sqrt{\frac{2\pi f_{lock}}{(3600)C_2}} \qquad (9\text{–}7)$$

The 3600 is the value of the internal filter resistor in ohms. You can see that the capture range is dependent on the filter bandwidth as determined by the internal resistor and the external capacitor C_2.

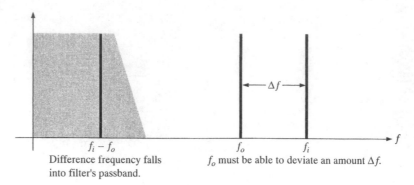

Difference frequency falls into filter's passband.

f_o must be able to deviate an amount Δf.

(a)

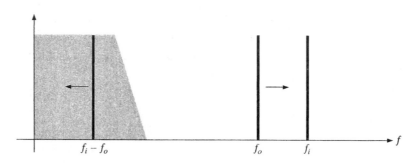

(b) $f_i - f_o$ decreases as f_o deviates towards f_i.

FIGURE 9–52
Illustration of the conditions for a PLL to acquire lock.

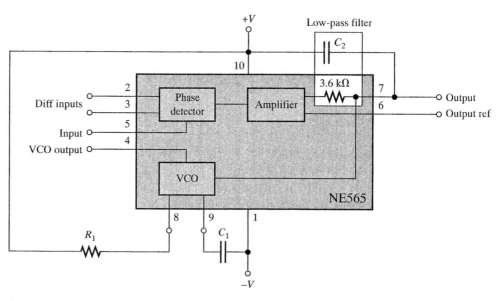

FIGURE 9–53
Block diagram of the NE565 PLL.

The PLL as an FM Demodulator

As you have seen, the VCO control voltage in a PLL depends on the deviation of the incoming frequency. The PLL will produce a voltage proportional to the frequency of the incoming signal which, in the case of FM, is the original modulating signal.

Figure 9–54 shows a typical connection for the NE565 as an FM demodulator. If the IF input is frequency modulated by a sinusoidal signal, we get a sinusoidal signal on the output as indicated. Since the maximum operating frequency is 500 kHz, this device must be used in double-conversion FM receivers. A double-conversion FM receiver is one in which essentially two mixers are used to first convert the RF to a 10.7 MHz IF and then convert this to a 455 kHz IF.

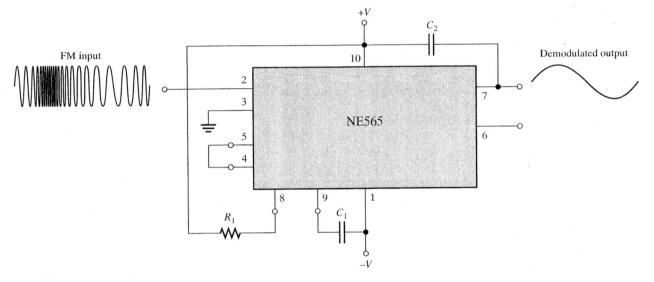

FIGURE 9–54
The NE565 as an FM demodulator.

The free-running frequency of the VCO is adjusted to approximately 455 kHz, which is the center of the modulated IF range. C_1 can be any value, but R_1 should be in the range from 2 kΩ to 20 kΩ. The input can be directly coupled as long as there is no dc voltage difference between pins 2 and 3. The VCO is connected to the phase detector by an external wire between pins 4 and 5. The capacitor between pins 7 and 8 is for eliminating possible oscillations. C_2 is the filter capacitor.

EXAMPLE 9–9   Determine the values for R_1, C_1, and C_2 for the NE565 in Figure 9–54 for a free-running frequency of 455 kHz and a capture range of ±10 kHz. The dc supply voltages are ±6 V.

Solution C_1 is calculated by using Equation (9–5). Choose $R_1 = 4.7$ kΩ.

$$f_o \cong \frac{1.2}{4R_1C_1}$$

$$C_1 \cong \frac{1.2}{4R_1f_o} = \frac{1.2}{4(4700 \ \Omega)(455 \times 10^3 \ \text{Hz})} = 140 \times 10^{-12} \ \text{F} = \textbf{140 pF}$$

The lock range and capture range must be determined before C_2 can be calculated. The lock range is

$$f_{lock} = \pm\frac{8f_o}{V_{CC}} = \pm\frac{8(455 \text{ kHz})}{12 \text{ V}} = \pm303 \text{ kHz}$$

The capture range is

$$f_{cap} \cong \pm\frac{1}{2\pi}\sqrt{\frac{2\pi f_{lock}}{(3600)C_2}}$$

$$f_{cap}^2 \cong \left(\frac{1}{2\pi}\right)^2\frac{2\pi f_{lock}}{(3600)C_2}$$

Therefore,

$$C_2 \cong \left(\frac{1}{2\pi}\right)^2\frac{2\pi f_{lock}}{(3600)f_{cap}^2} = \left(\frac{1}{2\pi}\right)^2\frac{2\pi(303 \times 10^3 \text{ Hz})}{(3600)(10 \times 10^3 \text{ Hz})^2} = 0.134 \times 10^{-6} \text{ F} = \textbf{0.134 } \boldsymbol{\mu}\textbf{F}$$

Practice Exercise What can you do to increase the capture range from ±10 kHz to ±15 kHz?

9–8 REVIEW QUESTIONS

1. List the three basic components in a phase-locked loop.
2. What is another circuit used in some PLLs other than the three listed in Question 1?
3. What is the basic function of a PLL?
4. What is the difference between the lock range and the capture range of a PLL?
5. Basically, how does a PLL track the incoming frequency?

9–9 ■ A SYSTEM APPLICATION

The DCE (data communications equipment) system introduced at the opening of this chapter includes an FSK (frequency shift keying) modem (modulator/demodulator). FSK is one method for modulating digital data for transmission over voice phone lines and is basically a form of frequency modulation. In this system application, the focus is on the low-speed modulator/demodulator (modem) board, which is implemented with a VCO for transmitting FSK signals and a PLL for receiving FSK signals.

After completing this section, you should be able to

❑ Apply what you have learned in this chapter to a system application
   ❑ Describe how a VCO and a PLL can be used in a communications system
   ❑ Discuss how FSK is used to send digital information over phone lines
   ❑ Translate between a printed circuit board and a schematic
   ❑ Analyze the modem circuitry
   ❑ Troubleshoot some common problems

A Brief Description of the System

The FSK **modem** interfaces a computer with the telephone network so that digital data, which are incompatible with the phone system because of bandwidth limitations, can be transmitted and received over regular phone lines, thus allowing computers to communicate with each other. Figure 9–55 shows a diagram of a simple data communications system in which a modem at each end of the phone line provides interfacing for a computer.

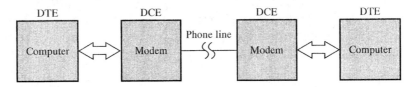

FIGURE 9–55
A data communications system.

The modem (DCE) consists of three basic functional blocks as shown in Figure 9–56: the FSK modem circuits, the phone line interface circuits, and the timing and control circuits. The dual polarity power supply is not shown. Although the focus of this system application is the FSK modem board, we will briefly look at each of the other parts to give you a basic idea of the overall system function.

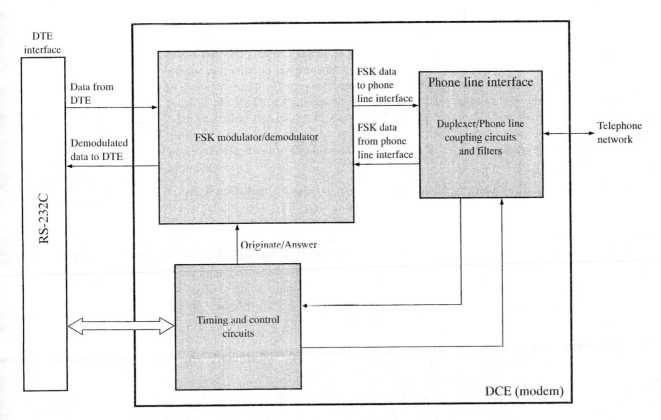

FIGURE 9–56
Basic block diagram of a modem.

The Phone Line Interface The main purposes of this circuitry are to couple the phone line to the modem by proper impedance matching, to provide necessary filtering, and to accommodate full-duplex transmission of data. *Full-duplex* means essentially that information can be going both ways on a single phone line at the same time. This allows a computer, connected to a modem, to be sending data and receiving data simultaneously without the transmitted data interfering with the received data. Full-duplexing is implemented by assigning the transmitted data one bandwidth and the received data another separate bandwidth within the 300 Hz to 3 kHz overall bandwidth of the phone network.

Timing and Control One basic function of the timing and control circuits is to determine the proper mode of operation for the modem. The two modes are the originate mode and the answer mode. Another function is to provide a standard interface (such as RS-232C) with the DTE (computer). The RS-232C standard requires certain defined command and control signals, data signals, and voltage levels for each signal.

Digital Data Before we get into FSK, let's briefly review digital data. A detailed knowledge of binary numbers is not necessary for this system application. Information is represented in digital form by 1s and 0s, which are the binary digits or bits. In terms of voltage waveforms, a 1 is generally represented by a high level and a 0 by a low level. A stream of serial data consists of a sequence of bits as illustrated by an example in Figure 9–57(a).

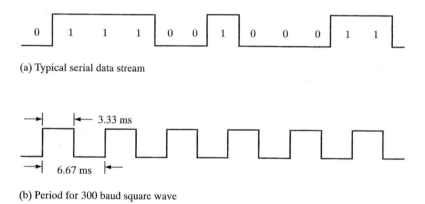

(a) Typical serial data stream

(b) Period for 300 baud square wave

FIGURE 9–57
A serial stream of digital data.

Baud Rate A low-speed modem, such as the one we are focusing on, sends and receives digital data at a rate of 300 bits/s or 300 baud.[1] For example, if we have an alternating sequence of 1s and 0s (highs and lows), as indicated in Figure 9–57(b), each bit takes 3.33 ms. Since it takes two bits, a 1 and a 0, to make up the period of this particular waveform, the fundamental frequency of this format is 1/6.67 ms = 150 Hz. This is the maximum frequency of a 300 baud data stream because normally there may be several consecutive 1s and/or several consecutive zeros in a sequence, thus reducing the frequency. As mentioned earlier, the telephone network has a 300 Hz minimum frequency response, so the fundamental frequency of the 300 baud data stream will fall outside of the telephone bandwidth. This prevents sending digital data in its pure form over the phone lines.

---

[1] Technically, bit rate and baud rate are not the same. Baud rate indicates how many frequency shifts are sent per second. Each frequency shift can represent more than one bit; thus, a 14,400 bits/s modem actually transmits at 2400 baud.

Frequency-Shift Keying (FSK)   FSK is one method used to overcome the bandwidth limitation of the telephone system so that digital data can be sent over the phone lines. The basic idea of FSK is to represent 1s and 0s by two different frequencies within the telephone bandwidth. By the way, any frequency within the telephone bandwidth is an audible tone. The standard frequencies for a full-duplex 300 baud modem in the originate mode are 1070 Hz for a 0 (called a space) and 1270 Hz for a 1 (called a mark). In the answer mode, 2025 Hz is a 0 and 2225 Hz is a 1. The relationship of these FSK frequencies and the telephone bandwidth is illustrated in Figure 9–58. Signals in both the originate and answer bands can exist at the same time on the phone line and not interfere with each other because of the frequency separation.

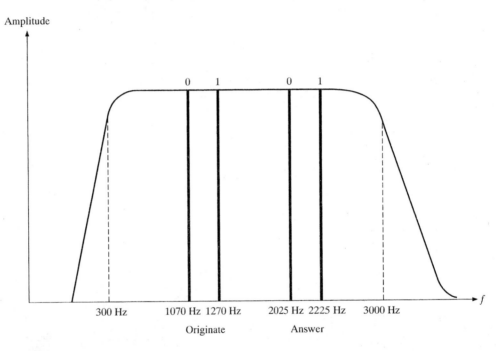

FIGURE 9–58
Frequencies for 300 baud, full-duplex data transmission.

An example of a digital data stream converted to FSK by a modem is shown in Figure 9–59.

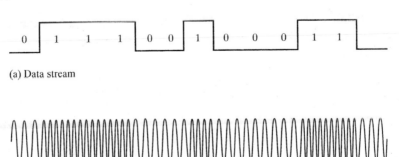

FIGURE 9–59
Example of FSK data.

Modem Circuit Operation

The FSK modem circuits, shown in Figure 9–60, contain an NE565 PLL and a VCO integrated circuit. The VCO can be a device such as the 4046 (not covered specifically in this chapter), which is a PLL device in which the VCO portion can be used by itself because all of the necessary inputs and outputs are available. The VCO in the NE565 cannot be used independently of the PLL because there is no input pin for the control voltage.

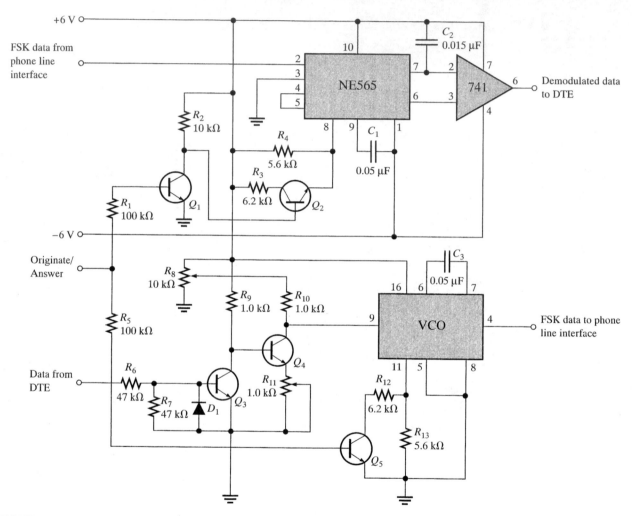

FIGURE 9–60
FSK modulator/demodulator circuit.

The function of the VCO is to accept digital data from a DTE and provide FSK modulation. The VCO is always the transmitting device. The digital data come in on the control voltage input (pin 9) of the VCO via a level-shifting circuit formed by Q_3 and Q_4. This circuit is used because the data from the RS-232C interface are dual polarity with a positive voltage representing a 0 and a negative voltage representing a 1. Potentiometer R_8 is for adjusting the high level of the control voltage and R_{11} is for adjusting the low level for the purpose of fine-tuning the frequency. Transistor Q_5 provides for originate/answer mode frequency selection by changing the value of the frequency-selection resistance from pin 11 to ground. Transistors Q_1 and Q_2 perform a similar function for the PLL.

When the digital data are at high levels, corresponding to logic 0s, the VCO oscillates at 1070 Hz in the originate mode and 2025 Hz in the answer mode. When the digital data are at low levels, corresponding to logic 1s, the VCO oscillates at 1270 Hz in the originate mode and 2225 Hz in the answer mode. An example of the originate mode is when a DTE issues a request for data and transmits that request to another DTE. An example of the answer mode is when the receiving DTE responds to a request and sends data back to the originating DTE.

The function of the PLL is to accept incoming FSK-modulated data and convert it to a digital data format for use by the DTE. The PLL is always a receiving device. When the modem is in the originate mode, the PLL is receiving answer-mode data from the other modem. When the modem is in the answer mode, the PLL is receiving originate-mode data from the other modem. The 741 is connected as a comparator that changes the data levels from the PLL to a dual-polarity format for compatibility with the RS-232C interface.

Now, so that you can take a closer look at the FSK modem board, let's take it out of the system and put it on the test bench.

ON THE TEST BENCH

■ ACTIVITY 1   Relate the PC Board to the Schematic

Locate and identify each component and each input/output pin on the PC board in Figure 9–61 using the schematic in Figure 9–60. Verify that the board and the schematic agree. If they do not agree, indicate the problem.

FIGURE 9–61

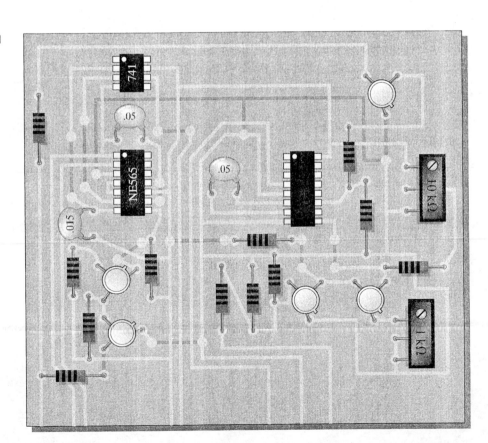

▨ ACTIVITY 2   Analyze the Circuits

For this application, the free-running frequencies of both the PLL and the VCO circuits are determined by the formula in Equation (9–5).

Step 1:   Verify that the free-running frequency for the PLL IC is approximately 1070 Hz in the originate mode and approximately 1270 Hz in the answer mode.

Step 2:   Repeat Step 1 for the VCO.

Step 3:   Determine the approximate minimum and maximum output voltages for the 741 comparator.

Step 4:   Determine the maximum high-level voltage on pin 9 of the VCO.

Step 5:   If a 300 Hz square wave that varies from +5 V to −5 V is applied to the data from the DTE input, what should you observe on pin 4 of the VCO?

Step 6:   When the data from the DTE are low, pin 9 of the VCO is at approximately 0 V. At this level, the VCO oscillates at 1070 Hz or 2025 Hz. When the data from the DTE go high, to what value should the voltage at pin 9 be adjusted to produce a 1270 Hz or 2225 Hz frequency if the transfer function of the VCO is 50 Hz/V?

▨ ACTIVITY 3   Write a Technical Report

Describe the overall operation of the FSK modem board. Specify how each circuit works and what its purpose is. Identify the function of each component. Use the results of Activity 2 as appropriate.

▨ ACTIVITY 4   Troubleshoot the System for Each of the Following Problems By Stating the Probable Cause or Causes

1. There is no demodulated data output voltage when there are verified FSK data from the phone line interface.

2. The NE565 properly demodulates 1070 Hz and 1270 Hz FSK data but does not properly demodulate 2025 Hz and 2225 Hz data.

3. The VCO produces no FSK output.

4. The VCO produces a continuous 1070 Hz tone in the originate mode and a continuous 2025 Hz tone in the answer mode when there are proper data from the DTE.

9–9 REVIEW QUESTIONS

1. The originate/answer input to the modem is *low.* In what mode is the system?

2. What is the purpose of diode D_1 in the FSK modem circuit?

3. The VCO is transmitting 1070 Hz and 1270 Hz FSK signals. To what frequencies does the PLL respond from another modem?

4. If the VCO is transmitting a constant 2225 Hz tone, what does this correspond to in terms of digital data? In what mode is the modem?

▨ SUMMARY

- ▨ In amplitude modulation (AM), the amplitude of a higher-frequency carrier signal is varied by a lower-frequency modulating signal (usually an audio signal).
- ▨ A basic superheterodyne AM receiver consists of an RF amplifier (not always), a mixer, a local oscillator, an IF (intermediate frequency) amplifier, an AM detector, and audio and power amplifiers.
- ▨ The IF in a standard AM receiver is 455 kHz.
- ▨ The AGC (automatic gain control) in a receiver tends to keep the signal strength constant within the receiver to compensate for variations in the received signal.
- ▨ In frequency modulation (FM), the frequency of a carrier signal is varied by a modulating signal.
- ▨ A superheterodyne FM receiver is basically the same as an AM receiver except that it requires a limiter to keep the IF amplitude constant, a different kind of detector or discriminator, and a de-emphasis network. The IF is 10.7 MHz.
- ▨ A four-quadrant linear multiplier can handle any combination of voltage polarities on its inputs.
- ▨ Amplitude modulation is basically a multiplication process.
- ▨ The multiplication of sinusoidal signals produces sum and difference frequencies.
- ▨ The output spectrum of a balanced modulator includes upper-side and lower-side frequencies, but no carrier frequency.
- ▨ The output spectrum of a standard amplitude modulator includes upper-side and lower-side frequencies and the carrier frequency.
- ▨ A linear multiplier is used as the mixer in receiver systems.
- ▨ A mixer converts the RF signal down to the IF signal. The radio frequency varies over the AM or FM band. The intermediate frequency is constant.
- ▨ One type of AM demodulator consists of a multiplier followed by a low-pass filter.
- ▨ The audio and power amplifiers boost the output of the detector or discriminator and drive the speaker.
- ▨ A voltage-controlled oscillator (VCO) produces an output frequency that can be varied by a control voltage. Its operation is based on a variable reactance.
- ▨ A VCO is a basic frequency modulator when the modulating signal is applied to the control voltage input.
- ▨ A phase-locked loop (PLL) is a feedback circuit consisting of a phase detector, a low-pass filter, a VCO, and sometimes an amplifier.
- ▨ The purpose of a PLL is to lock onto and track incoming frequencies.
- ▨ A linear multiplier can be used as a phase detector.
- ▨ A modem is a modulator/demodulator.
- ▨ DTE stands for digital terminal equipment.
- ▨ DCE stands for digital communications equipment.

▨ GLOSSARY

These terms are included in the end-of-book glossary.

Amplitude modulation (AM) A communication method in which a lower-frequency signal modulates (varies) the amplitude of a higher-frequency signal (carrier).

Audio Related to the range of frequencies that can be heard by the human ear and generally considered to be in the 200 Hz to 20 kHz range.

Balanced modulation A form of amplitude modulation in which the carrier is suppressed; sometimes known as *suppressed-carrier modulation*.

Demodulation The process in which the information signal is recovered from the IF carrier signal; the reverse of modulation.

Discriminator   A type of FM demodulator.

Freqency modulation (FM)   A communication method in which a lower-frequency intelligence-carrying signal modulates (varies) the frequency of a higher-frequency signal.

Mixer   A nonlinear circuit that combines two signals and produces the sum and difference frequencies.

Modem   A device that converts signals produced by one type of device to a form compatible with another; *mo*dulator/*dem*odulator.

Modulation   The process in which a signal containing information is used to modify the amplitude, frequency, or phase of a much higher-frequency signal called the carrier.

Multiplier   A linear device that produces an output voltage proportional to the product of two input voltages.

Phase-locked loop (PLL)   A device for locking onto and tracking the frequency of an incoming signal.

Voltage-controlled oscillator (VCO)   An oscillator for which the output frequency is dependent on a controlling input voltage.

■ KEY FORMULAS

(9–1) $V_{OUT} = KV_X V_Y$ Multiplier output voltage

(9–2) $v_1 v_2 = \dfrac{V_{1(p)} V_{2(p)}}{2} \cos 2\pi(f_1 - f_2)t$ Sum and difference frequencies

$$-\dfrac{V_{1(p)} V_{2(p)}}{2} \cos 2\pi(f_1 + f_2)t$$

(9–3) $V_{out} = V^2_{c(p)} \sin 2\pi f_c t$ Standard AM

$$+\dfrac{V_{c(p)} V_{m(p)}}{2} \cos 2\pi(f_c - f_m)t$$

$$-\dfrac{V_{c(p)} V_{m(p)}}{2} \cos 2\pi(f_c + f_m)t$$

(9–4) $V_c = \dfrac{V_i V_o}{2} \cos \theta_e$ PLL control voltage

(9–5) $f_o \cong \dfrac{1.2}{4R_1 C_1}$ Output frequency NE565

(9–6) $f_{lock} = \pm \dfrac{8f_o}{V_{CC}}$ Lock range NE565

(9–7) $f_{cap} \cong \pm \dfrac{1}{2\pi} \sqrt{\dfrac{2\pi f_{lock}}{(3600)C_2}}$ Capture range NE565

■ SELF-TEST

1. In amplitude modulation, the pattern produced by the peaks of the carrier signal is called the
   (a) index   (b) envelope   (c) audio signal   (d) upper-side frequency

2. Which of the following is not a part of an AM superheterodyne receiver?
   (a) Mixer   (b) IF amplifier   (c) DC restorer
   (d) Detector   (e) Audio amplifier   (f) Local oscillator

3. In an AM receiver, the local oscillator always produces a frequency that is above the incoming RF by
   (a) 10.7 kHz   (b) 455 MHz   (c) 10.7 MHz   (d) 455 kHz

4. An FM receiver has an IF frequency that is
   (a) in the 88 MHz to 108 MHz range
   (b) in the 540 kHz to 1640 kHz range
   (c) 455 kHz
   (d) greater than the IF in an AM receiver

5. The detector or discriminator in an AM or an FM receiver
   (a) detects the difference frequency from the mixer
   (b) changes the RF to IF
   (c) recovers the audio signal
   (d) maintains a constant IF amplitude

6. In order to handle all combinations of input voltage polarities, a multiplier must have
   (a) four-quadrant capability   (b) three-quadrant capability
   (c) four inputs   (d) dual-supply voltages

7. The internal attenuation of a multiplier is called the
   (a) transconductance   (b) scale factor   (c) reduction factor

8. When the two inputs of a multiplier are connected together, the device operates as a
   (a) voltage doubler   (b) square root circuit
   (c) squaring circuit   (d) averaging circuit

9. Amplitude modulation is basically a
   (a) summing of two signals   (b) multiplication of two signals
   (c) subtraction of two signals   (d) nonlinear process

10. The frequency spectrum of a balanced modulator contains
    (a) a sum frequency   (b) a difference frequency   (c) a carrier frequency
    (d) answers (a), (b), and (c)   (e) answers (a) and (b)   (f) answers (b) and (c)

11. The IF in a receiver is the
    (a) sum of the local oscillator frequency and the RF carrier frequency
    (b) local oscillator frequency
    (c) difference of the local oscillator frequency and the carrier RF frequency
    (d) difference of the carrier frequency and the audio frequency

12. When a receiver is tuned from one RF frequency to another,
    (a) the IF changes by an amount equal to the LO (local oscillator) frequency
    (b) the IF stays the same
    (c) the LO frequency changes by an amount equal to the audio frequency
    (d) both the LO and the IF frequencies change

13. The output of the AM detector goes directly to the
    (a) IF amplifier   (b) mixer   (c) audio amplifier   (d) speaker

14. If the control voltage to a VCO increases, the output frequency
    (a) decreases   (b) does not change   (c) increases

15. A PLL maintains lock by comparing
    (a) the phase of two signals
    (b) the frequency of two signals
    (c) the amplitude of two signals

■ PROBLEMS

SECTION 9–1 Basic Receivers

1. Label each block in the AM receiver in Figure 9–62.
2. Label each block in the FM receiver in Figure 9–63.

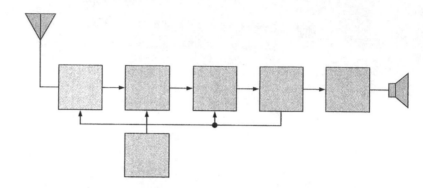

FIGURE 9–62

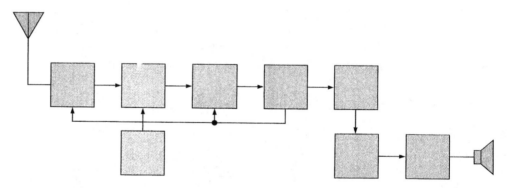

FIGURE 9–63

3. An AM receiver is tuned to a transmitted frequency of 680 kHz. What is the local oscillator (LO) frequency?
4. An FM receiver is tuned to a transmitted frequency of 97.2 MHz. What is the LO frequency?
5. The LO in an FM receiver is running at 101.9 MHz. What is the incoming RF? What is the IF?

SECTION 9–2 The Linear Multiplier

6. From the graph in Figure 9–64, determine the multiplier output voltage for each of the following pairs of input voltages.
   (a) $V_X = -4$ V, $V_Y = +6$ V (b) $V_X = +8$ V, $V_Y = -2$ V
   (c) $V_X = -5$ V, $V_Y = -2$ V (d) $V_X = +10$ V, $V_Y = +10$ V

7. How much pin 3 current is there for the multiplier in Figure 9–65? The potentiometer is set at 2.8 kΩ.

8. Determine the scale factor for the multiplier in Figure 9–65.

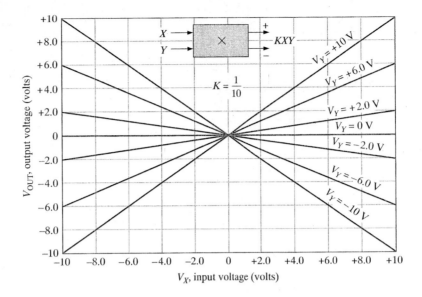

FIGURE 9–64

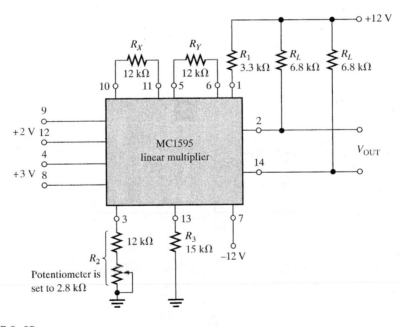

FIGURE 9–65

9. If a certain multiplier has a scale factor of 0.8 and the inputs are +3.5 V and −2.9 V, what is the output voltage?

10. Show the connections for the multiplier in Figure 9–65 in order to implement a squaring circuit.

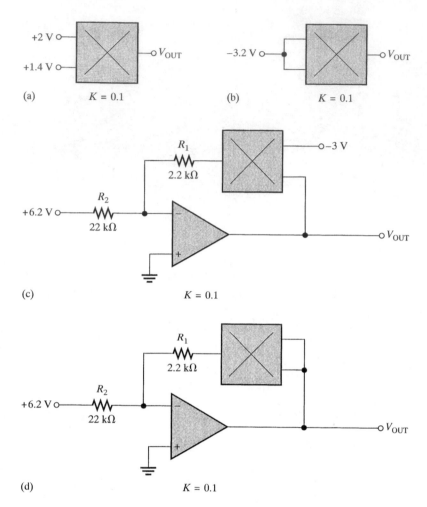

FIGURE 9–66

11. Determine the output voltage for each circuit in Figure 9–66.

SECTION 9–3   Amplitude Modulation

12. If a 100 kHz signal and a 30 kHz signal are applied to a balanced modulator, what frequencies will appear on the output?

13. What are the frequencies on the output of the balanced modulator in Figure 9–67?

FIGURE 9–67

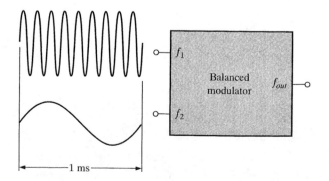

14. If a 1000 kHz signal and a 3 kHz signal are applied to a standard amplitude modulator, what frequencies will appear on the output?

15. What are the frequencies on the output of the standard amplitude modulator in Figure 9–68?

FIGURE 9–68

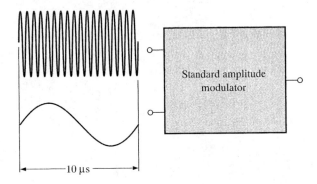

16. The frequency spectrum in Figure 9–69 is for the output of a standard amplitude modulator. Determine the carrier frequency and the modulating frequency.

17. The frequency spectrum in Figure 9–70 is for the output of a balanced modulator. Determine the carrier frequency and the modulating frequency.

18. A voice signal ranging from 300 Hz to 3 kHz amplitude modulates a 600 kHz carrier. Develop the frequency spectrum.

FIGURE 9–69

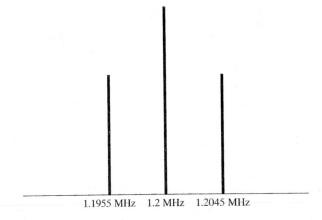

FIGURE 9–70

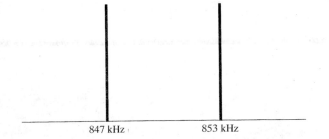

SECTION 9–4 The Mixer

19. Determine the output expression for a multiplier with one sinusoidal input having a peak voltage of 0.2 V and a frequency of 2200 kHz and the other input having a peak voltage of 0.15 V and a frequency of 3300 kHz.

20. Determine the output frequency of the IF amplifier for the frequencies shown in Figure 9–71.

FIGURE 9–71

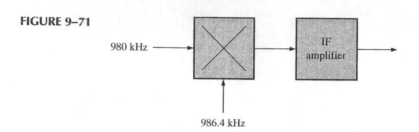

SECTION 9–5 AM Demodulation

21. The input to a certain AM receiver consists of a 1500 kHz carrier and two side frequencies separated from the carrier by 20 kHz. Determine the frequency spectrum at the output of the mixer amplifier.

22. For the same conditions stated in Problem 21, determine the frequency spectrum at the output of the IF amplifier.

23. For the same conditions stated in Problem 21, determine the frequency spectrum at the output of the AM detector (demodulator).

SECTION 9–6 IF and Audio Amplifiers

24. For a carrier frequency of 1.2 MHz and a modulating frequency of 8.5 kHz, list all of the frequencies on the output of the mixer in an AM receiver.

25. In a certain AM receiver, one amplifier has a passband from 450 kHz to 460 kHz and another has a passband from 10 Hz to 5 kHz. Identify these amplifiers.

26. Determine the maximum and minimum output voltages for the audio power amplifier in Figure 9–72.

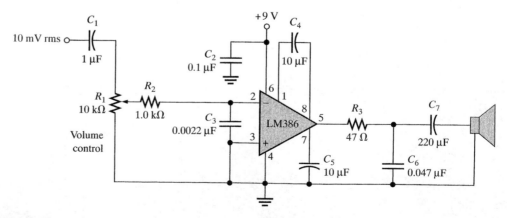

FIGURE 9–72

SECTION 9–7   Frequency Modulation

27. Explain how a VCO is used as a frequency modulator.

28. How does an FM signal differ from an AM signal?

29. What is the variable reactance element shown in the MC2833 diagram in Figure 9–44?

SECTION 9–8   The Phase-Locked Loop (PLL)

30. Label each block in the PLL diagram of Figure 9–73.

31. A PLL is locked onto an incoming signal with a peak amplitude of 250 mV and a frequency of 10 MHz at a phase angle of 30°. The 400 mV peak VCO signal is at a phase angle of 15°.
    (a) What is the VCO frequency?
    (b) What is the value of the control voltage being fed back to the VCO at this point?

FIGURE 9–73

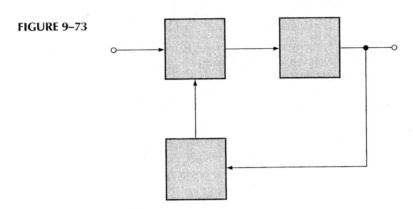

32. What is the conversion gain of a VCO if a 0.5 V increase in the control voltage causes the output frequency to increase by 3.6 kHz?

33. If the conversion gain of a certain VCO is 1.5 kHz per volt, how much does the frequency change if the control voltage increases 0.67 V?

34. Name two conditions for a PLL to acquire lock.

35. Determine the free-running frequency, the lock range, and the capture range for the PLL in Figure 9–74.

FIGURE 9–74

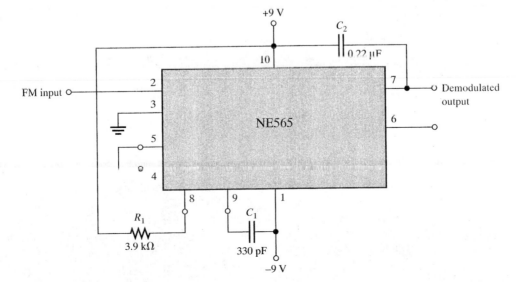

■ ANSWERS TO REVIEW QUESTIONS

Section 9–1

1. AM is amplitude modulation. FM is frequency modulation.
2. In AM, the modulating signal varies the amplitude of a carrier. In FM, the modulating signal varies the frequency of a carrier.
3. AM: 540 kHz to 1640 kHz; FM: 88 MHz to 108 MHz

Section 9–2

1. A four-quadrant multiplier can handle any combination (4) of positive and negative inputs. A one-quadrant multiplier can only handle two positive inputs, for example.
2. $K = 0.1$. K must be 1 for an output of 5 V.
3. Connect the two inputs together and apply a single input variable.

Section 9–3

1. Amplitude modulation is the process of varying the amplitude of a carrier signal with a modulating signal.
2. Balanced modulation produces no carrier frequency on the output, whereas standard AM does.
3. The carrier signal is the modulated signal and has a sufficiently high frequency for transmission. The modulating signal is a lower-frequency signal that contains information and varies the carrier amplitude according to its waveshape.
4. The upper-side frequency is the sum of the carrier frequency and the modulating frequency. The lower-side frequency is the difference of the carrier frequency and the modulating frequency.
5. By summing the peak carrier voltage and the modulating signal before mixing with the carrier signal

Section 9–4

1. The mixer produces (among other frequencies) a signal representing the difference between the incoming carrier frequency and the local oscillator frequency. This is called the intermediate frequency.
2. The mixer multiplies the carrier and the local oscillator signals.
3. 1000 kHz + 350 kHz = 1350 kHz, 1000 kHz − 350 kHz = 650 kHz

Section 9–5

1. The filter removes all frequencies except the audio.
2. Only the 1 kHz

Section 9–6

1. To amplify the 455 kHz amplitude modulated IF coming from the mixer
2. The IF center frequency is 455 kHz.
3. The 10 kHz bandwidth allows the upper-side and lower-side frequencies that contain the information to pass.
4. The audio amplifier follows the detector because the detector is the circuit that recovers the audio from the modulated IF.
5. The IF has a response of approximately 455 kHz ± 5 kHz. The typical audio amplifier has a maximum bandwidth from tens of hertz up to about 15 kHz although for many amplifiers, the bandwidth can be much less than this typical maximum.

Section 9–7

1. The frequency variation of an FM signal bears the information.

2. VCO is voltage-controlled oscillator.

3. VCOs are based on the principle of voltage-variable reactance.

Section 9–8

1. Phase detector, low-pass filter, and VCO

2. Sometimes a PLL uses an amplifier in the loop.

3. A PLL locks onto and tracks a variable incoming frequency.

4. The lock range specifies how much a lock-on frequency can deviate without the PLL losing lock. The capture range specifies how close the incoming frequency must be from the free-running VCO frequency in order for the PLL to lock.

5. The PLL detects a change in the phase of the incoming signal compared to the VCO signal that indicates a change in frequency. The positive feedback then causes the VCO frequency to change along with the incoming frequency.

Section 9–9

1. A *low* on the originate/answer input puts the modem in the originate mode.

2. The diode clips excess negative voltage to protect the base-emitter junction of the transistor.

3. The PLL responds to 2025 Hz and 2225 Hz because the other modem is transmitting in the answer mode.

4. A constant 2225 Hz represents a continuous string of 1s; answer mode

---

■ ANSWERS TO PRACTICE EXERCISES FOR EXAMPLES

9–1 -3.6 V from the graph in Figure 9–9

9–2 0.728 V

9–3 Modulate the carrier with a higher-frequency signal.

9–4 The balanced modulator output has the same side frequencies but does not have a carrier frequency.

9–5 $V_p = 0.025$ mV, $f = 455$ kHz

9–6 455 kHz

9–7 0.172 V

9–8 A decrease of 6 kHz; 244 kHz

9–9 Decrease C_2 to 0.0595 μF.

10

DATA CONVERSION CIRCUITS

Courtesy Hewlett-Packard Company

■ CHAPTER OBJECTIVES

☐ Explain analog switches and identify each type
☐ Discuss the operation of sample-and-hold amplifiers
☐ Explain analog and digital quantities and general interfacing considerations
☐ Describe the operation of digital-to-analog converters (DACs)
☐ Describe A/D conversion
☐ Discuss the operation of analog-to-digital converters (ADCs)
☐ Discuss the basic operation of V/F converters and F/V converters
☐ Troubleshoot DACs and ADCs
☐ Apply what you have learned in this chapter to a system application

Data conversion circuits make interfacing between analog and digital systems possible. Most things in nature occur in analog form. For example, your voice is analog, time is analog, temperature and pressure are analog, and the speed of your vehicle is analog. These quantities and others are first sensed or measured with analog (linear) circuits and are then frequently converted to digital form to facilitate storage, processing, or display.

Also, in many applications, information in digital form must be converted back to analog form. An example of this is digitized music that is stored on a CD. Before you can hear the sounds, the digital information must be converted to its original analog form.

In this chapter, we will study several basic types of circuits found in applications requiring data conversion.

This particular system consists of four large solar cell arrays (see Figure 10–61) that can be individually positioned for proper orientation to the sun's rays. Both the azimuth and the elevation of each array are controlled by stepping motors. The angular position in both azimuth and elevation is sensed by position potentiometers that produce voltages proportional to the angular positions. Many systems use synchros or resolvers as angular transducers, but our system uses potentiometers for simplicity.

The electronic control circuits utilize an analog multiplexer to obtain the analog position from each solar array. The output of the analog multiplexer is then converted to digital form by the analog-to-digital converter for processing by the digital controller. The computer in the digital controller determines how much each array must be rotated in both azimuth and elevation based on stored information about the sun's position. The digital controller then sends the appropriate number of pulses to step the motors to the proper position. Basically, the system controls the solar arrays so that they track the sun each day to maintain an approximate 90° angle to the sun's rays.

This chapter's system application focuses on the analog multiplexer and ADC circuits. For the system application in Section 10–9, in addition to the other topics, be sure you understand
☐ The basic operation of an analog switch
☐ The basic operation of an analog multiplexer
☐ The basic operation of analog-to-digital converters (ADCs)

10–1 ■ ANALOG SWITCHES

Analog switches are important in many types of electronic systems where it is necessary to switch signals on and off electronically. Major applications of analog switches are in signal selection, routing, and processing. Analog switches usually incorporate a FET as the basic switching element.

After completing this section, you should be able to

❑ Explain analog switches and identify each type
    ❑ Identify a single-pole–single-throw analog switch
    ❑ Identify a single-pole–double-throw analog switch
    ❑ Identify a double-pole–double-throw analog switch
    ❑ Describe the ADG202A analog switch IC
    ❑ Discuss multiple-channel analog switches

Types of Analog Switches

Three basic types of analog switches in terms of their functional operation are

❑ Single pole–single throw (SPST)

❑ Single pole–double throw (SPDT)

❑ Double pole–single throw (DPST)

Figure 10–1 illustrates these three basic types of analog switches. As you can see, the analog switch consists of a control element and one or more input-to-output paths called *switch channels*.

FIGURE 10–1

Basic types of analog switches.

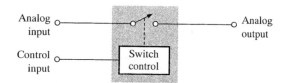

(a) Single pole–single throw (SPST)

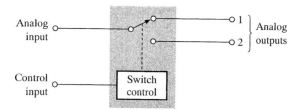

(b) Single pole–double throw (SPDT)

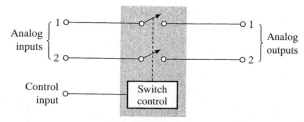

(c) Double pole–single throw (DPST)

An example of an analog switch is the ADG202A. This IC device has four independently operated SPST switches as shown in Figure 10–2(a). Typical packages are shown in Figure 10–2(b).

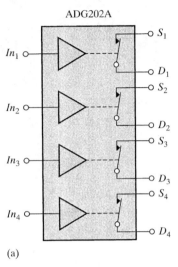

ADG202A

(a)

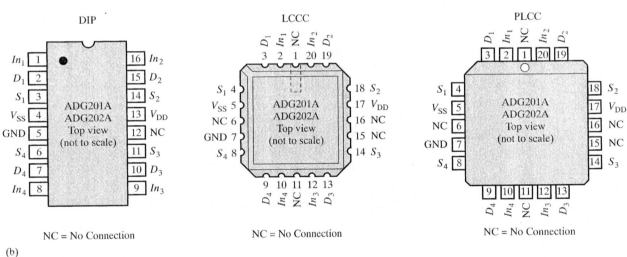

(b)

FIGURE 10–2
The ADG202A Quad SPST switches.

When the control input is at a high-level voltage (at least 2.4 V for the ADG202A), the switch is closed (on). When the control input is at a low-level voltage (no greater than 0.8 V for this device), the switch is open (off). The switches themselves are typically implemented with MOSFETs.

EXAMPLE 10–1   Determine the output waveform of the analog switch in Figure 10–3(a) for the control voltage and analog input voltage shown.

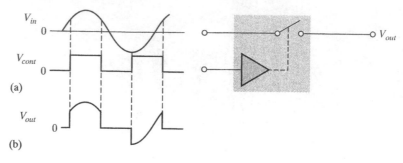

(a)

(b)

FIGURE 10–3

Solution   When the control voltage is high, the switch is closed and the analog input passes through to the output. When the control voltage is low, the switch is open and there is no output voltage. The output waveform is shown in Figure 10–3(b) in relation to the other voltages.

Practice Exercise   What will be the output waveform in Figure 10–3 if the frequency of the control voltage is doubled but keeping the same duty cycle?

Multiple Channel Analog Switches

In data acquisition systems where inputs from several different sources must be independently converted to digital form for processing, a technique called *multiplexing* is used. A separate analog switch is used for each analog source as illustrated in Figure 10–4 for a four-channel system. In this type of application, all of the outputs of the analog switches are connected together to form a common output and only one switch can be closed at a given time. The common switch outputs are connected to the input of a voltage follower as indicated.

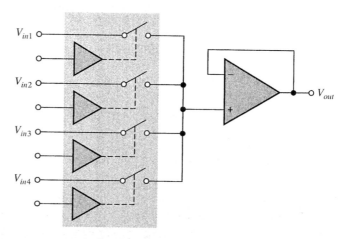

FIGURE 10–4
A four-channel analog multiplexer.

A good example of an IC analog multiplexer is the AD9300 shown in Figure 10–5. This device contains four analog switches that are controlled by a channel decoder. The inputs A_0 and A_1 determine which one of the four switches is on. If A_0 and A_1 are both low, input In_1 is selected. If A_0 is high and A_1 is low, input In_2 is selected. If A_0 is low and A_1 is high, input In_3 is selected. If A_0 and A_1 are both high, input In_4 is selected. The enable input controls the switch that connects or disconnects the output. The AD9300 is capable of switching 4 channels of video for applications including video routing, medical imaging, radar systems, and data acquisition systems.

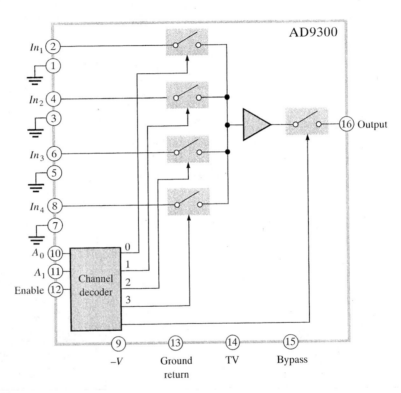

FIGURE 10–5
The AD9300 analog multiplexer.

EXAMPLE 10–2

Determine the output waveform of the analog multiplexer in Figure 10–6 for the control inputs and the analog inputs shown.

Solution   When a control input is a high level, the corresponding switch is closed and the analog voltage on its input is switched through to the output. Notice that only one control voltage is high at a time. The inputs to the switches are sinusoidal waves, each having a different frequency. The resulting output is a sequence of different sinusoidal waves that last for one second and that are separated by a one-second interval, as indicated in Figure 10–7.

FIGURE 10–6

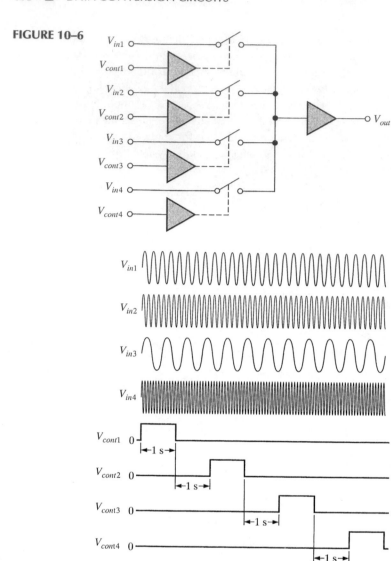

FIGURE 10–7

Practice Exercise How is the output waveform in Figure 10–7 affected if the time interval between the control voltage pulses is decreased?

10–1 REVIEW QUESTIONS

1. What is the purpose of an analog switch?
2. What is the basic function of an analog multiplexer?

10–2 ■ SAMPLE-AND-HOLD AMPLIFIERS

A sample-and-hold amplifier samples an analog input voltage at a certain point in time and retains or holds the sampled voltage for an extended time after the sample is taken. The sample-and-hold process keeps the sampled analog voltage constant for the length of time necessary to allow an analog-to-digital converter (ADC) to convert the voltage to digital form.

After completing this section, you should be able to

❑ Discuss the operation of sample-and-hold amplifiers
   ❑ Describe tracking in a sample-and-hold amplifier
   ❑ Define *aperture time, aperture jitter, acquisition time, droop,* and *feedthrough*
   ❑ Describe the AD582 sample-and-hold amplifier

A Basic Sample-and-Hold Circuit

A basic sample-and-hold circuit consists of an analog switch, a capacitor, and input and output buffer amplifiers as shown in Figure 10–8. The analog switch **samples** the analog input voltage through the input buffer amplifier, the capacitor (C_H) stores or holds the sampled voltage for a period of time, and the output buffer amplifier provides a high input impedance to prevent the capacitor from discharging quickly.

FIGURE 10–8
A basic sample-and-hold circuit.

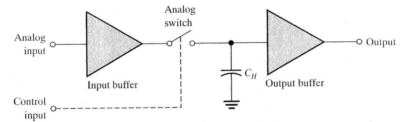

As illustrated in Figure 10–9, a relatively narrow control voltage pulse closes the analog switch and allows the capacitor to charge to the value of the input voltage. The switch then opens, and the capacitor holds the voltage for a long period of time because of the very high impedance discharge path through the op-amp input. So basically, the

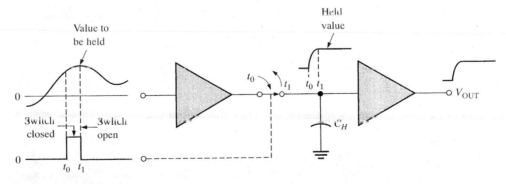

FIGURE 10–9
Basic action of a sample-and-hold.

sample-and-hold circuit converts an instantaneous value of the analog input voltage to a dc voltage.

Tracking During Sample Time

Perhaps a more appropriate designation for a sample-and-hold amplifier is sample/track-and-hold because the circuit actually tracks the input voltage during the sample interval. As indicated in Figure 10–10, the output follows the input during the time that the control voltage is high; and when the control voltage goes low, the last voltage is held until the next sample interval.

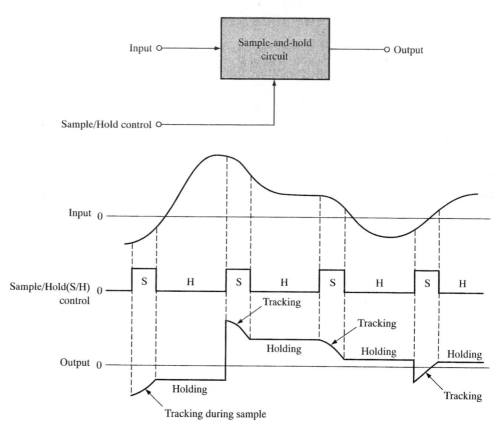

FIGURE 10–10
Example of tracking during a sample-and-hold sequence.

EXAMPLE 10–3

Determine the output voltage waveform for the sample/track-and-hold amplifier in Figure 10–11, given the input and control voltage waveforms.

Solution During the time that the control voltage is high, the analog switch is closed and the circuit is tracking the input. When the control voltage goes low, the analog switch opens; and the last voltage value is held at a constant level until the next time the control voltage goes high. This is shown in Figure 10–12.

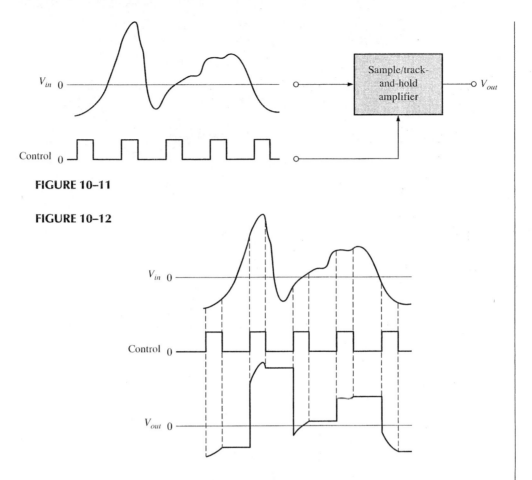

FIGURE 10–11

FIGURE 10–12

Practice Exercise   Sketch the output voltage waveform for Figure 10–11 if the control voltage frequency is reduced by half.

Performance Specifications

In addition to specifications similar to those of a closed-loop op-amp that were discussed in Chapter 2, several specifications are peculiar to sample-and-hold amplifiers. These include the aperture time, aperture jitter, acquisition time, droop, and feedthrough.

❑ **Aperture time**—the time for the analog switch to fully open after the control voltage switches from its sample level to its hold level. Aperture time produces a delay in the effective sample point.

❑ **Aperture jitter**—the uncertainty in the aperture time.

❑ **Acquisition time**—the time required for the device to reach its final value when the control voltage switches from its hold level to its sample level.

❑ **Droop**—the change in voltage from the sampled value during the hold interval because of charge leaking off of the hold capacitor.

❑ **Feedthrough**—the component of the output voltage that follows the input signal after the analog switch is opened. The inherent capacitance from the input to the output of the switch causes feedthrough.

Each of these parameters is illustrated in Figure 10–13 for an example input voltage waveform.

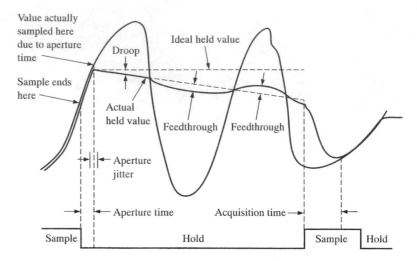

FIGURE 10–13
Sample and hold amplifier specifications. The effects are exaggerated for clarity. The black curve is input voltage waveform; the blue curve is output voltage.

A Specific Device

A good example of a basic sample-and-hold amplifier is the AD582. The circuit and pin configuration are shown in Figure 10–14. As shown in the figure, this particular device consists of two buffer amplifiers and an analog switch that is controlled by a logic gate. The hold capacitor must be connected externally to pin 6, and its value is selected depending on the application requirements.

FIGURE 10–14
The AD582 sample-and-hold amplifier.

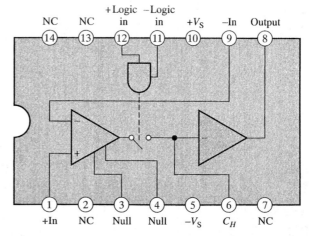

The control voltage for establishing the sample/hold intervals is applied between pins 11 and 12. The input signal to be sampled is applied to pin 1. A potentiometer for nulling out the offset voltage can be connected between pins 3 and 4, and the overall gain of the device can be set with external feedback connections. Two typical configurations are shown in Figure 10–15.

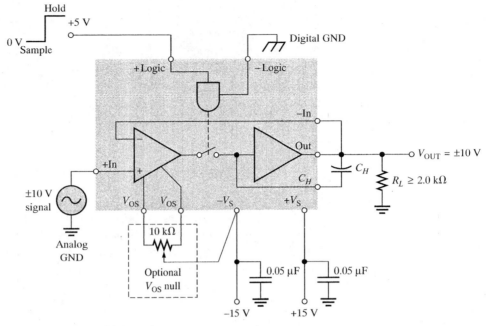

(a) Sample and hold with $A_v = +1$

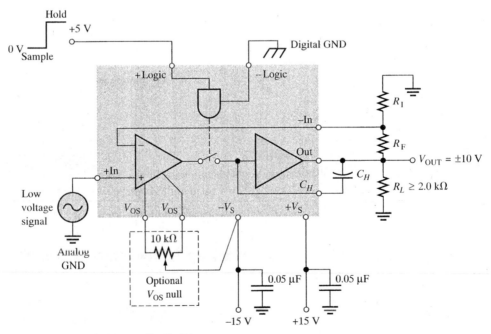

(b) Sample and hold with $A_v = (1 + R_F/R_I)$

FIGURE 10–15

Two configurations of the AD582 sample-and-hold amplifier.

10–2 REVIEW QUESTIONS

1. What is the basic function of a sample-and-hold amplifier?
2. In reference to the output of a sample-and-hold amplifier, what does droop mean?
3. Define *aperture time*.
4. What is acquisition time?

10–3 ■ INTERFACING THE ANALOG AND DIGITAL WORLDS

Analog quantities are sometimes called real-world quantities because most physical quantities are analog in nature. Many applications of computers and other digital systems require the input of real-world quantities, such as temperature, speed, position, pressure, and force. Real-world quantities can even include graphic images. Also, digital systems are often used to control real-world quantities. A basic familiarity with the binary number system is assumed for this and the next sections.

After completing this section, you should be able to

❑ Discuss digital and analog quantities and general interfacing considerations
   ❑ Describe an analog quantity
   ❑ Describe a digital quantity
   ❑ Discuss examples of real-world analog/digital interfacing

Digital and Analog Signals

An analog quantity is one that has a continuous set of values over a given range, as contrasted with discrete values for the digital case. Almost any measurable quantity is analog in nature, such as temperature, pressure, speed, and time. To further illustrate the difference between an analog and a digital representation of a quantity, let's take the case of a voltage that varies over a range from 0 V to +15 V. The analog representation of this quantity takes in all values between 0 and +15 V of which there is an infinite number.

In the case of *digital* representation using a 4-bit binary code, only sixteen values can be defined. More values between 0 and +15 can be represented by using more bits in the digital code. So an analog quantity can be represented to a high degree of accuracy with a digital code that specifies discrete values within the range. This concept is illustrated in Figure 10–16, where the analog function shown is a smoothly changing curve that takes on values between 0 V and +15 V. If a 4-bit code is used to represent this curve, each binary number represents a discrete point on the curve.

In Figure 10–16 the voltage on the analog curve is measured, or sampled, at each of thirty-five equal intervals. The voltage at each of these intervals is represented by a 4-bit code as indicated. At this point, we have a series of binary numbers representing various voltage values along the analog curve. This is the basic idea of analog-to-digital (A/D) conversion.

An approximation of the analog function in Figure 10–16 can be reconstructed from the sequence of digital numbers that has been generated. Obviously, there will be some

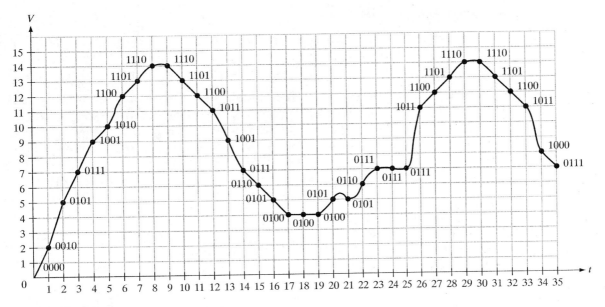

FIGURE 10–16
Discrete (digital) points on an analog curve.

error in the reconstruction because only certain values are represented (thirty-six in this example) and not the continuous set of values. If the digital values at all of the thirty-six points are graphed as shown in Figure 10–17, we have a reconstructed function. As you can see, the graph only approximates the original curve because values between the points are not known.

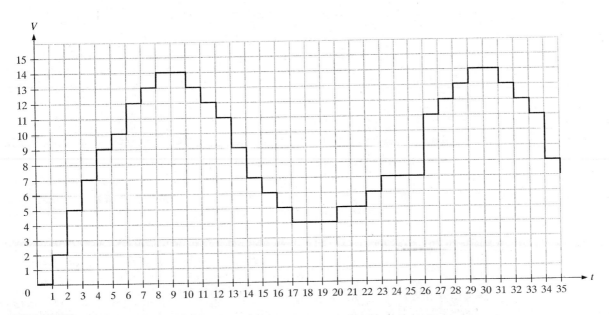

FIGURE 10–17
A rough digital reproduction of an analog curve.

Interfacing Applications

To interface between the digital and analog worlds, two basic processes are required. These are analog-to-digital (A/D) conversion and digital-to-analog (D/A) conversion. The following two system examples illustrate the application of these conversion processes.

An Electronic Thermostat

A simplified block diagram of a microprocessor-based electronic thermostat is shown in Figure 10–18. The room temperature sensor produces an analog voltage that is proportional to the temperature. This voltage is increased by the linear amplifier and applied to the **analog-to-digital converter (ADC),** where it is converted to a digital code and periodically sampled by the microprocessor. For example, suppose the room temperature is 67°F. A specific voltage value corresponding to this temperature appears on the ADC input and is converted to an 8-bit binary number, 01000011.

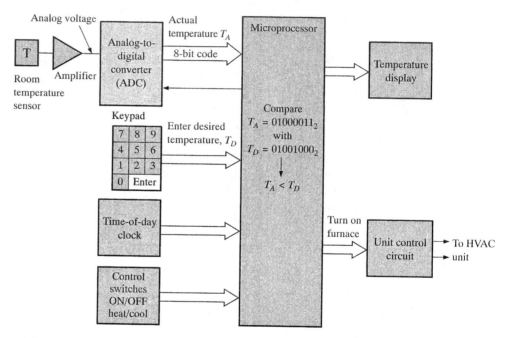

FIGURE 10–18
An electronic thermostat that uses an ADC.

Internally, the microprocessor compares this binary number with a binary number representing the desired temperature (say 01001000 for 72°F). This desired value has been previously entered from the keypad and stored in a register. The comparison shows that the actual room temperature is less than the desired temperature. As a result, the microprocessor instructs the unit control circuit to turn the furnace on. As the furnace runs, the microprocessor continues to monitor the actual temperature via the ADC. When the actual temperature equals or exceeds the desired temperature, the microprocessor turns the furnace off.

A Digital Audio Tape (DAT) Player/Recorder

Another system example that includes both A/D and D/A conversion is the DAT player/recorder. A basic block diagram is shown in Figure 10–19.

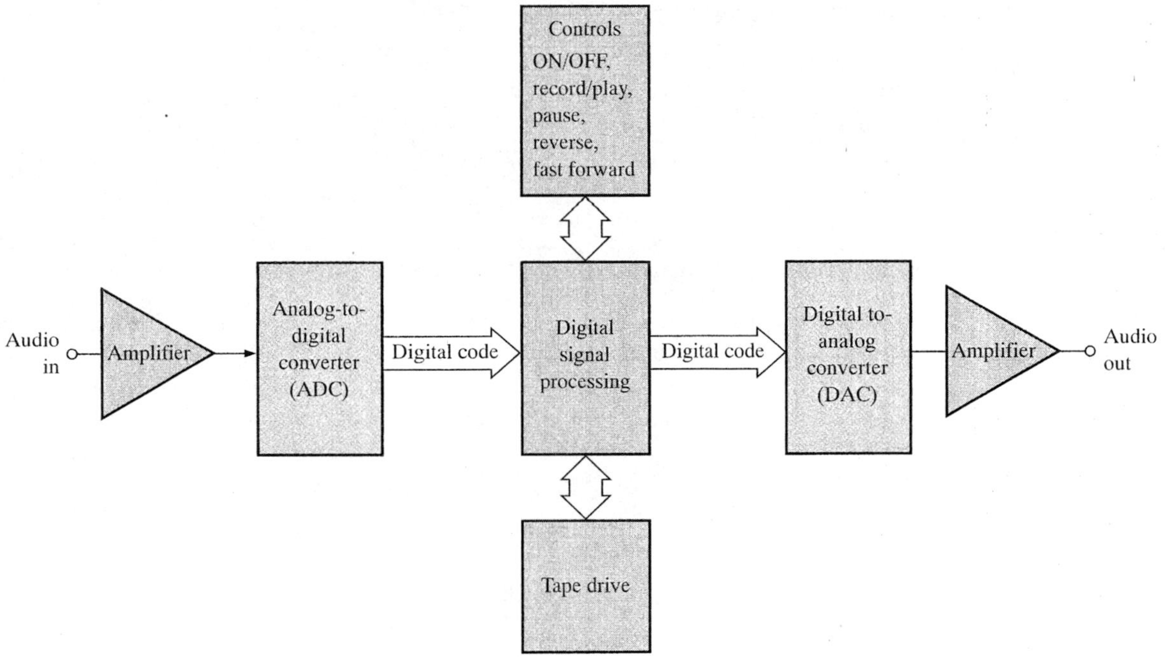

FIGURE 10–19
Basic block diagram of a DAT system.

An audio signal, of course, is an analog quantity. In the record mode, sound is picked up, amplified, and converted to digital form by the ADC. The digital codes representing the audio signal are processed and recorded on the tape.

In the play mode, the digitized audio signal is read from the tape, processed, and converted back to analog form by the **digital-to-analog converter (DAC).** It is then amplified and sent to the speaker system.

10–3 REVIEW QUESTIONS

1. In what form do quantities appear naturally?
2. Explain the basic purpose of A/D conversion.
3. Explain the basic purpose of D/A conversion.

10–4 ■ DIGITAL-TO-ANALOG (D/A) CONVERSION

D/A conversion is an important part of many systems. In this section, we will examine two basic types of digital-to-analog converters (DACs) and learn about their performance characteristics. The binary-weighted-input DAC was introduced in Chapter 4 as an example of a scaling adder application and is covered more thoroughly in this section. Also, a more commonly used configuration called the R/2R ladder DAC is introduced.

After completing this section, you should be able to

❑ Describe the operation of digital-to-analog converters (DACs)
   ❑ Describe the binary-weighted-input DAC
   ❑ Describe the *R/2R* ladder DAC
   ❑ Discuss resolution, accuracy, linearity, monotonicity, and settling time

Binary-Weighted-Input Digital-to-Analog Converter

The binary-weighted-input DAC uses a resistor network with resistance values that represent the binary weights of the input bits of the digital code. Figure 10–20 shows a 4-bit DAC of this type. Each of the input resistors will either have current or have no current, depending on the input voltage level. If the input voltage is zero (binary 0), the current is also zero. If the input voltage is high (binary 1), the amount of current depends on the input resistor value and is different for each input resistor, as indicated by the meters.

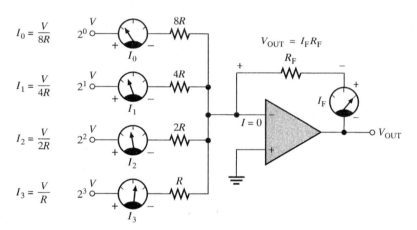

FIGURE 10–20
A 4-bit DAC with binary-weighted inputs.

    Since there is practically no current at the op-amp inverting ($-$) input, all of the input currents sum together and go through R_F. Since the inverting input is at 0 V (virtual ground), the drop across R_F is equal to the output voltage, so $V_{OUT} = I_F R_F$.

    The values of the input resistors are chosen to be inversely proportional to the binary weights of the corresponding input bits. The lowest-value resistor (R) corresponds to the highest binary-weighted input (2^3). The other resistors are multiples of R (that is, $2R$, $4R$, and $8R$) and correspond to the binary weights 2^2, 2^1, and 2^0, respectively. The input currents are also proportional to the binary weights. Thus, the output voltage is proportional to the sum of the binary weights because the sum of the currents is through R_F.

    One of the disadvantages of this type of DAC is the number of different resistor values. For example, an 8-bit converter requires eight resistors, ranging from some value of R to $128R$ in binary-weighted steps. This range of resistors requires tolerances of one part in 255 (less than 0.5%) to accurately convert the input, making this type of DAC very difficult to mass-produce.

EXAMPLE 10–4   Determine the output of the DAC in Figure 10–21(a) if the waveforms representing a sequence of 4-bit binary numbers in Figure 10–21(b) are applied to the inputs. Input D_0 is the least significant bit (LSB).

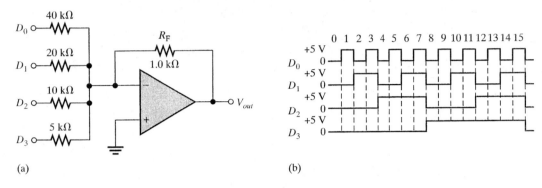

(a)   (b)

FIGURE 10–21

Solution   First, determine the current for each of the weighted inputs. Since the inverting $(-)$ input of the op-amp is at 0 V (virtual ground) and a binary 1 corresponds to $+5$ V, the current through any of the input resistors is 5 V divided by the resistance value.

$$I_0 = \frac{5\text{ V}}{40\text{ k}\Omega} = 0.125\text{ mA}$$

$$I_1 = \frac{5\text{ V}}{20\text{ k}\Omega} = 0.25\text{ mA}$$

$$I_2 = \frac{5\text{ V}}{10\text{ k}\Omega} = 0.5\text{ mA}$$

$$I_3 = \frac{5\text{ V}}{5\text{ k}\Omega} = 1.0\text{ mA}$$

There is essentially no current at the inverting op-amp input because of its extremely high impedance. Therefore, assume that all of the current goes through the feedback resistor R_F. Since one end of R_F is at 0 V (virtual ground), the drop across R_F equals the output voltage, which is negative with respect to virtual ground.

$$V_{OUT(D0)} = (1.0\text{ k}\Omega)(-0.125\text{ mA}) = -0.125\text{ V}$$
$$V_{OUT(D1)} = (1.0\text{ k}\Omega)(-0.25\text{ mA}) = -0.25\text{ V}$$
$$V_{OUT(D2)} = (1.0\text{ k}\Omega)(-0.5\text{ mA}) = -0.5\text{ V}$$
$$V_{OUT(D3)} = (1.0\text{ k}\Omega)(-1.0\text{ mA}) = -1.0\text{ V}$$

From Figure 10–21(b), the first binary input code is 0000, which produces an output voltage of 0 V. The next input code is 0001, which produces an output voltage of -0.125 V. The next code is 0010, which produces an output voltage of -0.25 V. The next code is 0011, which produces an output voltage of -0.125 V $+$ -0.25 V $=$ -0.375 V. Each successive binary code increases the output voltage by -0.125 V, so for this particular straight binary sequence on the inputs, the output is a stairstep waveform going from 0 V to -1.875 V in -0.125 V steps. This is shown in Figure 10–22.

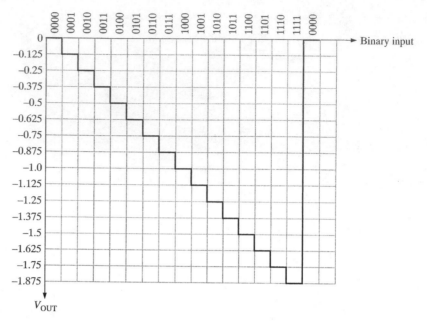

FIGURE 10–22
Output of the DAC in Figure 10–21.

Practice Exercise What size are the output steps of the DAC if the feedback resistance is changed to 2.0 kΩ?

The *R*/2*R* Ladder Digital-to-Analog Converter

Another method of D/A conversion is the *R*/2*R* ladder, as shown in Figure 10–23 for four bits. It overcomes one of the problems in the binary-weighted-input DAC in that it requires only two resistor values.

Start by assuming that the D_3 input is at a high level (+ 5 V) and the others are at a low level (ground, 0 V). This condition represents the binary number 1000. A circuit analysis will show that this reduces to the equivalent form shown in Figure 10–24(a). Essen-

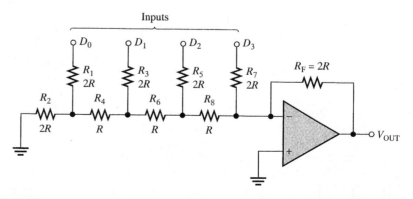

FIGURE 10–23
An R/2R ladder DAC.

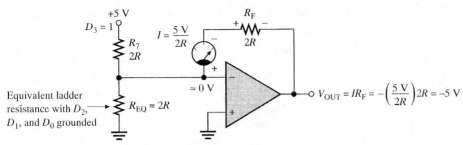

Equivalent ladder resistance with D_2, D_1, and D_0 grounded →

(a) Equivalent circuit for $D_3 = 1, D_2 = 0, D_1 = 0, D_0 = 0$

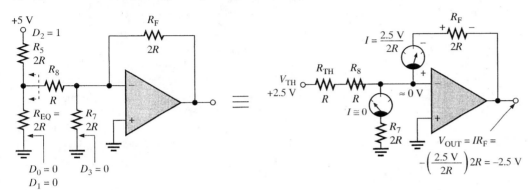

(b) Equivalent circuit for $D_3 = 0, D_2 = 1, D_1 = 0, D_0 = 0$

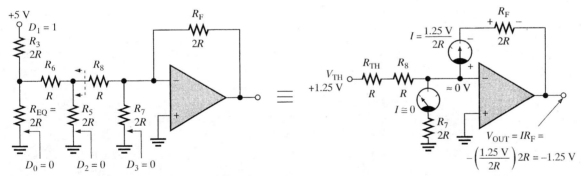

(c) Equivalent circuit for $D_3 = 0, D_2 = 0, D_1 = 1, D_0 = 0$

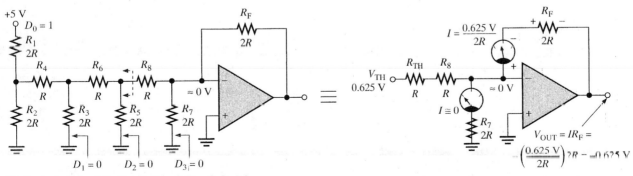

(d) Equivalent circuit for $D_3 = 0, D_2 = 0, D_1 = 0, D_0 = 1$

FIGURE 10–24

Analysis of the R/2R ladder DAC.

tially no current goes through the $2R$ equivalent resistance because the inverting input is at virtual ground. Thus, all of the current ($I = 5$ V$/2R$) through R_7 also goes through R_F, and the output voltage is -5 V.

Figure 10–24(b) shows the equivalent circuit when the D_2 input is at $+5$ V and the others are at ground. This condition represents 0100. If we thevenize looking from R_8, we get 2.5 V in series with R, as shown in part (b).[1] This results in a current through R_F of $I = 2.5$ V$/2R$, which gives an output voltage of -2.5 V. Keep in mind that there is no current at the op-amp inverting input and that there is no current through the equivalent resistance to ground because it has 0 V across it, due to the virtual ground.

Figure 10–24(c) shows the equivalent circuit when the D_1 input is at $+5$ V and the others are at ground. Again thevenizing looking from R_8, we get 1.25 V in series with R as shown. This results in a current through R_F of $I = 1.25$ V$/2R$, which gives an output voltage of -1.25 V.

In part (d) of Figure 10–24, the equivalent circuit representing the case where D_0 is at $+5$ V and the other inputs are at ground is shown. Thevenizing from R_8 gives an equivalent of 0.625 V in series with R as shown. The resulting current through R_F is $I = 0.625$ V$/2R$, which gives an output voltage of -0.625 V.

Notice that each successively lower-weighted input produces an output voltage that is halved, so that the output voltage is proportional to the binary weight of the input bits.

Performance Characteristics of Digital-to-Analog Converters

The performance characteristics of a DAC include resolution, accuracy, linearity, monotonicity, and settling time, each of which is discussed in the following list:

❑ **Resolution.** The resolution of a DAC is the reciprocal of the maximum number of discrete steps in the output. Resolution, of course, is dependent on the number of input bits. For example, a 4-bit DAC has a resolution of one part in $2^4 - 1$ (one part in fifteen). Expressed as a percentage, this is $(1/15)100 = 6.67\%$. The total number of discrete steps equals $2^n - 1$, where n is the number of bits. Resolution can also be expressed as the number of bits that are converted.

❑ **Accuracy.** Accuracy is a comparison of the actual output of a DAC with the expected output. It is expressed as a percentage of a full-scale, or maximum, output voltage. For example, if a converter has a full-scale output of 10 V and the accuracy is $\pm 0.1\%$, then the maximum error for any output voltage is $(10$ V$)(0.001) = 10$ mV. Ideally, the accuracy should be no worse than $\pm \frac{1}{2}$ of a least significant bit (LSB). For an 8-bit converter, the least significant bit is 0.39% of full scale. The accuracy should be approximately $\pm 0.2\%$.

❑ **Linearity.** A linear error is a deviation from the ideal straight-line output of a DAC. A special case is an offset error, which is the amount of output voltage when the input bits are all zeros.

❑ **Monotonicity.** A DAC is monotonic if it does not miss any steps when it is sequenced over its entire range of input bits.

❑ **Settling time.** Settling time is normally defined as the time it takes a DAC to settle within $\pm \frac{1}{2}$ LSB of its final value when a change occurs in the input code.

---

[1] Section 1–3 describes the Thevenin equivalent circuit and how to thevenize.

EXAMPLE 10–5

Determine the resolution, expressed as a percentage, of **(a)** an 8-bit DAC and **(b)** a 12-bit DAC.

Solution

(a) For the 8-bit converter,

$$\frac{1}{2^8 - 1} \times 100 = \frac{1}{255} \times 100 = \textbf{0.392\%}$$

(b) For the 12-bit converter,

$$\frac{1}{2^{12} - 1} \times 100 = \frac{1}{4095} \times 100 = \textbf{0.0244\%}$$

Practice Exercise Determine the percent resolution for an 18-bit converter.

10–4 REVIEW QUESTIONS

1. What is the disadvantage of the DAC with binary-weighted inputs?
2. What is the resolution of a 4-bit DAC?

10–5 ■ BASIC CONCEPTS OF ANALOG-TO-DIGITAL (A/D) CONVERSION

As you have seen, analog-to-digital conversion is the process by which an analog quantity is converted to digital form. A/D conversion is necessary when measured quantities must be in digital form for processing in a computer or for display or storage. Basic concepts of A/D conversion including resolution, conversion time, sampling theory, and quantization error are introduced in this section.

After completing this section, you should be able to

❑ Describe A/D conversion
   ❑ Define *resolution*
   ❑ Explain conversion time
   ❑ Discuss sampling theory
   ❑ Define *quantization error*

Resolution

An analog-to-digital converter (ADC) translates a continuous analog signal into a series of binary numbers. Each binary number represents the value of the analog signal at the time of conversion. The **resolution** of an ADC can be expressed as the number of bits (binary digits) used to represent each value of the analog signal. A 4-bit ADC can represent sixteen different values of an analog signal because $2^4 = 16$. An 8-bit ADC can represent 256 different values of an analog signal because $2^8 = 256$. A 12-bit ADC can represent 4096 different values of the analog signal because $2^{12} = 4096$. The more bits, the more accurate is the conversion and the greater is the resolution because more values of a given analog signal can be represented.

Resolution is basically illustrated in Figure 10–25 using the analog voltage ramp in part (a). For the case of 3-bit resolution as shown in part (b), only eight values of the voltage ramp can be represented by binary numbers. D/A reconstruction of the ramp using the eight binary values results in the stair-step approximation shown. For the case of 4-bit resolution as shown in part (c), sixteen values can be represented, and D/A reconstruction results in a more accurate 16-step approximation as shown. For the case of 5-bit resolution as shown in part (d), D/A reconstruction produces an even more accurate 32-step approximation of the ramp.

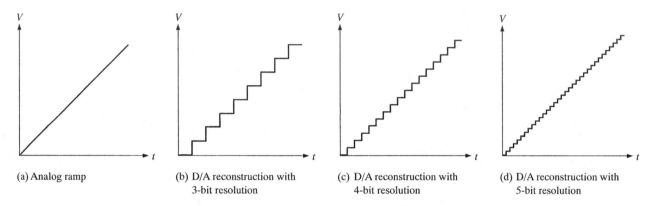

(a) Analog ramp   (b) D/A reconstruction with 3-bit resolution   (c) D/A reconstruction with 4-bit resolution   (d) D/A reconstruction with 5-bit resolution

FIGURE 10–25
Illustration of the effect of resolution on the representation of an analog signal (a ramp in this case).

Conversion Time

In addition to resolution, another important characteristic of ADCs is conversion time. The conversion of a value on an analog waveform into a digital quantity is not an instantaneous event, but it is a process that takes a certain amount of time. The conversion time can range from microseconds for fast converters to milliseconds for slower devices. Conversion time is illustrated in a basic way in Figure 10–26. As you can see, the value of the analog voltage to be converted occurs at time t_0 but the conversion is not complete until time t_1.

Sampling Theory

In A/D conversion, an analog waveform is sampled at a given point and the sampled value is then converted to a binary number. Since it takes a certain interval of time to accomplish the conversion, the number of samples of an analog waveform during a given period of time is limited. For example, if a certain ADC can make one conversion in 1 ms, it can make 1000 conversions in one second. That is, it can convert 1000 different analog values to digital form in a one-second interval.

In order to represent an analog waveform, the minimum sample rate must be greater than twice the maximum frequency component of the analog signal. This minimum sampling rate is known as the **Nyquist rate.** At the Nyquist rate, an analog signal is sampled and converted more than two times per cycle, which establishes the fundamental frequency

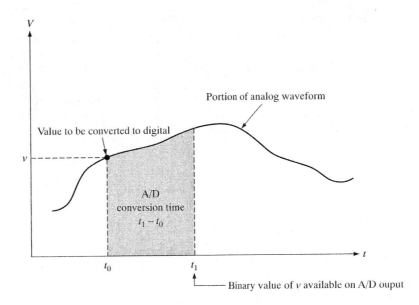

FIGURE 10–26
An illustration of A/D conversion time.

of the analog signal. Filtering can be used to obtain a facsimile of the original signal after D/A conversion. Obviously, a greater number of conversions per cycle of the analog signal results in a more accurate representation of the analog signal. This is illustrated in Figure 10–27 for two different sample rates. The lower waveforms are the D/A reconstructions for various sample rates.

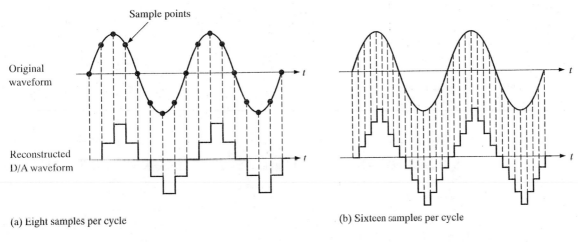

(a) Eight samples per cycle

(b) Sixteen samples per cycle

FIGURE 10–27
Illustration of two sampling rates.

Quantization Error

The term **quantization** in this context refers to determining a value for an analog quantity. Ideally, we would like to determine a value at a given instant and convert it immediately to digital form. This is, of course, impossible because of the conversion time of ADCs. Since an analog signal may change during a conversion time, its value at the end of the conversion time may not be the same as it was at the beginning (unless the input is a constant dc). This change in the value of the analog signal during the conversion time produces what is called the **quantization error,** as illustrated in Figure 10–28.

One way to avoid or at least minimize quantization error is to use a sample-and-hold circuit at the input to the ADC. As you learned in Section 10–2, a sample-and-hold amplifier quickly samples the analog input and then holds the sampled value for a certain time.

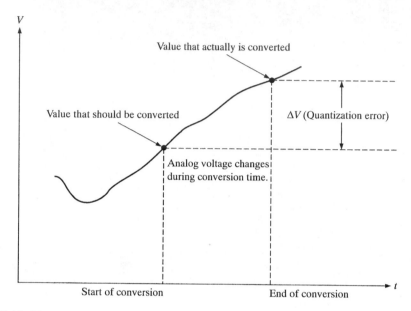

FIGURE 10–28

Illustration of quantization error in A/D conversion.

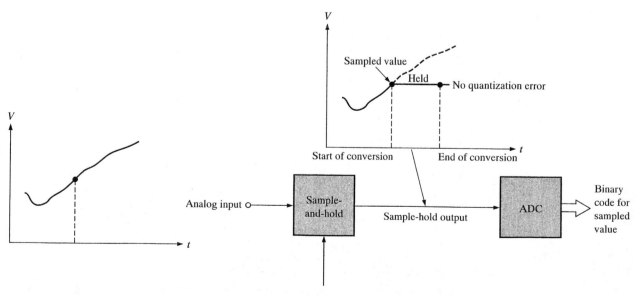

FIGURE 10–29

Using a sample-and-hold amplifier to avoid quantization error.

When used in conjunction with an ADC, the sample-and-hold is held constant for the duration of the conversion time. This allows the ADC to convert a constant value to digital form and avoids the quantization error. A basic illustration of this process is shown in Figure 10–29. When compared to the conversion in Figure 10–28, you can see that a more accurate representation of the analog input at the desired sample point is achieved.

10–5 REVIEW QUESTIONS

1. What is conversion time?
2. According to sampling theory, what is the minimum sampling rate for a 100 Hz sine wave?
3. Basically, how does a sample-and-hold circuit avoid quantization error in A/D conversion?

10–6 ■ ANALOG-TO-DIGITAL (A/D) CONVERSION METHODS

Now that you are familiar with some basic A/D conversion concepts, we will look at several methods for A/D conversion. These methods are flash (simultaneous), stair-step ramp, tracking, single-slope, dual-slope, and successive approximation. The flash and dual-slope methods were introduced in Chapter 4 as examples of op-amp applications. Some of that material is reviewed and expanded upon in this section.

After completing this section, you should be able to

❑ Discuss the operation of analog-to-digital converters (ADCs)
   ❑ Describe the flash ADC
   ❑ Describe the stairstep-ramp ADC
   ❑ Describe the tracking ADC
   ❑ Describe the single-slope ADC
   ❑ Describe the dual-slope ADC
   ❑ Describe the successive-approximation ADC

Flash (Simultaneous) Analog-to-Digital Converter

The **flash** (simultaneous) method utilizes comparators that compare reference voltages with the analog input voltage. When the analog voltage exceeds the reference voltage for a given comparator, a high-level output is generated. Figure 10–30 shows a 3-bit converter that uses seven comparator circuits; a comparator is not needed for the all-0s condition. A 4-bit converter of this type requires fifteen comparators. In general, $2^n - 1$ comparators are required for conversion to an n-bit binary code. The large number of comparators necessary for a reasonable-sized binary number is one of the disadvantages of the flash ADC. Its chief advantage is that it provides a fast conversion time.

    The reference voltage for each comparator is set by the resistive voltage-divider network. The output of each comparator is connected to an input of the priority encoder. The encoder is sampled by a pulse on the Enable input, and a 3-bit binary code representing the

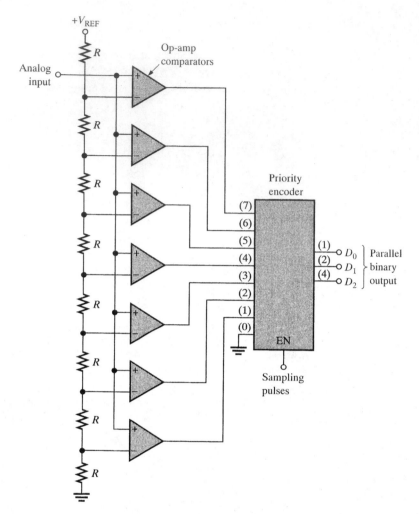

FIGURE 10–30
A 3-bit flash ADC.

value of the analog input appears on the encoder's outputs. The binary code is determined by the highest-order input having a high level.

The sampling rate determines the accuracy with which the sequence of digital codes represents the analog input of the ADC. The more samples taken in a given unit of time, the more accurately the analog signal is represented in digital form.

The following example illustrates the basic operation of the flash ADC in Figure 10–30.

EXAMPLE 10–6 Determine the binary code output of the 3-bit flash ADC for the analog input signal in Figure 10–31 and the sampling pulses (encoder Enable) shown. For this example, $V_{REF} = +8$ V.

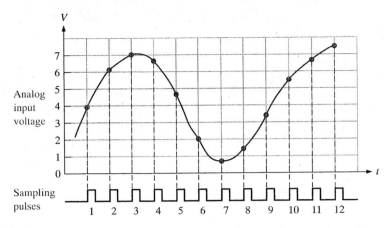

FIGURE 10–31

Sampling of values on an analog waveform for conversion to digital form.

Solution   The resulting A/D output sequence is listed as follows and shown in the waveform diagram of Figure 10–32 in relation to the sampling pulses.

$$100, 110, 111, 110, 100, 010, 000, 001, 011, 101, 110, 111$$

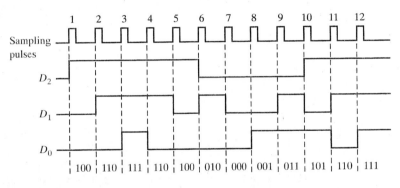

FIGURE 10–32

Resulting digital outputs for sampled values. Output D_0 is the least significant bit (LSB).

Practice Exercise   If the amplitude of the analog voltage in Figure 10–31 is reduced by half, what will the A/D output sequence be?

Stairstep-Ramp Analog-to-Digital Converter

The stairstep-ramp method of A/D conversion is also known as the *digital-ramp* or the *counter* method. It employs a DAC and a binary counter to generate the digital value of an analog input. Figure 10–33 shows a diagram of this type of converter.

Assume that the counter begins in the reset state (all 0s) and the output of the DAC is zero. Now assume that an analog voltage is applied to the input. When it exceeds the reference voltage (output of DAC), the comparator switches to a high-level output state and enables the AND gate. The clock pulses begin advancing the counter through its binary states, producing a stairstep reference voltage from the DAC. The counter continues to ad-

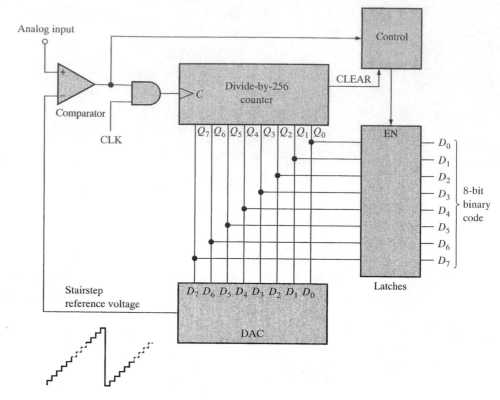

FIGURE 10–33
Stairstep-ramp ADC (8 bits).

vance from one binary state to the next, producing successively higher steps in the reference voltage. When the stairstep reference voltage reaches the analog input voltage, the comparator output will go to its low level and disable the AND gate, thus cutting off the clock pulses to stop the counter. The binary state of the counter at this point equals the number of steps in the reference voltage required to make the reference equal to or greater than the analog input. This binary number, of course, represents the value of the analog input. The control logic loads the binary count into the latches and resets the counter, thus beginning another count sequence to sample the input value.

The stairstep-ramp method is slower than the flash method because, in the worst case of maximum input, the counter must sequence through its maximum number of states before a conversion occurs. For an 8-bit conversion, this means a maximum of 256 counter states. Figure 10–34 illustrates a conversion sequence for a 4-bit conversion. Notice that for each sample, the counter must count from zero up to the point at which the stairstep reference voltage reaches the analog input voltage. The conversion time varies, depending on the analog voltage.

Tracking Analog-to-Digital Converter

The tracking method uses an up/down counter (a counter that can go either way in a binary sequence) and is faster than the stairstep-ramp method because the counter is not reset after each sample, but rather tends to track the analog input. Figure 10–35 shows a typical 8-bit tracking ADC.

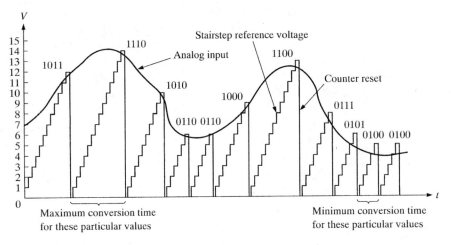

FIGURE 10–34
Example of a 4-bit conversion, showing an analog input and the stairstep reference voltage.

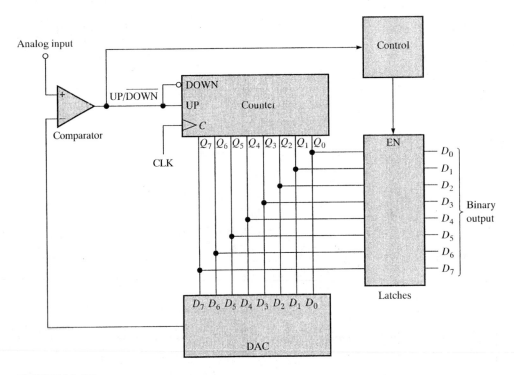

FIGURE 10–35
An 8-bit tracking ADC.

As long as the DAC reference voltage is less than the analog input, the comparator output level is high, putting the counter in the up mode, which causes it to produce an up sequence of binary counts. This causes an increasing stairstep reference voltage out of the DAC, which continues until the stairstep reaches the value of the input voltage.

When the reference voltage equals the analog input, the comparator's output switches to its low level and puts the counter in the down mode, causing it to back up one count. If the analog input is decreasing, the counter will continue to back down in its sequence and effectively track the input. If the input is increasing, the counter will back down one count after the comparison occurs and then will begin counting up again. When the input is constant, the counter backs down one count when a comparison occurs. The reference output is now less than the analog input, and the comparator output goes to its high level, causing the counter to count up. As soon as the counter increases one state, the reference voltage becomes greater than the input, switching the comparator to its low-output state. This causes the counter to back down one count. This back-and-forth action continues as long as the analog input is a constant value, thus causing an oscillation between two binary states in the ADC output. This is a disadvantage of this type of converter.

Figure 10–36 illustrates the tracking action of this type of ADC for a 4-bit conversion.

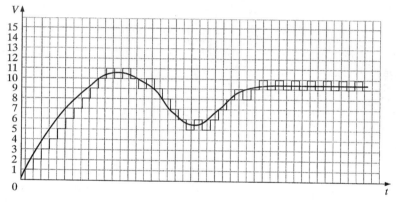

FIGURE 10–36
Tracking action of an ADC.

Single-Slope Analog-to-Digital Converter

Unlike the previous two methods, the single-slope converter does not require a DAC. It uses a linear ramp generator to produce a constant-slope reference voltage. A diagram is shown in Figure 10–37.

At the beginning of a conversion cycle, the counter is in the reset state and the ramp generator output is 0 V. The analog input is greater than the reference voltage at this point and therefore produces a high-level output from the comparator. This high level enables the clock to the counter and starts the ramp generator.

Assume that the slope of the ramp is 1 V/ms. The ramp will increase until it equals the analog input; at this point the ramp is reset, and the binary count is stored in the latches by the control logic. Let's assume that the analog input is 2 V at the point of comparison. This means that the ramp is also 2 V and has been running for 2 ms. Since the comparator output has been at its high level for 2 ms, 200 clock pulses have been allowed to pass through the gate to the counter (assuming a clock frequency of 100 kHz). At the point of comparison, the counter is in the binary state that represents decimal 200. With proper scaling and decoding, this number can be displayed as 2.00 V. This basic concept is used in some digital voltmeters.

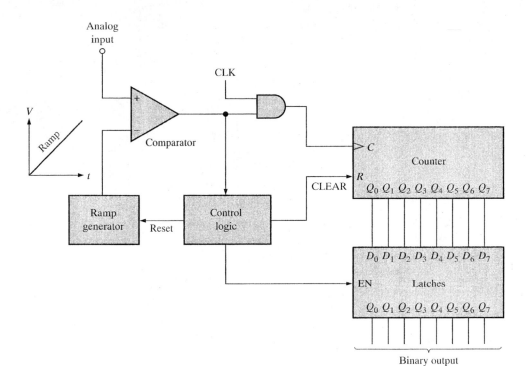

FIGURE 10–37
Single-slope ADC.

Dual-Slope Analog-to-Digital Converter

The operation of the dual-slope ADC is similar to that of the single-slope type except that a variable-slope ramp and a fixed-slope ramp are both used. This type of converter is common in digital voltmeters.

A ramp generator (integrator), A_1, is used to produce the dual-slope characteristic. A block diagram of a dual-slope ADC is shown in Figure 10–38 for reference.

Figure 10–39 illustrates dual-slope conversion. Let's assume that the counter is reset and the output of the integrator is zero. Now assume that a positive input voltage is applied to the input through the switch (SW) as selected by the control logic. Since the inverting (−) input of A_1 is at virtual ground, and assuming that V_{in} is constant for a period of time, there will be constant current through the input resistor R and therefore through the capacitor C. Capacitor C will charge linearly because the current is constant, and as a result, there will be a negative-going linear voltage ramp on the output of A_1, as illustrated in Figure 10–39(a).

When the counter reaches a specified count, it will be reset, and the control logic will switch the negative reference voltage $(-V_{REF})$ to the input of A_1, as shown in Figure 10–39(b). At this point the capacitor is charged to a negative voltage $(-V)$ proportional to the input analog voltage.

Now the capacitor discharges linearly because of the constant current from the $-V_{REF}$, as shown in Figure 10–39(c). This linear discharge produces a positive-going ramp on the A_1 output, starting at $-V$ and having a constant slope that is independent of the charge voltage. As the capacitor discharges, the counter advances from its reset state. The

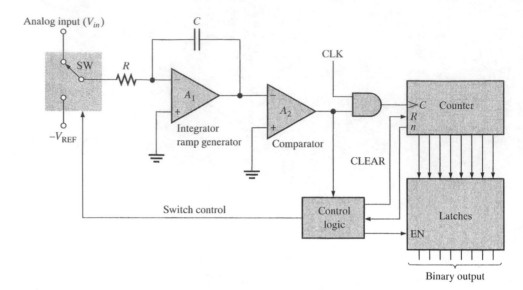

FIGURE 10–38
Dual-slope ADC.

time it takes the capacitor to discharge to zero depends on the initial voltage $-V$ (proportional to V_{in}) because the discharge rate (slope) is constant. When the integrator (A_1) output voltage reaches zero, the comparator (A_2) switches to its low state and disables the clock to the counter. The binary count is latched, thus completing one conversion cycle. The binary count is proportional to V_{in} because the time it takes the capacitor to discharge depends only on $-V$, and the counter records this interval of time.

Successive-Approximation Analog-to-Digital Converter

Perhaps the most widely used method of A/D conversion is **successive approximation.** It has a much faster conversion time than the other methods with the exception of the flash method. It also has a fixed conversion time that is the same for any value of the analog input.

Figure 10–40 shows a basic block diagram of a 4-bit successive-approximation ADC. It consists of a DAC, a successive-approximation register (SAR), and a comparator. The basic operation is as follows. The bits of the DAC are enabled one at a time, starting with the most significant bit (MSB). As each bit is enabled, the comparator produces an output that indicates whether the analog input voltage is greater or less than the output of the DAC. If the DAC output is greater than the analog input, the comparator's output is low, causing the bit in the register to reset. If the DAC output is less than the analog input, the bit is retained in the register. The system does this with the MSB first, then the next most significant bit, then the next, and so on. After all the bits of the DAC have been tried, the conversion cycle is complete.

In order to better understand the operation of the successive-approximation ADC, let's take a specific example of a 4-bit conversion. Figure 10–41 illustrates the step-by-step conversion of a given analog input voltage (5 V in this case). Let's assume that the DAC has the following output characteristic: $V_{OUT} = 8$ V for the 2^3 bit (MSB), $V_{OUT} = 4$ V for the 2^2 bit, $V_{OUT} = 2$ V for the 2^1 bit, and $V_{OUT} = 1$ V for the 2^0 bit (LSB).

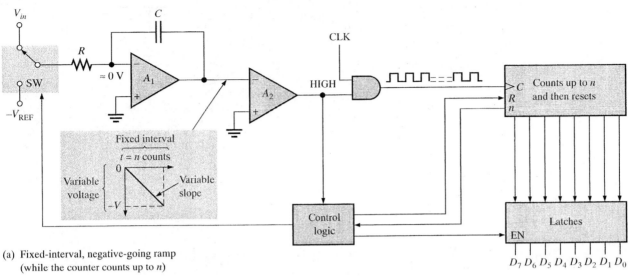

(a) Fixed-interval, negative-going ramp
(while the counter counts up to n)

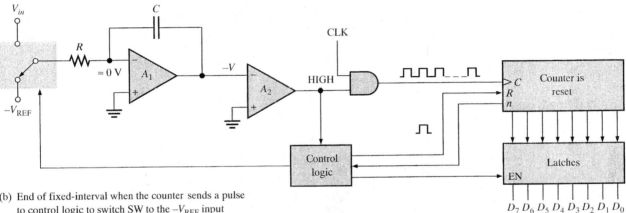

(b) End of fixed-interval when the counter sends a pulse
to control logic to switch SW to the $-V_{REF}$ input

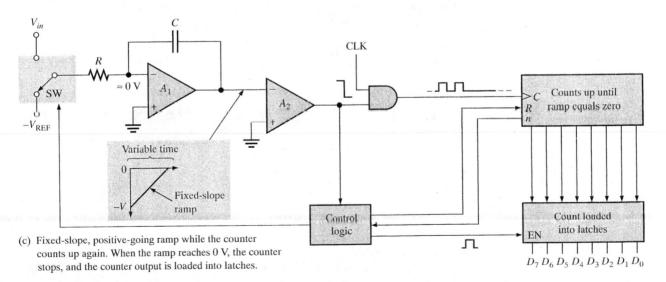

(c) Fixed-slope, positive-going ramp while the counter
counts up again. When the ramp reaches 0 V, the counter
stops, and the counter output is loaded into latches.

FIGURE 10–39
Dual-slope conversion.

FIGURE 10–40
Successive-approximation ADC.

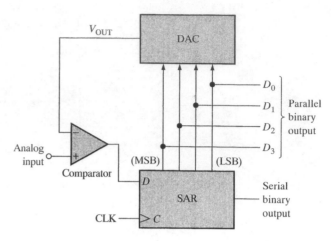

Figure 10–41(a) shows the first step in the conversion cycle with the MSB = 1. The output of the DAC is 8 V. Since this is greater than the analog input of 5 V, the output of the comparator is low, causing the MSB in the SAR to be reset to a 0.

Figure 10–41(b) shows the second step in the conversion cycle with the 2^2 bit equal to a 1. The output of the DAC is 4 V. Since this is less than the analog input of 5 V, the output of the comparator switches to its high level, causing this bit to be retained in the SAR.

Figure 10–41(c) shows the third step in the conversion cycle with the 2^1 bit equal to a 1. The output of the DAC is 6 V because there is a 1 on the 2^2 bit input and on the 2^1 bit input; 4 V + 2 V = 6 V. Since this is greater than the analog input of 5 V, the output of the comparator switches to its low level, causing this bit to be reset to a 0.

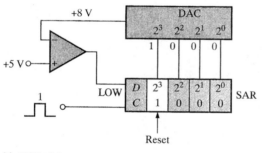

(a) MSB trial

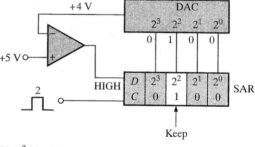

(b) 2^2-bit trial

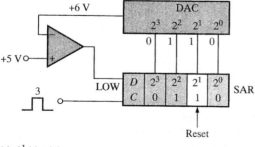

(c) 2^1-bit trial

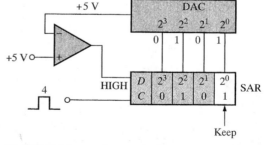

(d) LSB trial (conversion complete)

FIGURE 10–41
Successive-approximation conversion process.

Figure 10–41(d) shows the fourth and final step in the conversion cycle with the 2^0 bit equal to a 1. The output of the DAC is 5 V because there is a 1 on the 2^2 bit input and on the 2^0 bit input; 4 V + 1 V = 5 V.

The four bits have all been tried, thus completing the conversion cycle. At this point the binary code in the register is 0101, which is the binary value of the analog input of 5 V. Another conversion cycle now begins, and the basic process is repeated. The SAR is cleared at the beginning of each cycle.

A Specific Analog-to-Digital Converter

The ADC0804 is an example of a successive-approximation ADC. A block diagram is shown in Figure 10–42. This device operates from a +5 V supply and has a resolution of eight bits with a conversion time of 100 μs. Also, it has guaranteed monotonicity and an on-chip clock generator. The data outputs are tristate, so that they can be interfaced with a microprocessor bus system.

FIGURE 10–42
The ADC0804 analog-to-digital converter.

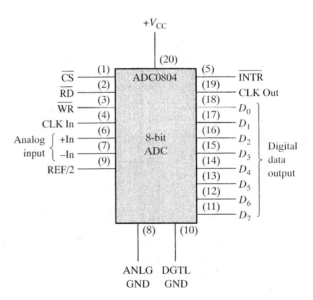

A detailed logic diagram of the ADC0804 is shown in Figure 10–43, and the basic operation of the device is as follows. The ADC0804 contains the equivalent of a 256-resistor DAC network. The successive-approximation logic sequences the network to match the analog differential input voltage $(+V_{IN} - (-V_{IN}))$ with an output from the resistive network. The MSB is tested first. After eight comparisons (sixty-four clock periods), an 8-bit binary code is transferred to an output latch, and the interrupt (INTR) output goes low. The device can be operated in a free-running mode by connecting the $\overline{INTR}$ output to the write ($\overline{WR}$) input and holding the conversion start ($\overline{CS}$) low. To ensure start-up under all conditions, a low $\overline{WR}$ input is required during the power-up cycle. Taking CS low anytime after that will interrupt the conversion process.

When the $\overline{WR}$ input goes low, the internal successive-approximation register (SAR) and the 8-bit shift register are reset. As long as both $\overline{CS}$ and $\overline{WR}$ remain low, the analog-to-digital converter remains in a reset state. Conversion starts one to eight clock periods after $\overline{CS}$ or $\overline{WR}$ makes a low-to-high transition.

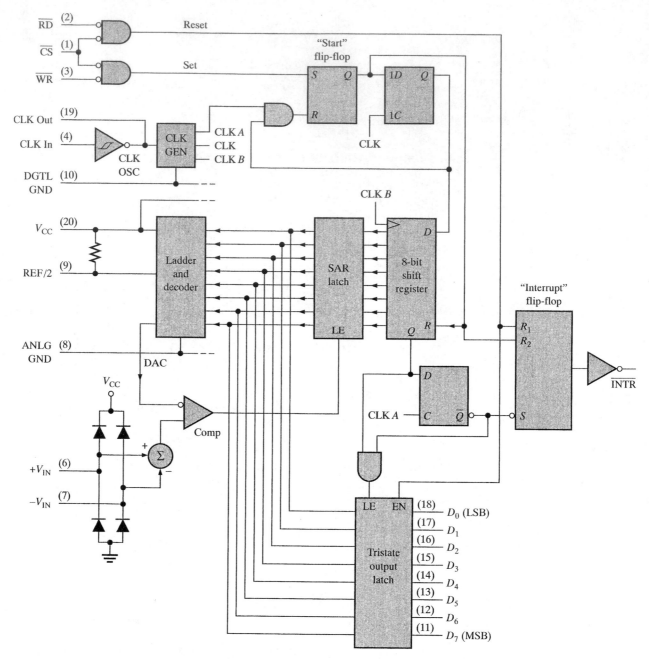

FIGURE 10–43
Logic diagram of the ADC0804 ADC.

When the $\overline{CS}$ and $\overline{WR}$ inputs are low, the start flip-flop is set, and the interrupt flip-flop and 8-bit register are reset. The high is ANDed with the next clock pulse, which puts a high on the reset input of the start flip-flop. If either $\overline{CS}$ or $\overline{WR}$ has gone high, the set signal to the start flip-flop is removed, causing it to be reset. A high is placed on the D input of the 8-bit shift register, and the conversion process is started. If the $\overline{CS}$ and $\overline{WR}$ inputs are still low, the start flip-flop, the 8-bit shift register, and the SAR remain reset. This action allows for wide $\overline{CS}$ and $\overline{WR}$ inputs, with conversion starting from one to eight clock periods after one of the inputs has gone high.

When the high input has been clocked through the 8-bit shift register, completing the SAR search, it is applied to an AND gate controlling the output latches and to the *D* input of a flip-flop. On the next clock pulse, the digital word is transferred to the tristate output latches, and the interrupt flip-flop is set. The output of the interrupt flip-flop is inverted to provide an $\overline{INTR}$ output that is high during conversion and low when conversion is complete.

When a low is at both the $\overline{CS}$ and $\overline{RD}$ inputs, the tristate output latch is enabled, the output code is applied to the D_0 through D_7 lines, and the interrupt flip-flop is reset. When either the $\overline{CS}$ or the $\overline{RD}$ input returns to a high, the D_0 through D_7 outputs are disabled. The interrupt flip-flop remains reset.

Several additional IC analog-to-digital converters are listed in Table 10–1.

TABLE 10–1
Several popular ADCs.

Device	Description	Resolution	Conversion Time	Supply Voltages
MC14433P	Dual Slope	3½ digits	40 ms	+5 V, +8 V
MC14443P	Single Slope	8 bits	300 ms	+5 V, +8 V
MC14447P	Single Slope	8 bits	300 ms	−5 V, +8 V
MC14559BCP	Successive Approximation	8 bits	—	±3 V to ±18 V
ADC0803	Successive Approximation	8 bits	100 μs	+5 V
ADC0809	Successive Approximation	8 bits	100 μs	+5 V
ADC0820	Flash Conversion	8 bits	1.18 μs	+5 V
TLC5540	Flash Conversion	8 bits	25 ns	+5 V
TLC320AD57	Sigma Delta	16/18 bits	20 μs	+5 V

10–6 REVIEW QUESTIONS

1. What is the fastest method of analog-to-digital conversion?
2. Which A/D conversion method uses an up/down counter?
3. The successive-approximation converter has a fixed conversion time. (True or false)

10–7 ■ VOLTAGE-TO-FREQUENCY (*V/F*) AND FREQUENCY-TO-VOLTAGE (*F/V*) CONVERTERS

Voltage-to-frequency converters convert an analog input voltage to a pulse stream or square wave in such a way that there is a linear relationship between the analog voltage and the frequency of the pulse stream. Frequency-to-voltage converters perform the inverse operation by converting a pulse stream to a voltage that is proportional to the pulse stream frequency. Actually, V/F and F/V converters can be used as ADCs and DACs in certain applications. In other applications, V/F and F/V converters are used, for example, in high-noise immunity digital transmission and in digital voltmeters.

After completing this section, you should be able to

❑ Discuss the basic operation of *V/F* and *F/V* converters
    ❑ Describe the AD650 *V/F* converter
    ❑ Discuss *V/F* and *F/V* applications

A Basic Voltage-to-Frequency Converter

The concept of voltage-to-frequency converters is shown in Figure 10–44. An analog voltage on the input is converted to a pulse signal with a frequency that is directly proportional to the amplitude of the input voltage. There are several ways to implement a *V/F* converter. For example, the VCO (voltage-controlled oscillator) with which you are already familiar can be used as one type of *V/F* converter. In this section, we will look at a relatively common implementation called the *charge-balance V/F converter.*

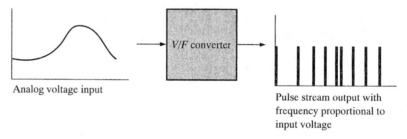

Analog voltage input

Pulse stream output with frequency proportional to input voltage

FIGURE 10–44
The basic V/F concept.

Figure 10–45 shows the diagram of a basic charge-balance *V/F* converter. It consists of an integrator, a comparator, a one-shot, a current source, and an electronic switch. The input resistor R_{in}, the integration capacitor C_{int}, and the one-shot timing capacitor C_{os} are components whose values are selected based on desired performance.

The basic operation of the *V/F* converter in Figure 10–46 is as follows. A positive input voltage produces an input current ($I_{in} = V_{in}/R_{in}$) which charges the capacitor C_{int}, as indicated in Figure 10–46(a). During this integrate mode, the integrator output voltage is a downward ramp, as shown. When the integrator output voltage reaches zero, the comparator triggers the one-shot. The one-shot produces a pulse with a fixed width, t_{os}, that switches the 1 mA current source to the input of the integrator and initiates the reset mode.

During the reset mode, current through the capacitor is in the opposite direction from the integrate mode, as indicated in Figure 10–46(b). This produces an upward ramp on the integrator output as indicated. After the one-shot times out, the current source is switched back to the integrator output, initiating another integrate mode and the cycle repeats.

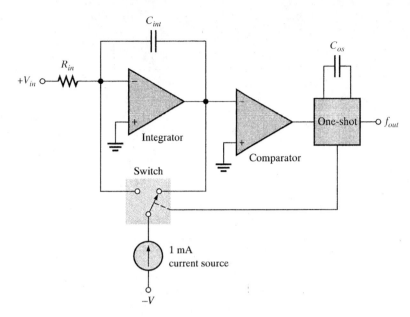

FIGURE 10–45
A basic voltage-to-frequency converter.

If the input voltage is held constant, the output waveform of the integrator is as shown in Figure 10–47(a), where the amplitude and the integrate time remain constant. The final output of the *V/F* converter is taken off the one-shot, as indicated in Figure 10–46. As long as the input voltage is constant, the output pulse stream has a constant frequency as indicated in Figure 10–47(b).

When the Input Voltage Increases An increase in the input voltage, V_{in}, causes the input current, I_{in}, to increase. In the basic relationship $I_C = (V_C/t)C$, the term V_C/t is the slope of the capacitor voltage. If the current increases, V_C/t also increases since C is constant. As applied to the *V/F* converter, this means that if the input current (I_{in}) increases, then the slope of the integrator output during the integrate mode will also increase and reduce the period of the final output voltage. Also, during the reset mode, the opposite current through the capacitor, 1 mA $- I_{in}$, is smaller, thus decreasing the slope of the upward ramp and reducing the amplitude of the integrator output voltage. This is illustrated in the waveform diagram of Figure 10–48 where the input voltage, and thus the input current, takes a step increase from one value to another. Notice that during reset, the positive-going slope of the integrator voltage is less, so it reaches a smaller amplitude during the time t_{os}. Remember, t_{os} does not change. Notice also that during integration, the negative-going slope of the integrator voltage is greater, so it reaches zero quicker. The net result of this increase in input voltage is that the output frequency increases an amount proportional to the increase in the input voltage. So, as the input voltage varies, the output frequency varies proportionally.

The AD650 Integrated Circuit *V/F* Converter

The AD650 is a good example of a *V/F* converter very similar to the basic device we just discussed. The main differences in the AD650 are the output transistor and the comparator

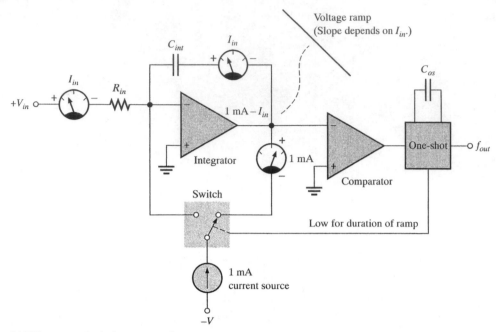

(a) *V/F converter in the integrate mode*

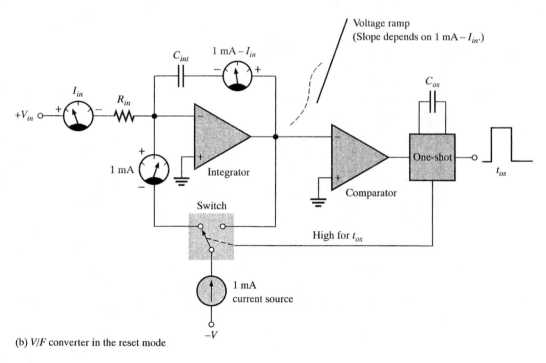

(b) *V/F converter in the reset mode*

FIGURE 10–46

Basic operation of a V/F converter for a constant input voltage.

FIGURE 10–47
V/F converter waveforms for a constant input voltage.

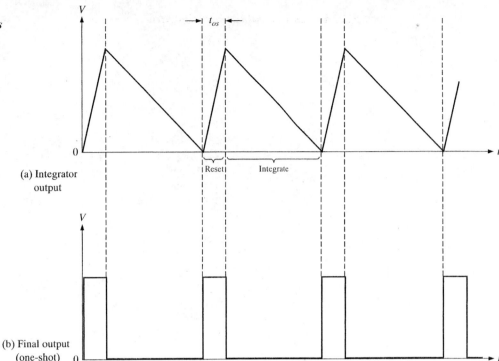

(a) Integrator output

(b) Final output (one-shot)

FIGURE 10–48
The output frequency increases when the input voltage increases.

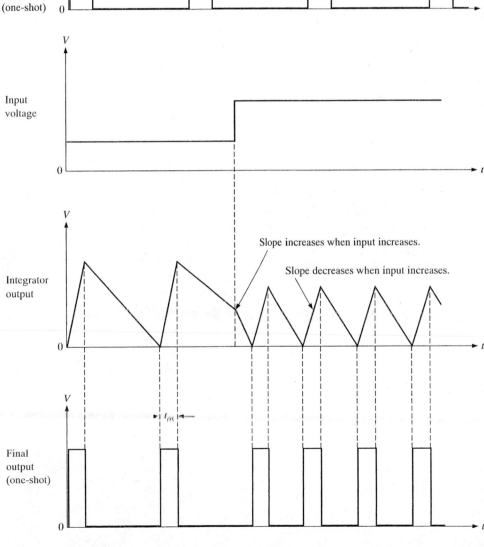

Input voltage

Integrator output

Slope increases when input increases.

Slope decreases when input increases.

Final output (one-shot)

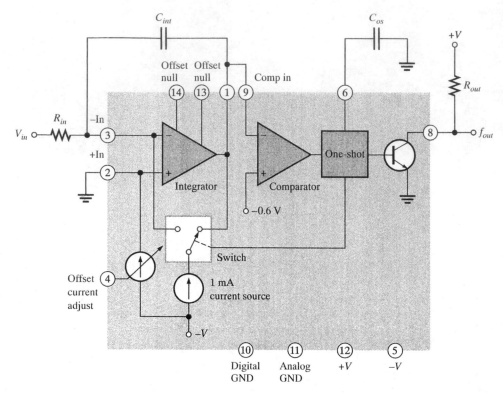

FIGURE 10–49
The AD650 V/F converter.

threshold voltage of −0.6 V instead of ground, as shown in Figure 10–49. The input resistor, integrating capacitor, one-shot capacitor, and the output pull-up resistor are external components, as indicated.

The values of the external components determine the operating characteristics of the device. The pulse width of the one-shot output is set by the following formula:

$$t_{os} = C_{os}(6.8 \times 10^3 \text{ s/F}) + 3 \times 10^{-7} \text{ s} \tag{10–1}$$

During the reset interval, the integrator output voltage increases by an amount expressed as

$$\Delta V = \frac{(1 \text{ mA} - I_{in})t_{os}}{C_{int}} \tag{10–2}$$

The duration of the integrate interval when the integrator output is sloping downward is

$$t_{int} = \frac{\Delta V}{I_{in}/C_{int}} = \frac{t_{os}(1 \text{ mA} - I_{in})/C_{int}}{I_{in}/C_{int}}$$

$$t_{int} = \left(\frac{1 \text{ mA}}{I_{in}} - 1\right)t_{os} \tag{10–3}$$

The period of a full cycle consists of the reset interval plus the integrate interval.

$$T = t_{os} + t_{int} = t_{os} + \left(\frac{1\ mA}{I_{in}} - 1\right)t_{os} = \left(1 + \frac{1\ mA}{I_{in}} - 1\right)t_{os} = \left(\frac{1\ mA}{I_{in}}\right)t_{os}$$

Therefore, the output frequency can be expressed as

$$f_{out} = \frac{I_{in}}{t_{os}(1\ mA)} \tag{10-4}$$

As you can see in Equation (10–4), the output frequency is directly proportional to the input current; and since $I_{in} = V_{in}/R_{in}$, it is also directly proportional to the input voltage and inversely proportional to the input resistance. The output frequency is also inversely proportional to t_{os}, which depends on the value of C_{os}.

EXAMPLE 10–7   Determine the output frequency for the AD650 *V/F* converter in Figure 10–50 when a constant input voltage of 5 V is applied.

FIGURE 10–50

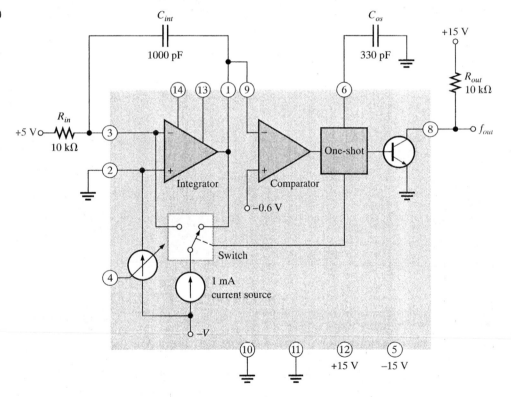

Solution

$$t_{os} = C_{os}(6.8 \times 10^3\ s/F) + 3 \times 10^{-7}\ s$$
$$= 330\ pF(6.8 \times 10^3\ s/F) + 3 \times 10^{-7}\ s = 2.5\ \mu s$$

$$I_{in} = \frac{V_{in}}{R_{in}} = \frac{5\ V}{10\ k\Omega} = 500\ \mu A$$

$$f_{out} = \frac{I_{in}}{t_{os}(1\ mA)} = \frac{500\ \mu A}{(2.5\ \mu s)(1\ mA)} = \textbf{200 kHz}$$

Practice Exercise What are the minimum and maximum output frequencies for the *V/F* converter in Figure 10–50 when a triangular wave with a minimum peak value of 1 V and a maximum peak value of 6 V is applied to the input?

A related device to the voltage-to-frequency converter is the 3-pin TSL235 light-to-frequency converter from Texas Instruments. The TSL235 combines a light sensor with a current-to-frequency converter on the same IC. It has a large dynamic range (1,000,000:1). Applications include transmission and reflective measurements such as proximity detectors, paper detection, and diagnostic equipment. The TSL245 is a similar IC but works with infrared; it can be used in conjunction with infrared sources.

A Basic *F/V* Converter

Figure 10–51 shows a basic frequency-to-voltage converter. The elements are the same as those in the voltage-to-frequency converter of Figure 10–45, but they are connected differently.

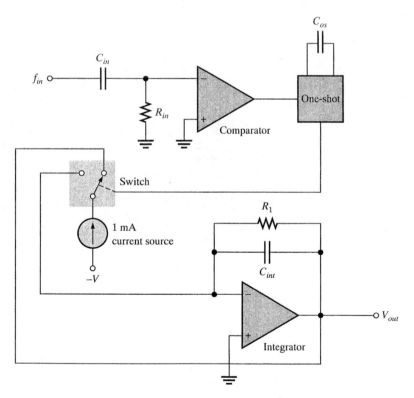

FIGURE 10–51
A basic frequency-to-voltage (F/V) converter.

When an input frequency is applied to the comparator input, it triggers the one-shot which produces a fixed pulse width (t_{os}) determined by C_{os}. This switches the 1 mA current source to the integrator input and C_{int} charges. Between one-shot pulses, C_{int} discharges through R_1. The higher the input frequency, the closer the one-shot pulses are together and the less C_{int} discharges. This causes the integrator output to increase as input frequency increases and to decrease as the input frequency decreases. The integrator output is the final voltage output of the *F/V* converter. *F/V* conversion action is illustrated by the waveforms in Figure 10–52. C_{int} and R_1 act as a filter and tend to smooth out the ripples on the integrator output as indicated by the dashed curve.

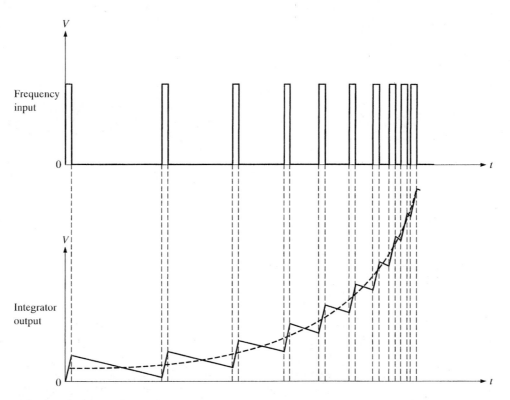

FIGURE 10–52
An example of frequency-to-voltage conversion.

Figure 10–53 shows the AD650 connected to function as a frequency-to-voltage converter. Compare this configuration with the voltage-to-frequency connection in Figure 10–49.

An Application

One application of *V/F* and *F/V* converters is in the remote sensing of a quantity (temperature, pressure, level) that is converted to an analog voltage by a transducer. The analog voltage is then converted to a pulse frequency by a *V/F* converter which is then transmitted by some method (radio link, fiber-optical link, telemetry) to a base unit receiver that includes an *F/V* converter. This basic application of *V/F* and *F/V* conversion is illustrated in Figure 10–54.

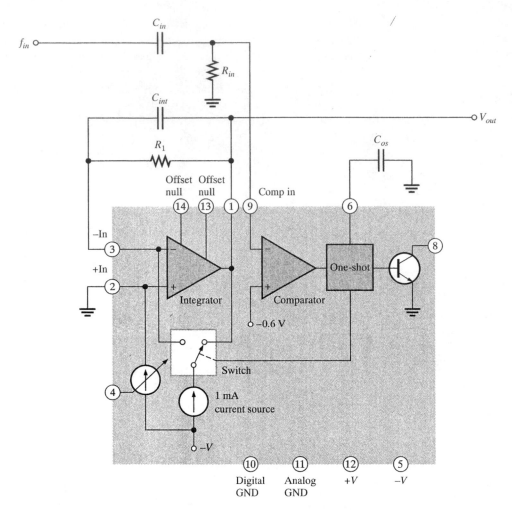

FIGURE 10–53
The AD650 connected as a frequency-to-voltage (F/V) converter.

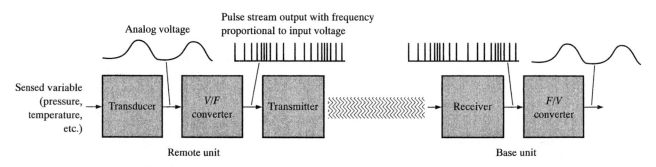

FIGURE 10–54
Basic application of V/F and F/V conversion.

10–7 REVIEW QUESTIONS

1. List the basic components in a typical V/F converter.

2. In a V/F converter, if the input voltage changes from 1 V to 6.5 V, what happens to the output?

3. Describe the basic differences between a V/F and a F/V converter in terms of inputs and outputs.

10–8 ■ TROUBLESHOOTING

Basic testing of DACs and ADCs includes checking their performance characteristics, such as monotonicity, offset, linearity, and gain, and checking for missing or incorrect codes. In this section, the fundamentals of testing these analog interfaces are introduced.

After completing this section, you should be able to

❑ Troubleshoot DACs and ADCs
   ❑ Identify D/A conversion errors
   ❑ Identify A/D conversion errors

Testing Digital-to-Analog Converters

The concept of DAC testing is illustrated in Figure 10–55. In this basic method, a sequence of binary codes is applied to the inputs, and the resulting output is observed. The binary code sequence extends over the full range of values from 0 to $2^n - 1$ in ascending order, where n is the number of bits.

FIGURE 10–55
Basic test setup for a DAC.

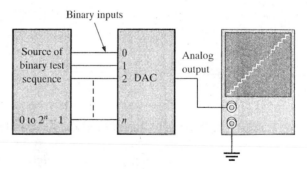

The ideal output is a straight-line stairstep as indicated. As the number of bits in the binary code is increased, the resolution is improved. That is, the number of discrete steps increases, and the output approaches a straight-line linear ramp.

D/A Conversion Errors

Several D/A conversion errors to be checked for are shown in Figure 10–56, which uses a 4-bit conversion for illustration purposes. A 4-bit conversion produces fifteen discrete steps. Each graph in the figure includes an ideal stairstep ramp for comparison with the faulty outputs.

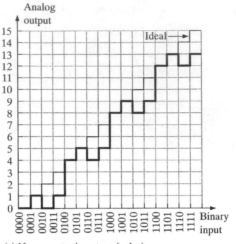

(a) Nonmonotonic output (color)

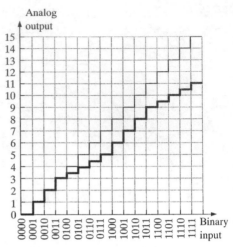

(b) Differential nonlinearity (color)

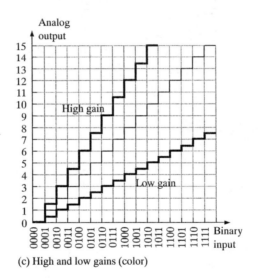

(c) High and low gains (color)

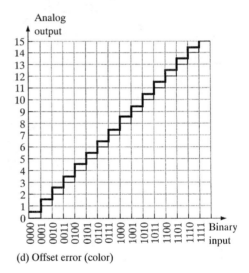

(d) Offset error (color)

FIGURE 10–56

Illustrations of several D/A conversion errors.

Nonmonotonicity The step reversals in Figure 10–56(a) indicate **nonmonotonic** performance, which is a form of nonlinearity. In this particular case, the error occurs because the 2^1 bit in the binary code is interpreted as a constant 0. That is, a short is causing the bit input line to be stuck in the low state.

Differential Nonlinearity Figure 10–56(b) illustrates differential nonlinearity in which the step amplitude is less than it should be for certain input codes. This particular output could be caused by the 2^2 bit having an insufficient weight, perhaps because of a faulty input resistor. We could also see steps with amplitudes greater than normal if a particular binary weight were greater than it should be.

Low or High Gain Output errors caused by low or high gain are illustrated in Figure 10–56(c). In the case of low gain, all of the step amplitudes are less than ideal. In the case of high gain, all of the step amplitudes are greater than ideal. This situation may be caused by a faulty feedback resistor in the op-amp circuit.

Offset Error An offset error is illustrated in Figure 10–56(d). Notice that when the binary input is 0000, the output voltage is nonzero and that this amount of offset is the same for all steps in the conversion. A faulty op-amp may be the culprit in this situation.

EXAMPLE 10–8

The DAC output in Figure 10–57 is observed when a straight 4-bit binary sequence is applied to the inputs. Identify the type of error, and suggest an approach to isolate the fault.

FIGURE 10–57

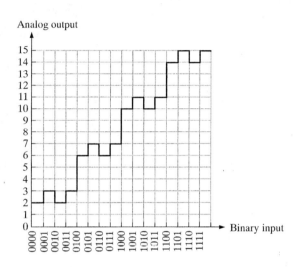

Solution The DAC in this case is nonmonotonic. Analysis of the output reveals that the device is converting the following sequence, rather than the actual binary sequence applied to the inputs.

0010, 0011, 0010, 0011, 0110, 0111, 0110, 0111, 1010, 1011, 1010, 1011, 1110, 1111, 1110, 1111

Apparently, the 2^1 bit (second from right) is stuck in the high (1) state. To find the problem, first monitor the bit input pin of the device. If it is changing states, the fault is internal, most likely an open. If the external pin is not changing states and is always high, check for an external short to $+V$ that may be caused by a solder bridge somewhere on the circuit board. If no problem is found here, disconnect the source output from the DAC input pin, and see if the output signal is correct. If these checks produce no results, the fault is most likely internal to the DAC, perhaps a short to the supply voltage.

Practice Exercise Graph the output of a DAC if a straight 4-bit binary sequence is applied to the inputs and the most significant bit input of the DAC is stuck high.

Testing Analog-to-Digital Converters

One method for testing ADCs is shown in Figure 10–58. A DAC is used as part of the test setup to convert the ADC output back to analog form for comparison with the test input.

A test input in the form of a linear ramp is applied to the input of the ADC. The resulting binary output sequence is then applied to the DAC test unit and converted to a stairstep ramp. The input and output ramps are compared for any deviation.

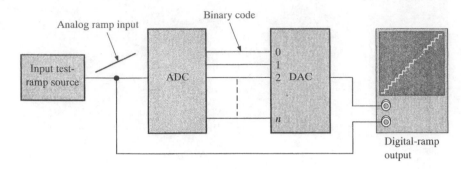

FIGURE 10–58
A method for testing ADCs.

A/D Conversion Errors

Again, a 4-bit conversion is used to illustrate the principles. Let's assume that the test input is an ideal linear ramp.

Missing Code The stairstep output in Figure 10–59(a) indicates that the binary code 1001 does not appear on the output of the ADC. Notice that the 1000 value stays for two intervals and then the output jumps to the 1010 value.

In a flash ADC, for example, a failure of one of the comparators can cause a missing-code error.

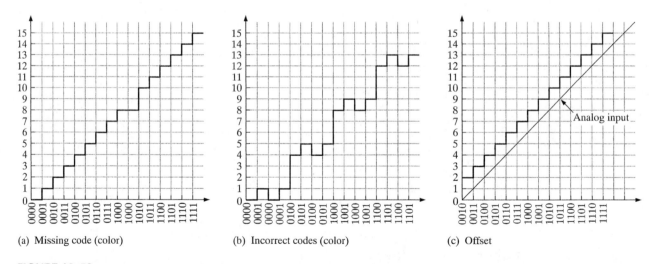

(a) Missing code (color)   (b) Incorrect codes (color)   (c) Offset

FIGURE 10–59
Illustrations of A/D conversion errors.

Incorrect Codes The stairstep output in Figure 10–59(b) indicates that several of the binary code words coming out of the ADC are incorrect. Analysis indicates that the 2^1-bit line is stuck in the low state in this particular case.

Offset Offset conditions are shown in 10–59(c). In this situation, the ADC interprets the analog input voltage as greater than its actual value. This error is probably due to a faulty comparator circuit.

EXAMPLE 10–9   A 4-bit flash ADC is shown in Figure 10–60(a). It is tested with a setup like the one in Figure 10–58. The resulting reconstructed analog output is shown in Figure 10–60(b). Identify the problem and the most probable fault.

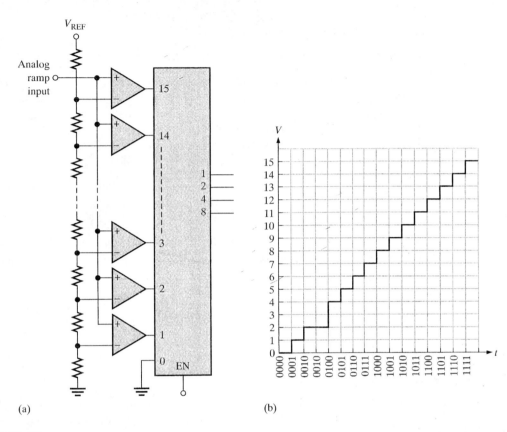

(a)   (b)

FIGURE 10–60

Solution   The binary code 0011 is missing from the ADC output, as indicated by the missing step. Most likely, the output of comparator 3 is stuck in its inactive state (low).

Practice Exercise   If the output of comparator 15 is stuck in the high state, what will be the reconstructed analog output when the ADC is tested in a setup like the one in Figure 10–58?

10–8 REVIEW QUESTIONS

1. How do you detect nonmonotonic behavior in a DAC?
2. What effect does low gain have on a DAC output?
3. Name two types of output errors in an ADC.

10–9 ■ A SYSTEM APPLICATION

The solar panel control system introduced at the opening of this chapter includes analog multiplexers and ADCs. The analog multiplexers accept position information from the potentiometers located on each solar unit and send it to the ADCs where the analog position information is converted to digital form. The digital outputs of the ADCs go to the digital controller for processing and control signals are sent back to the solar units to keep them properly positioned.

After completing this section, you should be able to

❑ Apply what you have learned in this chapter to a system application
   ❑ Describe how an analog multiplexer can be used to collect analog data
   ❑ Explain how an ADC is used in a system application
   ❑ Translate between a printed circuit board and a schematic
   ❑ Analyze the analog board
   ❑ Troubleshoot some common problems

A Brief Description of the System

The solar panel control system maintains each solar panel at approximately a 90° angle with respect to the sun's rays. Two angular positions are required to properly align the panels. The azimuth position is along a curve from east to west parallel with the horizon. The elevation position is from the horizon to directly overhead. These angular movements for a solar unit are illustrated in Figure 10–61. There are two stepping motors that drive the solar panel to the proper position, one for azimuth and one for elevation. Also, there are two position transducers (potentiometers are used as the sensors in this case) that produce voltages proportional to the positions of the panel as determined by the angular positions of the motor shafts.

In this application, the movement of the solar panel is very slow. The azimuth angle begins at an easterly orientation at sunrise and turns through approximately 180° by sunset. The elevation angle tracks the arc of the sun each day and also adjusts for seasonal varia-

FIGURE 10–61

A solar panel with position controls and sensors for azimuth and elevation.

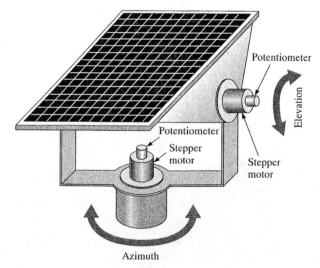

Potentiometer

Elevation

Potentiometer

Stepper motor

Stepper motor

Azimuth

tion of the sun's relative position in the sky. On the first day of winter it tracks through the lowest arc, and on the first day of summer it tracks through the highest arc.

A basic block diagram of the control electronics is shown in Figure 10–62. There are two 4-input analog multiplexers, one for the azimuth and one for the elevation. Each input has a dc voltage coming from the associated position potentiometer. Each voltage is proportional to the current angular position of the solar panel as measured by the potentiometer. Every five minutes the digital controller causes the multiplexer to quickly sequence through the four azimuth and the four elevation position voltages so that one voltage at a time is applied to the ADC. Each resulting digital code is processed by comparing the angle it represents to angular information about solar position based on time of day and date that is permanently stored in the digital controller's memory. Based on its computations, the digital controller issues the proper control signal to the appropriate stepping motor to update its position.

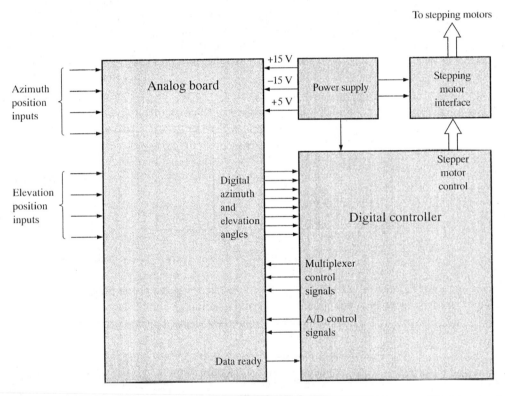

FIGURE 10–62
Basic block diagram of the solar panel control system.

The Potentiometer In this application, the potentiometer is used as a position transducer to convert the angular shaft position of the stepping motor to a proportional dc voltage. The potentiometer is mechanically linked to the associated motor so as the motor shaft turns, the wiper of the potentiometer slides along the resistive element. It is calibrated so that the smallest angle produces the smallest voltage. In Figure 10–63, a simplified diagram shows the basic construction and the schematic indicates minimum and maximum angles corresponding to the wiper position.

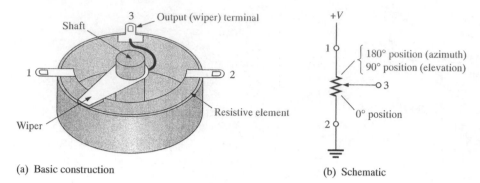

(a) Basic construction   (b) Schematic

FIGURE 10–63
The potentiometer as an angular transducer.

Stepping Motors   Although our focus is on the analog electronics in this system, a basic familiarization with stepping motors will help you have a better understanding of the overall system operation. A stepping motor is one in which the rotor shaft can be rotated in a series of incremental moves called steps or step angles. Stepping motors are available with step angles ranging from 0.9° to 30°. The motors used in this system are assumed to have a 2° step angle (180 steps/revolution). Rotation is achieved by applying a series of pulses to the windings. Generally, one pulse will rotate the shaft by one step angle (2° in our case). The sequence in which the pulses are applied to the windings determine the direction of rotation (CW or CCW). The speed of the motor is controlled by the rate at which the stepping pulses are applied. Obviously, in this system the motors move very slowly so speed is not a consideration.

Basic Operation of the Analog Circuits

A block diagram of the analog board is shown in Figure 10–64. DC voltages from the four azimuth potentiometers (remember there are four solar panels) are applied to the inputs of the upper AD9300, and dc voltages from the four elevation potentiometers are applied to the inputs of the lower AD9300.

The digital controller issues a high-level enable signal to the azimuth multiplexer. While the Enable input is high, the digital controller sequentially switches each of the four inputs to the output by applying a 2-bit binary code to the channel select inputs. As each input voltage appears on the multiplexer output, the digital controller issues a convert signal to the AD673 ADC to initiate a conversion. After the conversion is complete, the ADC then sends a data ready signal to the digital controller. The controller responds with a data enable signal, which places the 8-bit binary code on the outputs allowing the binary number to be processed by the controller.

After the first conversion, the digital controller advances to the next azimuth multiplexer input and repeats the operation. After all four azimuth inputs have been converted to digital and processed, the controller disables the azimuth multiplexer and enables the elevation multiplexer for conversion of its four inputs to digital. The 200 Ω variable resistors are used to adjust the input voltage to the ADC so that the highest output code (all 1s) corresponds to the maximum input voltage. Therefore, the full range of azimuth and elevation angles are converted into 256 discrete values.

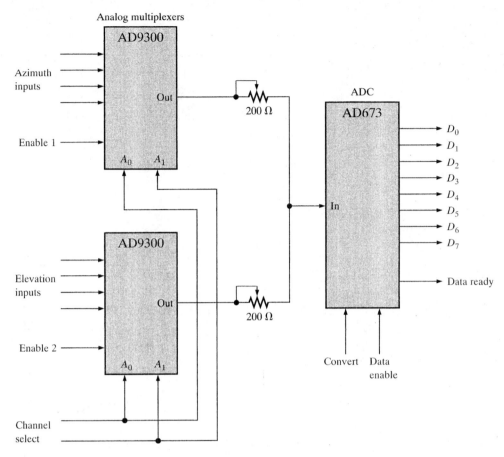

FIGURE 10–64
Simplified block diagram of the analog board.

The digital controller repeats the conversion sequence every eight minutes in this system. The basic timing diagram in Figure 10–65 graphically illustrates the operation.

Now, so that you can take a closer look at the analog board, let's take it out of the system and put it on the test bench.

ON THE TEST BENCH

▪ ACTIVITY 1 Relate the PC Board to the Schematic

Using the pin configuration diagrams on the data sheets for the AD9300 and the AD673 in Appendix A, complete the diagram in Figure 10–64 by adding pin numbers, ground connections, and supply voltage connections. Use the completed schematic to locate and identify the components and the input and output functions on the analog board in Figure 10–66.

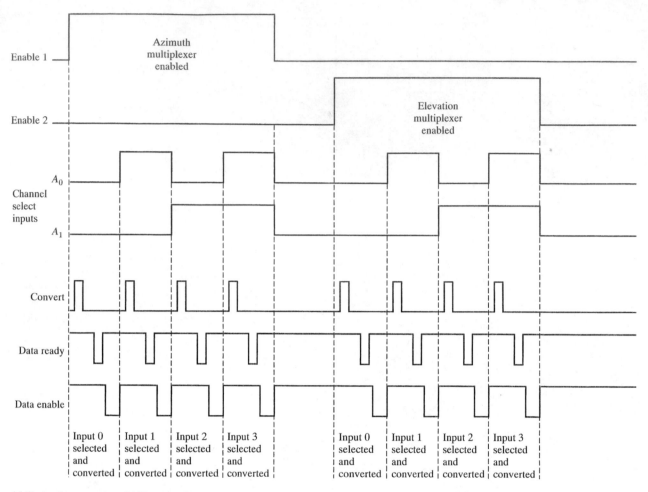

(a) Timing for one series of A/D conversions

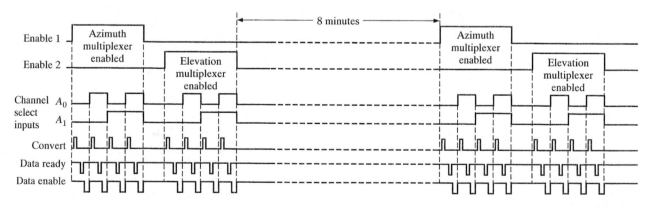

(b) A series of A/D conversions occurs every eight minutes.

FIGURE 10–65

Timing diagram for the A/D conversion sequence.

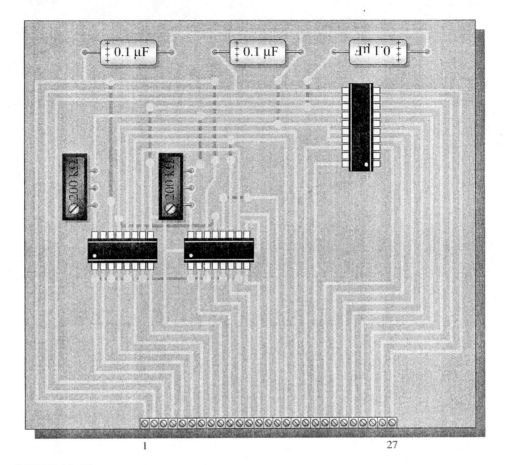

FIGURE 10–66

■ ACTIVITY 2   Analyze the System

For this application, assume that, at its limits, the azimuth potentiometer produces 0 V for a 0° orientation (due east) and 10 V for a 180° orientation (due west). At its limits, the elevation potentiometer produces 0 V for a 0° orientation (horizontal) and 10 V for a 90° orientation (vertical). The potentiometers rotate in 2° increments.

Step 1: Calculate the voltage increment that will occur on the azimuth multiplexer input when the azimuth motor makes one step.

Step 2: Calculate the voltage increment that will occur on the elevation multiplexer input when the elevation motor makes one step.

Step 3: Determine the resolution in degrees to which the ADC can represent the input voltage range for both azimuth and elevation. Is this resolution adequate? Explain.

■ ACTIVITY 3   Write a Technical Report

Describe the overall operation of the solar panel control system. List the functional requirements for each block in the system. Discuss the operation of the analog board and its role in the overall system. Use the results of Activity 2 if appropriate.

■ **ACTIVITY 4   Troubleshoot the System for Each of the Following
Problems By Stating the Probable Cause or Causes**

1. No voltage on one of the azimuth input lines to the analog multiplexer when there are
   2.22 V on each of the other three inputs.

2. All four solar units are stuck at a given elevation position, but their azimuth position advances properly.

3. One solar unit is stuck at a given azimuth position, but its elevation position advances
   properly.

10–9 REVIEW QUESTIONS

1. What is the purpose of the three capacitors on the analog board?
2. How many discrete values of the analog position voltage can be converted to digital?
3. Explain why this system repeats the conversion sequence every eight minutes.

■ **SUMMARY**

- There are three basic types of analog switches: single pole–single throw (SPST), single pole–double throw (SPDT), and double pole–double throw (DPDT).

- An analog switch is typically a MOSFET that is opened or closed with a control input.

- A sample-and-hold amplifier samples a voltage at a certain point in time and retains or holds that voltage for an interval of time.

- An analog quantity is one that has a continuous set of values over time.

- A digital quantity is one that has a set of discrete values over time.

- Two basic types of digital-to-analog converters (DACs) are the binary-weighted-input converter and the $R/2R$ ladder converter.

- The $R/2R$ ladder DAC is easier to implement because only two resistor values are required compared to a different value for each input in the binary-weighted-input DAC.

- The number of bits in an analog-to-digital converter (ADC) determines its resolution.

- The minimum sampling rate for A/D conversion is twice the maximum frequency component of the analog signal.

- The flash or simultaneous method of A/D conversion is the fastest.

- The successive-approximation method of A/D conversion is the most widely used.

- Other common methods of A/D conversion are single-slope, dual-slope, tracking, and stairstep ramp (counter method).

- In a voltage-to-frequency converter (V/F), the output frequency is directly proportional to the amplitude of the analog input voltage.

- In a frequency-to-voltage converter (F/V), the amplitude of the output voltage is directly proportional to the input frequency.

- Types of D/A conversion errors include nonmonotonicity, differential nonlinearity, low or high gain, and offset error.

- Types of A/D conversion errors include missing code, incorrect code, and offset.

■ GLOSSARY

These terms are included in the end-of-book glossary.

Accuracy   In relation to DACs or ADCs, a comparison of the actual output with the expected output, expressed as a percentage.

Acquisition time   In an analog switch, the time required for the device to reach its final value when switched from hold to sample.

Analog-to-digital converter (ADC)   A device used to convert an analog signal to a sequence of digital codes.

Aperture jitter   In an analog switch, the uncertainty in the aperture time.

Aperture time   In an analog switch, the time to fully open after being switched from sample to hold.

Digital-to-analog converter (DAC)   A device in which information in digital form is converted to an analog form.

Droop   In an analog switch, the change in the sampled value during the hold interval.

Feedthrough   In an analog switch, the component of the output voltage which follows the input voltage after the switch opens.

Flash   A method of A/D conversion.

Linearity   A straight-line relationship. A linear error is a deviation from the ideal straight-line output of a DAC.

Monotonicity   In relation to DACs, the presence of all steps in the output when sequenced over the entire range of input bits.

Nonmonotonicity   In relation to DACs, a step reversal or missing step in the output when sequenced over the entire range of input bits.

Nyquist rate   In sampling theory, the minimum rate at which an analog voltage can be sampled for A/D conversion. The sample rate must be equal to more than twice the maximum frequency component of the input signal.

Quantization   The determination of a value for an analog quantity.

Quantization error   The error resulting from the change in the analog voltage during the A/D conversion time.

Resolution   In relation to DACs or ADCs, the number of bits involved in the conversion. Also, for DACs, the reciprocal of the maximum number of discrete steps in the output.

Sample   The process of taking the instantaneous value of a quantity at a specific point in time.

Settling time   The time it takes a DAC to settle within $\pm\frac{1}{2}$ LSB of its final value when a change occurs in the input code.

Successive approximation   A method of A/D conversion.

■ KEY FORMULAS

V/F Converters

(10–1) $t_{os} = C_{os}(6.8 \times 10^3 \text{ s/F}) + 3 \times 10^{-7} \text{ s}$ One-shot time

(10–2) $\Delta V = \dfrac{(1 \text{ mA} - I_{in})t_{os}}{C_{in}}$ Integrator output increase in reset interval

(10–3) $t_{int} = \left(\dfrac{1 \text{ mA}}{I_{in}} - 1\right)t_{os}$ Integrate interval

(10–4) $f_{out} = \dfrac{I_{in}}{t_{os}(1 \text{ mA})}$ Output frequency

■ **SELF-TEST**

1. An analog switch
   (a) changes an analog signal to digital
   (b) connects or disconnects an analog signal to the output
   (c) stores the value of an analog voltage at a certain point
   (d) combines two or more analog signals onto a single line

2. An analog multiplexer
   (a) produces the sum of several analog voltages on an output line
   (b) connects two or more analog signals to an output at the same time
   (c) connects two or more analog signals to an output one at a time in sequence
   (d) distributes two or more analog signals to different outputs in sequence

3. A basic sample-and-hold circuit contains
   (a) an analog switch and an amplifier
   (b) an analog switch, a capacitor, and an amplifier
   (c) an analog multiplexer and a capacitor
   (d) an analog switch, a capacitor, and input and output buffer amplifiers

4. In a sample/track-and-hold amplifier,
   (a) the voltage at the end of the sample interval is held
   (b) the voltage at the beginning of the sample interval is held
   (c) the average voltage during the sample interval is held
   (d) the output follows the input during the sample interval
   (e) answers (a) and (d)

5. In an analog switch, the aperture time is the time it takes for the switch to
   (a) fully open after the control switches from hold to sample
   (b) fully close after the control switches from sample to hold
   (c) fully open after the control switches from sample to hold
   (d) fully close after the control switches from hold to sample

6. In a binary-weighted-input digital-to-analog converter (DAC),
   (a) all of the input resistors are of equal value
   (b) there are only two input resistor values required
   (c) the number of different input resistor values equals the number of inputs

7. In a 4-bit binary-weighted-input DAC, if the lowest-valued input resistor is 1 kΩ, the highest-valued input resistor is
   (a) 2 kΩ   (b) 4 kΩ   (c) 8 kΩ   (d) 16 kΩ

8. The advantage of an $R/2R$ ladder DAC is
   (a) it is more accurate   (b) it uses only two resistor values
   (c) it uses only one resistor value   (d) it can handle more inputs

9. In a DAC, monotonicity means that
   (a) the accuracy is within one-half of a least significant bit
   (b) there are no missing steps in the output
   (c) there is one bit missing from the input
   (d) there are no linear errors

10. An 8-bit analog-to-digital converter (ADC) can represent
    (a) 144 discrete values of an analog input
    (b) 4096 discrete values of an analog input
    (c) a continuous set of values of an analog input
    (d) 256 discrete values of an analog input

11. An analog signal must be sampled at a minimum rate greater than
    (a) twice the maximum frequency
    (b) twice the minimum frequency
    (c) the maximum frequency
    (d) the minimum frequency

12. Quantization error in an ADC is due to
    (a) poor resolution
    (b) nonlinearity of the input
    (c) a missing bit in the output
    (d) a change in the input voltage during the conversion time

13. Quantization error can be avoided by
    (a) using a higher resolution ADC
    (b) using a sample-and-hold prior to the ADC
    (c) shortening the conversion time
    (d) using a flash ADC

14. The type of ADC with the fastest conversion time is the
    (a) dual-slope   (b) single-slope
    (c) simultaneous   (d) successive-approximation

15. The output of a *V/F* converter
    (a) has an amplitude proportional to the frequency of the input
    (b) is a digital reproduction of the input voltage
    (c) has a frequency that is inversely proportional to the amplitude of the input
    (d) has a frequency that is directly proportional to the amplitude of the input

16. An element not found in the typical *V/F* converter is
    (a) a linear amplifier   (b) a one-shot
    (c) an integrator   (d) a comparator

■ PROBLEMS

SECTION 10–1 Analog Switches

1. Determine the output waveform for the analog switch in Figure 10–67(a) for each set of waveforms in parts (b), (c), and (d).

2. Determine the output of the 4-channel analog multiplexer in Figure 10–68 for the signal and control inputs shown.

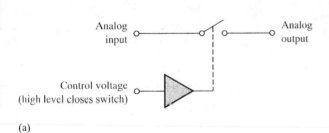

(a)

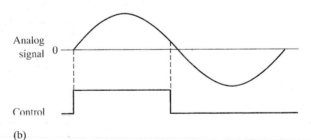

(b)

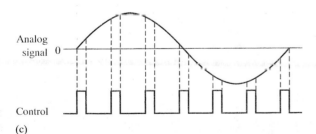

(c)

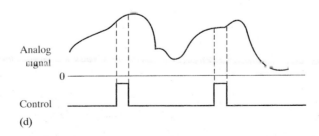

(d)

FIGURE 10–67

FIGURE 10–68

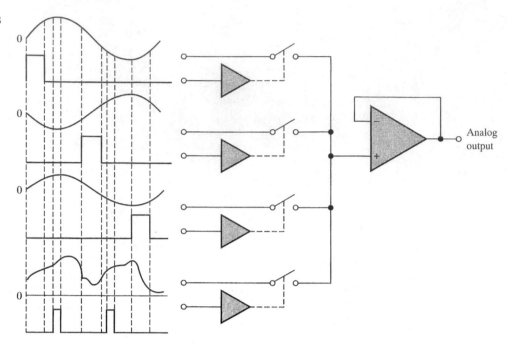

SECTION 10–2   Sample-and-Hold Amplifiers

3. Determine the output voltage waveform for the sample/track-and-hold amplifier in Figure 10–69 given the analog input and the control voltage waveforms shown. Sample is the high control level.

4. Repeat Problem 3 for the waveforms in Figure 10–70.

5. Determine the gain of each AD582 sample-and-hold amplifier in Figure 10–71.

FIGURE 10–69

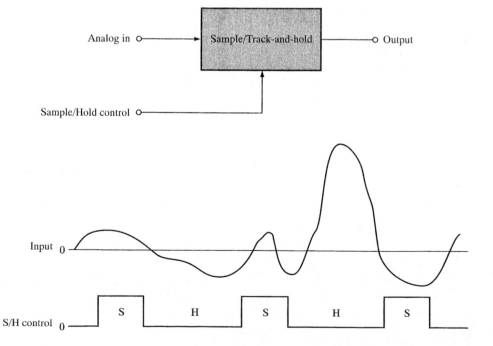

FIGURE 10–70

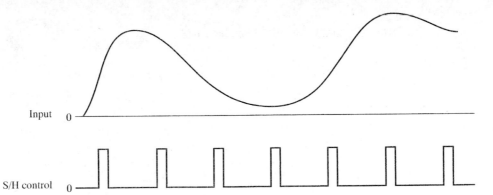

FIGURE 10–71

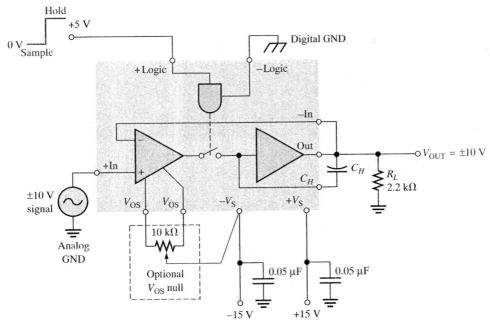

(a)

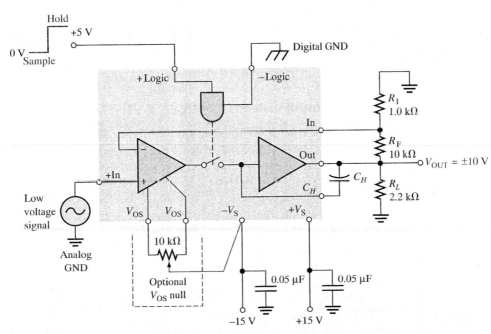

(b)

SECTION 10–3   Interfacing the Analog and Digital Worlds

6. The analog signal in Figure 10–72 is sampled at 2 ms intervals. Represent the signal by a series of 4-bit binary numbers.

7. Sketch the digital reproduction of the analog curve represented by the series of binary numbers developed in Problem 6.

8. Graph the analog signal represented by the following sequence of binary numbers: 1111, 1110, 1101, 1100, 1010, 1001, 1000, 0111, 0110, 0101, 0100, 0101, 0110, 0111, 1000, 1001, 1010, 1011, 1100, 1100, 1011, 1010, 1010.

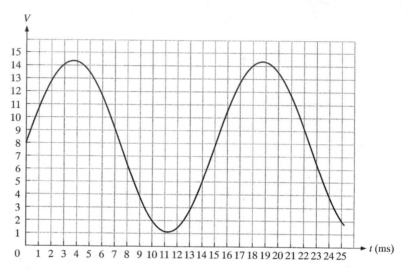

FIGURE 10–72

SECTION 10–4   Digital-to-Analog (D/A) Conversion

9. In a certain 4-bit DAC, the lowest-weighted resistor has a value of 10 kΩ. What should the values of the other input resistors be?

10. Determine the output of the DAC in Figure 10–73(a) if the sequence of 4-bit numbers in part (b) is applied to the inputs.

11. Repeat Problem 10 for the inputs in Figure 10–74.

12. Determine the resolution expressed as a percentage for each of the following DACs:
    (a) 3-bit   **(b)** 10-bit   **(c)** 18-bit

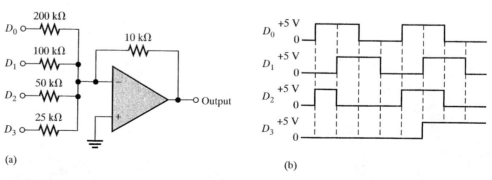

(a)

(b)

FIGURE 10–73

FIGURE 10–74

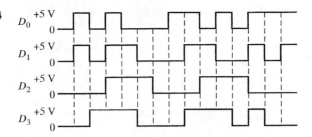

SECTION 10–5   Basic Concepts of Analog-to-Digital (A/D) Conversion

13. How many discrete values of an analog signal can each of the following ADCs represent?
    (a) 4-bit   **(b)** 5-bit   **(c)** 8-bit   **(d)** 16-bit

14. Determine the Nyquist rate for sinusoidal voltages with each of the following periods:
    (a) 10 s   **(b)** 1 ms   **(c)** 30 μs   **(d)** 1000 ns

15. What is the quantization error expressed in volts of an ADC with a sample-and-hold input for a sampled value of 3.2 V if the sample-and-hold has a droop of 100 mV/s? Assume that the conversion time of the ADC is 10 ms.

SECTION 10–6   Analog-to-Digital (A/D) Conversion Methods

16. Determine the binary output sequence of a 3-bit flash ADC for the analog input signal in Figure 10–75. The sampling rate is 100 kHz and $V_{REF} = 8V$.

17. Repeat Problem 16 for the analog waveform in Figure 10–76.

18. For a certain 4-bit successive-approximation ADC, the maximum ladder output is +8 V. If a constant +6 V is applied to the analog input, determine the sequence of binary states for the SAR.

FIGURE 10–75

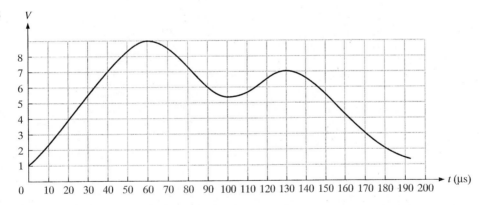

FIGURE 10–76

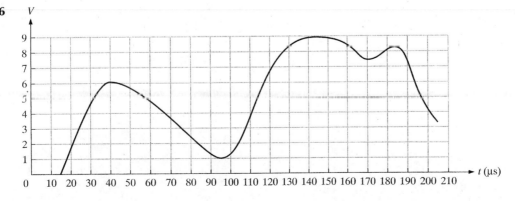

SECTION 10–7 Voltage-to-Frequency (*V/F*) and Frequency-to-Voltage (*F/V*) Converters

19. The analog input to a *V/F* converter increases from 0.5 V to 3.5 V. Does the output frequency increase, decrease, or remain unchanged?

20. Assume that when the input to a certain *V/F* converter is 0 V, there is no output signal (0 Hz). Also, when a constant +2 V is applied to the input, the corresponding output frequency is 1 kHz. Now, if the input takes a step up to +4 V, what is the output frequency?

21. Calculate the value of the timing capacitor required to produce a 5 μs pulse width in an AD650 *V/F* converter.

22. Determine the increase in the integrator output voltage during the reset interval in the AD650 shown in Figure 10–77.

23. Determine the minimum and maximum output frequencies of the AD6450 in Figure 10–77 for the portion of the input voltage shown within the shaded area in Figure 10–78.

FIGURE 10–77

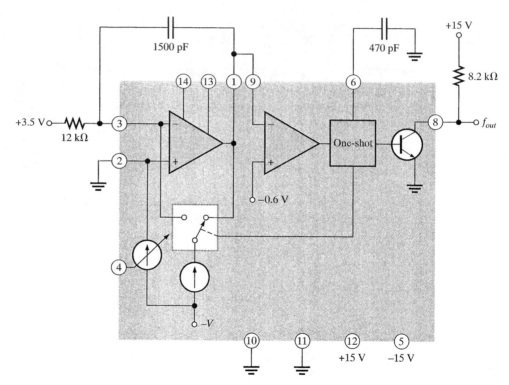

FIGURE 10–78

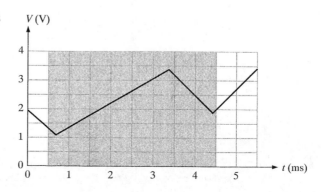

SECTION 10–8   Troubleshooting

24. A 4-bit DAC has failed in such a way that the MSB is stuck at 0. Draw the analog output when a straight binary sequence is applied to the inputs.

25. A straight binary sequence is applied to a 4-bit DAC and the output in Figure 10–79 is observed. What is the problem?

26. An ADC produces the following sequence of binary numbers when a certain analog signal is applied to its input: 0000, 0001, 0010, 0011, 0100, 0101, 0110, 0111, 0110, 0101, 0100, 0011, 0010, 0001, 0000.

    (a) Reconstruct the input from the digital codes as a DAC would.

    (b) If the ADC failed so that the code 0111 were missing from the output, what would the reconstructed output look like?

FIGURE 10–79

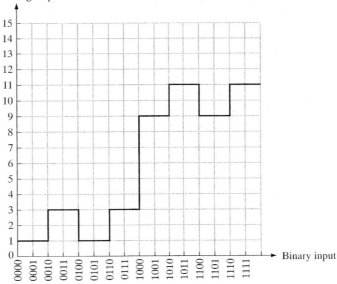

ANSWERS TO REVIEW QUESTIONS

Section 10–1

1. To switch analog signals on or off electronically.

2. An analog multiplexer switches analog voltages from several lines onto a common output line in a time sequence.

Section 10–2

1. A sample-and-hold retains the value of an analog signal taken at a given point.

2. Droop is the decrease in the held voltage due to capacitor leakage.

3. Aperture time is the time required for an analog switch to fully open at the end of a sample pulse.

4. Acquisition time is the time required for the device to reach final value at the start of the sample pulse.

Section 10–3

1. Natural quantities are in analog form.

2. A/D conversion changes an analog quantity into digital form.

3. D/A conversion changes a digital quantity into analog form.

Section 10–4

1. Each input resistor must have a different value.
2. 6.67%

Section 10–5

1. The time for a sampled analog value to be converted to digital is the conversion time.
2. Greater than 200 Hz.
3. The sample-and-hold keeps the sampled value constant during conversion.

Section 10–6

1. Flash is the fastest method of A/D conversion.
2. Tracking A/D conversion uses an up-down counter.
3. True

Section 10–7

1. V/F components: integrator, comparator, one-shot, current source, and switch
2. The output frequency increases proportionally.
3. The V/F converter has a voltage input and a frequency output. The F/V converter has a frequency input and a voltage output.

Section 10–8

1. Nonmonotonicity is indicated by a step reversal.
2. Step amplitudes are less than ideal.
3. Missing code and incorrect code

Section 10–9

1. The capacitors are for power supply decoupling.
2. $2^8 = 256$ values can be converted to digital.
3. Eight minutes is the minimum interval between 2-degree steps required to track the sun through a 180° arc for 12 hours.

■ **ANSWERS TO PRACTICE EXERCISES FOR EXAMPLES**

10–1 See Figure 10–80.
10–2 The same tones will be closer together.
10–3 See Figure 10–81.
10–4 0.25 V
10–5 0.00038%
10–6 010, 011, 011, 011, 010, 001, 000, 000, 001, 010, 011, 011
10–7 $f_{min} = 40$ kHz, $f_{max} = 240$ kHz
10–8 See Figure 10–82.
10–9 A constant 15 V output.

FIGURE 10–80

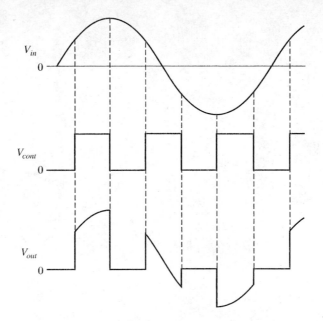

FIGURE 10–81

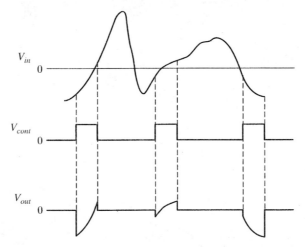

FIGURE 10–82

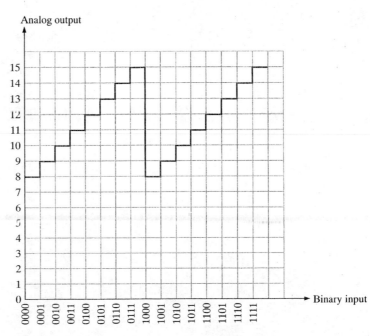

11

MEASUREMENT AND CONTROL CIRCUITS

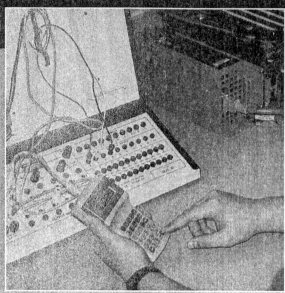

Courtesy Yuba College

■ **CHAPTER OBJECTIVES**

☐ Describe the basic operation of rms-to-dc converters
☐ Discuss angle measurement using a synchro
☐ Discuss the operation of three types of temperature-measuring circuits
☐ Describe methods of measuring strain, pressure, and motion
☐ Describe the operation of the zero-voltage switch
☐ Apply what you have learned in this chapter to a system application

This chapter introduces several types of transducers and related circuits for measuring basic physical analog parameters such as angular position, temperature, strain, pressure, and flow rate.

A **transducer** is a device that converts a physical parameter into another form. Transducers can be used at the input (a microphone) or output (a speaker) of a system. With electronic-measuring systems, the input transducer converts a quantity to be measured (temperature, humidity, flow rate, weight) into an electrical parameter (voltage, current, resistance, capacitance) that can be processed by an electronic instrument or system.

Transducers and their associated circuits are important in many applications. The measurement of angular position is critical in robotics, radar, and industrial machine control. Temperature-measuring and pressure-measuring circuits are used in industry for monitoring temperatures and pressures of various fluids or gases in tanks and pipes, and they are used in automotive applications for measuring temperatures and pressures in various parts of the automobile. Strain measurement is important for testing the strength of materials under stress in such areas as aircraft design. Also, zero-voltage switching, an important technique in controlling various functions, is introduced in this chapter.

The system application in Section 11–6 focuses on the measurement of wind speed and direction. The input from the wind speed measurement part of the system (anemometer) is generated by a type of flow meter in the form of a propeller arrangement mounted on a wind vane. The wind causes the flow meter blades to rotate on a shaft at a rate proportional to the wind speed. A magnetic device senses each rotation and the circuitry produces a pulse. The frequency of the pulses indicates the speed of the wind. The input from the wind direction measurement part of the system is generated by the wind vane attached to the shaft of a resolver. The wind vane aligns itself with the direction of the wind, and the resolver produces electrical signals proportional to the angular position. Figure 11–50 shows a simplified diagram for the wind-measuring system that will be the focus of the system application. Notice that some circuits from previous chapters are utilized in this application.

For the system application in Section 11–6, in addition to the other topics, be sure you understand

☐ Basic resolver and RDC operation
☐ The 555 timer, frequency-to-voltage converters, and ADCs

11–1 ■ RMS-TO-DC CONVERTERS

RMS-to-dc converters are used in several basic areas. One important application is in noise measurement for determining thermal noise, transistor, and switch contact noise. Another application is in the measurement of signals from mechanical phenomena such as strain, vibration, and expansion or contraction. RMS-to-dc converters are also useful for accurate measurements of low-frequency, low duty-cycle pulse trains.

After completing this section, you should be able to

❑ Describe the basic operation of rms-to-dc converters
   ❑ Define *rms*
   ❑ Explain the rms-to-dc conversion process
   ❑ List the basic circuits in an rms-to-dc converter
   ❑ Discuss the difference between explicit and implicit rms-to-dc converters
   ❑ Give examples of rms-to-dc converter applications

Definition of RMS

RMS stands for **root mean square** and is related to the amplitude of an ac signal. In practical terms, the rms value of an ac voltage is equal to the value of a dc voltage required to produce the same heating effect in a resistance. For this reason, it is sometimes called the *effective* value of an ac voltage. For example, an ac voltage with an rms value of 1 V produces the same amount of heat in a given resistor as a 1 V dc voltage. Mathematically, the rms value is found by taking the square root of the average (**mean**) of the signal voltage squared, as expressed in the following formula:

$$V_{rms} = \sqrt{\text{avg}(V_{in}^2)} \qquad\qquad (11\text{–}1)$$

RMS-to-DC Conversion

RMS-to-dc converters are electronic circuits that continuously compute the square of the input signal voltage, average it, and take the square root of the result. The output of an rms-to-dc converter is a dc voltage that is proportional to the rms value of the input signal. The block diagram in Figure 11–1 illustrates the basic conversion process.

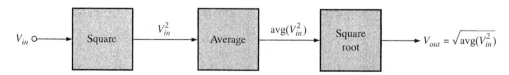

FIGURE 11–1
The rms-to-dc conversion process.

The Squaring Circuit   The squaring circuit is generally a linear multiplier with the signal applied to both inputs as shown in Figure 11–2. Linear multipliers were introduced in Chapter 9.

FIGURE 11–2
The squaring circuit is a linear multiplier.

V_{in} o

$V_{out} = V_{in} \times V_{in} = V_{in}^2$

The Averaging Circuit   The simplest type of averaging circuit is a single-pole low-pass filter on the input of an op-amp voltage-follower, as shown in Figure 11–3. The *RC* filter passes only the dc component (average value) of the squared input voltage. The overbar designates an average value.

FIGURE 11–3
A basic averaging circuit.

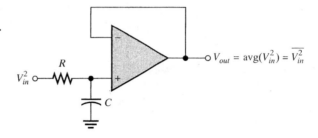

V_{in}^2 o

R

C

$V_{out} = \text{avg}(V_{in}^2) = \overline{V_{in}^2}$

The Square Root Circuit   Recall from Chapter 9 that a square root circuit uses a linear multiplier in an op-amp feedback loop as shown in Figure 11–4.

FIGURE 11–4
A square root circuit.

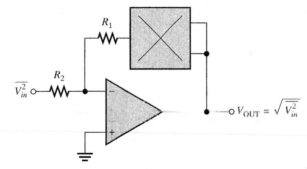

R_1

R_2

$\overline{V_{in}^2}$ o

$V_{OUT} = \sqrt{\overline{V_{in}^2}}$

A Complete RMS-to-DC Converter

Figure 11–5 shows the three functional circuits combined to form an rms-to-dc converter. This combination is often referred to as an *explicit* rms-to-dc converter because of the straightforward method used in determining the rms value.

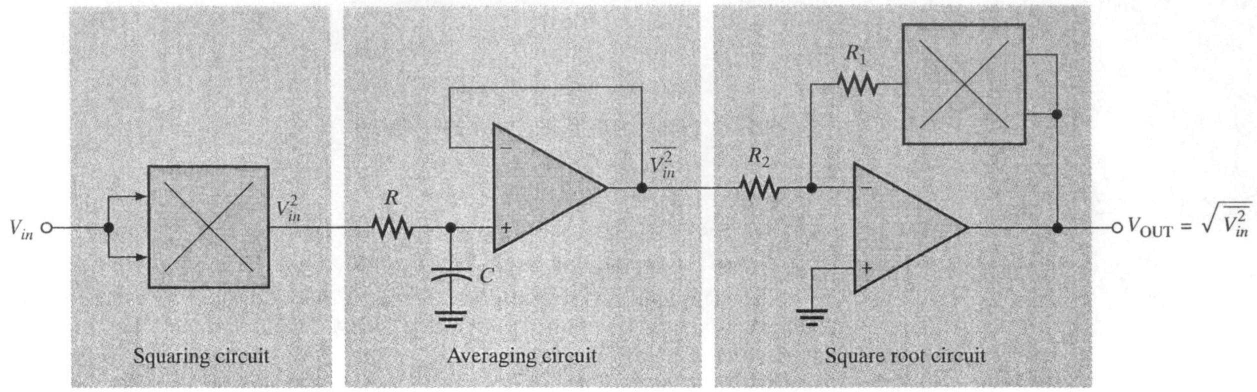

FIGURE 11–5
Explicit type of rms-to-dc converter.

Another method for achieving rms-to-dc conversion, sometimes called the *implicit* method, uses feedback to perform the square root operation. A basic circuit is shown in Figure 11–6. The first block squares the input voltage and divides by the output voltage. The averaging circuit produces the final dc output voltage, which is fed back to the squarer/divider circuit.

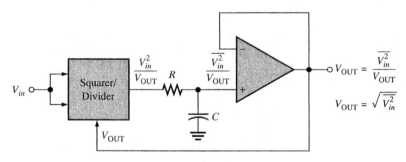

FIGURE 11–6
Implicit type of rms-to-dc converter.

The operation of the circuit in Figure 11–6 can be understood better by going through the mathematical steps performed by the circuit as follows. The expression for the output of the squarer/divider is

$$\frac{V_{in}^2}{V_{OUT}}$$

The voltage at the noninverting $(+)$ input to the voltage-follower is

$$V_{in(NI)} = \frac{\overline{V_{in}^2}}{V_{OUT}}$$

where the overbar indicates average value. The final output voltage is

$$V_{OUT} = \frac{\overline{V_{in}^2}}{V_{OUT}}$$

$$V_{OUT}^2 = \overline{V_{in}^2}$$

$$V_{OUT} = \sqrt{\overline{V_{in}^2}} \qquad (11-2)$$

The AD637 IC RMS-to-DC Converter

As an example of a specific IC device, let's look at the AD637 rms-to-dc converter. This device is essentially an implicit type of converter except that it has an absolute value circuit at the input and it uses an inverting low-pass filter for averaging, as indicated in Figure 11–7. The averaging capacitor, C_{avg}, is an external component that can be selected to provide a minimum averaging error under various input conditions.

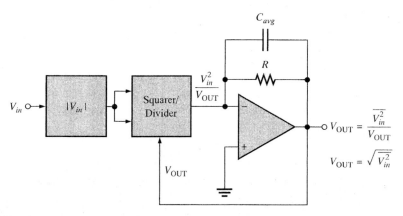

FIGURE 11–7
Basic diagram of the AD637 rms-to-dc converter.

The absolute value circuit in the first block is simply a full-wave rectifier that changes all the negative portions of an input voltage to positive. The squarer/divider circuit is actually implemented with log and antilog circuits as shown in Figure 11–8.

Notice that the second block in Figure 11–8 produces the log of the square of the input by taking the log of V_{in} and then multiplying it by two.

$$2 \log V_{in} = \log V_{in}^2$$

This equation is based on the fundamental rule of logarithms that states the log of a variable squared is equal to twice the log of the variable. The third block is a subtracter that subtracts the logarithm of the output voltage from the log of the input squared.

$$2 \log V_{in} - \log V_{OUT} = \log V_{in}^2 - \log V_{OUT} = \log\left(\frac{V_{in}^2}{V_{OUT}}\right)$$

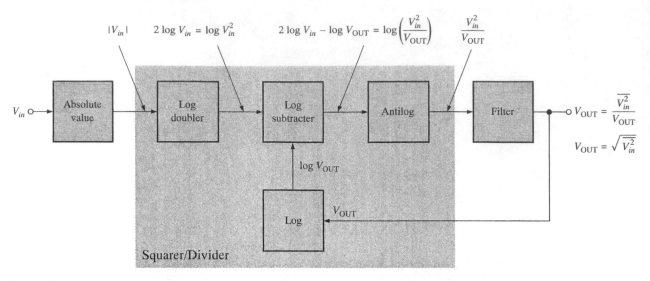

FIGURE 11–8
Internal function diagram of the AD637.

This equation is based on the fundamental rule of logarithms that states that the difference of two logarithmic terms equals the logarithm of the quotient of the two terms. The antilog circuit takes the antilog of $\log(V_{in}^2/V_{OUT})$ and produces an output equal to V_{in}^2/V_{OUT}, as indicated in the figure. The low-pass filter averages the output of the antilog circuit and produces the final output.

Examples of RMS-to-DC Converter Applications

In addition to the measurement applications mentioned in the section introduction, rms-to-dc converters are used in a variety of system applications. A couple of typical applications are in automatic gain control (AGC) circuits and rms voltmeters. Let's look at each of these in a general way.

AGC Circuits Figure 11–9 shows a general diagram of an AGC circuit that incorporates an rms-to-dc converter. AGC circuits are used in audio systems to keep the output amplitude constant when the input signal level varies over a certain range. They are also used in signal generators to keep the output amplitude constant with variations in waveform, duty cycle, and frequency.

RMS Voltmeters Figure 11–10 basically illustrates the rms-to-dc converter in an rms voltmeter. The rms-to-dc converter produces a dc output that is the rms value of the input signal. This rms value is then converted to digital form by an ADC and displayed.

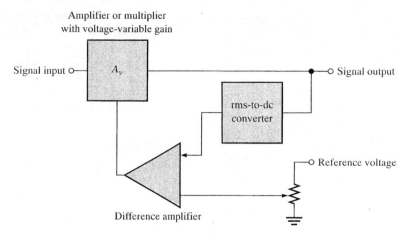

FIGURE 11–9
A simplified AGC circuit using an rms-to-dc converter.

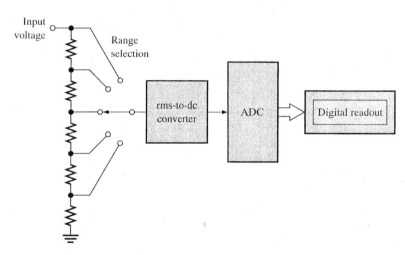

FIGURE 11–10
A simplified rms voltmeter.

11–1 REVIEW QUESTIONS

1. What is the basic purpose of an rms-to-dc converter?
2. What are the three internal functions performed by an rms-to-dc converter?

11–2 ■ ANGLE-MEASURING CIRCUITS

In many applications, the angular position of a shaft or other mechanical mechanism must be measured and converted to an electrical signal for processing or display. Examples of this mechanical-to-electrical interfacing are found in radar and satellite antennas, wind vanes, solar systems, industrial machines including robots, and military fire control systems, to name a few. In this section, the circuits for interfacing angular position transducers, called synchros and resolvers, are introduced. Transducers, in general, are devices that convert a physical parameter from one form to another. Before we get into the circuits used in angular measurements, a brief introduction to synchros will provide some background.

After completing this section, you should be able to

❑ Discuss angle measurement using a synchro
    ❑ Define *synchro* and explain the basic operation
    ❑ Define *resolver* and explain the basic operation
    ❑ Discuss synchro-to-digital converters and resolver-to-digital converters
    ❑ Describe the basic operation of an RDC
    ❑ Show how angles can be represented by digital codes
    ❑ Discuss an RDC application

Synchros

A **synchro** is an electromechanical transducer used for shaft angle measurement and positioning. There are several different types of synchros, but all can be thought of basically as rotating transformers. In physical appearance, a synchro resembles a small ac motor as shown in Figure 11–11(a) with a diameter ranging from about 0.5 in. to about 4 in.

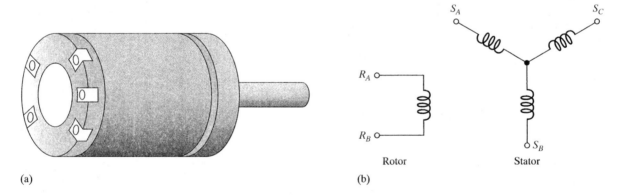

(a)   (b)

FIGURE 11–11
A typical synchro and its basic winding structure.

The basic synchro consists of a **rotor,** which can revolve within a fixed **stator** assembly. A shaft is connected to the rotor so that when the shaft rotates, the rotor also rotates. In most synchros, there is a rotor winding and three stator windings. The stator windings are connected as shown in Figure 11–11(b) and are separated by 120° around the stator. The windings are brought out to a terminal block at one end of the housing.

Synchro Voltages When a reference sinusoidal voltage is applied across the rotor winding, the voltage induced across any one of the stator windings is proportional to the sine of the angle (θ) between the rotor winding and the stator winding. The angle θ is dependent on the shaft position.

The voltage induced across any two stator windings (between any two stator terminals) is the sum or difference of the two stator voltages. These three voltages, called *syn-*

chro format voltages, are represented in Figure 11–12 and are derived using a basic trigonometric identity. The important thing is that each of the three synchro format voltages is a function of the shaft angle, θ, and can be used to determine the angular position at any time. As the shaft rotates, the format voltages change proportionally.

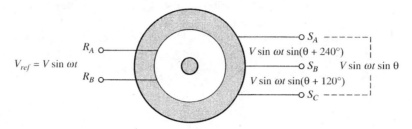

FIGURE 11–12
Synchro format voltages with a reference voltage applied to the rotor.

Resolvers

The **resolver** is a particular type of synchro that is often used in rotational systems to transduce the angular position. Resolvers differ from regular synchros in that the rotor and two stator windings are separated from each other by 90° rather than by 120°. The basic winding configuration of a simple resolver is shown in Figure 11–13.

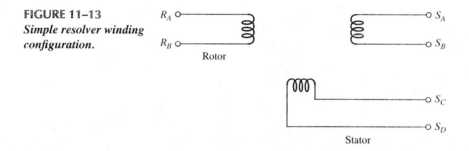

FIGURE 11–13
Simple resolver winding configuration.

Resolver Voltages   If a reference sinusoidal voltage is applied to the rotor winding, the resulting voltages across the stator windings are as given in Figure 11–14. These voltages are a function of the shaft angle θ and are called *resolver format voltages.* One of the voltages is proportional to the sine of θ and the other voltage is proportional to the cosine of θ. Notice that the resolver has a four-terminal output compared to the three-terminal output of the standard synchro.

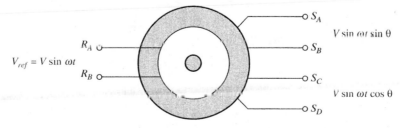

FIGURE 11–14
Resolver format voltages with a reference voltage applied to the rotor.

Basic Operation of Synchro-to-Digital and Resolver-to-Digital Converters

Electronic circuits known as **synchro-to-digital converters (SDCs)** and **resolver-to-digital converters (RDCs)** are used to convert the format voltages from a synchro or resolver to a digital format. These devices may be considered a very specialized form of analog-to-digital converter.

All converters, both SDCs and RDCs, operate internally with resolver format voltages. Therefore, the output format voltages of a synchro must first be transformed into resolver format by a special type of transformer called the *Scott-T transformer,* as illustrated in Figure 11–15.

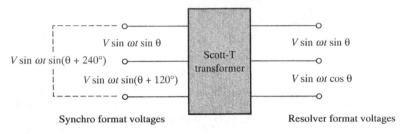

FIGURE 11–15
Inputs and outputs of a Scott-T transformer.

Some SDCs have internal Scott-T transformers, but others require a separate transformer. Other than the transformer, the basic operation and internal circuitry of SDCs and RDCs are the same, so let's focus on RDCs. A simplified block diagram of a tracking RDC is shown in Figure 11–16.

The two resolver format voltages, $V_1 = V \sin \omega t \sin \theta$ and $V_2 = V \sin \omega t \cos \theta$, are applied to the RDC inputs as indicated in Figure 11–16 (θ is the current shaft angle of the resolver). These resolver voltages go through buffers to special multiplier circuits. Let's assume that the current state of the up/down counter represents some angle, ϕ. The digital code representing ϕ is applied to the multiplier circuits along with the resolver voltages. The cosine multiplier takes the cosine of ϕ and multiplies it times the resolver voltage V_1. The sine multiplier takes the sine of ϕ and multiplies it times the resolver voltage V_2. The resulting output of the cosine multiplier is

$$V_1 \cos \phi = V \sin \omega t \sin \theta \cos \phi$$

The resulting output of the sine multiplier is

$$V_2 \sin \phi = V \sin \omega t \cos \theta \sin \phi$$

These two voltages are subtracted by the error amplifier to produce the following error voltage:

$$V \sin \omega t \sin \theta \cos \phi - V \sin \omega t \cos \theta \sin \phi = V \sin \omega t (\sin \theta \cos \phi - \cos \theta \sin \phi)$$

A basic trigonometric identity reduces the error voltage expression to

$$V \sin \omega t \sin(\theta - \phi)$$

The phase-sensitive detector produces a dc error voltage proportional to $\sin(\theta - \phi)$, which is applied to the integrator. The output of the integrator drives a voltage-controlled

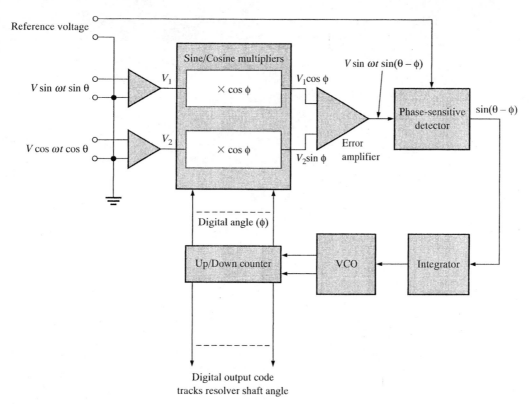

FIGURE 11–16
Simplified diagram of a resolver-to-digital converter (RDC).

oscillator (VCO), which provides clock pulses to the up/down counter. When the counter reaches the value of the current shaft angle θ, then $\phi = \theta$ and

$$\sin(\theta - \phi) = 0$$

If the sine is zero, then the difference of the angles is $0°$.

$$\theta - \phi = 0°$$

At this point, the angle stored in the counter equals the resolver shaft angle.

$$\phi = \theta$$

When the shaft angle changes, the counter will count up or down until its count equals the new shaft angle. Therefore, the RDC will continuously track the resolver shaft angle and produce an output digital code that equals the angle at all times.

Representation of Angles with a Digital Code

The most common method of representing an angular measurement with a digital code is given in Table 11–1 for word lengths up to 16 bits. A binary 1 in any bit position means that the corresponding angle is included, and a 0 means that the corresponding angle is not included.

TABLE 11–1
Bit weights for resolver-to-digital conversion.

Bit Position	Angle (Degrees)
1 (MSB)	180.00000
2	90.00000
3	45.00000
4	22.50000
5	11.25000
6	5.62500
7	2.81250
8	1.40625
9	0.70313
10	0.35156
11	0.17578
12	0.08790
13	0.04395
14	0.02197
15	0.01099
16	0.00549

EXAMPLE 11–1

A certain RDC has an 8-bit digital output. What is the angle being measured if the output code is 01001101? The left-most bit is the MSB.

Solution

Bit Position	Bit	Angle (Degrees)
1	0	0
2	1	90.00000
3	0	0
4	0	0
5	1	11.25000
6	1	5.62500
7	0	0
8	1	1.40625

To get the angle represented, add all the included angles (as indicated by the presence of a 1 in the output code).

$$90.00000° + 11.25000° + 5.62500° + 1.40625° = \mathbf{108°}$$

Although more digits are carried in the calculation to show the process, the answer is rounded to the nearest degree. An 8-bit code has a resolution of 1.4°.

Practice Exercise   What is the angular shaft position measured by a 12-bit RDC when it has a binary code of 100000100001 on its outputs?

A Specific Resolver-to-Digital Converter

To illustrate a typical IC device, let's look at Analog Devices 1S20, which is a 12-bit converter. The diagram for this device is shown in Figure 11–17, and as you can see, it is basically the same as the general RDC in Figure 11–16 with some additions. Additional circuits include the latch and output buffers for controlling the data transfer to and inter-

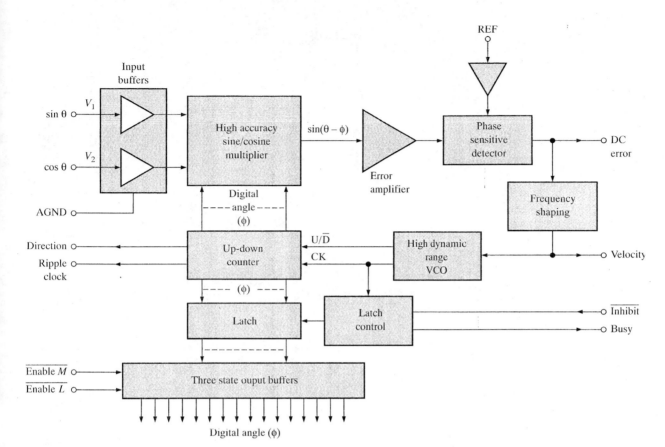

FIGURE 11–17
Diagram of the 1S20 resolver-to-digital converter.

facing with other digital systems. These additional circuits do not affect the conversion process.

Additional outputs include the Busy signal that indicates when the counter is in transition from one code to another as a result of the angle changing. The Direction output indicates the direction of rotation of the resolver shaft. The Ripple clock output indicates when the counter resets from all 1s to all 0s or vice versa, indicating that the shaft has completed a full revolution (360°). The Velocity output is proportional to the rate of change of the input angle.

Additional inputs include the $\overline{\text{Inhibit}}$ signal that is used to control the transfer of data from the counter to the latch. The two $\overline{\text{Enable}}$ inputs are used to enable the buffer outputs.

A Simple Resolver-to-Digital Converter Application

The measurement of wind direction is one example of how an RDC can be used. As shown in Figure 11–18, a wind vane is fixed to the shaft of a resolver. As the wind vane moves to align with the direction of the wind, the resolver shaft rotates and its angle indicates the wind direction. The resolver output is applied to an RDC and the resulting digital output code, which represents the wind direction, drives a digital readout.

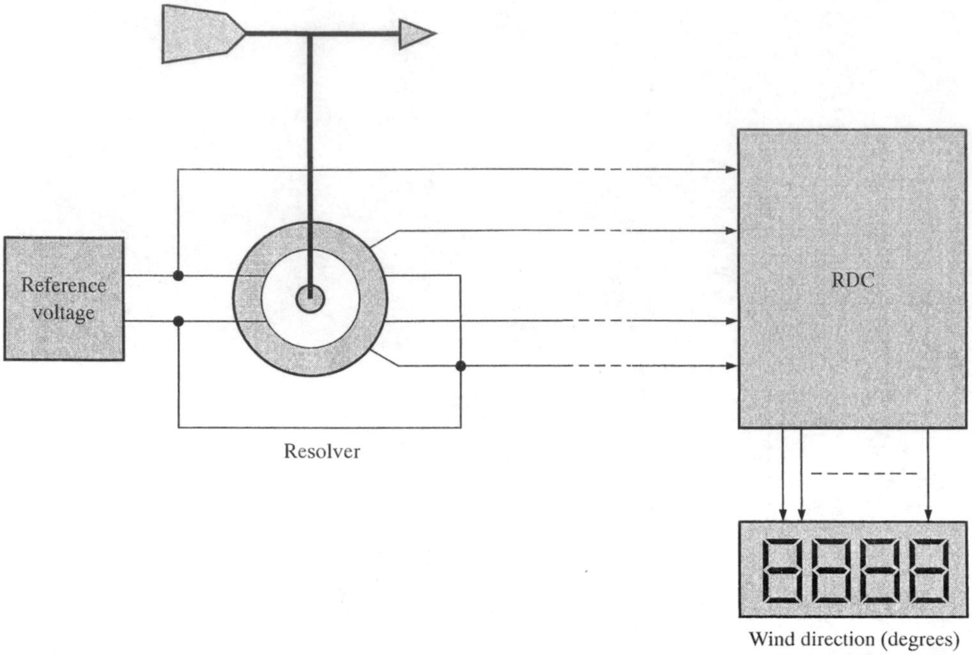

FIGURE 11–18
Measurement and display of wind direction with a resolver and an RDC.

11–2 REVIEW QUESTIONS

1. What is a transducer that converts a mechanical shaft position into electrical signals called?
2. What type of input does an RDC accept?
3. What type of output does an RDC produce?
4. What is the function of an RDC?

11–3 ■ TEMPERATURE-MEASURING CIRCUITS

Temperature is perhaps the most common physical parameter that is measured and converted to electrical form. Several types of temperature sensors respond to temperature and produce a corresponding indication by a change or alteration in a physical characteristic that can be detected by an electronic circuit. Common types of temperature sensors are thermocouples, resistance temperature detectors (RTDs), and thermistors. In this section, we will look at each of these sensors and at signal conditioning circuits that are required to interface the transducers to electronic equipment.

After completing this section, you should be able to

❑ Discuss the operation of three types of temperature-measuring circuits
    ❑ Describe the thermocouple and how to interface it with an electronic circuit
    ❑ Discuss a thermocouple signal conditioner

❑ Describe the resistance temperature detector (RTD) and circuit interfacing
❑ Describe the thermistor and circuit interfacing

The Thermocouple

The **thermocouple** is formed by joining two dissimilar metals. A small voltage, called the *Seebeck voltage,* is produced across the junction of the two metals when heated, as illustrated in Figure 11–19. The amount of voltage produced is dependent on the types of metals and is directly proportional to the temperature of the junction (positive temperature coefficient); however, this voltage is generally much less than 100 mV. The voltage versus temperature characteristic of thermocouples is somewhat nonlinear, but the amount of nonlinearity is predictable. Thermocouples are widely used in certain industries because they have a wide temperature range and can be used to measure very high temperatures.

FIGURE 11–19

A voltage proportional to temperature is generated when a thermocouple is heated.

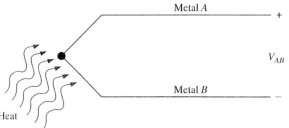

Metal *A* +

V_{AB}

Metal *B* −

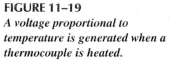

Heat

Some common metal combinations used in commercial thermocouples are chromel-alumel (chromel is a nickel-chromium alloy and alumel is a nickel-aluminum alloy), iron-constantan (constantan is a copper-nickel alloy), chromel-aluminum, tungsten-rhenium alloys, and platinum-10% Rh/Pt. Each of these types of thermocouple has a different temperature range, coefficient, and voltage characteristic and is designated by the letter *E, J, K, W,* and *S,* respectively. The overall temperature range covered by thermocouples is from −250°C to 2000°C. Each type covers a different portion of this range, as shown in Figure 11–20.

Thermocouple-to-Electronics Interface   When a thermocouple is connected to a signal-conditioning circuit, as illustrated in Figure 11–21, an *unwanted* thermocouple is effectively created at the point(s) where one or both of the thermocouple wires connect to the circuit terminals made of a dissimilar metal. The unwanted thermocouple junction is sometimes referred to as a **cold junction** in some references because it is normally at a significantly lower temperature than that being measured by the measuring thermocouple. These unwanted thermocouples can have an unpredictable effect on the overall voltage that is sensed by the circuit because the voltage produced by the unwanted thermocouple opposes the measured thermocouple voltage and its value depends on ambient temperature.

Example of a Thermocouple-to-Electronics Interface   As shown in Figure 11–22, a copper/constantan thermocouple (known as type *T*) is used, in this case, to measure the temperature in an industrial temperature chamber. The copper thermocouple wire is connected to a copper terminal on the circuit board and the constantan wire is also connected to a copper terminal on the circuit board. The copper-to-copper connection is no

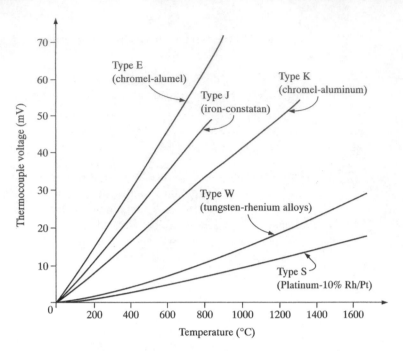

FIGURE 11–20
Output of some common thermocouples with 0°C as the reference temperature.

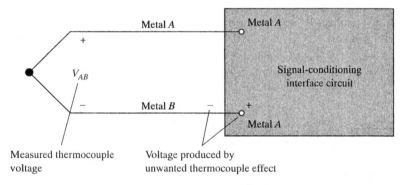

FIGURE 11–21
Creation of an unwanted thermocouple in a thermocouple-to-electronics interface.

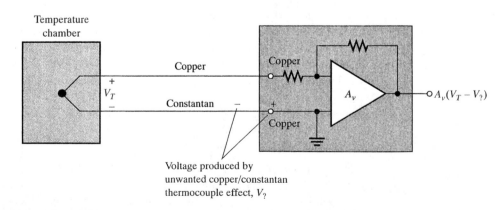

FIGURE 11–22
A simplified temperature-measuring circuit with an unwanted thermocouple at the junction of the constantan wire and the copper terminal.

problem because the metals are the same. The constantan-to-copper connection acts as an unwanted thermocouple that will produce a voltage in opposition to the thermocouple voltage because the metals are dissimilar.

Since the unwanted thermocouple connection is not at a fixed temperature, its effects are unpredictable and it will introduce inaccuracy into the measured temperature. One method for eliminating an unwanted thermocouple effect is to add a reference thermocouple at a constant known temperature (usually 0°C). Figure 11–23 shows that by using a reference thermocouple that is held at a constant known temperature, the unwanted thermocouple at the circuit terminal is eliminated because both contacts to the circuit terminals are now copper-to-copper. The voltage produced by the reference thermocouple is a known constant value and can be compensated for in the circuitry.

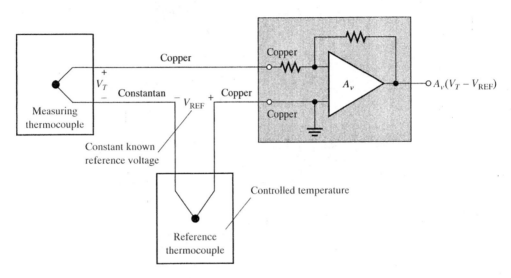

FIGURE 11–23
Using a reference thermocouple in a temperature-measuring circuit.

EXAMPLE 11–2

Suppose the thermocouple in Figure 11–22 is measuring 200°C in an industrial oven. The circuit board is in an area where the ambient temperature can vary from 15°C to 35°C. Using Table 11–2 for a type-*T* (copper/constantan) thermocouple, determine the voltage across the circuit input terminals at the ambient temperature extremes. What is the maximum percent error in the voltage at the circuit input terminals?

TABLE 11–2
Type-T thermocouple voltage.

Temperature (°C)	Output (mV)
−200	−5.603
−100	−3.378
0	0.000
+100	4.277
+200	9.286
+300	14.860
+400	20.869

Solution From Table 11–2, you know that the measuring thermocouple is producing 9.286 mV. To determine the voltage that the unwanted thermocouple is creating at 15°C, you must interpolate from the table. Since 15°C is 15% of 100°C, a linear interpolation gives the following voltage:

$$0.15(4.277 \text{ mV}) = 0.642 \text{ mV}$$

Since 35°C is 35% of 100°C, the voltage is

$$0.35(4.277 \text{ mV}) = 1.497 \text{ mV}$$

The voltage across the circuit input terminals at 15°C is

$$9.286 \text{ mV} - 0.642 \text{ mV} = 8.644 \text{ mV}$$

The voltage across the circuit input terminals at 35°C is

$$9.286 \text{ mV} - 1.497 \text{ mV} = 7.789 \text{ mV}$$

The maximum percent error in the voltage at the circuit input terminals is

$$\left(\frac{9.286 \text{ mV} - 7.789 \text{ mV}}{9.286 \text{ mV}} \right) 100\% = \mathbf{16.1\%}$$

You can never be sure how much it is off because you have no control over the ambient temperature. Also, the linear interpolation may or may not be accurate depending on the linearity of the temperature characteristic of the unwanted thermocouple.

Practice Exercise In the case of the circuit in Figure 11–22, if the temperature being measured goes up to 300°C, what is the maximum percent error in the voltage across the circuit input terminals?

EXAMPLE 11–3

Now, refer to the thermocouple circuit in Figure 11–23. Suppose the thermocouple is measuring 200°C. Again, the circuit board is in an area where the ambient temperature can vary from 15°C to 35°C. The reference thermocouple is held at exactly 0°C. Determine the voltage across the circuit input terminals at the ambient temperature extremes.

Solution From Table 11–2 in Example 11–2, the thermocouple voltage is 0 V at 0°C. Since the reference thermocouple produces no voltage at 0°C and is completely independent of ambient temperature, there is no error in the measured voltage over the ambient temperature range. Therefore, the voltage across the circuit input terminals at both temperature extremes equals the measuring thermocouple voltage, which is **9.286 mV**.

Practice Exercise If the reference thermocouple were held at −100°C instead of 0°C, what would be the voltage across the circuit input terminals if the measuring thermocouple were at 400°C?

Compensation It is bulky and expensive to maintain a reference thermocouple at a fixed temperature (usually an ice bath is required). Another approach is to compensate for the unwanted thermocouple effect by adding a compensation circuit as shown in Figure 11–24. This is sometimes referred to as *cold-junction compensation*. The compensation circuit consists of a resistor and an integrated circuit temperature sensor with a temperature coefficient that matches that of the unwanted thermocouple.

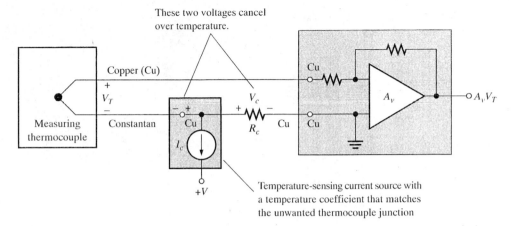

FIGURE 11–24
Simplified circuit for compensation of unwanted thermocouple effect.

The current source in the temperature sensor produces a current that creates a voltage drop, V_c, across the compensation resistor, R_c. The resistance is adjusted so that this voltage drop is equal and opposite the voltage produced by the unwanted thermocouple at a given temperature. When the ambient temperature changes, the current changes proportionally, so that the voltage across the compensation resistor is always approximately equal to the unwanted thermocouple voltage. Since the compensation voltage, V_c, is opposite in polarity to the unwanted thermocouple voltage, the unwanted voltage is effectively cancelled.

A Thermocouple Signal Conditioner

The functions shown in the circuit of Figure 11–24 plus others are available in single-package hybrid IC circuits, known as *thermocouple signal conditioners*. The 2B50 is a good example of this type of circuit. It is designed for interfacing a thermocouple with various types of electronic systems and provides gain, compensation, isolation, common-mode rejection, and other features in one package. A diagram showing the simplified internal structure package pin numbers is shown in Figure 11–25(a), and the physical package configuration is shown in Figure 11–25(b).

A basic description of the 2B50 shown in Figure 11–26 follows. The thermocouple can be connected directly to the two screw terminals or to circuit board pins 1 and 2. An input filtering and protection circuit precedes the op-amp. The voltage gain can be set anywhere from 50 to 1000 by an external resistor, R_G, between pins 3 and 5 as shown. The internal isolated power supply (part of the isolation amplifier) provides convenient dc voltages for use with external input offset adjustment as shown in the figure. The cold-junction compensation circuit provides for compensation of the thermocouple connection at the circuit input terminals. Compensating circuits for J-, K-, and T-type thermocouples are built into the device and are configured using connections to pins 41 or 42. (Refer to the pin designations that were shown in part (a) of Figure 11–25.) An external resistor connected to the X input (pin 40) configures the circuit for other types of thermocouples. The value of R_X is selected for the particular type of thermocouple based on the manufacturer's recommendation. The op-amp is followed by an isolation amplifier whose output is filtered

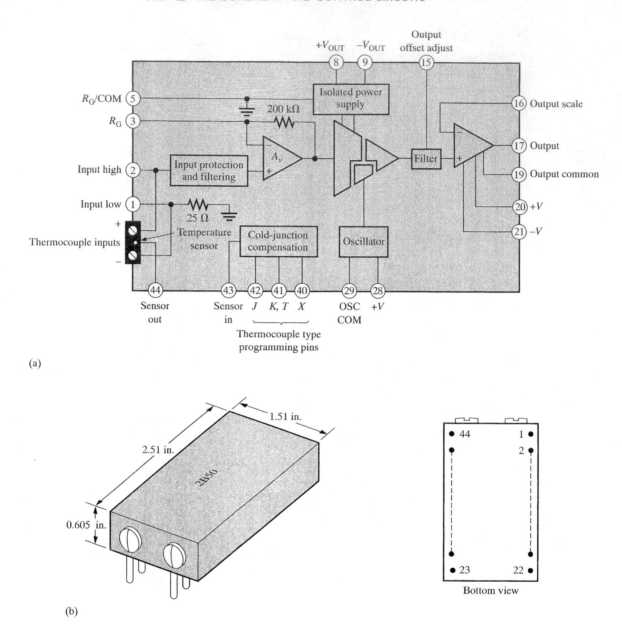

(a)

(b)

FIGURE 11–25
The 2B50 thermocouple signal conditioner.

and is applied to an output op-amp. If the output scale (pin 16) is connected to the output (pin 17), the output op-amp is a voltage-follower and the overall gain of the 2B50 is equal to the gain set by R_G. In this configuration, the full-scale output range is ±5 V. If a greater output range is required, the overall gain can be doubled by connecting external resistors as shown in Figure 11–26. This gives the output op-amp a gain of two and an output range of ±10 V. The oscillator shown is associated with the isolation amplifier (refer to Chapter 8).

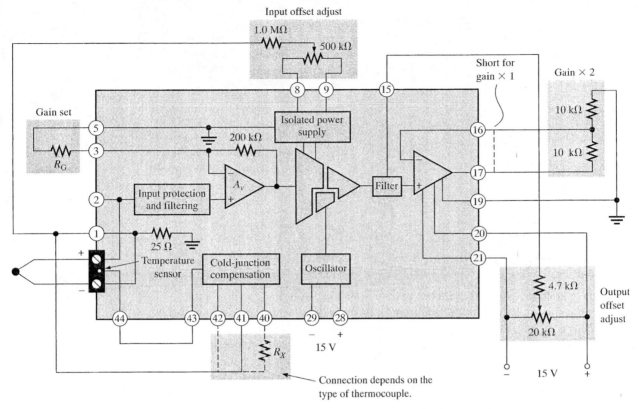

FIGURE 11–26
The 2B50 thermocouple signal conditioner externally configured for a typical application.

EXAMPLE 11–4

Determine the overall gain and the type of thermocouple in the signal conditioner of Figure 11–27.

Solution The input op-amp is connected in a noninverting configuration. Recall from Chapter 2 that the voltage gain of this type of configuration is

$$A_v = \frac{R_f}{R_i} + 1$$

In this case, R_f is the 200 kΩ internal resistor and R_i is the 1.0 kΩ external gain resistor, R_G. The gain of the input op-amp is, therefore,

$$A_v = \frac{200 \text{ k}\Omega}{1 \text{ k}\Omega} + 1 = 201$$

The output op-amp is connected as a voltage-follower; its gain is 1. The overall gain of the device is

$$A_v \times 1 = 201$$

The external connection from pin 1 to pin 42 indicates that this device is set up for a type-*J* thermocouple.

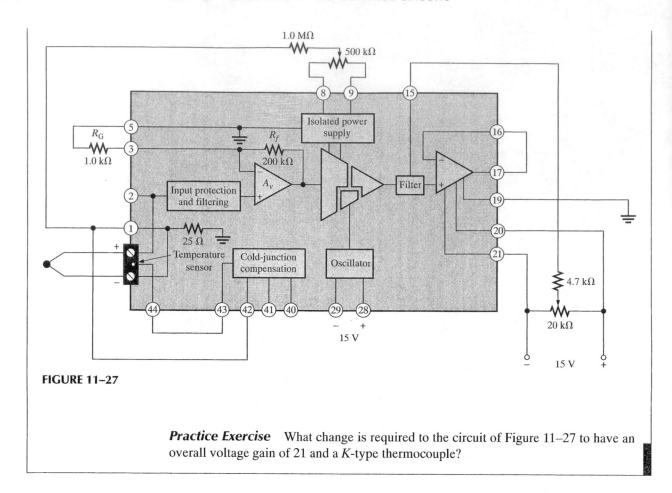

FIGURE 11-27

Practice Exercise   What change is required to the circuit of Figure 11–27 to have an overall voltage gain of 21 and a *K*-type thermocouple?

Resistance Temperature Detectors (RTDs)

A second major type of temperature transducer is the **resistance temperature detector (RTD).** The RTD is a resistive device in which the resistance changes directly with temperature (positive temperature coefficient). The RTD is more nearly linear than the thermocouple. RTDs are constructed in either a wire-wound configuration or by a metal-film technique. The most common RTDs are made of either platinum, nickel, or nickel alloys.

Generally, RTDs are used to sense temperature in two basic ways. First, as shown in Figure 11–28(a), the RTD is driven by a current source and, since the current is constant, the change in voltage across it is proportional (by Ohm's law) to the change in its resistance with temperature. Second, as shown in Figure 11–28(b), the RTD is connected in a 3-wire bridge circuit; and the bridge output voltage is used to sense the change in the RTD resistance and, thus, the temperature.

Theory of the 3-wire Bridge   To avoid subjecting the three bridge resistors to the same temperature that the RTD is sensing, the RTD is usually remotely located to the point where temperature variations are to be measured and connected to the rest of the bridge by long wires. The resistance of the three bridge resistors must remain constant. The long extension wires to the RTD have resistance that can affect the accurate operation of the bridge.

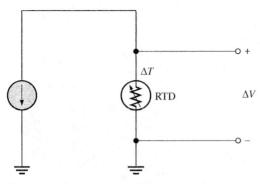

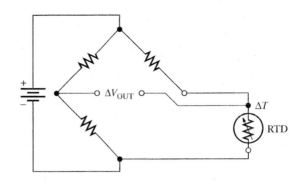

(a) A change in temperature, ΔT, produces a change in voltage, ΔV, across the RTD proportional to the change in RTD resistance when the current is constant.

(b) A change in temperature, ΔT, produces a change in bridge output voltage, ΔV, proportional to the change in RTD resistance.

FIGURE 11–28

Basic methods of employing an RTD in a temperature-sensing circuit.

Figure 11–29(a) shows the RTD connected in the bridge with a 2-wire configuration. Notice that the resistance of both of the long connecting wires appear in the same leg of the bridge as the RTD. Recall from your study of basic circuits that $V_{OUT} = 0$ V and the bridge is balanced when $R_{RTD} = R_3$ if $R_1 = R_2$. The wire resistances will throw the bridge off balance when $R_{RTD} = R_3$ and will cause an error in the output voltage for any value of the RTD resistance because they are in series with the RTD in the same leg of the bridge.

The 3-wire configuration in Figure 11–29(b) overcomes the wire resistance problem. By connecting a third wire to one end of the RTD as shown, the resistance of wire A is now placed in the same leg of the bridge as R_3 and the resistance of wire B is placed in the same leg of the bridge as the RTD. Because the wire resistances are now in opposite legs of the bridge, their effects will cancel if both wire resistances are the same (equal lengths of same type of wire). The resistance of the third wire has no effect; essentially no current goes through it because the output terminals of the bridge are open or are connected across a very high impedance. The balance condition is expressed as

$$R_{RTD} + R_B = R_3 + R_A$$

If $R_A = R_B$, then they cancel in the equation and the balance condition is completely independent of the wire resistances.

$$R_{RTD} = R_3$$

The method described here is important in many measurements that use a sensitive transducer and a bridge. It is often used in strain-gage measurements (described in Section 11–4).

Basic RTD Temperature-Sensing Circuits

Two simplified RTD measurement circuits are shown in Figure 11–30. The circuit in part (a) is one implementation of an RTD driven by a constant current. The operation is as follows. From your study of basic op-amp circuits, recall that the input current and the current through the feedback path are essentially equal because the input impedance of the op-amp is ideally infinite. Therefore, the constant current through the RTD is set by the constant input voltage, V_{IN}, and the input resistance, R_1, because the inverting input is at virtual

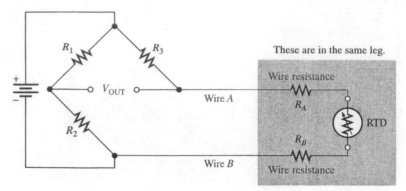

(a) Two-wire bridge connection

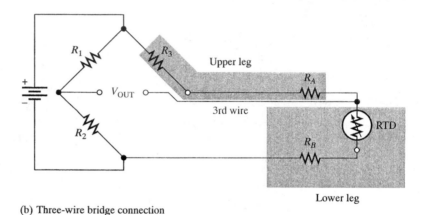

(b) Three-wire bridge connection

FIGURE 11–29
Comparison of 2-wire and 3-wire bridge connections in an RTD circuit.

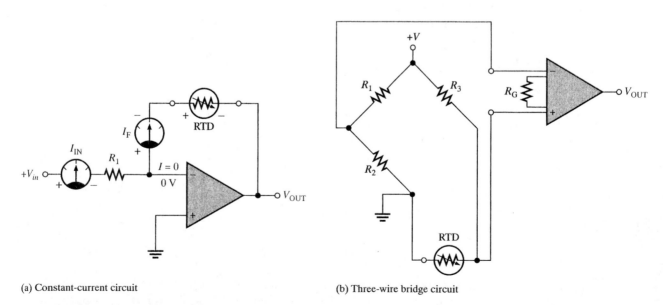

(a) Constant-current circuit

(b) Three-wire bridge circuit

FIGURE 11–30
Basic RTD temperature-measuring circuits.

ground. The RTD is in the feedback path and, therefore, the output voltage of the op-amp is equal to the voltage across the RTD. As the resistance of the RTD changes with temperature, the voltage across the RTD also changes because the current is constant.

The circuit in Figure 11–30(b) shows a basic circuit in which an instrumentation amplifier is used to amplify the voltage across the 3-wire bridge circuit. The RTD forms one leg of the bridge; and as its resistance changes with temperature, the bridge output voltage also changes proportionally. The bridge is adjusted for balance (V_{OUT} = 0 V) at some reference temperature, say 0°C. This means that R_3, is selected to equal the resistance of the RTD at this reference temperature.

EXAMPLE 11–5

Determine the output voltage of the instrumentation amplifier in the RTD circuit in Figure 11–31 if the resistance of the RTD is 1320 Ω at the temperature being measured.

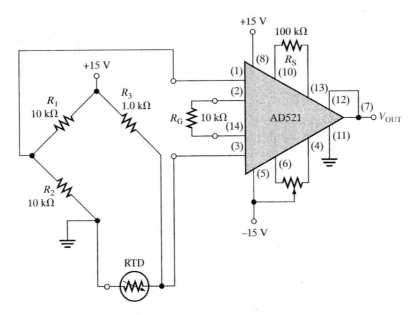

FIGURE 11–31

Solution The bridge output voltage is

$$V_{OUT} = \left(\frac{R_{RTD}}{R_3 + R_{RTD}}\right)15\text{ V} - \left(\frac{R_2}{R_1 + R_2}\right)15\text{ V} = \left(\frac{1320\text{ }\Omega}{2320\text{ }\Omega}\right)15\text{ V} - \left(\frac{10\text{ k}\Omega}{20\text{ k}\Omega}\right)15\text{ V}$$

$$= 8.53\text{ V} - 7.5\text{ V} = 1.03\text{ V}$$

Using Equation (8–3), the voltage gain of the AD521 instrumentation amplifier is

$$A_v = \frac{R_S}{R_G} = \frac{100\text{ k}\Omega}{10\text{ k}\Omega} = 10$$

The output voltage from the amplifier is

$$V_{OUT} = (10)(1.03\text{ V}) = \textbf{10.3 V}$$

Practice Exercise What must be the nominal resistance of the RTD in Figure 11–31 to balance the bridge at 25°C? What is the amplifier output voltage when the bridge is balanced?

An RTD Signal Conditioner

The 2B31 is a good example of a hybrid integrated circuit device that provides for either 3-wire bridge or current-source RTD measurements. Figure 11–32 is a simplified diagram of the 2B31. It includes an input instrumentation amplifier followed by a buffer amplifier that drives an active 2 kHz low-pass filter having a Bessel characteristic. The amplifier gain is set by external resistors, and the filter response can be externally adjusted up to 5 kHz. The filter eliminates 60 Hz or any noise picked up on the long RTD lines. The adjustable bridge excitation circuit is basically a precision dc power supply used to generate accurate bridge voltage or a constant current depending on the RTD sensing method.

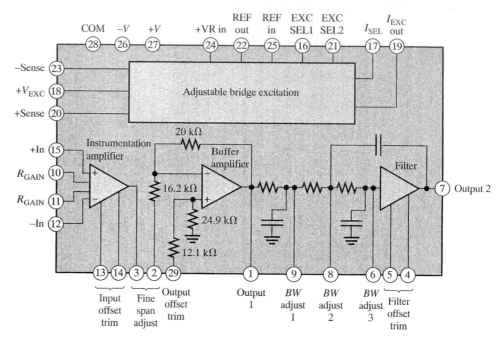

FIGURE 11–32
The 2B31 RTD signal conditioner.

Typical connections for the signal conditioner are shown in Figure 11–33. Part (a) is a constant-current configuration. The bridge excitation circuit is connected to operate as a constant-current source to supply current to the external RTD, and the resulting voltage across the RTD is applied to the inputs of the device. Voltage gain is set by R_G and R_F. Also, input and output offset adjustments can be implemented but are left off for simplicity.

Figure 11–33(b) shows a 3-wire RTD bridge connection. In this case, the bridge excitation circuit is connected to operate as a precision voltage supply. The bridge voltage can be externally adjusted by the value of the external resistor.

Voltage Gain of the 2B31   The voltage gain of the 2B31 is determined by the gain of the instrumentation amplifier and the gain of the buffer amplifier that follows it. The voltage gain for the instrumentation amplifier in this particular device is determined by a 94 kΩ internal resistance and the external resistor R_G.

$$A_{v(inst)} = \frac{94 \text{ k}\Omega}{R_G} + 1$$

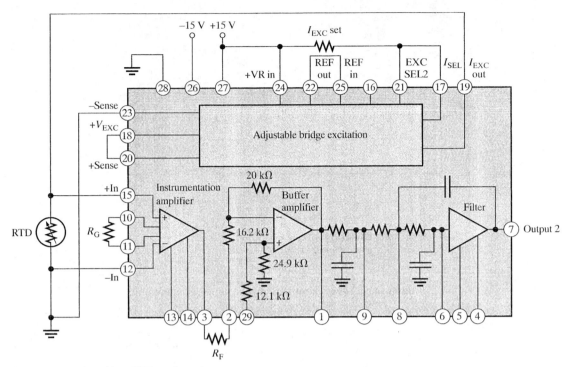

(a) Constant-current driven RTD configuration

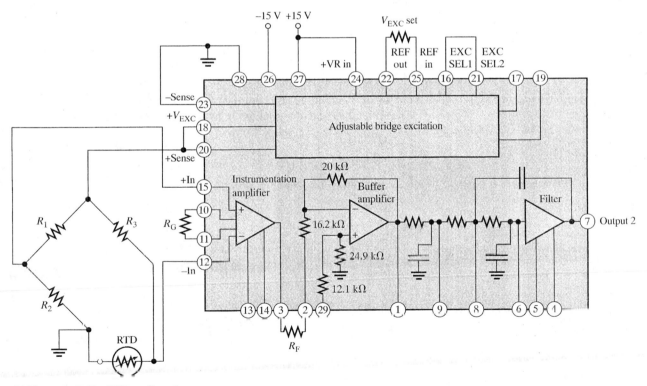

(b) Three-wire bridge RTD configuration

FIGURE 11–33

Two basic RTD temperature-measuring circuits using the 2B31 signal conditioner.

The buffer amplifier is an inverting op-amp configuration. Its gain is determined by the 16.2 kΩ and the 20 kΩ internal resistors and the external resistor R_F.

The active filter circuit has a unity gain. Therefore, the overall voltage gain of the device is

$$A_v = \left(\frac{94 \text{ k}\Omega}{R_G} + 1\right)\left(\frac{20 \text{ k}\Omega}{R_F + 16.2 \text{ k}\Omega}\right)$$

EXAMPLE 11–6

Determine the voltage gain of a 2B31 for $R_G = 4.7$ kΩ and $R_F = 10$ kΩ.

Solution
$$A_v = \left(\frac{94 \text{ k}\Omega}{4.7 \text{ k}\Omega} + 1\right)\left(\frac{20 \text{ k}\Omega}{10 \text{ k}\Omega + 16.2 \text{ k}\Omega}\right) = (21)(0.763) = \mathbf{16.0}$$

Practice Exercise   If pins 2 and 3 of the 2B31 are shorted together, how is the gain affected for $R_G = 4.7$ kΩ?

Thermistors

A third major type of temperature transducer is the **thermistor,** which is a resistive device made from a semiconductive material such as nickel oxide or cobalt oxide. The resistance of a thermistor changes inversely with temperature (negative temperature coefficient). The temperature characteristic for thermistors is more nonlinear than that for thermocouples or RTDs; in fact, a thermistor's temperature characteristic is essentially logarithmic. Also, like the RTD, the temperature range of a thermistor is more limited than that of a thermocouple. Thermistors have the advantage of a greater sensitivity than either thermocouples or RTDs and are generally less expensive. This means that their change in resistance per degree change in temperature is greater. Since they are both variable-resistance devices, the thermistor and the RTD can be used in similar circuits.

Like the RTD, thermistors can be used in constant-current-driven configurations or in bridges. In Figure 11–34, the general response of a thermistor in a constant-current

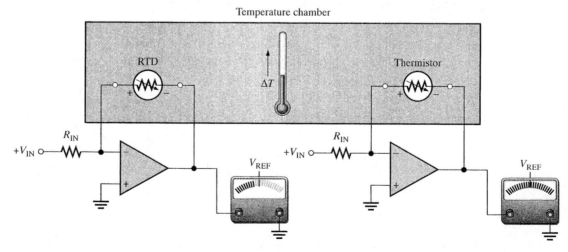

FIGURE 11–34
General comparison of the responses of a thermistor circuit to a similar RTD circuit.

op-amp circuit is compared to that of an RTD in a similar circuit. Both the RTD and the thermistor are exposed to the same temperature environment as indicated. It is assumed that at some reference temperature, the RTD and the thermistor have the same resistance and produce the same output voltage. In the RTD circuit, as the temperature increases from the reference value (becomes more negative), the op-amp's output voltage decreases from the reference value because the resistance of the RTD increases. In the thermistor circuit, as the temperature increases, the op-amp's output voltage increases from the reference value (becomes less negative) because the thermistor's resistance decreases due to its negative temperature coefficient. Also, for the same temperature change, the change in the output voltage of the thermistor circuit is greater than the corresponding change in the output voltage of the RTD circuit because of the greater sensitivity of the thermistor.

11–3 REVIEW QUESTIONS

1. What is a thermocouple?
2. How can temperature be measured with a thermocouple?
3. What is an RTD and how does its operation differ from a thermocouple?
4. What is the primary operational difference between an RTD and a thermistor?
5. Of the three devices introduced in this section, which one would most likely be used to measure extremely high temperatures?

11–4 ■ STRAIN-MEASURING, PRESSURE-MEASURING, AND MOTION-MEASURING CIRCUITS

In this section, methods of measuring three types of force-related parameters (strain, pressure, and motion) are examined. A variety of applications require the measurement of these three parameters. Also, other parameters, such as the flow rate of a fluid, can be measured indirectly by measuring strain, pressure, or motion.

After completing this section, you should be able to

❑ Describe methods of measuring strain, pressure, and motion
   ❑ Explain how a strain gage operates
   ❑ Discuss strain gage circuits
   ❑ Explain how pressure transducers work
   ❑ Discuss pressure-measuring circuits
   ❑ List several pressure transducer applications
   ❑ Explain displacement transducers, velocity transducers, and acceleration transducers

The Strain Gage

Strain is the deformation, either expansion or compression, of a material due to a force acting on it. For example, a metal rod or bar will lengthen slightly when an appropriate force is applied as illustrated in Figure 11–35(a). Also, if a metal plate is bent, there is an expansion of the upper surface, called *tensile strain,* and a compression of the lower surface, called *compressive strain,* as shown in Figure 11–35(b).

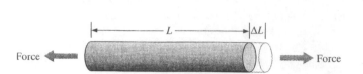

(a) Strain occurs as length changes from L to $L + \Delta L$ when force is applied.

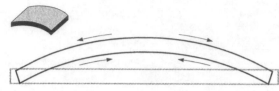

(b) Strain occurs when the flat plate is bent, causing the upper surface to expand and the lower surface to contract.

FIGURE 11–35
Examples of strain.

Strain gages are based on the principle that the resistance of a material increases if its length increases and decreases if its length decreases. This is expressed by the following formula (which you should recall from your dc/ac circuits course).

$$R = \frac{\rho L}{A} \qquad\qquad (11\text{–}3)$$

This formula states that the resistance of a material, such as a length of wire, depends directly on the resistivity (ρ) and the length (L) and inversely on the cross-sectional area (A).

A **strain gage** is basically a long very thin strip of resistive material that is bonded to the surface of an object on which strain is to be measured, such as a wing or tail section of an airplane under test. When a force acts on the object to cause a slight elongation, the strain gage also lengthens proportionally and its resistance increases. Most strain gages are formed in a pattern similar to that in Figure 11–36(a) to achieve enough length for a sufficient resistance value in a smaller area. It is then placed along the line of strain as indicated in Figure 11–36(b).

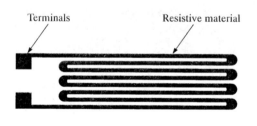

(a) Typical strain gage configuration.

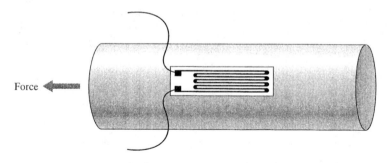

(b) The strain gage is bonded to the surface to be measured along the line of force. When the surface lengthens, the strain gage stretches.

FIGURE 11–36
A simple strain gage and its placement.

The Gage Factor of a Strain Gage

An important characteristic of strain gages is the **gage factor (*GF*),** which is defined as the ratio of the fractional change in resistance to the fractional change in length along the axis of the gage. For metallic strain gages, the *GF*s are typically around 2. The concept of gage factor is illustrated in Figure 11–37 and expressed in Equation (11–4) where R is the nominal resistance and ΔR is the change in re-

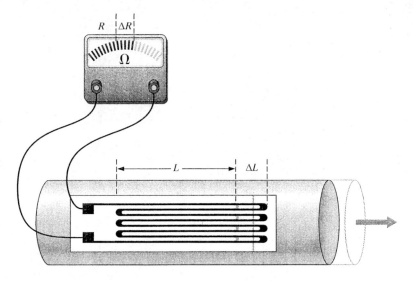

FIGURE 11–37
Illustration of gage factor. The ohmmeter symbol is not intended to represent a practical method for measuring ΔR.

sistance due to strain. The fractional change in length ($\Delta L/L$) is designated strain (ϵ) and is usually expressed in parts per million, called *microstrain* (designated $\mu\epsilon$).

$$GF = \frac{\Delta R/R}{\Delta L/L} \qquad\qquad (11\text{–}4)$$

EXAMPLE 11–7

A certain material being measured under stress undergoes a strain of 5 parts per million (5 $\mu\epsilon$). The strain gage has a nominal (unstrained) resistance of 320 Ω and a gage factor of 2.0. Determine the resistance change in the strain gage.

Solution

$$GF = \frac{\Delta R/R}{\Delta L/L} = \frac{\Delta R/R}{\epsilon}$$

$$\Delta R = (GF)(R)(\epsilon) = 2.0(320\ \Omega)(5 \times 10^{-6}) = \textbf{3.2 m}\boldsymbol{\Omega}$$

Practice Exercise If the strain in this example is 8 $\mu\epsilon$, how much does the resistance change?

Basic Strain Gage Circuits

Because a strain gage exhibits a resistance change when the quantity it is sensing changes, it is typically used in circuits similar to those used for RTDs. The basic difference is that strain instead of temperature is being measured. Therefore, strain gages are usually applied in bridge circuits or in constant-current-driven circuits, as shown in Figure 11–38, just as RTDs and thermistors are. In fact, the 2B31 signal conditioner is designed for use with both strain gages and RTDs. Refer to Section 11–3 for the operation of the basic circuits.

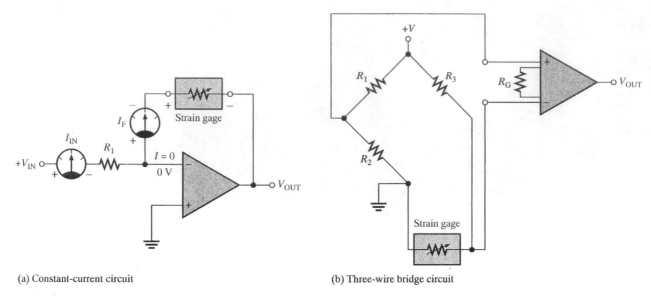

(a) Constant-current circuit   (b) Three-wire bridge circuit

FIGURE 11–38
Basic strain-measuring circuits.

Pressure Transducers

Pressure transducers are devices that exhibit a change in resistance proportional to a change in pressure. Basically, pressure sensing is accomplished using a strain gage bonded to a flexible diaphragm as shown in Figure 11–39(a). Figure 11–39(b) shows the diaphragm with no net pressure exerted on it. When a net positive pressure exists on one side of the diaphragm, as shown in Figure 11–39(c), the diaphragm is pushed upward and its surface expands. This expansion causes the strain gage to lengthen and its resistance to increase.

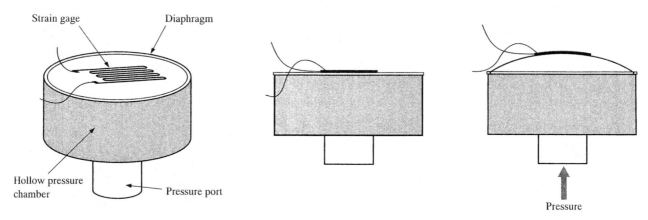

(a) Basic pressure gage construction

(b) With no net pressure on diaphragm, strain gage resistance is at its nominal value (side view).

(c) Net pressure forces diaphragm to expand, causing elongation of the strain gage and thus an increase in its resistance.

FIGURE 11–39
A simplified pressure sensor constructed with a strain gage bonded to a flexible diaphragm.

Pressure transducers typically are manufactured using a foil strain gage bonded to a stainless steel diaphragm or by integrating semiconductor strain gages (resistors) in a silicon diaphragm. Either way, the basic principle remains the same.

Pressure transducers come in three basic configurations in terms of relative pressure measurement. The absolute pressure transducer measures applied pressure relative to a vacuum as illustrated in Figure 11–40(a). The gage pressure transducer measures applied pressure relative to the pressure of the surroundings (ambient pressure) as illustrated in Figure 11–40(b). The differential pressure transducer measures one applied pressure relative to another applied pressure as shown in Figure 11–40(c). Some transducer configurations include circuitry such as bridge completion circuits and op-amps within the same package as the sensor itself, as indicated.

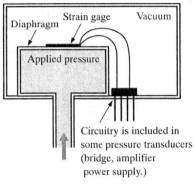

(a) Absolute pressure transducer

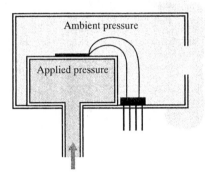

(b) Gage pressure transducer

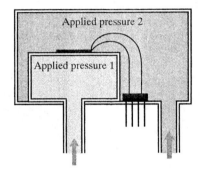

(c) Differential pressure transducer

FIGURE 11–40
Three basic types of pressure transducers.

Pressure-Measuring Circuits

Because pressure transducers are devices in which the resistance changes with the quantity being measured, they are usually in a bridge configuration as shown by the basic op-amp bridge circuit in Figure 11–41(a). In some cases, the complete circuitry is built into the transducer package, and in other cases the circuitry is external to the sensor. The symbols in parts (b) through (d) of Figure 11–41 are sometimes used to represent the complete pressure transducer with an amplified output. The symbol in part (b) represents the absolute pressure transducer, the symbol in part (c) represents the gage pressure transducer, and the symbol in part (d) represents the differential pressure transducer.

Flow Rate Measurement One common method of measuring the flow rate of a fluid through a pipe is the differential-pressure method. A flow restriction device such as a Venturi section (or other type of restriction such as an orifice) is placed in the flow stream. The Venturi section is formed by a narrowing of the pipe, as indicated in Figure 11–42. Although the velocity of the fluid increases as it flows through the narrow channel, the volume of fluid per minute (volumetric flow rate) is constant throughout the pipe.

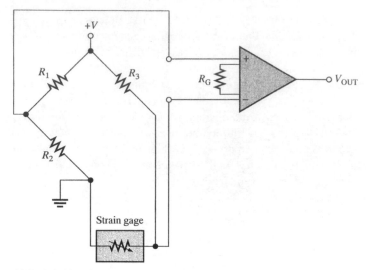

(a) Basic bridge circuit

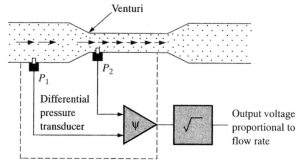

(b) Absolute pressure transducer   (c) Gage pressure transducer   (d) Differential pressure transducer

FIGURE 11–41
A basic pressure transducer circuit and symbols.

FIGURE 11–42
A basic method of flow rate measurement.

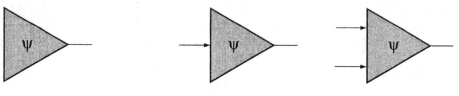

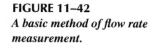

Because the velocity of the fluid increases as it goes through the restricted area, the pressure also increases. If pressure is measured at a wide point and at a narrow point, the flow rate can be determined because flow rate is proportional to the square root of the differential pressure, as shown in Figure 11–42.

Pressure Transducer Applications Pressure transducers are used anywhere there is a need to determine the pressure of a substance. In medical applications, pressure transducers are used for blood pressure measurement; in aircraft, pressure transducers are used for altitude pressure, cabin pressure, and hydraulic pressure; in automobiles, pressure transducers are used for fuel flow, oil pressure, brake line pressure, manifold pressure, and steering system pressure, to name a few applications.

Motion-Measuring Circuits

Displacement Transducers *Displacement* is a quantity that indicates the change in position of a body or point. Angular displacement refers to a rotation that can be measured in degrees or radians. Displacement transducers can be either contacting or noncontacting.

Contacting transducers typically use a sensing shaft with a coupling device to follow the position of the measured quantity. A contacting type of displacement sensor that relates a change in inductance to displacement is the linear variable differential transformer (LVDT). The sensing shaft is connected to a moving magnetic core inside a specially wound transformer. A typical LVDT is shown in Figure 11–43. The primary of the transformer is in line and located between two identical secondaries. The primary winding is excited with ac (usually in the range of 1 to 5 kHz). When the core is centered, the voltage induced in each secondary is equal. As the core moves off center, the voltage in one secondary will be greater than the other. With the demodulator circuit shown, the polarity of the output changes as the core passes the center position. The transducer has excellent sensitivity, linearity, and repeatability.

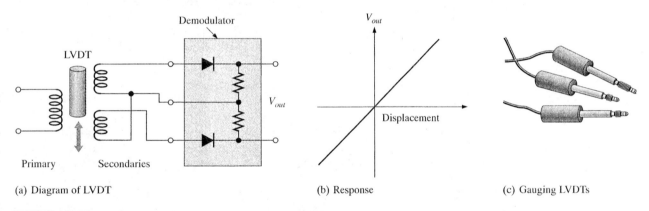

(a) Diagram of LVDT   (b) Response   (c) Gauging LVDTs

FIGURE 11–43
LVDT displacement transducers.

Noncontacting displacement transducers include optical and capacitive transducers. Photocells can be arranged to observe light through holes in an encoding disk or to count fringes painted on the surface to be measured. Optical systems are fast; but noise, including background light sources, can produce spurious signals in optical sensors. It is useful to build hysteresis into the system if noise is a problem (see Section 4–1).

Fiber-optic sensors make excellent proximity detectors for close ranges. Reflective sensors use two fiber bundles, one for transmitting light and the other for receiving light from a reflective surface, as illustrated in Figure 11–44. Light is transmitted in the fiber bundle without any significant attenuation. When it leaves the transmitting fiber bundle, it forms a spot on the target that is inversely proportional to the square of the distance. The receiving bundle is aimed at the spot and collects the reflected light to an optical sensor. The light intensity detected by the receiving bundle depends on the physical size and arrangement of the fibers as well as the distance to the spot and the reflecting surface, but

FIGURE 11–44
Fiber-optic proximity detector.

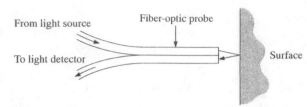

the technique can respond to distances approaching 1 microinch. The major disadvantage is limited dynamic range.

Capacitive sensors can be made into very sensitive displacement and proximity transducers. The capacitance is varied by moving one of the plates of a capacitor with respect to the second plate. The moving plate can be any metallic surface such as the diaphragm of a capacitive microphone or a surface that is being measured. The capacitor can be used to control the frequency of a resonant circuit to convert the capacitive change into a usable electrical output.

Velocity Transducers
Velocity is defined as the rate of change of displacement. It follows that velocity can be determined indirectly with a displacement sensor and measuring the time between two positions. A direct measurement of velocity is possible with certain transducers that have an output proportional to the velocity to be measured. These transducers can respond to either linear or angular velocity. Linear velocity transducers can be constructed using a permanent magnet inside a concentric coil, forming a simple motor by generating an emf proportional to the velocity. Either the coil or the magnet can be fixed and the other moved with respect to the fixed component. The output is taken from the coil.

There are a variety of transducers that are designed to measure angular velocity. Tachometers, a class of angular velocity transducers, provide a dc or ac voltage output. A dc tachometer is basically a small generator with a coil that rotates in a constant magnetic field. A voltage is induced in the coil as it rotates in the magnetic field. The average value of the induced voltage is proportional to the speed of rotation, and the polarity is indicative of the direction of rotation, an advantage with dc tachometers. AC tachometers can be designed as generators that provide an output frequency that is proportional to the rotational speed.

Another technique for measuring angular velocity is to rotate a shutter over a photosensitive element. The shutter interrupts a light source from reaching the photocells, causing the output of the photocells to vary at a rate proportional to the rotational speed.

Acceleration Transducers
Acceleration is usually measured by use of a spring-supported seismic mass, mounted in a suitable enclosure as shown in Figure 11–45. Damping is provided by a dashpot, which is a mechanical device to reduce the vibration. The relative motion between the case and the mass is proportional to the acceleration. A secondary transducer such as a resistive displacement transducer or an LVDT is used to convert the relative motion to an electrical output. Ideally, the mass does not move when the case accelerates because of its inertia; in practice, it does because of forces applied to it through the spring. The accelerometer has a natural frequency, the period of which should be shorter than the time required for the measured acceleration to change. Accelerometers used to measure vibration should also be used at frequencies less than the natural frequency.

An accelerometer that uses the basic principle of the LVDT can be constructed to measure vibration. The mass is made from a magnet that is surrounded with a coil. Voltage induced in the coil is a function of the acceleration.

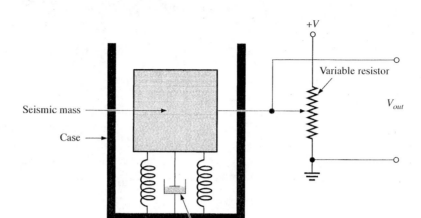

FIGURE 11–45
A basic accelerometer. Motion is converted to a voltage by the variable resistor.

Another type of accelerometer uses a piezoelectric crystal in contact with the seismic mass. The crystal generates an output voltage in response to forces induced by the acceleration of the mass. Piezoelectric crystals are small in size and have a natural frequency that is very high; they can be used to measure high-frequency vibration. The drawback to piezoelectric crystals is that the output is very low and the impedance of the crystal is high, making it subject to problems from noise.

11–4 REVIEW QUESTIONS

1. Describe a basic strain gage.
2. Describe a basic pressure gage.
3. List three types of pressure gages.
4. **(a)** What is an LVDT? **(b)** What does it measure?

11–5 ■ POWER-CONTROL CIRCUITS

Many types of integrated circuits are used for controlling power to a load. In this section, you will learn about the zero-voltage switch that is used for driving an SCR or triac to control the amount of power to a load. The main advantage of the zero-voltage switching method is the elimination of large switching transients that occur when an SCR or triac is turned on near the peak of the ac voltage. Zero-voltage switches are used in the control of heaters, lamps, valves, motors, to name a few applications.

After completing this section, you should be able to

❑ Describe the operation of the zero-voltage switch
   ❑ Discuss the CA3079 zero-voltage switch

The Zero-Voltage Switch

A zero-voltage switch is used to control the amount of power to a load and to switch the power on only when the line voltage crosses the zero-voltage axis. This operation is illustrated in Figure 11–46 where the zero-voltage switch triggers an SCR in order to turn power on to a load. An SCR is turned on by a trigger pulse at the gate and turns off when the current falls to a low value. The load might be a resistive heating element, and the power is typically turned on for several cycles of the ac and then turned off for several cycles to maintain a certain temperature. The zero-voltage switch uses a sensing circuit to determine when to turn power on. In the case of a heat-control circuit, the sensing can be accomplished by a thermistor in a bridge network.

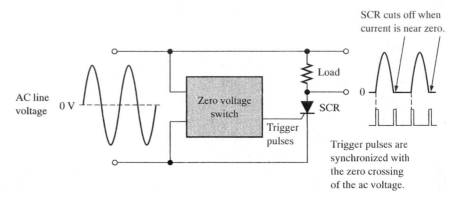

FIGURE 11–46
Basic operation of a zero-voltage switch.

The Reason for Switching at Zero Voltage The key feature of a zero-voltage switch is that it generates trigger pulses for the control device only at the points where the ac line voltage crosses zero. The main reason for synchronizing the switching with the zero crossings is to prevent **radio frequency interference (RFI).**

As demonstrated in Figure 11–47, if the SCR or triac were switched on somewhere near the peak of the ac cycle, for example, there would be a sudden inrush of current to the load. When there is a sudden transition of voltage or current, many high-frequency components are generated. (The rising and falling edges of a pulse contain high frequencies.) By switching the SCR or triac on when the voltage across it is zero, the sudden increase in current is prevented because the current will increase sinusoidally with the ac voltage. Zero-voltage switching also prevents thermal shock to the load which, depending on the type of load, may shorten its life.

A Specific Zero-Voltage Switch

The CA3079 is an example of an IC zero-voltage switch. A block diagram of this device is shown in Figure 11–48. The limiter and the power supply allow the CA3079 to be operated directly from the ac power line with no external dc power required. The zero-crossing detector determines when the ac voltage crosses zero and produces synchronized pulses at each zero crossing. The on/off sensing amplifier, which is basically a comparator, accepts inputs from external sensing circuits (such as a temperature sensor) and provides an output to an AND gate that either enables or inhibits the zero-crossing pulses. The driver provides

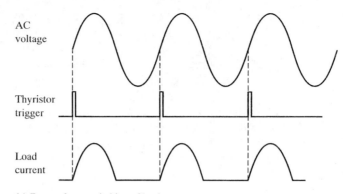

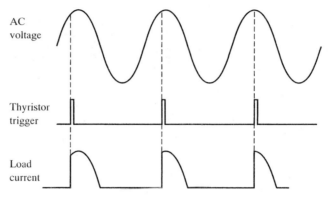

(a) Zero-voltage switching of load current

(b) Unsynchronized switching of load current produces current transients
    that cause RFI.

FIGURE 11–47

Comparison of zero-voltage switching to unsynchronized switching of power to a load.

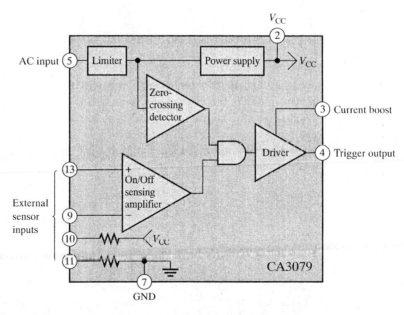

FIGURE 11–48

The CA3079 zero-voltage switch.

sufficient current to the gate of an external thyristor. The amount of drive current can be increased by connecting the current boost pin to the dc supply voltage pin.

A CA3079 connected for a typical application is shown in Figure 11–49. The ac line voltage is applied across the triac and load and to the input of the CA3079A through a limiting resistor, R_{in}. The manufacturer recommends a 2 W, 10 kΩ resistor for R_{in} if the ac voltage is 120 V rms. A bridge circuit with a thermistor in one leg is the temperature sensor in this particular application. The variable resistor in the bridge, R_{set}, adjusts the temperature level. Notice that the other part of the bridge is formed by the internal resistors R_1 and R_2, which are connected using pins 9, 10, and 11. The output circuit in this case drives a triac which, when turned on, conducts current through a resistive heating element.

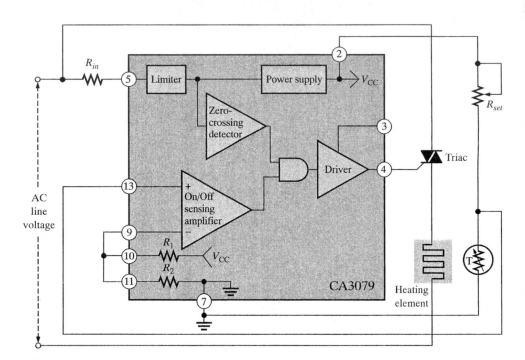

FIGURE 11–49
An example of a CA3079 application as a temperature controller.

As long as the temperature of the heating element is below the set point, the triac is triggered on at each zero crossing of the ac voltage and turns off near the end of each half cycle when the load current drops to near zero. The load continues to get current on each half-cycle until the increasing temperature causes the thermistor resistance to decrease to a value where the bridge voltage at pin 13 is less than the voltage at pin 9. At this point, the output of the on/off sensing amplifier goes low and inhibits the AND gate, thus preventing trigger pulses from getting to the triac. The triac remains off until the temperature of the load drops enough for the thermistor resistance to increase back to a point where the voltage at pin 13 is greater than the voltage at pin 9. At this point, the on/off sensing amplifier switches back to its high output state and enables the AND gate so that the trigger pulses again turn on the triac. Examine the circuit in Figure 11–49 carefully until you see how the input bridge circuit is formed.

11-5 REVIEW QUESTIONS

1. Explain the basic purpose in zero-voltage switching.
2. How does an SCR differ from a triac in terms of delivering power to a load in a circuit such as in Figure 11–49?

11-6 ■ A SYSTEM APPLICATION

The wind speed and direction measurement system presented at the beginning of the chapter is a representation of a type of instrument that is typically found at a meteorological data-gathering facility. This system is actually two systems in one because it measures two parameters, wind speed and wind direction, independently. In this system, you will find circuits that you learned about in this chapter and some that were studied in previous chapters. The measurement of wind direction was briefly mentioned in Section 11–2.

After completing this section, you should be able to

❑ Apply what you have learned in this chapter to a system application
   ❑ Explain how resolvers and RDCs are used to measure wind direction
   ❑ Descibe how wind speed can be measured
   ❑ Translate between the printed circuit board and a schematic
   ❑ Analyze the measurement circuit board
   ❑ Troubleshoot some common problems

A Brief Description of the System

As you can see in the system block diagram in Figure 11–50, there are two transducers—the anemometer flow meter and the resolver. The flow meter used in this system is basically a propeller-type instrument in which the blades revolve as the wind blows across them. The faster the wind blows, the faster the blades revolve. A magnetic sensor detects each time one revolution is completed and produces a short duration pulse that triggers the 555 one-shot. The frequency of the pulse train produced by the one-shot increases as the wind speed increases. A frequency-to-voltage converter produces an output voltage that is proportional to the frequency and thus the wind speed. This voltage is converted to digital form and the resulting digital code goes to the microprocessor, which translates it to a binary number corresponding to the wind speed and produces a digital readout. A resolver and a resolver-to-digital converter (RDC) are used to measure the wind direction. The digital output of the RDC goes to the microprocessor where it is translated to an appropriate binary number and displayed.

As indicated in Figure 11–50, one circuit board in this system contains the measurement circuitry and another board contains the microprocessor, display circuitry, and power supply. Our focus in this section is on the measurement circuit board.

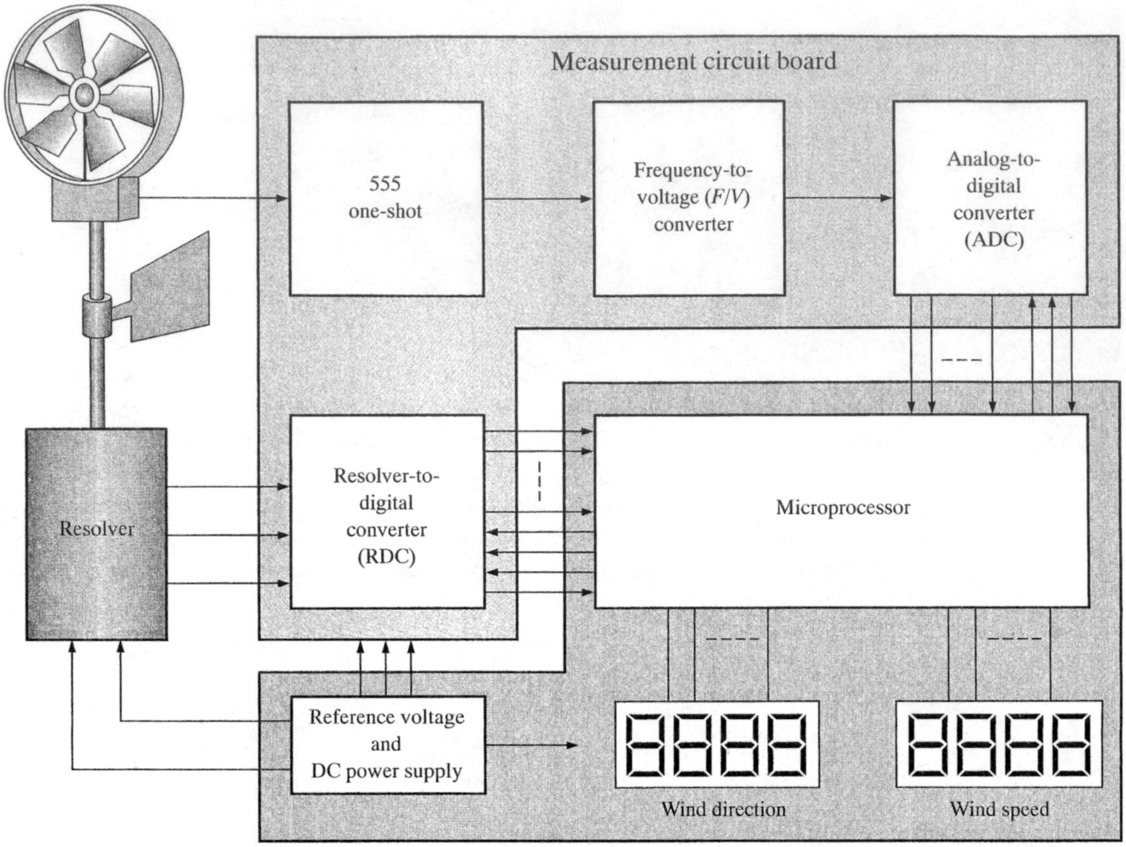

FIGURE 11–50
Block diagram of the wind-measuring system.

Now, so that you can take a closer look at the measurement circuit board, let's take it out of the system and put it on the test bench.

ON THE TEST BENCH

■ ACTIVITY 1 Relate the PC Board to the Schematic

Locate and identify each component and each input/output pin on the PC board in Figure 11–51 after all of the inputs and outputs on the schematic in Figure 11–52 have been identified. Verify that the board and the schematic agree.

■ ACTIVITY 2 Write a Technical Report

Describe the overall operation of the measurement circuit board. Specify how each circuit works and its purpose.

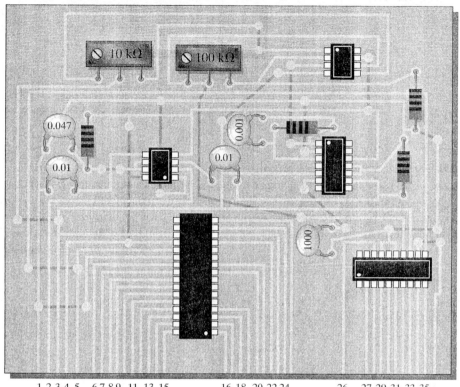

FIGURE 11–51

■ **ACTIVITY 3   Troubleshoot the Circuit Board for Each of the Following Problems by Stating the Probable Cause or Causes**

1. No pulses out of the one-shot.

2. There is a 100 mV level out of the F/V converter, but the output of the ADC indicates zero.

3. There are pulses out of the one-shot but no voltage out of the F/V converter.

4. For one complete revolution of the resolver shaft, the maximum angle represented by the RDC output code is 180.

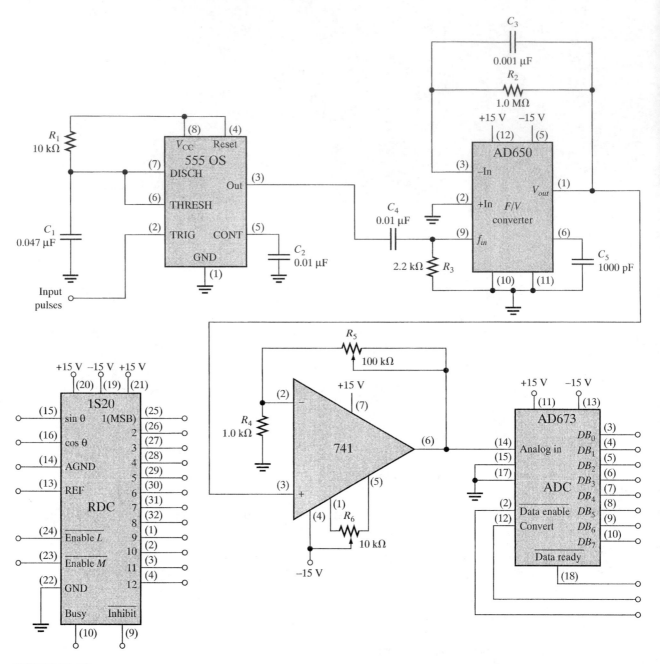

FIGURE 11–52

11–6 REVIEW QUESTIONS

1. Which components determine the pulse width of the one-shot?
2. What is the purpose of the 741 op-amp circuit?

■ SUMMARY

- An rms-to-dc converter performs three basic functions: squaring, averaging, and taking the square root.
- Squaring is usually implemented with a linear multiplier.
- A simple averaging circuit is a low-pass filter that passes only the dc component of the input.
- A square root circuit utilizes a linear multiplier in the feedback loop of an op-amp.
- A synchro is a shaft angle transducer having three stator windings.
- A resolver is a type of synchro which, in its simplest form, has two stator windings.
- The output voltages of a synchro or resolver are called *format voltages* and are proportional to the shaft angle.
- A resolver-to-digital converter (RDC) converts resolver format voltages to a digital code that represents the angular position of the shaft.
- A thermocouple is a type of temperature transducer formed by the junction of two dissimilar metals.
- When the thermocouple junction is heated, a voltage is generated across the junction that is proportional to the temperature.
- Thermocouples can be used to measure very high temperatures.
- The resistance temperature detector (RTD) is a temperature transducer in which the resistance changes directly with temperature. It has a positive temperature coefficient.
- RTDs are typically used in bridge circuits or in constant-current circuits to measure temperature. They have a more limited temperature range than thermocouples.
- The thermistor is a temperature transducer in which the resistance changes inversely with temperature. It has a negative temperature coefficient.
- Thermistors are more sensitive than RTDs or thermocouples, but their temperature range is limited.
- The strain gage is based on the fact that the resistance of a material increases when its length increases.
- The gage factor of a strain gage is the fractional change in resistance to the fractional change in length.
- Pressure transducers are constructed with strain gages bonded to a flexible diaphragm.
- An absolute pressure transducer measures pressure relative to a vacuum.
- A gage pressure transducer measures pressure relative to ambient pressure.
- A differential pressure transducer measures one pressure relative to another pressure.
- The flow rate of a liquid can be measured using a differential pressure gage.
- A zero-voltage switch generates pulses at the zero crossings of an ac voltage for triggering a thyristor used in power control.
- Motion measuring circuits include LVDT displacement transducers, velocity transducers, and accelerometers.

■ GLOSSARY

These terms are included in the end-of-book glossary.

Cold junction   A reference thermocouple held at a fixed temperature and used for compensation in thermocouple circuits.

Gage factor (*GF*)   The ratio of the fractional change in resistance to the fractional change in length along the axis of the gage.

Mean   Average value.

Radio frequency interference (RFI) High frequencies produced when high values of current and voltage are rapidly switched on and off.

Resistance temperature detector (RTD) A type of temperature transducer in which resistance is directly proportional to temperature.

Resolver A type of synchro.

Resolver-to-digital converter (RDC) An electronic circuit that converts resolver voltages to a digital format which represents the angular position of the rotor shaft.

Root mean square (RMS) The value of an ac voltage that corresponds to a dc voltage that produces the same heating effect in a resistance.

Rotor The part of a synchro that is attached to the shaft and rotates. The rotor winding is located on the rotor.

Stator The part of a synchro that is fixed. The stator windings are located on the stator.

Strain The expansion or compression of a material caused by stress forces acting on it.

Strain gage A transducer formed by a resistive material in which a lengthening or shortening due to stress produces a proportional change in resistance.

Synchro An electromechanical transducer used for shaft angle measurement and control.

Synchro-to-digital converter (SDC) An electronic circuit that converts synchro voltages to a digital format which represents the angular position of the rotor shaft.

Thermistor A type of temperature transducer in which resistance is inversely proportional to temperature.

Thermocouple A type of temperature transducer formed by the junction of two dissimilar metals which produces a voltage proportional to temperature.

Transducer A device that converts a physical parameter into an electrical quantity.

■ KEY FORMULAS

(11–1) $V_{rms} = \sqrt{\mathrm{avg}(V_{in}^2)}$ Root-mean-square value

(11–2) $V_{OUT} = \sqrt{\overline{V_{in}^2}}$ RMS-to-dc converter output

(11–3) $R = \dfrac{\rho L}{A}$ Resistance of a material

(11–4) $GF = \dfrac{\Delta R/R}{\Delta L/L}$ Gage factor of a strain gage

---

■ SELF-TEST

1. The rms value of an ac signal is equal to
   (a) the peak value
   (b) the dc value that produces the same heating effect
   (c) the square root of the average value
   (d) answers (b) and (c)

2. An explicit type of rms-to-dc converter contains
   (a) a squaring circuit   (b) an averaging circuit
   (c) a square root circuit   (d) a squarer/divider circuit
   (e) all of the above   (f) answers (a), (b), and (c) only

3. A synchro produces
   (a) three format voltages   (b) two format voltages
   (c) one format voltage   (d) one reference voltage

4. A resolver produces
(a) three format voltages   (b) two format voltages
(c) one format voltage   (d) none of these

5. A Scott-T transformer is used for
(a) coupling the reference voltage to a synchro or resolver
(b) changing resolver format voltages to synchro format voltages
(c) changing synchro format voltages to resolver format voltages
(d) isolating the rotor winding from the stator windings

6. The output of an RDC is a
(a) sine wave with an amplitude proportional to the angular position of the resolver shaft
(b) digital code representing the angular position of the stator housing
(c) digital code representing the angular position of the resolver shaft
(d) sine wave with a frequency proportional to the angular position of the resolver shaft

7. A thermocouple
(a) produces a change in resistance for a change in temperature
(b) produces a change in voltage for a change in temperature
(c) is made of two dissimilar metals
(d) answers (b) and (c)

8. In a thermocouple circuit, where each of the thermocouple wires is connected to a copper circuit board terminal,
(a) an unwanted thermocouple is produced   (b) compensation is required
(c) a reference thermocouple must be used   (d) all of these
(e) answers (a) and (c)

9. A typical thermocouple signal conditioner includes
(a) an isolation amplifier   (b) an instrumentation amplifier
(c) cold-junction compensation   (d) none of these
(e) answers (a) and (c)

10. An RTD
(a) produces a change in resistance for a change in temperature
(b) has a negative temperature coefficient
(c) has a wider temperature range than a thermocouple
(d) all of these

11. The purpose of a 3-wire bridge is to eliminate
(a) nonlinearity of an RTD
(b) the effects of wire resistance in an RTD circuit
(c) noise from the RTD resistance
(d) none of these

12. A thermistor has
(a) less sensitivity than an RTD
(b) a greater temperature range than a thermocouple
(c) a negative temperature coefficient
(d) a positive temperature coefficient

13. Both RTDs and thermistors are used in
(a) circuits that measure resistance   (b) circuits that measure temperature
(c) bridge circuits   (d) constant-current-driven circuits
(e) answers (b), (c), and (d)   (f) answers (b) and (c) only

14. When the length of a strain gage increases,
(a) it produces more voltage   (b) its resistance increases
(c) its resistance decreases   (d) it produces an open circuit

15. A higher gage factor indicates that the strain gage is
   (a) less sensitive to a change in length
   (b) more sensitive to a change in length
   (c) has more total resistance
   (d) made of a physically larger conductor

16. Many types of pressure transducers are made with
   (a) thermistors   (b) RTDs
   (c) strain gages   (d) none of these

17. Gage pressure is measured relative to
   (a) ambient pressure   (b) a vacuum
   (c) a reference pressure

18. The flow rate of a liquid can be measured
   (a) with a string
   (b) with a temperature sensor
   (c) with an absolute pressure transducer
   (d) with a differential pressure transducer

19. Zero-voltage switching is commonly used in
   (a) determining thermocouple voltage   (b) SCR and triac power control circuits
   (c) in balanced bridge circuits   (d) RFI generation

20. A major disadvantage of unsynchronized switching of power to a load is
   (a) lack of efficiency   (b) possible damage to the thyristor
   (c) RFI generation

■ PROBLEMS

SECTION 11–1   RMS-to-DC Converters

1. A 5 V dc voltage is applied across a 1.0 kΩ resistor. To achieve the same power in the 1.0 kΩ resistor as produced by the dc voltage, what must be the rms value of a sinusoidal voltage?

2. Based on the fundamental definition of rms, determine the rms value for a symmetrical square wave with an amplitude of ± 1 V.

SECTION 11–2   Angle-Measuring Circuits

3. A certain RDC has an 8-bit digital output. What is the angle that is being measured if the output code is 10000111?

4. Repeat Problem 3 for an RDC output code of 00010101.

5. What is the purpose of the $\overline{\text{Enable } M}$ and $\overline{\text{Enable } L}$ inputs on a 1S20 RDC?

6. Explain the Direction and Velocity outputs on a 1S20 RDC.

SECTION 11–3   Temperature-Measuring Circuits

7. Three identical thermocouples are each exposed to a different temperature as follows: Thermocouple A is exposed to 450°C, thermocouple B is exposed to 420°C, and thermocouple C is exposed to 1200°C. Which thermocouple produces the most voltage?

8. You have two thermocouples. One is a K type and the other is a T type. In general, what do these letter designations tell you?

9. Determine the output voltage of the op-amp in Figure 11–53 if the thermocouple is measuring a temperature of 400°C and the circuit itself is at 25°C. Refer to Table 11–2.

10. What should be the output voltage in Problem 9 if the circuit is properly compensated?

11. Determine the overall gain of the 2B50 thermocouple signal conditioner in Figure 11–54.

12. What type of thermocouple is the 2B50 in Figure 11–54 set up for?

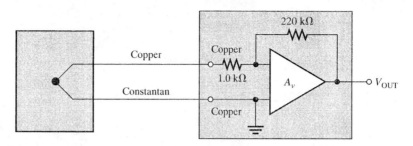

FIGURE 11–53

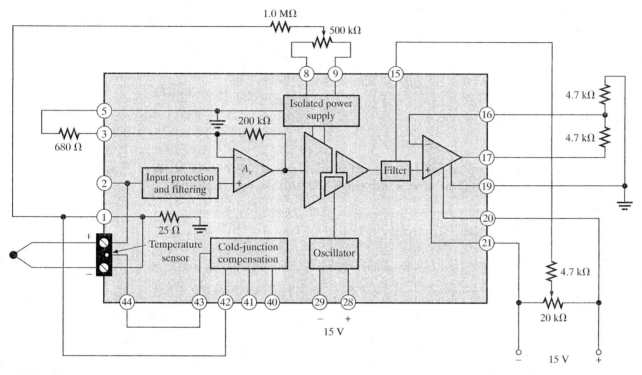

FIGURE 11–54

13. At what resistance value of the RTD will the bridge circuit in Figure 11–55 be balanced if the wires running to the RTD each have a resistance of 10 Ω?

FIGURE 11–55

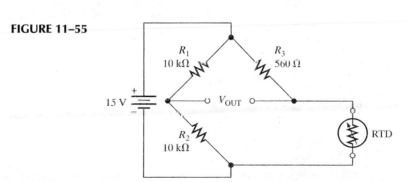

14. At what resistance value of the RTD will the bridge circuit in Figure 11–56 be balanced if the wires running to the RTD each have a resistance of 10 Ω?

15. Explain the difference in the results of Problems 13 and 14.

FIGURE 11–56

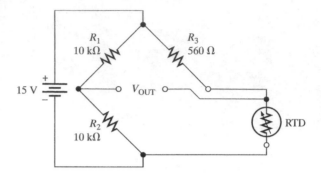

16. Determine the output voltage of the instrumentation amplifier in Figure 11–57 if the resistance of the RTD is 697 Ω at the temperature being measured.

FIGURE 11–57

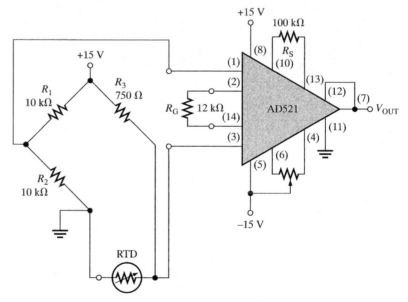

17. Determine the voltage gain of the 2B31 in Figure 11–58.

SECTION 11–4 Strain-Measuring, Pressure-Measuring, and Motion-Measuring Circuits

18. A certain material being measured undergoes a strain of 3 parts per million. The strain gage has a nominal resistance of 600 Ω and a gage factor of 2.5. Determine the resistance change in the strain gage.

19. Explain how a strain gage can be used to measure pressure.

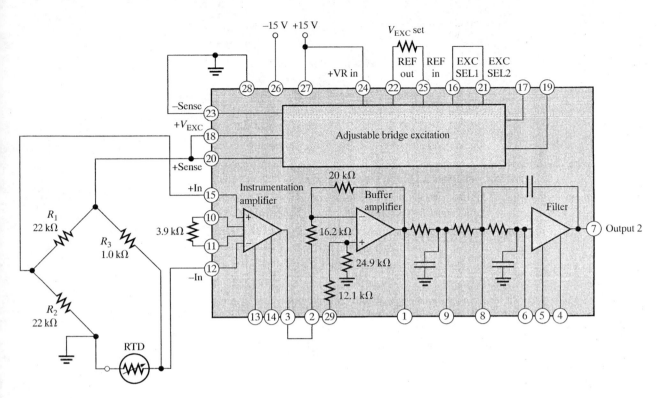

FIGURE 11–58

20. Identify and compare the three symbols in Figure 11–59.

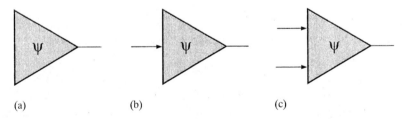

(a)   (b)   (c)

FIGURE 11–59

SECTION 11–5 Power-Control Circuits

21. Show how to connect the CA3079 zero-voltage switch to a 3-wire RTD bridge in order to control a triac-driven heating element.

22. Assume the RTD in Problem 21 has a resistance of 200 Ω at 50°C, 400 Ω at 100°C, and 600 Ω at 150°C. To what value must you set the other external bridge resistor to maintain a temperature of 125°C?

■ ANSWERS TO REVIEW QUESTIONS

Section 11–1

1. An rms-to-dc converter produces a dc output voltage that is equal to the rms value of the ac input voltage.

2. Internally, an rms-to-dc converter squares, averages, and takes the square root.

Section 11–2

1. Synchro

2. An RDC accepts resolver format voltages on its inputs.

3. An RDC produces a digital code representing the angular shaft position of the resolver.

4. An RDC converts the angular shaft position of a resolver into a digital code.

Section 11–3

1. A thermocouple is a temperature transducer formed by the junction of two dissimilar metals.

2. A thermocouple produces a voltage proportional to the temperature at the junction.

3. An RTD is a resistance temperature detector in which the resistance is proportional to the temperature, whereas the thermocouple produces a voltage.

4. An RTD has a positive temperature coefficient and a thermistor has a negative temperature coefficient.

5. The thermocouple has a greater temperature range than the RTD or thermistor.

Section 11–4

1. Basically, a strain gage is a resistive element whose dimensions can be altered by an applied force to produce a change in resistance.

2. Basically, a pressure gage is a strain gage bonded to a flexible diaphragm.

3. Absolute, gage, and differential

4. (a) A linear variable differential transformer
   (b) Displacement

Section 11–5

1. Zero-voltage switching eliminates fast transitions in the current to a load, thus reducing RFI emissions and thermal shock to the load element.

2. An SCR is unidirectional and therefore allows current through the load only during half of the ac cycle. A triac is bidirectional and allows current during the complete cycle.

Section 11–6

1. R_1 and C_1 set the pulse width of the one-shot.

2. The noninverting op-amp provides gain for the output of the F/V converter and permits adjustment of the input to the ADC for calibration purposes.

■ **ANSWERS TO PRACTICE EXERCISES FOR EXAMPLES**

11–1 183°

11–2 10.1%

11–3 24.247 mV

11–4 Change R_G to 10 kΩ and move wire from pin 42 to pin 41.

11–5 1.0 kΩ; 0 V

11–6 The gain increases to 25.9.

11–7 5.12 mΩ

A DATA SHEETS

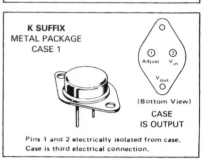

LM117
LM217
LM317

THREE-TERMINAL ADJUSTABLE POSITIVE VOLTAGE REGULATORS

SILICON MONOLITHIC
INTEGRATED CIRCUIT

THREE-TERMINAL ADJUSTABLE OUTPUT POSITIVE VOLTAGE REGULATORS

The LM117/217/317 are adjustable 3-terminal positive voltage regulators capable of supplying in excess of 1.5 A over an output voltage range of 1.2 V to 37 V. These voltage regulators are exceptionally easy to use and require only two external resistors to set the output voltage. Further, they employ internal current limiting, thermal shutdown and safe area compensation, making them essentially blow-out proof.

The LM117 series serve a wide variety of applications including local, on card regulation. This device can also be used to make a programmable output regulator, or by connecting a fixed resistor between the adjustment and output, the LM117 series can be used as a precision current regulator.

- Output Current in Excess of 1.5 Ampere in K and T Suffix Packages
- Output Current in Excess of 0.5 Ampere in H Suffix Package
- Output Adjustable between 1.2 V and 37 V
- Internal Thermal Overload Protectiion
- Internal Short-Circuit Current Limiting Constant with Temperature
- Output Transistor Safe-Area Compensation
- Floating Operation for High Voltage Applications
- Standard 3-lead Transistor Packages
- Eliminates Stocking Many Fixed Voltages

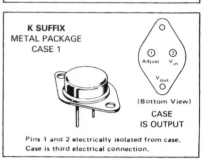

K SUFFIX
METAL PACKAGE
CASE 1

(Bottom View)
CASE
IS OUTPUT

Pins 1 and 2 electrically isolated from case.
Case is third electrical connection.

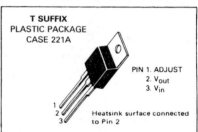

T SUFFIX
PLASTIC PACKAGE
CASE 221A

PIN 1. ADJUST
2. V_{out}
3. V_{in}

Heatsink surface connected to Pin 2

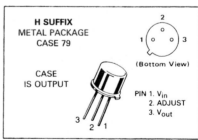

H SUFFIX
METAL PACKAGE
CASE 79

CASE
IS OUTPUT

(Bottom View)

PIN 1. V_{in}
2. ADJUST
3. V_{out}

STANDARD APPLICATION

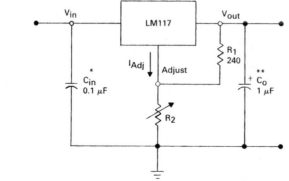

* = C_{in} is required if regulator is located an appreciable distance from power supply filter.
** = C_O is not needed for stability, however it does improve transient response.

$$V_{out} = 1.25 \text{ V} \left(1 + \frac{R_2}{R_1}\right) + I_{Adj} R_2$$

Since I_{Adj} is controlled to less than 100 μA, the error associated with this term is negligible in most applications.

ORDERING INFORMATION

Device	Tested Operating Temperature Range	Package
LM117H LM117K	$T_J = -55°C$ to $+150°C$	Metal Can Metal Power
LM217H LM217K	$T_J = -25°C$ to $+150°C$	Metal Can Metal Power
LM317H LM317K LM317T	$T_J = 0°C$ to $+125°C$	Metal Can Metal Power Plastic Power
LM317BT#	$T_J = -40°C$ to $+125°C$	Plastic Power

#Automotive temperature range selections are available with special test conditions and additional tests.
Contact your local Motorola sales office for information.

LM117, LM217, LM317

MAXIMUM RATINGS

Rating	Symbol	Value	Unit
Input-Output Voltage Differential	V_I-V_O	40	Vdc
Power Dissipation	P_D	Internally Limited	
Operating Junction Temperature Range LM117 LM217 LM317	T_J	 -55 to $+150$ -25 to $+150$ 0 to $+150$	°C
Storage Temperature Range	T_{stg}	-65 to $+150$	°C

ELECTRICAL CHARACTERISTICS

(V_I-V_O = 5.0 V; I_O = 0.5 A for K and T packages; I_O = 0.1 A for H package; T_J = T_{low} to T_{high} [see Note 1]; I_{max} and P_{max} per Note 2; unless otherwise specified.)

Characteristic	Figure	Symbol	LM117/217 Min	LM117/217 Typ	LM117/217 Max	LM317 Min	LM317 Typ	LM317 Max	Unit
Line Regulation (Note 3) T_A = 25°C, 3.0 V ≤ V_I-V_O ≤ 40 V	1	Reg$_{line}$	—	0.01	0.02	—	0.01	0.04	%/V
Load Regulation (Note 3) T_A = 25°C, 10 mA ≤ I_O ≤ I_{max} V_O ≤ 5.0 V V_O ≥ 5.0 V	2	Reg$_{load}$	 — —	 5.0 0.1	 15 0.3	 — —	 5.0 0.1	 25 0.5	 mV %/V_O
Thermal Regulation (T_A = +25°C) 20 ms Pulse		—	—	0.02	0.07	—	0.03	0.07	%/W
Adjustment Pin Current	3	I_{Adj}	—	50	100	—	50	100	μA
Adjustment Pin Current Change 2.5 V ≤ V_I-V_O ≤ 40 V 10 mA ≤ I_L ≤ I_{max}, P_D ≤ P_{max}	1,2	ΔI$_{Adj}$	—	0.2	5.0	—	0.2	5.0	μA
Reference Voltage (Note 4) 3.0 V ≤ V_I-V_O ≤ 40 V 10 mA ≤ I_O ≤ I_{max}, P_D ≤ P_{max}	3	V_{ref}	1.2	1.25	1.3	1.2	1.25	1.3	V
Line Regulation (Note 3) 3.0 V ≤ V_I-V_O ≤ 40 V	1	Reg$_{line}$	—	0.02	0.05	—	0.02	0.07	%/V
Load Regulation (Note 3) 10 mA ≤ I_O ≤ I_{max} V_O ≤ 5.0 V V_O ≥ 5.0 V	2	Reg$_{load}$	 — —	 20 0.3	 50 1.0	 — —	 20 0.3	 70 1.5	 mV %/V_O
Temperature Stability (T_{low} ≤ T_J ≤ T_{high})	3	T_S	—	0.7	—	—	0.7	—	%/V_O
Minimum Load Current to Maintain Regulation (V_I-V_O = 40 V)	3	I_{Lmin}	—	3.5	5.0	—	3.5	10	mA
Maximum Output Current V_I-V_O ≤ 15 V, P_D ≤ P_{max} K and T Packages H Package V_I-V_O = 40 V, P_D ≤ P_{max}, T_A = 25°C K and T Packages H Package	3	I_{max}	 1.5 0.5 0.25 —	 2.2 0.8 0.4 0.07	 — — — —	 1.5 0.5 0.15 —	 2.2 0.8 0.4 0.07	 — — — —	A
RMS Noise, % of V_O T_A = 25°C, 10 Hz ≤ f ≤ 10 kHz	—	N	—	0.003	—	—	0.003	—	%/V_O
Ripple Rejection, V_O = 10 V, f = 120 Hz (Note 5) Without C_{Adj} C_{Adj} = 10 μF	4	RR	 — 66	 65 80	 — —	 — 66	 65 80	 — —	dB
Long-Term Stability, T_J = T_{high} (Note 6) T_A = 25°C for Endpoint Measurements	3	S	—	0.3	1.0	—	0.3	1.0	%/1.0 k Hrs.
Thermal Resistance Junction to Case H Package K Package T Package	—	$R_{\theta JC}$	 — — —	 12 2.3 —	 15 3.0 —	 — — —	 12 2.3 5.0	 15 3.0 —	°C/W

NOTES: (1) T_{low} = -55°C for LM117 T_{high} = $+150$°C for LM117
      = -25°C for LM217   = $+150$°C for LM217
      = 0°C for LM317   = $+125$°C for LM317
   (2) I_{max} = 1.5 A for K and T Packages
      = 0.5 A for H Package
      P_{max} = 20 W for K Package
           = 20 W for T Package
           = 2.0 W for H Package
   (3) Load and line regulation are specified at constant junction temperature. Changes in V_O due to heating effects must

be taken into account separately. Pulse testing with low duty cycle is used.
(4) Selected devices with tightened tolerance reference voltage available.
(5) C_{ADJ}, when used, is connected between the adjustment pin and ground.
(6) Since Long-Term Stability cannot be measured on each device before shipment, this specification is an engineering estimate of average stability from lot to lot.

MC1455

TIMING CIRCUIT

SILICON MONOLITHIC INTEGRATED CIRCUIT

TIMING CIRCUIT

The MC1455 monolithic timing circuit is a highly stable controller capable of producing accurate time delays, or oscillation. Additional terminals are provided for triggering or resetting if desired. In the time delay mode of operation, the time is precisely controlled by one external resistor and capacitor. For astable operation as an oscillator, the free running frequency and the duty cycle are both accurately controlled with two external resistors and one capacitor. The circuit may be triggered and reset on falling waveforms, and the output structure can source or sink up to 200 mA or drive MTTL circuits.

- Direct Replacement for NE555 Timers
- Timing From Microseconds Through Hours
- Operates in Both Astable and Monostable Modes
- Adjustable Duty Cycle
- High Current Output Can Source or Sink 200 mA
- Output Can Drive MTTL
- Temperature Stability of 0.005% per °C
- Normally "On" or Normally "Off" Output

G SUFFIX
METAL PACKAGE
CASE 601

1. Ground
2. Trigger
3. Output
4. Reset

5. Control Voltage
6. Threshold
7. Discharge
8. V_{CC}

V_{CC}

P1 SUFFIX
PLASTIC PACKAGE
CASE 626

U SUFFIX
CERAMIC PACKAGE
CASE 693

D SUFFIX
PLASTIC PACKAGE
CASE 751
(SO-8)

FIGURE 1 — 22-SECOND SOLID-STATE TIME DELAY RELAY CIRCUIT

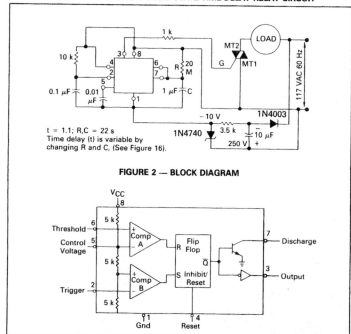

t = 1.1; R,C = 22 s
Time delay (t) is variable by
changing R and C, (See Figure 16).

FIGURE 2 — BLOCK DIAGRAM

ORDERING INFORMATION

Device	Alternate	Temperature Range	Package
MC1455G	—	0°C to +70°C	Metal Can
MC1455P1	NE555V		Plastic DIP
MC1455D	—		SO-8
MC1455U	—		Ceramic DIP
MC1455BP1	—	−40°C to +85°C	Plastic DIP

MC1455

MAXIMUM RATINGS (T_A = +25°C unless otherwise noted.)

Rating	Symbol	Value	Unit
Power Supply Voltage	V_{CC}	+18	Vdc
Discharge Current (Pin 7)	I_7	200	mA
Power Dissipation (Package Limitation)	P_D		
Metal Can		680	mW
Derate above T_A = +25°C		4.6	mW/°C
Plastic Dual In-Line Package		625	mW
Derate above T_A = +25°C		5.0	mW/°C
Operating Temperature Range (Ambient)	T_A		°C
MC1455B		–40 to +85	
MC1455		0 to +70	
Storage Temperature Range	T_{stg}	–65 to +150	°C

FIGURE 3 — GENERAL TEST CIRCUIT

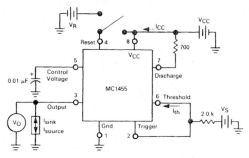

Test Circuit for Measuring dc Parameters: (to set output and measure parameters)
a) When V_S ⁻ 2 3 V_{CC}, V_O is low.
b) When V_S ⁻ 1 3 V_{CC}, V_O is high.
c) When V_O is low, pin 7 sinks current. To test for Reset, set V_O high, apply Reset voltage, and test for current flowing into pin 7. When Reset is not in use, it should be tied to V_{CC}

ELECTRICAL CHARACTERISTICS (T_A = +25°C, V_{CC} = +5.0 V to +15 V unless otherwise noted.)

Characteristics	Symbol	Min	Typ	Max	Unit
Operating Supply Voltage Range	V_{CC}	4.5	—	16	V
Supply Current	I_{CC}				mA
V_{CC} = 5.0 V, R_L = ∞		—	3.0	6.0	
V_{CC} = 15 V, R_L = ∞		—	10	15	
Low State, (Note 1)					
Timing Error (Note 2)					
R = 1.0 kΩ to 100 kΩ					
Initial Accuracy C = 0.1 μF		—	1.0	—	%
Drift with Temperature		—	50	—	PPM/°C
Drift with Supply Voltage		—	0.1	—	%/Volt
Threshold Voltage	V_{th}	—	2/3	—	xV_{CC}
Trigger Voltage	V_T				V
V_{CC} = 15 V		—	5.0	—	
V_{CC} = 5.0 V		—	1.67	—	
Trigger Current	I_T	—	0.5	—	μA
Reset Voltage	V_R	0.4	0.7	1.0	V
Reset Current	I_R	—	0.1	—	mA
Threshold Current (Note 3)	I_{th}	—	0.1	0.25	μA
Discharge Leakage Current (Pin 7)	I_{dis}	—	—	100	nA
Control Voltage Level	V_{CL}				V
V_{CC} = 15 V		9.0	10	11	
V_{CC} = 5.0 V		2.6	3.33	4.0	
Output Voltage Low	V_{OL}				V
(V_{CC} = 15 V)					
I_{sink} = 10 mA		—	0.1	0.25	
I_{sink} = 50 mA		—	0.4	0.75	
I_{sink} = 100 mA		—	2.0	2.5	
I_{sink} = 200 mA		—	2.5	—	
(V_{CC} = 5.0 V)					
I_{sink} = 8.0 mA		—	—	—	
I_{sink} = 5.0 mA		—	0.25	0.35	
Output Voltage High	V_{OH}				V
(I_{source} = 200 mA)					
V_{CC} = 15 V		—	12.5	—	
(I_{source} = 100 mA)					
V_{CC} = 15 V		12.75	13.3	—	
V_{CC} = 5.0 V		2.75	3.3	—	
Rise Time of Output	t_{OLH}	—	100	—	ns
Fall Time of Output	t_{OHL}	—	100	—	ns

NOTES:
1. Supply current when output is high is typically 1.0 mA less.
2. Tested at V_{CC} = 5.0 V and V_{CC} = 15 V. Monostable mode
3. This will determine the maximum value of R_A + R_B for 15 V operation. The maximum total R = 20 megohms.

**MC1741
MC1741C**

OPERATIONAL AMPLIFIER

SILICON MONOLITHIC
INTEGRATED CIRCUIT

INTERNALLY COMPENSATED, HIGH PERFORMANCE OPERATIONAL AMPLIFIERS

. . . designed for use as a summing amplifier, integrator, or amplifier with operating characteristics as a function of the external feedback components.

- No Frequency Compensation Required
- Short-Circuit Protection
- Offset Voltage Null Capability
- Wide Common-Mode and Differential Voltage Ranges
- Low-Power Consumption
- No Latch Up

MAXIMUM RATINGS (T_A = +25°C unless otherwise noted)

Rating	Symbol	MC1741C	MC1741	Unit
Power Supply Voltage	V_{CC} V_{EE}	+18 −18	+22 −22	Vdc Vdc
Input Differential Voltage	V_{ID}	±30		Volts
Input Common Mode Voltage (Note 1)	V_{ICM}	±15		Volts
Output Short Circuit Duration (Note 2)	t_S	Continuous		
Operating Ambient Temperature Range	T_A	0 to +70	−55 to +125	°C
Storage Temperature Range Metal and Ceramic Packages Plastic Packages	T_{stg}	−65 to +150 −55 to +125		°C

NOTES:
1. For supply voltages less than +15 V, the absolute maximum input voltage is equal to the supply voltage.
2. Supply voltage equal to or less than 15 V.

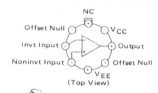

NC
Offset Null — V_{CC}
Invt Input — Output
Noninvt Input — Offset Null
V_{EE}
(Top View)

G SUFFIX
METAL PACKAGE
CASE 601

P1 SUFFIX
PLASTIC PACKAGE
CASE 626

U SUFFIX
CERAMIC PACKAGE
CASE 693

D SUFFIX
PLASTIC PACKAGE
CASE 751
(SO-8)

PIN CONNECTIONS

Offset Null — NC
Invt Input — V_{CC}
Noninvt Input — Output
V_{EE} — Offset Null
(Top View)

EQUIVALENT CIRCUIT SCHEMATIC

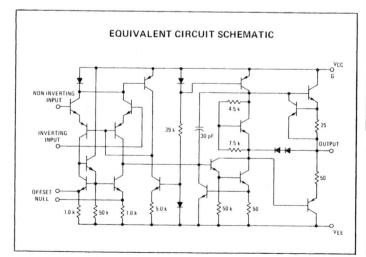

ORDERING INFORMATION

Device	Alternate	Temperature Range	Package
MC1741CD	—	0°C to +70°C	SO-8
MC1741CG	LM741CH, μA741HC		Metal Can
MC1741CP1	LM741CN, μA741TC		Plastic DIP
MC1741CU	—		Ceramic DIP
MC1741G	—	−55°C to +125°C	Metal Can
MC1741U	—		Ceramic DIP

MC1741, MC1741C

ELECTRICAL CHARACTERISTICS ($V_{CC} = +15$ V, $V_{EE} = -15$ V, $T_A = 25°C$ unless otherwise noted).

Characteristic	Symbol	MC1741 Min	MC1741 Typ	MC1741 Max	MC1741C Min	MC1741C Typ	MC1741C Max	Unit
Input Offset Voltage ($R_S \leqslant 10$ k)	V_{IO}	--	1.0	5.0	—	2.0	6.0	mV
Input Offset Current	I_{IO}	-	20	200	—	20	200	nA
Input Bias Current	I_{IB}	-	80	500	—	80	500	nA
Input Resistance	r_i	0.3	2.0	--	0.3	2.0	—	MΩ
Input Capacitance	C_i	-	1.4	—	—	1.4	—	pF
Offset Voltage Adjustment Range	V_{IOR}	—	± 15	—	—	± 15	—	mV
Common Mode Input Voltage Range	V_{ICR}	± 12	± 13	--	± 12	± 13	—	V
Large Signal Voltage Gain ($V_O = \pm 10$ V, $R_L \geqslant 2.0$ k)	A_v	50	200	-	20	200	—	V/mV
Output Resistance	r_o	-	75	.	—	75	—	Ω
Common Mode Rejection Ratio ($R_S \leqslant 10$ k)	CMRR	70	90	--	70	90	—	dB
Supply Voltage Rejection Ratio ($R_S \leqslant 10$ k)	PSRR		30	150	—	30	150	μV/V
Output Voltage Swing ($R_L \geqslant 10$ k) ($R_L \geqslant 2$ k)	V_O	 ± 12 ± 10	 ± 14 ± 13	 - -	 ± 12 ± 10	 ± 14 ± 13	 — —	V
Output Short-Circuit Current	I_{os}	-	20	—	—	20	—	mA
Supply Current	I_D	--	1.7	2.8	—	1.7	2.8	mA
Power Consumption	P_C	..	50	85	—	50	85	mW
Transient Response (Unity Gain — Non-Inverting) ($V_I = 20$ mV, $R_L \geqslant 2$ k, $C_L \leqslant 100$ pF) Rise Time ($V_I = 20$ mV, $R_L \geqslant 2$ k, $C_L \leqslant 100$ pF) Overshoot ($V_I = 10$ V, $R_L \geqslant 2$ k, $C_L \leqslant 100$ pF) Slew Rate	 t_{TLH} os SR		 0.3 15 0.5	 - -- -	 -- — —	 0.3 15 0.5	 — — —	 μs % V/μs

ELECTRICAL CHARACTERISTICS ($V_{CC} = +15$ V, $V_{EE} = -15$ V, $T_A = T_{low}$ to T_{high} unless otherwise noted).

Characteristic	Symbol	MC1741 Min	MC1741 Typ	MC1741 Max	MC1741C Min	MC1741C Typ	MC1741C Max	Unit
Input Offset Voltage ($R_S \leqslant 10$ kΩ)	V_{IO}	—	1.0	6.0	—	—	7.5	mV
Input Offset Current ($T_A = 125°C$) ($T_A = -55°C$) ($T_A = 0°C$ to $+70°C$)	I_{IO}	 -- -- —	 7.0 85 —	 200 500 —	 — — —	 — — —	 — — 300	nA
Input Bias Current ($T_A = 125°C$) ($T_A = -55°C$) ($T_A = 0°C$ to $+70°C$)	I_{IB}	 — — --	 30 300 —	 500 1500 --	 — — —	 — — —	 — — 800	nA
Common Mode Input Voltage Range	V_{ICR}	± 12	± 13	—	—	—	—	V
Common Mode Rejection Ratio ($R_S \leqslant 10$ k)	CMRR	70	90	—	—	—	—	dB
Supply Voltage Rejection Ratio ($R_S \leqslant 10$ k)	PSRR	—	30	150	--		—	μV/V
Output Voltage Swing ($R_L \geqslant 10$ k) ($R_L \geqslant 2$ k)	V_O	 ± 12 ± 10	 ± 14 ± 13	 — —	 -- ± 10	 — ± 13	 — —	V
Large Signal Voltage Gain ($R_L \geqslant 2$ k, $V_{out} = \pm 10$ V)	A_v	26	—	..	15	—	—	V/mV
Supply Currents ($T_A = 125°C$) ($T_A = -55°C$)	I_D	 --	 1.5 2.0	 2.5 3.3	 . —	 — --	 — --	mA
Power Consumption ($T_A = +125°C$) ($T_A = -55°C$)	P_C	 - --	 45 60	 75 100	 — —	 — —	 — --	mW

*$T_{high} = 125°C$ for MC1741 and $70°C$ for MC1741C
$T_{low} = -55°C$ for MC1741 and $0°C$ for MC1741C

MC1741, MC1741C

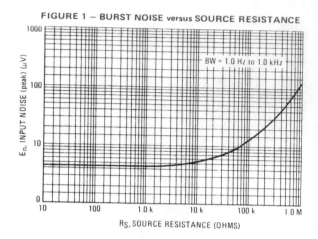

FIGURE 1 – BURST NOISE versus SOURCE RESISTANCE

BW = 1.0 Hz to 1.0 kHz

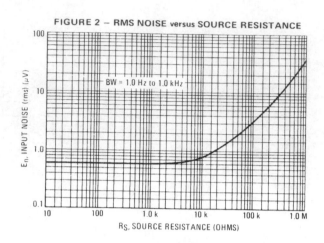

FIGURE 2 – RMS NOISE versus SOURCE RESISTANCE

BW = 1.0 Hz to 1.0 kHz

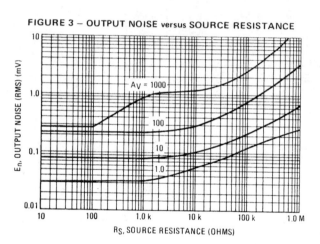

FIGURE 3 – OUTPUT NOISE versus SOURCE RESISTANCE

$A_V = 1000$

100

10

1.0

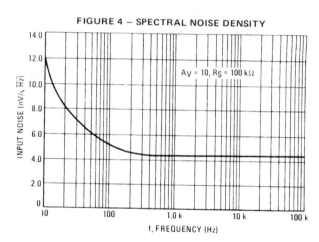

FIGURE 4 – SPECTRAL NOISE DENSITY

$A_V = 10$, $R_S = 100 \, k\Omega$

FIGURE 5 – BURST NOISE TEST CIRCUIT

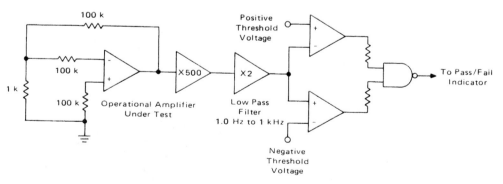

Unlike conventional peak reading or RMS meters, this system was especially designed to provide the quick response time essential to burst (popcorn) noise testing.

The test time employed is 10 seconds and the 20 μV peak limit refers to the operational amplifier input thus eliminating errors in the closed-loop gain factor of the operational amplifier under test.

MC1741, MC1741C

FIGURE 6 – POWER BANDWIDTH
(LARGE SIGNAL SWING versus FREQUENCY)

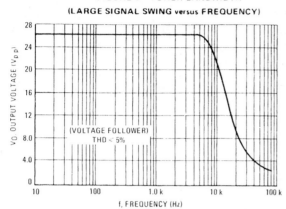

FIGURE 7 – OPEN LOOP FREQUENCY RESPONSE

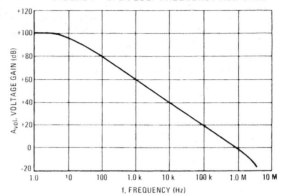

FIGURE 8 – POSITIVE OUTPUT VOLTAGE SWING
versus LOAD RESISTANCE

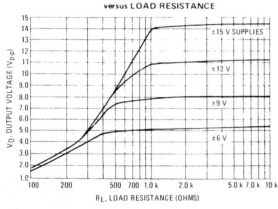

FIGURE 9 – NEGATIVE OUTPUT VOLTAGE SWING
versus LOAD RESISTANCE

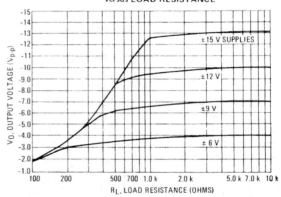

FIGURE 10 – OUTPUT VOLTAGE SWING versus
LOAD RESISTANCE (Single Supply Operation)

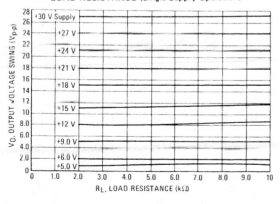

FIGURE 11 – SINGLE SUPPLY INVERTING AMPLIFIER

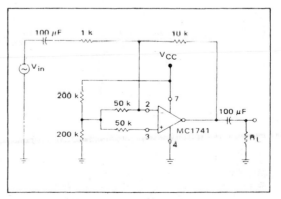

543

MC7800 Series

THREE-TERMINAL POSITIVE VOLTAGE REGULATORS

These voltage regulators are monolithic integrated circuits designed as fixed-voltage regulators for a wide variety of applications including local, on-card regulation. These regulators employ internal current limiting, thermal shutdown, and safe-area compensation. With adequate heatsinking they can deliver output currents in excess of 1.0 ampere. Although designed primarily as a fixed voltage regulator, these devices can be used with external components to obtain adjustable voltages and currents.

- Output Current in Excess of 1.0 Ampere
- No External Components Required
- Internal Thermal Overload Protection
- Internal Short-Circuit Current Limiting
- Output Transistor Safe-Area Compensation
- Output Voltage Offered in 2% and 4% Tolerance

THREE-TERMINAL POSITIVE FIXED VOLTAGE REGULATORS

SILICON MONOLITHIC INTEGRATED CIRCUITS

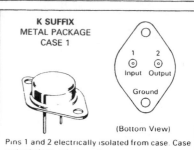

K SUFFIX
METAL PACKAGE
CASE 1

1 — Input
2 — Output
Ground

(Bottom View)

Pins 1 and 2 electrically isolated from case. Case is third electrical connection

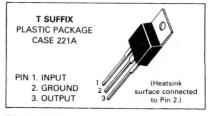

T SUFFIX
PLASTIC PACKAGE
CASE 221A

PIN 1. INPUT
2. GROUND
3. OUTPUT

(Heatsink surface connected to Pin 2.)

REPRESENTATIVE SCHEMATIC DIAGRAM

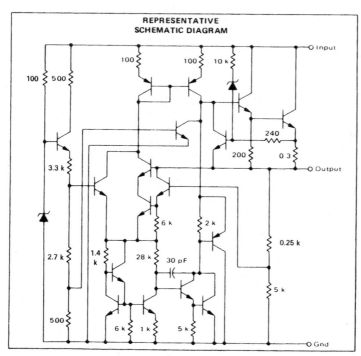

STANDARD APPLICATION

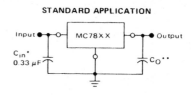

A common ground is required between the input and the output voltages. The input voltage must remain typically 2.0 V above the output voltage even during the low point on the input ripple voltage.

XX = these two digits of the type number indicate voltage.

* = C_{in} is required if regulator is located an appreciable distance from power supply filter.

** = C_O is not needed for stability; however, it does improve transient response.

XX indicates nominal voltage

ORDERING INFORMATION

Device	Output Voltage Tolerance	Tested Operating Junction Temp. Range	Package
MC78XXK MC78XXAK*	4% 2%	−55 to +150°C	Metal Power
MC78XXCK* MC78XXACK*	4% 2%	0 to +125°C	
MC78XXCT MC78XXACT	4% 2%		Plastic Power
MC78XXBT	4%	−40 to +125°C	

*2% regulators in Metal Power packages are available in 5, 12 and 15 volt devices.

TYPE NO./VOLTAGE

MC7805	5.0 Volts	MC7812	12 Volts
MC7806	6.0 Volts	MC7815	15 Volts
MC7808	8.0 Volts	MC7818	18 Volts
MC7809	9.0 Volts	MC7824	24 Volts

MC7800 Series

MAXIMUM RATINGS (T_A = +25°C unless otherwise noted.)

Rating	Symbol	Value	Unit
Input Voltage (5.0 V – 18 V)	V_{in}	35	Vdc
(24 V)		40	
Power Dissipation and Thermal Characteristics			
Plastic Package			
T_A = +25°C	P_D	Internally Limited	Watts
Derate above T_A = +25°C	$1/\theta_{JA}$	15.4	mW/°C
Thermal Resistance, Junction to Air	θ_{JA}	65	°C/W
T_C = +25°C	P_D	Internally Limited	Watts
Derate above T_C = +75°C (See Figure 1)	$1/\theta_{JC}$	200	mW/°C
Thermal Resistance, Junction to Case	θ_{JC}	5.0	°C/W
Metal Package			
T_A = +25°C	P_D	Internally Limited	Watts
Derate above T_A = +25°C	$1/\theta_{JA}$	22.5	mW/°C
Thermal Resistance, Junction to Air	θ_{JA}	45	°C/W
T_C = +25°C	P_D	Internally Limited	Watts
Derate above T_C = +65°C (See Figure 2)	$1/\theta_{JC}$	182	mW/°C
Thermal Resistance, Junction to Case	θ_{JC}	5.5	°C/W
Storage Junction Temperature Range	T_{stg}	– 65 to + 150	°C
Operating Junction Temperature Range	T_J		°C
MC7800,A		– 55 to + 150	
MC7800C,AC		0 to + 150	
MC7800B		– 40 to + 150	

DEFINITIONS

Line Regulation — The change in output voltage for a change in the input voltage. The measurement is made under conditions of low dissipation or by using pulse techniques such that the average chip temperature is not significantly affected

Load Regulation — The change in output voltage for a change in load current at constant chip temperature

Maximum Power Dissipation — The maximum total device dissipation for which the regulator will operate within specifications

Quiescent Current — That part of the input current that is not delivered to the load

Output Noise Voltage — The rms ac voltage at the output, with constant load and no input ripple, measured over a specified frequency range

Long Term Stability — Output voltage stability under accelerated life test conditions with the maximum rated voltage listed in the devices' electrical characteristics and maximum power dissipation

MC7800 Series

MC7805, B, C
ELECTRICAL CHARACTERISTICS (V_{in} = 10 V, I_O = 500 mA, T_J = T_{low} to T_{high} [Note 1] unless otherwise noted).

Characteristic	Symbol	MC7805 Min	Typ	Max	MC7805B Min	Typ	Max	MC7805C Min	Typ	Max	Unit
Output Voltage (T_J = +25°C)	V_O	4.8	5.0	5.2	4.8	5.0	5.2	4.8	5.0	5.2	Vdc
Output Voltage (5.0 mA ≤ I_O ≤ 1.0 A, P_O ≤ 15 W) 7.0 Vdc ≤ V_{in} ≤ 20 Vdc 8.0 Vdc ≤ V_{in} ≤ 20 Vdc	V_O	— 4.65	— 5.0	— 5.35	— 4.75	— 5.0	— 5.25	4.75 —	5.0 —	5.25 —	Vdc
Line Regulation (T_J = +25°C, Note 2) 7.0 Vdc ≤ V_{in} ≤ 25 Vdc 8.0 Vdc ≤ V_{in} ≤ 12 Vdc	Reg_{line}	— —	2.0 1.0	50 25	— —	7.0 2.0	100 50	— —	7.0 2.0	100 50	mV
Load Regulation (T_J = +25°C, Note 2) 5.0 mA ≤ I_O ≤ 1.5 A 250 mA ≤ I_O ≤ 750 mA	Reg_{load}	— —	25 8.0	100 25	— —	40 15	100 50	— —	40 15	100 50	mV
Quiescent Current (T_J = +25°C)	I_B	—	3.2	6.0	—	4.3	8.0	—	4.3	8.0	mA
Quiescent Current Change 7.0 Vdc ≤ V_{in} ≤ 25 Vdc 8.0 Vdc ≤ V_{in} ≤ 25 Vdc 5.0 mA ≤ I_O ≤ 1.0 A	ΔI_B	— — —	— 0.3 0.04	— 0.8 0.5	— — —	— 1.3 0.5	— — —	— — —	— — —	1.3 — 0.5	mA
Ripple Rejection 8.0 Vdc ≤ V_{in} ≤ 18 Vdc, f = 120 Hz	RR	68	75	—	—	68	—	—	68	—	dB
Dropout Voltage (I_O = 1.0 A, T_J = +25°C)	$V_{in} - V_O$	—	2.0	2.5	—	2.0	—	—	2.0	—	Vdc
Output Noise Voltage (T_A = +25°C) 10 Hz ≤ f ≤ 100 kHz	V_n	—	10	40	—	10	—	—	10	—	μV/V_O
Output Resistance f = 1.0 kHz	r_O	—	17	—	—	17	—	—	17	—	mΩ
Short-Circuit Current Limit (T_A = +25°C) V_{in} = 35 Vdc	I_{sc}	—	0.2	1.2	—	0.2	—	—	0.2	—	A
Peak Output Current (T_J = +25°C)	I_{max}	1.3	2.5	3.3	—	2.2	—	—	2.2	—	A
Average Temperature Coefficient of Output Voltage	TCV_O	—	±0.6	—	—	-1.1	—	—	-1.1	—	mV/°C

MC7805A, AC
ELECTRICAL CHARACTERISTICS (V_{in} = 10 V, I_O = 1.0 A, T_J = T_{low} to T_{high} [Note 1] unless otherwise noted).

Characteristics	Symbol	MC7805A Min	Typ	Max	MC7805AC Min	Typ	Max	Unit
Output Voltage (T_J = +25°C)	V_O	4.9	5.0	5.1	4.9	5.0	5.1	Vdc
Output Voltage (5.0 mA ≤ I_O ≤ 1.0 A, P_O ≤ 15 W) 7.5 Vdc ≤ V_{in} ≤ 20 Vdc	V_O	4.8	5.0	5.2	4.8	5.0	5.2	Vdc
Line Regulation (Note 2) 7.5 Vdc ≤ V_{in} ≤ 25 Vdc, I_O = 500 mA 8.0 Vdc ≤ V_{in} ≤ 12 Vdc 8.0 Vdc ≤ V_{in} ≤ 12 Vdc, T_J = +25°C 7.3 Vdc ≤ V_{in} ≤ 20 Vdc, T_J = +25°C	Reg_{line}	— — — —	2.0 3.0 1.0 2.0	10 10 4.0 10	— — — —	7.0 10 2.0 7.0	50 50 25 50	mV
Load Regulation (Note 2) 5.0 mA ≤ I_O ≤ 1.5 A, T_J = +25°C 5.0 mA ≤ I_O ≤ 1.0 A 250 mA ≤ I_O ≤ 750mA, T_J = +25°C 250 mA ≤ I_O ≤ 750 mA	Reg_{load}	— — — —	2.0 2.0 1.0 1.0	25 25 15 25	— — — —	25 25 — 8.0	100 100 — 50	mV
Quiescent Current T_J = +25°C	I_B	— —	— 3.2	5.0 4.0	— —	— 4.3	6.0 6.0	mA
Quiescent Current Change 8.0 Vdc ≤ V_{in} ≤ 25 Vdc, I_O = 500 mA 7.5 Vdc ≤ V_{in} ≤ 20 Vdc, T_J = +25°C 5.0 mA ≤ I_O ≤ 1.0 A	ΔI_B	— — —	0.3 0.2 0.04	0.5 0.5 0.2	— — —	— — —	0.8 0.8 0.5	mA
Ripple Rejection 8.0 Vdc ≤ V_{in} ≤ 18 Vdc, f = 120 Hz, T_J = +25°C 8.0 Vdc ≤ V_{in} ≤ 18 Vdc, f = 120 Hz, I_O = 500 mA	RR	68 68	75 75	— —	— 68	— —	— —	dB
Dropout Voltage (I_O = 1.0 A, T_J = +25°C)	$V_{in} - V_O$	—	2.0	2.5	—	2.0	—	Vdc
Output Noise Voltage (T_A = +25°C) 10 Hz ≤ f ≤ 100 kHz	V_n	—	10	40	—	10	—	μV/V_O
Output Resistance (f = 1.0 kHz)	r_O	—	2.0	—	—	17	—	mΩ
Short-Circuit Current Limit (T_A = +25°C) V_{in} = 35 Vdc	I_{sc}	—	0.2	1.2	—	0.2	—	A
Peak Output Current (T_J = +25°C)	I_{max}	1.3	2.5	3.3	—	2.2	—	A
Average Temperature Coefficient of Output Voltage	TCV_O	—	±0.6	—	—	-1.1	—	mV/°C

NOTES: 1. T_{low} = -55°C for MC78XX, A T_{high} = +150°C for MC78XX, A
= 0° for MC78XXC, AC = +125°C for MC78XXC, AC, B
= -40°C for MC78XXB

2. Load and line regulation are specified at constant junction temperature. Changes in V_O due to heating effects must be taken into account separately. Pulse testing with low duty cycle is used.

546

MC7900 Series

THREE-TERMINAL NEGATIVE VOLTAGE REGULATORS

The MC7900 Series of fixed output negative voltage regulators are intended as complements to the popular MC7800 Series devices. These negative regulators are available in the same seven-voltage options as the MC7800 devices. In addition, one extra voltage option commonly employed in MECL systems is also available in the negative MC7900 Series.

Available in fixed output voltage options from −5.0 to −24 volts, these regulators employ current limiting, thermal shutdown, and safe-area compensation — making them remarkably rugged under most operating conditions. With adequate heatsinking they can deliver output currents in excess of 1.0 ampere.

● No External Components Required
● Internal Thermal Overload Protection
● Internal Short-Circuit Current Limiting
● Output Transistor Safe-Area Compensation
● Available in 2% Voltage Tolerance (See Ordering Information)

THREE-TERMINAL NEGATIVE FIXED VOLTAGE REGULATORS

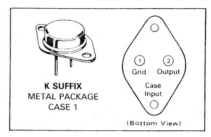

K SUFFIX
METAL PACKAGE
CASE 1

① Gnd ② Output
Case Input

(Bottom View)

T SUFFIX
PLASTIC PACKAGE
CASE 221A

PIN 1. GROUND
2. INPUT
3. OUTPUT

(Heatsink surface connected to Pin 2)

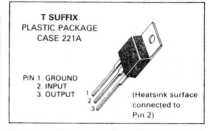

SCHEMATIC DIAGRAM

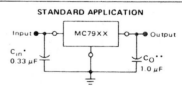

STANDARD APPLICATION

Input ● ─ ○ ─── MC79XX ─── ○ ─ ● Output

C_{in}*
0.33 µF

C_O**
1.0 µF

A common ground is required between the input and the output voltages. The input voltage must remain typically 2.0 V more negative even during the high point on the input ripple voltage.

XX = these two digits of the type number indicate voltage.

* = C_{in} is required if regulator is located an appreciable distance from power supply filter.

** = C_O improves stability and transient response.

ORDERING INFORMATION

Device	Output Voltage Tolerance	Tested Operating Junction Temp. Range	Package
MC79XXCK MC79XXACK*	4% 2%	$T_J = 0°C$ to $+125°C$	Metal Power**
MC79XXCT MC79XXACT*	4% 2%		Plastic Power
MC79XXBT#	4%	$T_J = -40°C$ to $+125°C$	

XX indicates nominal voltage.
*2% output voltage tolerance available in 6, 12 and 15 volt devices.
**Metal power package available in 5, 12 and 15 volt devices.
#Automotive temperature range selections are available with special test conditions and additional tests in 5, 12 and 15 volt devices. Contact your local Motorola sales office for information.

DEVICE TYPE/NOMINAL OUTPUT VOLTAGE

Device	Voltage	Device	Voltage
MC7905	5.0 Volts	MC7912	12 Volts
MC7905.2	5.2 Volts	MC7915	15 Volts
MC7906	6.0 Volts	MC7918	18 Volts
MC7908	8.0 Volts	MC7924	24 Volts

MC7900 Series

MAXIMUM RATINGS (T_A = +25°C unless otherwise noted.)

Rating	Symbol	Value	Unit
Input Voltage (–5.0 V $\geqslant$ V_O $\geqslant$ –18 V)	V_I	–35	Vdc
(24 V)		–40	
Power Dissipation			
Plastic Package			
T_A = +25°C	P_D	Internally Limited	Watts
Derate above T_A = +25°C	$1/R_{\theta JA}$	15.4	mW/°C
T_C = +25°C	P_D	Internally Limited	Watts
Derate above T_C = +95°C (See Figure 1)	$1/R_{\theta JC}$	200	mW/°C
Metal Package			
T_A = +25°C	P_D	Internally Limited	Watts
Derate above T_A = +25°C	$1/R_{\theta JA}$	22.2	mW/°C
T_C = +25°C	P_D	Internally Limited	Watts
Derate above T_C = +65°C	$1/R_{\theta JC}$	182	mW/°C
Storage Junction Temperature Range	T_{stg}	–65 to +150	°C
Junction Temperature Range	T_J	0 to +150	°C

THERMAL CHARACTERISTICS

Characteristic	Symbol	Max	Unit
Thermal Resistance, Junction to Ambient — Plastic Package	$R_{\theta JA}$	65	°C/W
— Metal Package		45	
Thermal Resistance, Junction to Case — Plastic Package	$R_{\theta JC}$	5.0	°C/W
— Metal Package		5.5	

MC7905C ELECTRICAL CHARACTERISTICS (V_I = –10 V, I_O = 500 mA, 0°C < T_J < +125°C unless otherwise noted.)

Characteristic	Symbol	Min	Typ	Max	Unit
Output Voltage (T_J = +25°C)	V_O	–4.8	–5.0	–5.2	Vdc
Line Regulation (Note 1)	Reg_{line}				mV
(T_J = +25°C, I_O = 100 mA)					
–7.0 Vdc $\geqslant$ V_I $\geqslant$ –25 Vdc		—	7.0	50	
–8.0 Vdc $\geqslant$ V_I $\geqslant$ –12 Vdc		—	2.0	25	
(T_J = +25°C, I_O = 500 mA)					
–7.0 Vdc $\geqslant$ V_I $\geqslant$ –25 Vdc		—	35	100	
–8.0 Vdc $\geqslant$ V_I $\geqslant$ –12 Vdc			8.0	50	
Load Regulation (T_J = +25°C) (Note 1)	Reg_{load}				mV
5.0 mA $\leqslant$ I_O $\leqslant$ 1.5 A		—	11	100	
250 mA $\leqslant$ I_O $\leqslant$ 750 mA		—	4.0	50	
Output Voltage	V_O	–4.75	—	–5.25	Vdc
–7.0 Vdc $\geqslant$ V_I $\geqslant$ –20 Vdc, 5.0 mA $\leqslant$ I_O $\leqslant$ 1.0 A, P $\leqslant$ 15 W					
Input Bias Current (T_J = +25°C)	I_{IB}	—	4.3	8.0	mA
Input Bias Current Change	ΔI_{IB}				mA
–7.0 Vdc $\geqslant$ V_I $\geqslant$ –25 Vdc		—	—	1.3	
5.0 mA $\leqslant$ I_O $\leqslant$ 1.5 A		—	—	0.5	
Output Noise Voltage (T_A = +25°C, 10 Hz $\leqslant$ f $\leqslant$ 100 kHz)	e_{on}	—	40	—	µV
Ripple Rejection (I_O = 20 mA, f = 120 Hz)	RR	—	70	—	dB
Dropout Voltage	$V_I - V_O$	—	2.0	—	Vdc
I_O = 1.0 A, T_J = +25°C					
Average Temperature Coefficient of Output Voltage	$\Delta V_O / \Delta T$	—	–1.0	—	mV/°C
I_O = 5.0 mA, 0°C $\leqslant$ T_J $\leqslant$ +125°C					

Note:
1. Load and line regulation are specified at constant junction temperature. Changes in V_O due to heating effects must be taken into account separately. Pulse testing with low duty cycle is used.

MC7900 Series

MC7912C ELECTRICAL CHARACTERISTICS (V_I = -19 V, I_O = 500 mA, 0°C < T_J < +125°C unless otherwise noted.)

Characteristic	Symbol	Min	Typ	Max	Unit
Output Voltage (T_J = +25°C)	V_O	-11.5	-12	-12.5	Vdc
Line Regulation (Note 1)	Reg_{line}				mV
(T_J = +25°C, I_O = 100 mA)					
-14.5 Vdc $\geq V_I \geq$ -30 Vdc		—	13	120	
-16 Vdc $\geq V_I \geq$ -22 Vdc		—	6.0	60	
(T_J = +25°C, I_O = 500 mA)					
-14.5 Vdc $\geq V_I \geq$ -30 Vdc		—	55	240	
-16 Vdc $\geq V_I \geq$ -22 Vdc		—	24	120	
Load Regulation (T_J = +25°C) (Note 1)	Reg_{load}				mV
5.0 mA $\leq I_O \leq$ 1.5 A		—	46	240	
250 mA $\leq I_O \leq$ 750 mA		—	17	120	
Output Voltage	V_O	-11.4	—	-12.6	Vdc
-14.5 Vdc $\geq V_I \geq$ -27 Vdc, 5.0 mA $\leq I_O \leq$ 1.0 A, P $\leq$ 15 W					
Input Bias Current (T_J = +25°C)	I_{IB}	—	4.4	8.0	mA
Input Bias Current Change	ΔI_{IB}				mA
-14.5 Vdc $\geq V_I \geq$ -30 Vdc		—	—	1.0	
5.0 mA $\leq I_O \leq$ 1.5 A		—	—	0.5	
Output Noise Voltage (T_A = +25°C, 10 Hz $\leq$ f $\leq$ 100 kHz)	e_{on}	—	75	—	μV
Ripple Rejection (I_O = 20 mA, f = 120 Hz)	RR	—	61	—	dB
Dropout Voltage	$V_I - V_O$	—	2.0	—	Vdc
I_O = 1.0 A, T_J = +25°C					
Average Temperature Coefficient of Output Voltage	$\Delta V_O / \Delta T$	—	-1.0	—	mV/°C
I_O = 5.0 mA, 0°C $\leq T_J \leq$ +125°C					

MC7912AC ELECTRICAL CHARACTERISTICS (V_I = -19 V, I_O = 500 mA, 0°C < T_J < +125°C unless otherwise noted.)

Characteristic	Symbol	Min	Typ	Max	Unit
Output Voltage (T_J = +25°C)	V_O	-11.75	-12	-12.25	Vdc
Line Regulation (Note 1)	Reg_{line}				mV
-16 Vdc $\geq V_I \geq$ -22 Vdc, I_O = 1.0 A, T_J = 25°C		—	6.0	60	
-16 Vdc $\geq V_I \geq$ -22 Vdc, I_O = 1.0 A		—	24	120	
-14.8 Vdc $\geq V_I \geq$ -30 Vdc, I_O = 500 mA		—	24	120	
-14.5 Vdc $\geq V_I \geq$ -27 Vdc, I_O = 1.0 A, T_J = 25°C		—	13	120	
Load Regulation (Note 1)	Reg_{load}				mV
5.0 mA $\leq I_O \leq$ 1.5 A, T_J = 25°C		—	46	150	
250 mA $\leq I_O \leq$ 750 mA		—	17	75	
5.0 mA $\leq I_O \leq$ 1.0 A		—	35	150	
Output Voltage	V_O	-11.5	—	-12.5	Vdc
-14.8 Vdc $\geq V_I \geq$ -27 Vdc, 5.0 mA $\leq I_O \leq$ 1.0 A, P $\leq$ 15 W					
Input Bias Current	I_{IB}	—	4.4	8.0	mA
Input Bias Current Change	ΔI_{IB}				mA
16 Vdc $\geq V_I \geq$ -30 Vdc		—	—	0.8	
5.0 mA $\leq I_O \leq$ 1.0 A		—	—	0.5	
5.0 mA $\leq I_O \leq$ 1.5 A, T_J = 25°C		—	—	0.5	
Output Noise Voltage (T_A = +25°C, 10 Hz $\leq$ f $\leq$ 100 kHz)	e_{on}	—	75	—	μV
Ripple Rejection (I_O = 20 mA, f = 120 Hz)	RR	—	61	—	dB
Dropout Voltage	$V_I - V_O$	—	2.0	—	Vdc
I_O = 1.0 A, T_J = +25°C					
Average Temperature Coefficient of Output Voltage	$\Delta V_O / \Delta T$	—	-1.0	—	mV/°C
I_O = 5.0 mA, 0°C $\leq T_J \leq$ +125°C					

Note
1. Load and line regulation are specified at constant junction temperature. Changes in V_O due to heating effects must be taken into account separately. Pulse testing with low duty cycle is used.

ANALOG DEVICES

FEATURES
Low Nonlinearity: ±0.012% max (AD295C)
Low Gain Drift: ±60ppm/°C max
Floating Input and Output Power: ±15V dc @ 5mA
3-Port Isolation: ±2500V CMV (Input to Output)
Complies with NEMA ICS1-111
Gain Adjustable: 1V/V to 1000V/V
User Configurable Input Amplifier

APPLICATIONS
Motor Controls
Process Signal Isolator
High Voltage Instrumentation Amplifier
Multichannel Data Acquisition Systems
Off Ground Signal Measurements

AD295 FUNCTIONAL BLOCK DIAGRAM

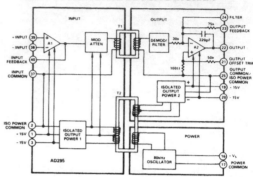

GENERAL DESCRIPTION
The AD295 is a high accuracy, high reliability hybrid isolation amplifier designed for industrial, instrumentation and medical applications. Three performance versions are available offering guaranteed nonlinearity error at 10V p-p output: ±0.05% max (AD295A), ±0.025% max (AD295B), ±0.012% max (AD295C). Using a pulse width modulation technique the AD295 provides 3-port isolation between input, output and power supply ports. Using this technique, the AD295 interrupts ground loops and leakage paths and minimizes the effect of high voltage transients. Additionally, floating (isolated) power ±15V dc @ 5mA is available at both the input and output. The AD295's gain can be programmed at the input, output or both sections allowing for user flexibility. An uncommitted input amplifier allows configuration as a buffer, inverter, subtractor or differential amplifier.

The AD295 is provided in an epoxy sealed ceramic 40-pin package that insures quality performance, high stability and accuracy. Input/output pin spacing complies with NEMA (ICS1-111) separation specifications required for many industrial applications.

WHERE TO USE THE MODEL AD295
Industrial: The AD295 is designed for measuring signals in harsh industrial environments. The AD295 provides high accuracy with complete galvanic isolation and protection from transients or where ground fault currents or high common-mode voltages are present. The AD295 can be applied in process controllers, current loop receivers, motor controls and weighing systems.

Instrumentation: In data acquisition systems the AD295 provides common-mode rejection for conditioning thermocouples, strain gauges or other low-level signals where high performance and system protection is required.

Medical: In biomedical and patient monitoring equipment like diagnostic systems and blood pressure monitors, the AD295 provides protection from lethal ground fault currents. Low level signal recording and monitoring is achieved with the AD295's low input noise (2μV p-p @ G = 1000V/V) and high CMR (106dB @ 60Hz).

DESIGN FEATURES AND USER BENEFITS
Isolated Power: Isolated power supply sections at the input and output provide ±15V dc @ 5mA. Isolated power is load regulated to 4%. This feature permits the AD295 to excite floating signal conditioners, front-end buffer amplifiers and remote transducers at the input and external circuitry at the output. This eliminates the need for a separate dc/dc converter.

Input Amplifier: The uncommitted input amplifier allows the user to configure the input as a buffer, inverter, subtractor or differential amplifier to meet the application need.

Adjustable Gain: Gain can be selected at the input, output or both. Thus, circuit response can be tailored to the user's application. The AD295 provides the user with flexibility for circuit optimization without requiring external active components.

Three-Port Isolation: Provides true galvanic isolation between input, output and power supply ports. Eliminates the need for power supply and output ports being returned through a common ground.

Wide Operating Temperature: The AD295 is designed to operate over the −40°C to +100°C temperature range with rated performance over −25°C to +85°C.

Leakage: The low coupling capacitance between input and output yields a ground leakage current of less than 2μA rms at 115V ac, 60Hz. The AD295 meets standards established by UL STD 544.

SPECIFICATIONS (typical @ +25°C, & V_S = +15V unless otherwise noted)

Model	AD295A	AD295B	AD295C
GAIN			
Range	1V/V to 1000V/V	*	*
Open Loop	100dB	*	*
Accuracy G = 1V/V	± 1.5%	*	*
vs. Temperature (− 25°C to + 85°C)			
G = 1V/V to 100V/V	± 60ppm/°C max	*	*
Nonlinearity (± 5V Swing) G = 1V–100V/V	± 0.05% max	± 0.025% max	± 0.012% max
INPUT VOLTAGE RATINGS			
Linear Differential Range	± 10V min	*	*
Max Safe Differential Input	± 15V	*	*
Max CMV (Input to Output)			
Continuous ac or dc	± 2500V peak	*	*
ac, 60Hz, 1 Minute Duration	2500V rms	*	*
Max CMV (Input to Power Common/Output to Power Common)			
Continuous ac or dc	± 2000V peak	*	*
ac, 60Hz, 1 Minute Duration	2000V rms	*	*
CMR, Input to Output 60Hz. G = 1V/V			
$R_S \leq 1k\Omega$ Balanced Source Impedance	106dB	*	*
$R_S \leq 1k$ Source Impedance Imbalance	103dB min	*	*
Max Leakage Current, Input to Output			
@ 115V ac, 60Hz	2μA rms max	*	*
INPUT IMPEDANCE			
Differential	$5 \cdot 10^7 \Omega$ 33pF	*	*
Common Mode	$10^8 \Omega$ 20pF	*	*
INPUT BIAS CURRENT			
Initial, @ + 25°C	5nA max	*	*
vs. Temperature (− 25°C to + 85°C)	− 25pA/°C max	*	*
INPUT DIFFERENCE CURRENT			
Initial, @ + 25°C	± 2nA max	*	*
vs. Temperature (− 25°C to + 85°C)	± 5pA/°C max	*	*
INPUT NOISE (Gain = 1000V/V)			
Voltage			
0.01Hz to 10Hz	2μV p-p	*	*
10Hz to 1kHz	1μV rms	*	*
Current			
0.01Hz to 10Hz	10pA p-p	*	*
FREQUENCY RESPONSE			
Small Signal (−3dB)			
G = 1V/V to 100V/V	4.5kHz	*	*
G = 1000V/V	600Hz	*	*
Full Power, 20V p-p Output			
G = 1V/V to 100V/V	1.4kHz	*	*
G = 1000V/V	200Hz	*	*
Slew Rate G = 1V/V to 100V/V	0.1V/μs	*	*
Settling Time G = 1V/V			
(to ± 0.1% for 10V Step)	550μs	*	*
(to ± 0.1% for 20V Step)	700μs	*	*
OFFSET VOLTAGE, REFERRED TO INPUT			
Initial @ + 25°C (Adjustable to Zero)	$\pm \left(3 + \frac{15}{G_{IN}}\right)$ mV max	*	
vs. Temperature (− 25°C to + 85°C)	$\pm \left(10 + \frac{450}{G_{IN}}\right)\frac{\mu V}{°C}$ max	$\pm \left(3 + \frac{300}{G_{IN}}\right)\frac{\mu V}{°C}$ max	$\pm \left(1.5 + \frac{150}{G_{IN}}\right)\frac{\mu V}{°C}$ max
vs. Supply	$\pm \left(1 + \frac{200}{G_{IN}}\right)\frac{\mu V}{\%}$	*	
RATED OUTPUT			
Voltage, 2kΩ Load	± 10V min	*	*
Output Impedance	2Ω (dc to 100Hz)	*	*
Output Ripple (10Hz to 10kHz)	6mV p-p	*	*
(10Hz to 100kHz)	40mV p-p	*	*
ISOLATED POWER SUPPLIES [1] (V_{ISO1} & V_{ISO2})			
Voltage	± 15V dc	*	*
Accuracy	± 5%	*	*
Current [2]	± 5mA max	*	*
Load Regulation (No Load to Full Load)	− 4%	*	*
Ripple, 100kHz BW	12mV p-p	*	
POWER SUPPLY (+ V_S)			
Voltage, Rated Performance	+ 15V dc ± 3%	*	
Voltage, Operating	+ 12V dc to + 16V dc	*	
Current, Quiescent (V_S = + 15V)	40mA	*	
With V_{ISO} Loaded	45mA	*	
TEMPERATURE RANGE			
Rated Performance	− 25°C to 85°C	*	*
Operating	− 40°C to + 100°C	*	*
Storage	− 40°C to + 100°C	*	*
CASE DIMENSIONS	2.7" × 0.88" × 0.375"	*	*

NOTES
[1] V_{ISO1} accuracy and regulation 10%
[2] = 10mA can be supplied by V_{ISO1} if V_{ISO2} is not used

*Specifications same as AD295A
Specifications subject to change without notice.

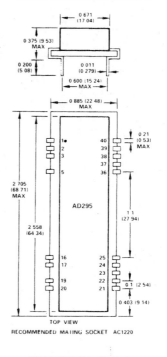

OUTLINE DIMENSIONS
Dimensions shown in inches and (mm).

TOP VIEW
RECOMMENDED MATING SOCKET AC1220

PIN DESIGNATIONS

PIN	FUNCTION	PIN	FUNCTION
1	+ 15V (+ V_{ISO1})	40	INPUT FEEDBACK
2	V_{ISO1} COM	39	+ INPUT
3	15V (− V_{ISO1})	38	− INPUT
		37	INPUT COM
5	NO CONNECTION	36	NO CONNECTION
16	+ V_S	25	OUTPUT COM V_{ISO2} COM
17	POWER COMMON	24	FILTER
		23	OUTPUT FEEDBACK
19	+ 15V (+ V_{ISO2})	22	OUTPUT
20	− 15V (+ V_{ISO2})	21	OUTPUT OFFSET TRIM

551

Integrated Circuit
Precision Instrumentation Amplifier

AD521

FEATURES
Programmable Gains from 0.1 to 1000
Differential Inputs
High CMRR: 110dB min
Low Drift: $2\mu V/^\circ C$ max (L)
Complete Input Protection, Power ON and Power OFF
Functionally Complete with the Addition of Two Resistors
Internally Compensated
Gain Bandwidth Product: 40MHz
Output Current Limited: 25mA
Very Low Noise: $0.5\mu V$ p-p, 0.1Hz to 10Hz, RTI @ G = 1000
Chips are Available

AD521 PIN CONFIGURATION

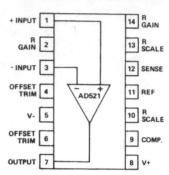

PRODUCT DESCRIPTION
The AD521 is a second generation, low cost, monolithic IC instrumentation amplifier developed by Analog Devices. As a true instrumentation amplifier, the AD521 is a gain block with differential inputs and an accurately programmable input/output gain relationship.

The AD521 IC instrumentation amplifier should not be confused with an operational amplifier, although several manufacturers (including Analog Devices) offer op amps which can be used as building blocks in variable gain instrumentation amplifier circuits. Op amps are general-purpose components which, when used with precision-matched external resistors, can perform the instrumentation amplifier function.

An instrumentation amplifier is a precision differential voltage gain device optimized for operation in a real world environment, and is intended to be used wherever acquisition of a useful signal is difficult. It is characterized by high input impedance, balanced differential inputs, low bias currents and high CMR.

As a complete instrumentation amplifier, the AD521 requires only two resistors to set its gain to any value between 0.1 and 1000. The ratio matching of these resistors does not affect the high CMRR (up to 120dB) or the high input impedance (3 X $10^9 \Omega$) of the AD521. Furthermore, unlike most operational amplifier-based instrumentation amplifiers, the inputs are protected against overvoltages up to ±15 volts beyond the supplies.

The AD521 IC instrumentation amplifier is available in four different versions of accuracy and operating temperature range. The economical "J" grade, the low drift "K" grade, and the lower drift, higher linearity "L" grade are specified from 0 to

+70°C. The "S" grade guarantees performance to specification over the extended temperature range: –55°C to +125°C.

PRODUCT HIGHLIGHTS
1. The AD521 is a true instrumentation amplifier in integrated circuit form, offering the user performance comparable to many modular instrumentation amplifiers at a fraction of the cost.

2. The AD521 has low guaranteed input offset voltage drift ($2\mu V/^\circ C$ for L grade) and low noise for precision, high gain applications.

3. The AD521 is functionally complete with the addition of two resistors. Gain can be preset from 0.1 to more than 1000.

4. The AD521 is fully protected for input levels up to 15V beyond the supply voltages and 30V differential at the inputs.

5. Internally compensated for all gains, the AD521 also offers the user the provision for limiting bandwidth.

6. Offset nulling can be achieved with an optional trim pot.

7. The AD521 offers superior dynamic performance with a gain-bandwidth product of 40MHz, full peak response of 100kHz (independent of gain) and a settling time of $5\mu s$ to 0.1% of a 10V step.

SPECIFICATIONS (typical @ $V_S = \pm15V$, $R_L = 2k\Omega$ and $T_A = 25°C$ unless otherwise specified)

MODEL	AD521JD	AD521KD	AD521LD	AD521SD (AD521SD/883B)
GAIN				
Range (For Specified Operation, Note 1)	1 to 1000	*	*	*
Equation	$G = R_S/R_G$ V/V	*	*	*
Error from Equation	$(\pm0.25 - 0.004G)\%$	*	*	*
Nonlinearity (Note 2)				
$1 \leqslant G \leqslant 1000$	0.2% max	*	0.1% max	*
Gain Temperature Coefficient	$\pm(3 \pm0.05G)$ppm/°C	*	*	$\pm(15 \pm0.4G)$ppm/°C
OUTPUT CHARACTERISTICS				
Rated Output	±10V, ±10mA min	*	*	*
Output at Maximum Operating Temperature	±10V @ 5mA min	*	*	*
Impedance	0.1Ω	*	*	*
DYNAMIC RESPONSE				
Small Signal Bandwidth (±3dB)				
G = 1	>2MHz	*	*	*
G = 10	300kHz	*	*	*
G = 100	200kHz	*	*	*
G = 1000	40kHz	*	*	*
Small Signal, ±1.0% Flatness				
G = 1	75kHz	*	*	*
G = 10	26kHz	*	*	*
G = 100	24kHz	*	*	*
G = 1000	6kHz	*	*	*
Full Peak Response (Note 3)	100kHz	*	*	*
Slew Rate, $1 \leqslant G \leqslant 1000$	10V/µs	*	*	*
Settling Time (any 10V step to within 10mV of Final Value)				
G = 1	7µs	*	*	*
G = 10	5µs	*	*	*
G = 100	10µs	*	*	*
G = 1000	35µs	*	*	*
Differential Overload Recovery (±30V Input to within 10mV of Final Value) (Note 4)				
G = 1000	50µs	*	*	*
Common Mode Step Recovery (30V Input to within 10mV of Final Value) (Note 5)				
G = 1000	10µs	*	*	*
VOLTAGE OFFSET (may be nulled)				
Input Offset Voltage (V_{OS_I})	3mV max (2mV typ)	1.5mV max (0.5mV typ)	1.0mV max (0.5mV typ)	**
vs. Temperature	15µV/°C max (7µV/°C typ)	5µV/°C max (1.5µV/°C typ)	2µV/°C max	**
vs. Supply	3µV/%	*	*	**
Output Offset Voltage (V_{OS_O})	400mV max (200mV typ)	200mV max (30mV typ)	100mV max	**
vs. Temperature	400µV/°C max (150µV/°C typ)	150µV/°C max (50µV/°C typ)	75µV/°C max	**
vs. Supply (Note 6)	$0.005 V_{OS_O}/\%$	*	*	*
INPUT CURRENTS				
Input Bias Current (either input)	80nA max	40nA max	**	**
vs. Temperature	1nA/°C max	500pA/°C max	**	**
vs. Supply	2%/V	*	*	*
Input Offset Current	20nA max	10nA max	**	**
vs. Temperature	250pA/°C max	125pA/°C max	**	**
INPUT				
Differential Input Impedance (Note 7)	$3 \times 10^9 \Omega \| 1.8pF$	*	*	*
Common Mode Input Impedance (Note 8)	$6 \times 10^{10} \Omega \| 3.0pF$	*	*	*
Input Voltage Range for Specified Performance (with respect to ground)	±10V	*	*	*
Maximum Voltage without Damage to Unit, Power ON or OFF Differential Mode (Note 9)	30V	*	*	*
Voltage at either input (Note 9)	$V_S \pm15V$	*	*	*
Common Mode Rejection Ratio, DC to 60Hz with 1kΩ source unbalance				
G = 1	70dB min (74dB typ)	74dB min (80dB typ)	**	**
G = 10	90dB min (94dB typ)	94dB min (100dB typ)	**	**
G = 100	100dB min (104dB typ)	104dB min (114dB typ)	**	**
G = 1000	100dB min (110dB typ)	110dB min (120dB typ)	**	**
NOISE				
Voltage RTO (p-p) @ 0.1Hz to 10Hz (Note 10)	$\sqrt{(0.5G)^2 + (225)^2}\,\mu V$	*	*	*
RMS RTO, 10Hz to 10kHz	$\sqrt{(1.2G)^2 + (50)^2}\,\mu V$	*	*	*
Input Current, rms, 10Hz to 10kHz	15pA (rms)	*	*	*
REFERENCE TERMINAL				
Bias Current	3µA	*	*	*
Input Resistance	10MΩ	*	*	*
Voltage Range	±10V	*	*	*
Gain to Output	1	*	*	*
POWER SUPPLY				
Operating Voltage Range	±5V to ±18V	*	*	*
Quiescent Supply Current	5mA max	*	*	*
TEMPERATURE RANGE				
Specified Performance	0 to +70°C	*	*	-55°C to +125°C
Operating	-25°C to +85°C	*	*	-55°C to +125°C
Storage	-65°C to +150°C	*	*	*
PACKAGE OPTION [1]				
Ceramic (D-14)	AD521JD	AD521KD	AD521LD	AD521SD

NOTES
[1] See Section 20 for package outline information
* Specifications same as AD521JD.
** Specifications same as AD521KD.
Specifications subject to change without notice.

Voltage-to-Frequency and Frequency-to-Voltage Converter

AD650

FEATURES
V/F Conversion to 1MHz
Reliable Monolithic Construction
Very Low Nonlinearity
   0.002% typ at 10kHz
   0.005% typ at 100kHz
   0.07% typ at 1MHz
Input Offset Trimmable to Zero
CMOS or TTL Compatible
Unipolar, Bipolar, or Differential V/F
V/F or F/V Conversion
Available in Surface Mount
MIL-STD-883-Compliant Versions Available

PIN CONFIGURATION

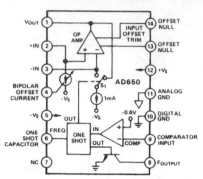

PRODUCT DESCRIPTION

The AD650 V/F/V (voltage-to-frequency or frequency-to-voltage converter) provides a combination of high frequency operation and low nonlinearity previously unavailable in monolithic form. The inherent monotonicity of the V/F transfer function makes the AD650 useful as a high-resolution analog-to-digital converter. A flexible input configuration allows a wide variety of input voltage and current formats to be used, and an open-collector output with separate digital ground allows simple interfacing to either standard logic families or opto-couplers.

The linearity error of the AD650 is typically 20ppm (0.002% of full scale) and 50ppm (0.005%) maximum at 10kHz full scale. This corresponds to approximately 14-bit linearity in an analog-to-digital converter circuit. Higher full-scale frequencies or longer count intervals can be used for higher resolution conversions. The AD650 has a useful dynamic range of six decades allowing extremely high resolution measurements. Even at 1MHz full scale, linearity is guaranteed less than 1000ppm (0.1%) on the AD650KN, KP, BD and SD grades.

In addition to analog-to-digital conversion, the AD650 can be used in isolated analog signal transmission applications, phased-locked-loop circuits, and precision stepper motor speed controllers. In the F/V mode, the AD650 can be used in precision tachometer and FM demodulator circuits.

The input signal range and full-scale output frequency are user-programmable with two external capacitors and one resistor. Input offset voltage can be trimmed to zero with an external potentiometer.

The AD650JN and AD650KN are offered in a plastic 14-pin DIP package. The AD650JP and AD650KP are available in a

20-pin plastic leaded chip carrier (PLCC). Both plastic packaged versions of the AD650 are specified for the commerical (0 to +70°C) temperature range. For industrial temperature range (-25°C to +85°C) applications, the AD650AD and AD650BD are offered in a ceramic package. The AD650SD is specified for the full -55°C to +125°C extended temperature range.

PRODUCT HIGHLIGHTS

1. In addition to very high linearity, the AD650 can operate at full scale output frequency up to 1MHz. The combination of these two features makes the AD650 an inexpensive solution for applications requiring high resolution monotonic A/D conversion.

2. The AD650 has a very versatile architecture that can be configured to accommodate bipolar, unipolar, or differential input voltages, or unipolar input currents.

3. TTL or CMOS compatibility is achieved using an open collector frequency output. The pullup resistor can be connected to voltages up to +30V, or +15V or +5V for conventional CMOS or TTL logic levels.

4. The same components used for V/F conversion can also be used for F/V conversion by adding a simple logic biasing network and reconfiguring the AD650.

5. The AD650 provides separate analog and digital grounds. This feature allows prevention of ground loops in real-world applications.

6. The AD650 is available in versions compliant with MIL-STD-883. Refer to the Analog Devices Military Products Databook or current AD650/883B data sheet for detailed specifications.

AD650—SPECIFICATIONS (@ +25°C with $V_S = \pm 15V$ unless otherwise noted)

Model	AD650J/AD650A Min	Typ	Max	AD650K/AD650B Min	Typ	Max	AD650S Min	Typ	Max	Units
DYNAMIC PERFORMANCE										
Full Scale Frequency Range			1			1			1	MHz
Nonlinearity[1] f_{max} = 10kHz		0.002	0.005		0.002	0.005		0.002	0.005	%
100kHz		0.005	**0.02**		0.005	**0.02**		0.005	**0.02**	%
500kHz		0.02	0.05		0.02	0.05		0.02	0.05	%
1MHz		0.1			0.05	**0.1**		0.05	**0.1**	%
Full Scale Calibration Error[2], 100kHz		±5			±5			±5		%
1MHz		±10			±10			±5		%
vs. Supply[3]	0.015		+0.015	−0.015		+0.015	−0.015		+0.015	% of FSR/V
vs. Temperature										
A, B, and S Grades										
at 10kHz			±75			±75			±75	ppm/°C
at 100kHz			±150			±150			±150	ppm/°C
J and K Grades										
at 10kHz		±75			±75					ppm/°C
at 100kHz		±150			±150					ppm/°C
BIPOLAR OFFSET CURRENT										
Activated by 1.24kΩ between pins 4 and 5	0.45	0.5	0.55	0.45	0.5	0.55	0.45	0.5	0.55	mA
DYNAMIC RESPONSE										
Maximum Settling Time for Full Scale										
Step Input	1 Pulse of New Frequency Plus 1μs			1 Pulse of New Frequency Plus 1μs			1 Pulse of New Frequency Plus 1μs			
Overload Recovery Time										
Step Input	1 Pulse of New Frequency Plus 1μs			1 Pulse of New Frequency Plus 1μs			1 Pulse of New Frequency Plus 1μs			
ANALOG INPUT AMPLIFIER (V/F Conversion)										
Current Input Range (Figure 1)	0		+0.6	0		+0.6	0		+0.6	mA
Voltage Input Range (Figure 5)	10		0	10		0	10		0	V
Differential Impedance	2MΩ‖10pF			2MΩ‖10pF			2MΩ‖10pF			
Common Mode Impedance	1000MΩ‖10pF			1000MΩ‖10pF			1000MΩ‖10pF			
Input Bias Current										
Noninverting Input		40	**100**		40	**100**		40	**100**	nA
Inverting Input		±8	**±20**		±8	**±20**		±8	**±20**	nA
Input Offset Voltage										
(Trimmable to Zero)			±4			±4			±4	mV
vs. Temperature (T_{min} to T_{max})		±30			±30			±30		μV/°C
Safe Input Voltage		±V_S			±V_S			±V_S		C
COMPARATOR (F/V Conversion)										
Logic "0" Level	V_S		1	V_S		1	V_S		+1	V
Logic "1" Level	0		+V_S	0		+V_S	0		+V_S	V
Pulse Width Range[4]	0.1		(0.3 · t_{OS})	0.1		(0.3 · t_{OS})	0.1		(0.3 · t_{OS})	μs
Input Impedance		250			250			250		kΩ
OPEN COLLECTOR OUTPUT (V/F Conversion)										
Output Voltage in Logic "0"										
I_{SINK} 8mA, T_{min} to T_{max}			0.4			0.4			0.4	V
Output Leakage Current in Logic "1"			**100**			**100**			**100**	nA
Voltage Range[5]	0		+36	0		+36	0		+36	V
AMPLIFIER OUTPUT (F/V Conversion)										
Voltage Range (1500Ω min load resistance)	0		+10	0		+10	0		+10	V
Source Current (750Ω max load resistance)	10			10			10			mA
Capacitive Load (Without Oscillation)			100			100			100	pF
POWER SUPPLY										
Voltage, Rated Performance	±9		±18	±9		±18	±9		±18	V
Quiescent Current		8			8			8		mA
TEMPERATURE RANGE										
Rated Performance – N Package	0		+70	0		+70				°C
D Package	25		+85	25		+85	55		+125	°C
Storage – N Package	25		+85	25		+85				°C
D Package	65		+150	65		+150	65		+150	°C
PACKAGE OPTIONS[6]										
PLCC (P-20A)	AD650JP			AD650KP						
Plastic DIP (N-14)	AD650JN			AD650KN						
Ceramic DIP (D-14)	AD650AD			AD650BD			AD650SD			

NOTES

[1]Nonlinearity is defined as deviation from a straight line from zero to full scale, expressed as a fraction of full scale.

[2]Full scale calibration error adjustable to zero.

[3]Measured at full scale output frequency of 100kHz.

[4]Refer to F/V conversion section of the text.

[5]Referred to digital ground.

[6]D = Ceramic DIP; N = Plastic DIP; P = Plastic Leaded Chip Carrier. For outline information see Package Information section.

Specifications subject to change without notice.

Specifications shown in boldface are tested on all production units at final electrical test. Results from those tests are used to calculate outgoing quality levels. All min and max specifications are guaranteed, although only those shown in boldface are tested on all production units.

ABSOLUTE MAXIMUM RATINGS

Total Supply Voltage $+V_S$ to $-V_S$ 36V
Storage Temperature Ceramic . . . . $-55°C$ to $+165°C$
               Plastic . . . . $-25°C$ to $+125°C$
Differential Input Voltage (Pins 2 & 3) . . . . $\pm 10V$
Maximum Input Voltage . . . . $\pm V_S$
Open Collector Output   Voltage Above Digital GND . . 36V
                      Current 50mA
Amplifier Short Ckt to Ground . . . . Indefinite
Comparator Input Voltage (Pin 9) . . . . $\pm V_S$

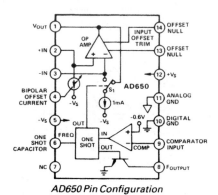

AD650 Pin Configuration

ORDERING GUIDE

Part[1] Number	Gain Tempco ppm/°C 100kHz	1MHz 100kHz Linearity	Specified Temperature Range °C	Package
AD650JN	150 typ	0.1% typ	0 to +70	Plastic DIP
AD650KN	150 typ	0.1% max	0 to +70	Plastic DIP
AD650JP	150 typ	0.1% typ	0 to +70	PLCC
AD650KP	150 typ	0.1% max	0 to +70	PLCC
AD650AD	150 max	0.1% typ	−25 to +85	Ceramic
AD650BD	150 max	0.1% max	−25 to +85	Ceramic
AD650SD	150 max	0.1% max	−55 to +125	Ceramic

NOTE
[1]For details on grade and package offerings screened in accordance with
MIL-STD-883, refer to the Analog Devices Military Products Databook or
current AD650/883B data sheet.

FEATURES
Complete 8-Bit A/D Converter with Reference, Clock
and Comparator
30µs Maximum Conversion Time
Full 8- or 16-Bit Microprocessor Bus Interface
Unipolar and Bipolar Inputs
No Missing Codes Over Temperature
Operates on +5V and −12V to −15V Supplies
MIL-STD-883 Compliant Version Available

FUNCTIONAL BLOCK DIAGRAM

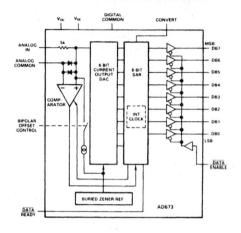

PRODUCT DESCRIPTION
The AD673 is a complete 8-bit successive approximation analog-to-digital converter consisting of a DAC, voltage reference, clock, comparator, successive approximation register (SAR) and 3 state output buffers–all fabricated on a single chip. No external components are required to perform a full accuracy 8-bit conversion in 20µs.

The AD673 incorporates advanced integrated circuit design and processing technologies. The successive approximation function is implemented with I²L (integrated injection logic). Laser trimming of the high stability SiCr thin film resistor ladder network insures high accuracy, which is maintained with a temperature compensated sub-surface Zener reference.

Operating on supplies of +5V and −12V to −15V, the AD673 will accept analog inputs of 0 to +10V or −5V to +5V. The trailing edge of a positive pulse on the CONVERT line initiates the 20µs conversion cycle. DATA READY indicates completion of the conversion.

The AD673 is available in two versions. The AD673J as specified over the 0 to +70°C temperature range and the AD673S guarantees ± ½LSB relative accuracy and no missing codes from −55°C to +125°C.

Two package configurations are offered. All versions are also offered in a 20-pin hermetically sealed ceramic DIP. The AD673J is also available in a 20-pin plastic DIP.

PRODUCT HIGHLIGHTS
1. The AD673 is a complete 8-bit A/D converter. No external components are required to perform a conversion.

2. The AD673 interfaces to many popular microprocessors without external buffers or peripheral interface adapters.

3. The device offers true 8-bit accuracy and exhibits no missing codes over its entire operating temperature range.

4. The AD673 adapts to either unipolar (0 to +10V) or bipolar (−5V to +5V) analog inputs by simply grounding or opening a single pin.

5. Performance is guaranteed with +5V and −12V or −15V supplies.

6. The AD673 is available in a version compliant with MIL-STD-883. Refer to the Analog Devices Military Products Databook or current /883B data sheet for detailed specifications.

AD673—SPECIFICATIONS

($T_A = 25°C$, $V+ = +5V$, $V- = -12V$ or $-15V$, all voltages measured with respect to digital common, unless otherwise indicated.)

Model	AD673J Min	AD673J Typ	AD673J Max	AD673S Min	AD673S Typ	AD673S Max	Units
RESOLUTION		8			8		Bits
RELATIVE ACCURACY,[1]			± 1/2			± 1/2	LSB
$T_A = T_{min}$ to T_{max}			± 1/2			± 1/2	LSB
FULL SCALE CALIBRATION[2]		± 2			± 2		LSB
UNIPOLAR OFFSET			± 1/2			± 1/2	LSB
BIPOLAR OFFSET			± 1/2			± 1/2	LSB
DIFFERENTIAL NONLINEARITY,[3]	8			8			Bits
$T_A = T_{min}$ to T_{max}	8			8			Bits
TEMPERATURE RANGE	0		+ 70	− 55		+ 125	°C
TEMPERATURE COEFFICIENTS							
Unipolar Offset			± 1			± 1	LSB
Bipolar Offset			± 1			± 1	LSB
Full Scale Calibration[2]			± 2			± 2	LSB
POWER SUPPLY REJECTION							
Positive Supply							
$+4.5 \leq V+ \leq +5.5V$			± 2			± 2	LSB
Negative Supply							
$-15.75V \leq V- \leq -14.25V$			± 2			± 2	LSB
$-12.6V \leq V- \leq -11.4V$			± 2			± 2	LSB
ANALOG INPUT IMPEDANCE	3.0	5.0	7.0	3.0	5.0	7.0	kΩ
ANALOG INPUT RANGES							
Unipolar	0		+ 10	0		+ 10	V
Bipolar	− 5		+ 5	− 5		+ 5	V
OUTPUT CODING							
Unipolar	Positive True Binary			Positive True Binary			
Bipolar	Positive True Offset Binary			Positive True Offset Binary			
LOGIC OUTPUT							
Output Sink Current							
($V_{OUT} = 0.4V$ max, T_{min} to T_{max})	3.2			3.2			mA
Output Source Current[4]							
($V_{OUT} = 2.4V$ min, T_{min} to T_{max})	0.5			0.5			mA
Output Leakage			± 40			± 40	μA
LOGIC INPUTS							
Input Current			± 100			± 100	μA
Logic "1"	2.0			2.0			V
Logic "0"			0.8			0.8	V
CONVERSION TIME, T_A and							
T_{min} to T_{max}	10	20	30	10	20	30	μs
POWER SUPPLY							
V +	+ 4.5	+ 5.0	+ 7.0	+ 4.5	+ 5.0	+ 7.0	V
V −	− 11.4	− 15	− 16.5	− 11.4	− 15	− 16.5	V
OPERATING CURRENT							
V +		15	20		15	20	mA
V −		9	15		9	15	mA

NOTES

[1]Relative accuracy is defined as the deviation of the code transition points from the ideal transfer point on a straight line from the zero to the full scale of the device.

[2]Full scale calibration is guaranteed trimmable to zero with an external 200Ω potentiometer in place of the 15Ω fixed resistor.
Full scale is defined as 10 volts minus 1LSB, or 9.961 volts.

[3]Defined as the resolution for which no missing codes will occur.

[4]The data output lines have active pull-ups to source 0.5mA. The DATA READY line is open collector with a nominal 6kΩ internal pull-up resistor.

Specifications subject to change without notice.

Specifications shown in boldface are tested on all production units at final electrical test. Results from those tests are used to calculate outgoing quality levels. All min and max specifications are guaranteed, although only those shown in boldface are tested on all production units.

ABSOLUTE MAXIMUM RATINGS

V + to Digital Common . . . . 0 to + 7V
V − to Digital Common . . . . 0 to − 16.5V
Analog Common to Digital Common . . . . ± 1V
Analog Input to Analog Common . . . . ± 15V
Control Inputs . . . . 0 to V +
Digital Outputs (High Impedance State) . . . . . 0 to V +
Power Dissipation 800mW

ORDERING GUIDE

Model	Temperature Range	Relative Accuracy	Package Options[1]
AD673JN	0 to + 70°C	± 1/2LSB max	Plastic DIP (N-20)
AD673JD	0 to + 70°C	± 1/2LSB max	Ceramic DIP (D-20)
AD673SD[2]	− 55°C to + 125°C	± 1/2LSB max	Ceramic DIP (D-20)
AD673JP	0 to + 70°C	± 1/2LSB max	PLCC (P-20A)

NOTES
[1]D = Ceramic DIP; N = Plastic DIP; P = Plastic Leaded Chip Carrier. For outline information see Package Information section.
[2]For details on grade and package offering screened in accordance with MIL-STD-883, refer to Analog Devices Military Products Databook.

FUNCTIONAL DESCRIPTION

A block diagram of the AD673 is shown in Figure 1. The positive CONVERT pulse must be at least 500ns wide. $\overline{DR}$ goes high within 1.5μs after the leading edge of the convert pulse indicating that the internal logic has been reset. The negative edge of the CONVERT pulse initiates the conversion. The internal 8-bit current output DAC is sequenced by the integrated injection logic (I^2L) successive approximation register (SAR) from its most significant bit to least significant bit to provide an output current which accurately balances the input signal current through the 5kΩ resistor. The comparator determines whether the addition of each successively weighted bit current causes the DAC current sum to be greater or less than the input current; if the sum is more, the bit is turned off. After testing all bits, the SAR contains a 8-bit binary code which accurately represents the input signal to within (0.05% of full scale).

The temperature compensated buried Zener reference provides the primary voltage reference to the DAC and ensures excellent stability with both time and temperature. The bipolar offset input controls a switch which allows the positive bipolar offset current (exactly equal to the value of the MSB less ½LSB) to be injected into the summing (+) node of the comparator to offset the DAC output. Thus the nominal 0 to + 10V unipolar input range becomes a − 5V to + 5V range. The 5kΩ thin film input resistor is trimmed so that with a full scale input signal, an input current will be generated which exactly matches the DAC output with all bits on.

UNIPOLAR CONNECTION

The AD673 contains all the active components required to perform a complete A/D conversion. Thus, for many applications, all that is necessary is connection of the power supplies (+ 5V and − 12V to − 15V), the analog input and the convert pulse. However, there are some features and special connections which should be considered for achieving optimum performance. The functional pin-out is shown in Figure 2.

The standard unipolar 0 to + 10V range is obtained by shorting the bipolar offset control pin (pin 16) to digital common (pin 17).

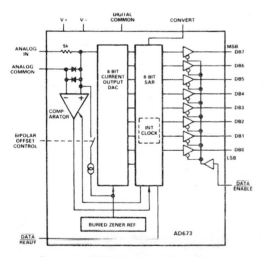

Figure 1. AD673 Functional Block Diagram

The SAR drives $\overline{DR}$ low to indicate that the conversion is complete and that the data is available to the output buffers. $\overline{DATA\ ENABLE}$ can then be activated to enable the 8-bits of data desired. $\overline{DATA\ ENABLE}$ should be brought high prior to the next conversion to place the output buffers in the high impedance state.

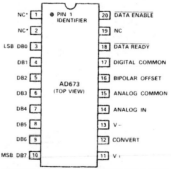

*PINS 1 & 2 ARE INTERNALLY CONNECTED TO TEST POINTS AND SHOULD BE LEFT FLOATING

Figure 2. AD673 Pin Connections

4 × 1 Wideband Video Multiplexer

AD9300

FEATURES
34MHz Full Power Bandwidth
±0.1dB Gain Flatness to 8MHz
72dB Crosstalk Rejection @ 10MHz
0.03°/0.01% Differential Phase/Gain
Cascadable for Switch Matrices
MIL-STD-883 Compliant Versions Available

APPLICATIONS
Video Routing
Medical Imaging
Electro-Optics
ECM Systems
Radar Systems
Data Acquisition

FUNCTIONAL BLOCK DIAGRAM
(Based on Cerdip)

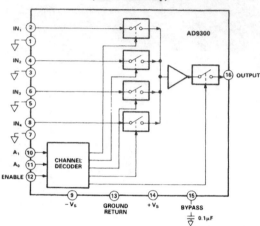

GENERAL DESCRIPTION
The AD9300 is a monolithic high-speed video signal multiplexer useable in a wide variety of applications.

Its four channels of video input signals can be randomly switched at megahertz rates to the single output. In addition, multiple devices can be configured in either parallel or cascade arrangements to form switch matrices. This flexibility in using the AD9300 is possible because the output of the device is in a high-impedance state when the chip is not enabled; when the chip is enabled, the unit acts as a buffer with a high input impedance and low output impedance.

An advanced bipolar process provides fast, wideband switching capabilities while maintaining crosstalk rejection of 72dB at 10MHz. Full power bandwidth is a minimum 27MHz. The device can be operated from ±10V to ±15V power supplies.

The AD9300K is available in a 16-pin ceramic DIP and a 20-pin PLCC and is designed to operate over the commercial temperature range of 0 to +70°C. The AD9300TQ is a hermetic 16-pin ceramic DIP for military temperature range (−55°C to +125°C) applications. This part is also available processed to MIL-STD-883. The AD9300 is available in a 20-pin LCC as the model AD9300TE, which operates over a temperature range of −55°C to +125°C.

The AD9300 Video Multiplexer is available in versions compliant with MIL-STD-883. Refer to the Analog Devices *Military Products Databook* or current AD9300/883B data sheet for detailed specifications.

PIN DESIGNATIONS

DIP

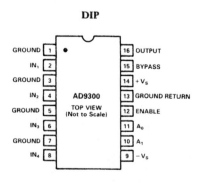

LCC and PLCC

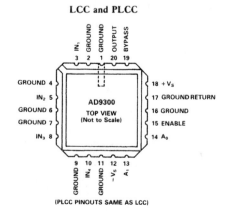

(PLCC PINOUTS SAME AS LCC)

AD9300—SPECIFICATIONS

ELECTRICAL CHARACTERISTICS ($\pm V_S = \pm 12V \pm 5\%$; $C_L = 10pF$; $R_L = 2k\Omega$, unless otherwise noted)

Parameter (Conditions)	Temp	Test Level	COMMERCIAL 0°C to +70°C AD9300KQ/KP Min	Typ	Max	Units
INPUT CHARACTERISTICS						
Input Offset Voltage	+25°C	I		3	10	mV
Input Offset Voltage	Full	VI			14	mV
Input Offset Voltage Drift[2]	Full	V		75		µV/°C
Input Bias Current	+25°C	I		15	37	µA
Input Bias Current	Full	VI			55	µA
Input Resistance	+25°C	V		3.0		MΩ
Input Capacitance	+25°C	V		2		pF
Input Noise Voltage (dc to 8MHz)	+25°C	V		16		µV rms
TRANSFER CHARACTERISTICS						
Voltage Gain[3]	+25°C	I	0.990	0.994		V/V
Voltage Gain[3]	Full	VI	0.985			V/V
DC Linearity[4]	+25°C	V		0.01		%
Gain Tolerance ($V_{IN} = \pm 1V$)						
dc to 5MHz	+25°C	I		0.05	0.1	dB
5MHz to 8MHz	+25°C	I		0.1	0.3	dB
Small-Signal Bandwidth	+25°C	V		350		MHz
($V_{IN} = 100mV$ p-p)						
Full Power Bandwidth[5]	+25°C	I	27	34		MHz
($V_{IN} = 2V$ p-p)						
Output Swing	Full	VI	±2			V
Output Current (Sinking @ = 25°C)	+25°C	V		5		mA
Output Resistance	+25°C	IV, V		9	15	Ω
DYNAMIC CHARACTERISTICS						
Slew Rate[6]	+25°C	I	170	215		V/µs
Settling Time						
(to 0.1% on ±2V Output)	+25°C	IV		70	100	ns
Overshoot						
To T-Step[7]	+25°C	V		<0.1		%
To Pulse[8]	+25°C	V		<10		%
Differential Phase[9]	+25°C	IV		0.03	0.1	°
Differential Gain[9]	+25°C	IV		0.01	0.1	%
Crosstalk Rejection						
Three Channels[10]	+25°C	IV	68	72		dB
One Channel[11]	+25°C	IV	70	76		dB
SWITCHING CHARACTERISTICS[12]						
A_X Input to Channel HIGH Time[13]	+25°C	I		40	50	ns
(t_{HIGH})						
A_X Input to Channel LOW Time[15]	+25°C	I		35	45	ns
(t_{LOW})						
Enable to Channel ON Time[15]	+25°C	I		35	45	ns
(t_{ON})						
Enable to Channel OFF Time[16]	+25°C	I		35	45	ns
(t_{OFF})						
Switching Transient[17]	+25°C	V		60		mV

EXPLANATION OF TEST LEVELS

Test Level I	–	100% production tested.
Test Level II	–	100% production tested at +25°C, and sample tested at specified temperatures.
Test Level III	–	Sample tested only.
Test Level IV	–	Parameter is guaranteed by design and characterization testing.
Test Level V	–	Parameter is a typical value only.
Test Level VI	–	All devices are 100% production tested at +25°C. 100% production tested at temperature extremes for military temperature devices; sample tested at temperature extremes for commercial/industrial devices.

Parameter (Conditions)	Temp	Test Level	COMMERCIAL 0°C to +70°C AD9300KQ/KP			Units
			Min	Typ	Max	
DIGITAL INPUTS						
Logic "1" Voltage	Full	VI	2			V
Logic "0" Voltage	Full	VI			0.8	V
Logic "1" Current	Full	VI			5	µA
Logic "0" Current	Full	VI			1	µA
POWER SUPPLY						
Positive Supply Current (+12V)	+25°C	I		13	16	mA
Positive Supply Current (+12V)	Full	VI		13	16	mA
Negative Supply Current (−12V)	+25°C	I		12.5	15	mA
Negative Supply Current (−12V)	Full	VI		12.5	16	mA
Power Supply Rejection Ratio ($\pm V_S = \pm 12V \pm 5\%$)	Full	VI	67	75		dB
Power Dissipation ($\pm 12V$)[19]	+25°C	V		306		mW

NOTES

[1]Permanent damage may occur if any one absolute maximum rating is exceeded. Functional operation is not implied, and device reliability may be impaired by exposure to higher-than-recommended voltages for extended periods of time.

[2]Measured at extremes of temperature range.

[3]Measured as slope of V_{OUT} versus V_{IN} with $V_{IN} = \pm 1V$.

[4]Measured as worst deviation from end-point fit with $V_{IN} = \pm 1V$.

[5]Full Power Bandwith (FPBW) based on Slew Rate (SR). $FPBW = SR \cdot 2\pi V_{PEAK}$

[6]Measured between 20% and 80% transition points of $\pm 1V$ output.

[7]T-Step = Sin^2X Step, when Step between 0V and +700mV points has 10%-to-90% risetime = 125ns.

[8]Measured with a pulse input having slew rate >250V µs.

[9]Measured at output between 0.28Vdc and 1.0Vdc with $V_{IN} = 284mV$ p-p at 3.58MHz and 4.43MHz.

[10]This specification is critically dependent on circuit layout. Value shown is measured with selected channel grounded and 10MHz 2V p-p signal applied to remaining three channels. If selected channel is grounded through 75Ω, value is approximately 6dB higher.

[11]This specification is critically dependent on circuit layout. Value shown is measured with selected channel grounded and 10MHz 2V p-p signal applied to one other channel. If selected channel is grounded through 75Ω, value is approximately 6dB higher. Minimum specification in ( ) applies to DIPs.

[12]Consult system timing diagram.

[13]Measured from address change to 90% point of −2V to +2V output LOW-to-HIGH transition.

[14]Measured from address change to 90% point of +2V to −2V output HIGH-to-LOW transition.

[15]Measured from 50% transition point of ENABLE input to 90% transition of 0V to −2V and 0V to +2V output.

[16]Measured from 50% transition point of ENABLE input to 10% transition of +2V to 0V and −2V to 0V output.

[17]Measured while switching between two grounded channels.

[18]Maximum power dissipation is a package-dependent parameter related to the following typical thermal impedances:

    16-Pin Ceramic $\theta_{JA} = 87°C/W$; $\theta_{JC} = 25°C/W$
    20-Pin LCC $\theta_{JA} = 74°C/W$; $\theta_{JC} = 10°C/W$
    20-Pin PLCC $\theta_{JA} = 71°C/W$; $\theta_{JC} = 26°C/W$

Specifications subject to change without notice.

ABSOLUTE MAXIMUM RATINGS[1]

Supply Voltages ($\pm V_S$) . . . . $\pm 16V$
Analog Input Voltage Each Input
  (IN_1 thru IN_4) . . . . $\pm 3.5V$
Differential Voltage Between Any Two
  Inputs (IN_1 thru IN_4) 5V
Digital Input Voltages (A_0, A_1, ENABLE) . −0.5V to +5.5V

Output Current
  Sinking 6.0mA
  Sourcing 6.0mA
Operating Temperature Range
  AD9300KQ/KP . . . . 0°C to +70°C
Storage Temperature Range . . . . −65°C to +150°C
Junction Temperature . . . . +175°C
Lead Soldering (10sec) . . . . +300°C

ORDERING INFORMATION

Device	Temperature Range	Description	Package Option[1]
AD9300KQ	0 to +70°C	16-Pin Cerdip, Commercial	Q-16
AD9300TE/883B[2]	−55°C to +125°C	20-Pin LCC, Military Temperature	E-20A
AD9300TQ/883B[2]	−55°C to +125°C	16-Pin Cerdip, Military Temperature	Q-16
AD9300KP	0 to +70°C	20-Pin PLCC, Commercial	P-20A

NOTES

[1]E = Ceramic Leadless Chip Carrier; P = Plastic Leaded Chip Carrier; Q = Cerdip. For outline information see Package Information section.

[2]For specifications, refer to Analog Devices *Military Products Databook*.

BURR-BROWN®

LOG100

Precision
LOGARITHMIC AND LOG RATIO AMPLIFIER

FEATURES

- **ACCURACY**
  0.37% FSO max Total Error
  Over 5 Decades
- **LINEARITY**
  0.1% max Log Conformity
  Over 5 Decades
- **EASY TO USE**
  Pin-selectable Gains
  Internal Laser-trimmed Resistors
- **WIDE INPUT DYNAMIC RANGE**
  6 Decades, 1nA to 1mA
- **HERMETIC CERAMIC DIP**

APPLICATIONS

- **LOG, LOG RATIO AND ANTILOG COMPUTATIONS**
- **ABSORBANCE MEASUREMENTS**
- **DATA COMPRESSION**
- **OPTICAL DENSITY MEASUREMENTS**
- **DATA LINEARIZATION**
- **CURRENT AND VOLTAGE INPUTS**

DESCRIPTION

The LOG100 uses advanced integrated circuit technologies to achieve high accuracy, ease of use, low cost, and small size. It is the logical choice for your logarithmic-type computations. The amplifier has guaranteed maximum error specifications over the full six-decade input range (1nA to 1mA) and for all possible combinations of I_1 and I_2. Total error is guaranteed so that involved error computations are not necessary.

The circuit uses a specially designed compatible thin-film monolithic integrated circuit which contains amplifiers, logging transistors, and low drift thin-film

resistors. The resistors are laser-trimmed for maximum precision. FET input transistors are used for the amplifiers whose low bias currents (1pA typical) permit signal currents as low as 1nA while maintaining guaranteed total errors of 0.37% FSO maximum.

Because scaling resistors are self-contained, scale factors of 1V, 3V or 5V per decade are obtained simply by pin selections. No other resistors are required for log ratio applications. The LOG100 will meet its guaranteed accuracy with no user trimming. Provisions are made for simple adjustments of scale factor, offset voltage, and bias current if enhanced performance is desired.

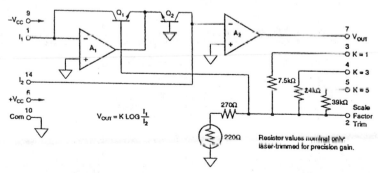

$$V_{OUT} = K \, LOG \frac{I_1}{I_2}$$

Resistor values nominal only laser-trimmed for precision gain.

International Airport Industrial Park • Mailing Address: PO Box 11400 • Tucson, AZ 85734 • Street Address: 6730 S. Tucson Blvd. • Tucson, AZ 85706
Tel: (520) 746-1111 • Twx: 910-952-1111 • Cable: BBRCORP • Telex: 066-6491 • FAX: (520) 889-1510 • Immediate Product Info: (800) 548-6132

PDS-437E

SPECIFICATIONS

ELECTRICAL

$T_A = +25°C$ and $\pm V_{CC} = \pm15V$, after 15 minute warm-up, unless otherwise specified.

PARAMETER	CONDITIONS	LOG100JP MIN	LOG100JP TYP	LOG100JP MAX	UNITS
TRANSFER FUNCTION					
Log Conformity Error[1]	Either I_1 or I_2		$V_{OUT} = K \, Log \, (I_1/I_2)$		
Initial	1nA to 100µA (5 decades)		0.04	0.1	%
	1nA to 1mA (6 decades)		0.15	0.25	%
Over Temperature	1nA to 100µA (5 decades)		0.002		%/°C
	1nA to 1mA (6 decades)		0.001		%/°C
K Range[2]			1, 3, 5		V/decade
Accuracy			0.3		%
Temperature Coefficient			0.03		%/°C
ACCURACY					
Total Error[3]	K = 1,[4] Current Input Operation				
Initial	$I_1, I_2 = 1mA$			±55	mV
	$I_1, I_2 = 100µA$			±30	mV
	$I_1, I_2 = 10µA$			±25	mV
	$I_1, I_2 = 1µA$			±20	mV
	$I_1, I_2 = 100nA$			±25	mV
	$I_1, I_2 = 10nA$			±30	mV
	$I_1, I_2 = 1nA$			±37	mV
vs Temperature	$I_1, I_2 = 1mA$		±0.20		mV/°C
	$I_1, I_2 = 100µA$		±0.37		mV/°C
	$I_1, I_2 = 10µA$		±0.28		mV/°C
	$I_1, I_2 = 1µA$		±0.033		mV/°C
	$I_1, I_2 = 100nA$		±0.28		mV/°C
	$I_1, I_2 = 10nA$		±0.51		mV/°C
	$I_1, I_2 = 1nA$		±1.26		mV/°C
vs Supply	$I_1, I_2 = 1mA$		±4.3		mV/V
	$I_1, I_2 = 100µA$		±1.5		mV/V
	$I_1, I_2 = 10µA$		±0.37		mV/V
	$I_1, I_2 = 1µA$		±0.11		mV/V
	$I_1, I_2 = 100nA$		±0.61		mV/V
	$I_1, I_2 = 10nA$		±0.91		mV/V
	$I_1, I_2 = 1nA$		±2.6		mV/V
INPUT CHARACTERISTICS (of Amplifiers A_1 and A_2)					
Offset Voltage					
Initial			±0.7	±5	mV
vs Temperature			±80		µV/°C
Bias Current					
Initial			1	5[5]	pA
vs Temperature			Doubles Every 10°C		
Voltage Noise	10Hz to 10kHz, RTI		3		µVrms
Current Noise	10Hz to 10kHz, RTI		0.5		pArms
AC PERFORMANCE					
3dB Response[6], $I_2 = 10µA$					
1nA	$C_C = 4500pF$		0.11		kHz
1µA	$C_C = 150pF$		38		kH
10µA	$C_C = 150pF$		27		kH
1mA	$C_C = 50pF$		45		kHz
Step Response[8]					
Increasing	$C_C = 150pF$				
1µA to 1mA			11		µs
100nA to 1µA			7		µs
10nA to 100nA			110		µs
Decreasing	$C_C = 150pF$				
1mA to 1µA			45		µs
1µA to 100nA			20		µs
100nA to 10nA			550		µs
OUTPUT CHARACTERISTICS					
Full Scale Output (FSO)					
Rated Output		±10			V
Voltage	$I_{OUT} = ±5mA$	±10			V
Current	$V_{OUT} = ±10V$	±5			mA
Current Limit					
Positive			12.5		mA
Negative			15		mA
Impedance			0.05		Ω

SPECIFICATIONS (CONT)

ELECTRICAL

$T_A = +25°C$ and $\pm V_{CC} = \pm 15V$, after 15 minute warm-up, unless otherwise specified.

| PARAMETER | CONDITIONS | LOG100JP | | | UNITS |
		MIN	TYP	MAX	
POWER SUPPLY REQUIREMENTS					
Rated Voltage			±15		VDC
Operating Range	Derated Performance	±12		±18	VDC
Quiescent Current			±7	±9	mA
AMBIENT TEMPERATURE RANGE					
Specification		0		+70	
Operating Range	Derated Performance	−25		+85	°C
Storage		−40		+85	°C

NOTES: (1) Log Conformity Error is the peak deviation from the best-fit straight line of the V_{OUT} vs Log I_{IN} curve expressed as a percent of peak-to-peak full scale output. (2) May be trimmed to other values. See Applications section. (3) The worst-case Total Error for any ratio of I_1/I_2 is the largest of the two errors when I_1 and I_2 are considered separately. (4) Total Error at other values of K is K times Total Error for K = 1. (5) Guaranteed by design. Not directly measurable due to amplifier's committed configuration. (6) 3dB and transient response are a function of both the compensation capacitor and the level of input current. See Typical Performance Curves.

ABSOLUTE MAXIMUM RATINGS

Supply .... ±18V
Internal Power Dissipation.. 600mV
Input Current 10mA
Input Voltage Range .... ±18V
Storage Temperature Range .... −40°C to +85°C
Lead Temperature (soldering, 10s) .... +300°C
Output Short-circuit Duration .... Continuous to ground
Junction Temperature .... 175°C

SCALE FACTOR PIN CONNECTIONS

K, V/DECADE	CONNECTIONS
5	5 to 7
3	4 to 7
1.9	4 and 5 to 7
1	3 to 7
0.85	3 and 5 to 7
0.77	3 and 4 to 7
0.68	3 and 4 and 5 to 7

FREQUENCY COMPENSATION

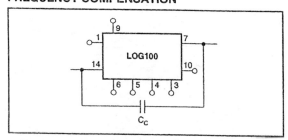

ORDERING INFORMATION

MODEL	PACKAGE	SPECIFIED TEMPERATURE RANGE
LOG100JP	14-Pin Hermetic Ceramic DIP	0°C to +70°C

PIN CONFIGURATION

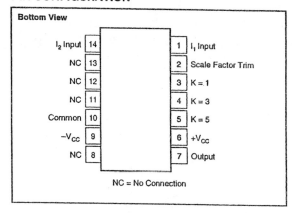

ELECTROSTATIC DISCHARGE SENSITIVITY

Any integral circuit can be damaged by ESD. Burr-Brown recommends that all integrated circuits be handled with appropriate precautions. Failure to observe proper handling and installation procedures can cause damage.

ESD damage can range from subtle performance degradation to complete device failure. Precision integrated circuits may be more susceptible to damage because very small parametric changes could cause the device not to meet published specifications.

PACKAGE INFORMATION

MODEL	PACKAGE	PACKAGE DRAWING NUMBER[1]
LOG100JP	14-Pin Hermetic Ceramic DIP	148[2]

NOTES: (1) For detailed drawing and dimension table, please see end of data sheet, or Appendix D of Burr-Brown IC Data Book. (2) During 1994 the package was changed from plastic to hermetic ceramic. Pinout, model number, and specifications remained unchanged. The metal lid of the new package is internally connected to common, pin 10.

LOG100

TYPICAL PERFORMANCE CURVES

$T_A = +25°C$, $V_{cc} = ±15VDC$, unless otherwise noted.

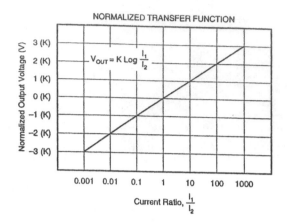

NORMALIZED TRANSFER FUNCTION

$$V_{OUT} = K \, Log \frac{I_1}{I_2}$$

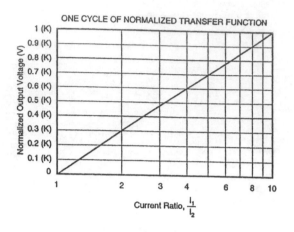

ONE CYCLE OF NORMALIZED TRANSFER FUNCTION

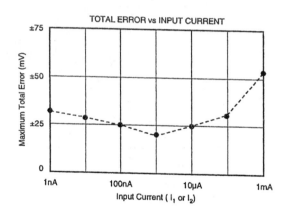

TOTAL ERROR vs INPUT CURRENT

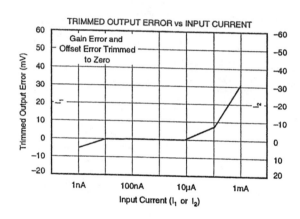

TRIMMED OUTPUT ERROR vs INPUT CURRENT

Gain Error and Offset Error Trimmed to Zero

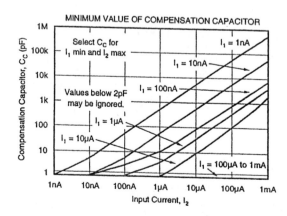

MINIMUM VALUE OF COMPENSATION CAPACITOR

Select C_C for I_1 min and I_2 max

Values below 2pF may be ignored.

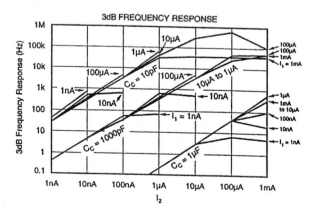

3dB FREQUENCY RESPONSE

Hybrid, Tracking Resolver-to-Digital Converters

1S20/1S40/1S60/1S61

FEATURES
Low Cost
32-Pin Hybrid
High Tracking Rate 170rps at 12 Bits
Velocity Output
DC Error Output
Logic Outputs for Extension Pitch Counter

APPLICATIONS
Numerical Control of Machine Tools
Robotics

1S20/1S40/1S60/1S61 FUNCTIONAL BLOCK DIAGRAM

GENERAL DESCRIPTION

The 1S20/40/60/61 are a series of low cost hybrid converters with a high tracking rate and all essential features for numerically controlled machine applications. These converters are housed in a 32-pin triple DIP ceramic package measuring $1.1'' \times 1.7'' \times 0.205''$ ($28 \times 43.2 \times 5.2$mm).

The 1S20/40/60/61 convert resolver format input signals into a parallel natural binary digital word. Typically, these signals would be obtained from a brushless resolver and the resolver/converter combination gives a parallel absolute angular output word similar to that provided by an absolute encoder. The ratiometric conversion principle of the 1S20/40/60/61 series ensures high noise immunity and tolerance of lead length when the converter is at a distance from the resolver.

The output word is in three-state digital logic form with a high and low byte enable input so that the converter can communicate with an 8- or 16-bit digital highway. In this series there are 12-, 14- and two 16-bit resolution (± 4 arc mins and ± 10 arc mins accuracy) models available.

Repeatability is 1LSB for all models under constant temperature conditions.

The 1S20/40/60/61 are available with three frequency options covering the range 400Hz to 10kHz.

Models Available
Four models are available in this range and three frequency options for each model.

1S20 is a 12-bit up to 170 revolutions per second
1S40 is a 14-bit up to 42.5 revolutions per second
1S60 is a 16-bit up to 10.5 revolutions per second
1S61 is a 16-bit up to 10.5 revolutions per second

APPLICATIONS/USER BENEFITS

The 1S20/40/60/61 has been specifically designed for the numerically controlled machine and robot industry. Using the type 2 servo loop tracking principle ideally suits these converters to the electrically noisy environment found in these industrial applications.

By using hybrid construction techniques, small size, low power and high reliability are further benefits offered by these converters. This small size with the three-state digital outputs makes these converters ideal for multichannel operation.

The layout of the connections simplifies the parallel connection to a digital highway.

The provision of the digital outputs of DIRECTION and RIPPLE CLOCK allow simple extension counters for multi-pitch operation to be implemented.

Analog outputs of velocity and dc error for control loop stabilization and bite (built in test) provide two more features required in these applications.

SPECIFICATIONS (typical @ +25°C, unless otherwise specified)

Models	1S20	1S40	1S60	1S61	Units
RESOLUTION	12	14	16	16	Bits
ACCURACY[1]	± 8.5	± 5.3	± 4.0	± 10	arc-mins
REPEATABILITY[2]	1	*	*	*	LSB
SIGNAL AND REFERENCE FREQUENCY[3]	400-10k	*	*	*	Hz
DIGITAL OUTPUT	Parallel natural binary				
Max Load	20	*	*	*	LSTTL
TRACKING RATE (min)					
400Hz – 2.6kHz	50	12.5	3.0	3.0	rps
2.6kHz – 5kHz	90	22.5	5.5	5.5	rps
5kHz – 10kHz	170	42.5	10.5	10.5	rps
SETTLING TIME					
400Hz – 2.6kHz	150	180	350	350	ms
2.6kHz – 5kHz	40	50	130	130	ms
5kHz – 10kHz	20	25	60	60	ms
ACCELERATION CONSTANT (K_a)					
400Hz – 2.6kHz	9,500	*	*	*	$\sec^{-2}$
2.6kHz – 5kHz	144,000	*	*	*	$\sec^{-2}$
5kHz – 10kHz	713,000	*	*	*	$\sec^{-2}$
SIGNAL VOLTAGE	2.0	*	*	*	V rms
SIGNAL INPUT IMPEDANCE	>10	*	*	*	MΩ
REFERENCE VOLTAGE	2.0	*	*	*	V rms
REFERENCE INPUT IMPEDANCE	125	*	*	*	kΩ
ALLOWABLE PHASE SHIFT[4] (Signal to Reference)	± 10	*	*	*	Degrees
BUSY OUTPUT[5]	Logic "Hi" when Busy				
Max Load	20	*	*	*	LSTTL
BUSY WIDTH	430	*	*	*	ns
ENABLE INPUTS	Logic "Lo" to ENABLE				
Load	1	*	*	*	LSTTL
ENABLE AND DISABLE TIMES	120 (typ)	*	*	*	ns
	220 (max)	*	*	*	ns
INHIBIT INPUT	Logic "Lo" to INHIBIT	*	*	*	
Load	1	*	*	*	LSTTL
DIRECTION OUTPUT (DIR)[5]	Logic "Hi" when counting up Logic "Lo" when counting down				
Max Load	20	*	*	*	LSTTL
RIPPLE CLOCK[5]	Negative pulse indicating when internal counters change from all "1's" to all "0's" or vice versa.				
Max Load	20	*	*	*	LSTTL
VELOCITY OUTPUT[6] (at specified min tracking rate).					
Polarity	positive for increasing angle	*	*	*	–
Output Voltage[7]	± 10	*	*	*	V dc
Accuracy	± 10	*	*	*	% FSD
Zero Offset	± 8	*	*	*	mV
DC ERROR OUTPUT VOLTAGE[6]	40	10	2.5	2.5	mV/LSB
POWER SUPPLIES					
+ V_S	+ 11.5 to + 16	*	*	*	V
– V_S	– 11.5 to – 16	*	*	*	V
+ 5V	+ 4.75 to + 5.25	*	*	*	V
POWER SUPPLY CONSUMPTION[7]					
+ V_S	20, 30 (max)	*	*	*	mA
– V_S	20, 30 (max)	*	*	*	mA
+ 5V	105, 125 (max)	*	*	*	mA
POWER DISSIPATION[7]	1.1, 1.5 (max)	*	*	*	W
TEMPERATURE RANGE					
Operating	0 to + 70	*	*	*	°C
Storage	– 55 to + 125	*	*	*	°C
PACKAGE OPTION[8]	DH-32E	*	*	*	
WEIGHT	1 (28)	*	*	*	oz. (grms)

NOTES

[1] Specified over the operating temperature range and for:
    a). ± 10% signal and reference amplitude variation.
    b). 10% signal and reference harmonic distortion.
    c). ± 10% on frequency range of option.
[2] Specified at constant temperature. Over the operating temperature range, worst case repeatability could be up to 1.5 arc mins for all models.
[3] See frequency range options.

[4] For no additional error with a static input, see "Dynamic Accuracy vs. Resolver Phase Shift"
[5] See timing diagram.
[6] These outputs should be connected via buffers or comparator inputs (max load 100pF).
[7] ± V_S = ± 15 volts.
[8] See Section 14 for package outline information.
*Specifications same as 1S20.
Specifications subject to change without notice.

High Performance, Economy Strain Gage/RTD Conditioners

2B30/2B31

FEATURES
Low Cost
Complete Signal Conditioning Function
Low Drift: 0.5μV/°C max ("L"); Low Noise: 1μV p-p max
Wide Gain Range: 1 to 2000V/V
Low Nonlinearity: 0.0025% max ("L")
High CMR: 140dB min (60Hz, G = 1000V/V)
Input Protected to 130V rms
Adjustable Low Pass Filter: 60dB/Decade Roll-Off (from 2Hz)
Programmable Transducer Excitation: Voltage (4V to 15V @ 100mA) or Current (100μA to 10mA)

APPLICATIONS
Measurement and Control of:
    Pressure, Temperature, Strain/Stress, Force, Torque
Instrumentation: Indicators, Recorders, Controllers
Data Acquisition Systems
Microcomputer Analog I/O

2B31 FUNCTIONAL BLOCK DIAGRAM

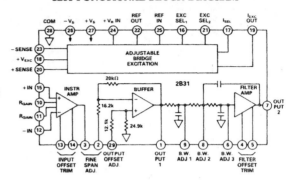

GENERAL DESCRIPTION
Models 2B30 and 2B31 are high performance, low cost, compact signal conditioning modules designed specifically for high accuracy interface to strain gage-type transducers and RTD's (resistance temperature detectors). The 2B31 consists of three basic sections: a high quality instrumentation amplifier; a three-pole low pass filter, and an adjustable transducer excitation. The 2B30 has the same amplifier and filter as the 2B31, but no excitation capability.

Available with low offset drift of 0.5μV/°C max (RTI, G = 1000V/V) and excellent linearity of 0.0025% max, both models feature guaranteed low noise performance (1μV p-p max) and outstanding 140dB common mode rejection (60Hz, CMV = ±10V, G = 1000V/V) enabling the 2B30/2B31 to maintain total amplifier errors below 0.1% over a 20°C temperature range. The low pass filter offers 60dB/decade roll-off from 2Hz to reduce normal-mode noise bandwidth and improve system signal-to-noise ratio. The 2B31's regulated transducer excitation stage features a low output drift (0.015%/°C max) and a capability of either constant voltage or constant current operation.

Gain, filter cutoff frequency, output offset level and bridge excitation (2B31) are all adjustable, making the 2B30/2B31 the industry's most versatile high-accuracy transducer-interface modules. Both models are offered in three accuracy selections, J/K/L, differing only in maximum nonlinearity and offset drift specifications.

APPLICATIONS
The 2B30/2B31 may be easily and directly interfaced to a wide variety of transducers for precise measurement and control of pressure, temperature, stress, force and torque. For ap-

plications in harsh industrial environments, such characteristics as high CMR, input protection, low noise, and excellent temperature stability make 2B30/2B31 ideally suited for use in indicators, recorders, and controllers.

The combination of low cost, small size and high performance of the 2B30/2B31 offers also exceptional quality and value to the data acquisition system designer, allowing him to assign a conditioner to each transducer channel. The advantages of this approach over low level multiplexers include significant improvements in system noise and resolution, and elimination of crosstalk and aliasing errors.

DESIGN FEATURES AND USER BENEFITS
High Noise Rejection: The true differential input circuitry with high CMR (140dB) eliminating common-mode noise pickup errors, input filtering minimizing RFI/EMI effects, output low pass filtering (f_c=2Hz) rejecting 50/60Hz line frequency pickup and series-mode noise.

Input and Output Protection: Input protected for shorts to power lines (130V rms), output protected for shorts to ground and either supply.

Ease of Use: Direct transducer interface with minimum external parts required, convenient offset and span adjustment capability.

Programmable Transducer Excitation: User-programmable adjustable excitation source-constant voltage (4V to 15V @ 100mA) or constant current (100μA to 10mA) to optimize transducer performance.

Adjustable Low Pass Filter: The three-pole active filter (f_c=2Hz) reducing noise bandwidth and aliasing errors with provisions for external adjustment of cutoff frequency.

SPECIFICATIONS (typical @ +25°C and $V_S = \pm15V$ unless otherwise noted)

MODEL	2B30J 2B31J	2B30K 2B31K	2B30L 2B31L
GAIN[1]			
Gain Range	1 to 2000V/V	*	*
Gain Equation	$G = (1 + 94k\Omega/R_G)[20k\Omega/(R_F + 16.2k\Omega)]$	*	*
Gain Equation Accuracy	±2%	*	*
Fine Gain (Span) Adjust Range	±20%		
Gain Temperature Coefficient	±25ppm/°C max (±10ppm/°C typ)	*	*
Gain Nonlinearity	±0.01% max	±0.005% max	±0.0025% max
OFFSET VOLTAGES[1]			
Total Offset Voltage, Referred to Input			
Initial @ +25°C	Adjustable to Zero (±0.5mV typ)	*	*
Warm-Up Drift, 10 Min., G = 1000	Within ±5μV (RTI) of Final Value	*	
vs. Temperature			
G = 1V/V	±150μV/°C max	±75μV/°C max	±50μV/°C max
G = 1000V/V	±3μV/°C max	±1μV/°C max	±0.5μV/°C max
At Other Gains	±(3 + 150/G)μV/°C max	±(1 + 75/G)μV/°C max	±(0.5 + 50/G)μV/°C max
vs. Supply, G = 1000V/V	±25μV/V	*	*
vs. Time, G = 1000V/V	±5μV/month		
Output Offset Adjust Range	±10V		
INPUT BIAS CURRENT			
Initial @ +25°C	+200nA max (100nA typ)	*	*
vs. Temperature (0 to +70°C)	-0.6nA/°C	*	*
INPUT DIFFERENCE CURRENT			
Initial @ +25°C	±5nA	*	*
vs. Temperature (0 to +70°C)	±40pA/°C	*	*
INPUT IMPEDANCE			
Differential	100MΩ‖47pF		
Common Mode	100MΩ‖47pF		
INPUT VOLTAGE RANGE			
Linear Differential Input	±10V		
Maximum Differential or CMV Input			
Without Damage	130V rms	*	*
Common Mode Voltage	±10V		
CMR, 1kΩ Source Imbalance			
G = 1V/V, dc to 60Hz[1]	90dB	*	*
G = 100V/V to 2000V/V, 60Hz[1]	140dB min	*	*
dc[2]	90dB min (112 typ)	*	*
INPUT NOISE			
Voltage, G = 1000V/V			
0.01Hz to 2Hz	1μV p-p max	*	*
10Hz to 100Hz[2]	1μV p-p	*	*
Current, G = 1000			
0.01Hz to 2Hz	70pA p-p	*	*
10Hz to 100Hz[2]	30pA rms	*	*
RATED OUTPUT[1]			
Voltage, 2kΩ Load[3]	±10V min	*	*
Current	±5mA min	*	*
Impedance, dc to 2Hz, G = 100V/V	0.1Ω	*	*
Load Capacitance	0.01μF max	*	*
DYNAMIC RESPONSE (Unfiltered)[2]			
Small Signal Bandwidth			
-3dB Gain Accuracy, G = 100V/V	30kHz	*	*
G = 1000V/V	5kHz	*	*
Slew Rate	1V/μs	*	*
Full Power	15kHz	*	*
Settling Time, G = 100, ±10V Output			
Step to 0.1%	30μs	*	*
LOW PASS FILTER (Bessel)			
Number of Poles	3	*	*
Gain (Pass Band)	+1	*	*
Cutoff Frequency (-3dB Point)	2Hz	*	*
Roll-Off	60dB/decade	*	*
Offset (at 25°C)	±5mV	*	*
Settling Time, G = 100V/V, ±10V			
Output Step to ±0.1%	600ms	*	*
BRIDGE EXCITATION (See Table 1)			
POWER SUPPLY[4]			
Voltage, Rated Performance	±15V dc	*	*
Voltage, Operating[5]	±(12 to 18)V dc	*	*
Current, Quiescent[6]	±15mA	*	*
TEMPERATURE RANGE			
Rated Performance	0 to +70°C	*	*
Operating	-25°C to +85°C	*	*
Storage	-55°C to +125°C	*	*
CASE SIZE	2" x 2" x 0.4" (51 x 51 x 10.2mm)	*	*

NOTES
*Specifications same as 2B30J/2B31J
[1] Specifications referred to output at pin 7 with 3.75k, 1%, 25ppm/°C fine span resistor installed and internally set 2Hz filter cutoff frequency.
[2] Specifications referred to the unfiltered output at pin 1.
[3] Protected for shorts to ground and/or either supply voltage.
[4] Recommended power supply ADI model 902-2 or model 2B35.
[5] Tracking power supplies.
[6] Does not include bridge excitation and load currents.
Specifications subject to change without notice.

OUTLINE DIMENSIONS
Dimensions shown in inches and (mm).

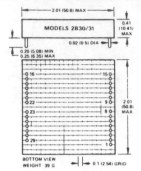

PIN DESIGNATIONS

PIN	FUNCTION	PIN	FUNCTION
1	OUTPUT 1 (UNFILTERED)	16	EXC SEL 1
2	FINE GAIN (SPAN) ADJ.	17	I SEL
3	FINE GAIN (SPAN) ADJ.	18	V_{EXC} OUT
4	FILTER OFFSET TRIM	19	I_{EXC} OUT
5	FILTER OFFSET TRIM	20	SENSE HIGH (+)
6	BANDWIDTH ADJ 3	21	EXC SEL 2
7	OUTPUT 2 (FILTERED)	22	REF OUT
8	BANDWIDTH ADJ. 2	23	SENSE LOW (-)
9	BANDWIDTH ADJ. 1	24	REGULATOR $-V_R$ IN
10	R_{GAIN}	25	REF IN
11	R_{GAIN}	26	$+V_S$
12	-INPUT	27	$-V_S$
13	INPUT OFFSET TRIM	28	COMMON
14	INPUT OFFSET TRIM	29	OUTPUT OFFSET TRIM
15	+ INPUT		

Note: Pins 16 thru 25 are not connected in Model 2B30.

AC1211/AC1213 MOUNTING CARDS

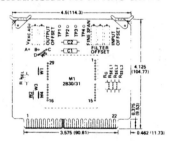

AC1211/AC1213 CONNECTOR DESIGNATIONS

PIN	FUNCTION	PIN	FUNCTION
A	REGULATOR $-V_R$ IN	1	EXC SEL 1
B	SENSE LOW (-)	2	I SEL
C	REF OUT	3	V_{EXC} OUT
D	REF IN	4	I_{EXC} OUT
E		5	SENSE HIGH (+)
F		6	EXC SEL 2
H		7	OUTPUT OFFSET TRIM
J		8	
K	$-V_S$	9	$-V_S$
L	$+V_S$	10	$-V_S$
M		11	
N	COMMON	12	COMMON
P		13	
R	FINE GAIN ADJ	14	
S	FINE GAIN ADJ	15	
T	FILTER OFFSET TRIM	16	
U	FILTER OFFSET TRIM	17	R_{GAIN}
V	OUTPUT 2 (FILTERED)	18	R_{GAIN}
W	- INPUT	19	OUTPUT 1 (UNFILTERED)
X	INPUT OFFSET TRIM	20	BANDWIDTH ADJ. 1
Y	INPUT OFFSET TRIM	21	BANDWIDTH ADJ. 3
Z	+ INPUT	22	BANDWIDTH ADJ. 2

The AC1211/AC1213 mounting card is available for the 2B30/2B31. The AC1211/AC1213 is an edge connector card with pin receptacles for plugging in the 2B30/2B31. In addition, it has provisions for installing the gain resistors and the bridge excitation, offset adjustment and filter cutoff programming components. The AC1211/AC1213 is provided with a Cinch 251-22-30-160 (or equivalent) edge connector. The AC1213 includes the adjustment pots; no pots are provided with the AC1211.

Isolated, Thermocouple Signal Conditioner

2B50

FEATURES
Accepts J, K, T, E, R, S or B Thermocouple Types
Internally Provided Cold Junction Compensation
High CMV Isolation: ±1500V pk
High CMR: 160dB min @ 60Hz
Low Drift: ±1μV/°C max (2B50B)
High Linearity: ±0.01% max (2B50B)
Input Protection and Filtering
Screw Terminal Input Connections

APPLICATIONS
Precision Thermocouple Signal Conditioning for:
    Process Control and Monitoring
    Industrial Automation
    Energy Management
    Data Acquisition Systems

2B50 FUNCTIONAL BLOCK DIAGRAM

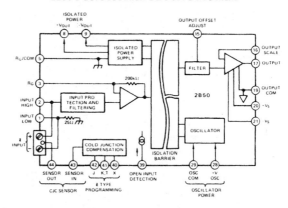

GENERAL DESCRIPTION
The model 2B50 is a high performance thermocouple signal conditioner providing input protection, isolation and common mode rejection, amplification, filtering and integral cold junction compensation in a single, compact package.

The 2B50 has been designed to condition low level analog signals, such as those produced by thermocouples, in the presence of high common mode voltages. Featuring direct thermocouple connection via screw terminals and internally provided reference junction temperature sensor, the 2B50 may be jumper programmed to provide cold junction compensation for thermocouple types J, K, T, and B, or resistor programmed for types E, R, and S.

The high performance of the 2B50 is accomplished by the use of reliable transformer isolation techniques. This assures complete input to output galvanic isolation (±1500V pk) and excellent common mode rejection (160dB @ 60Hz).

Other key features include: input protection (220V rms), filtering (NMR of 70dB @ 60Hz), low drift amplification (±1μV/°C max – 2B50B), and high linearity (±0.01% max – 2B50B).

APPLICATIONS
The 2B50 has been designed to provide thermocouple signal conditioning in data acquisition systems, computer interface systems, and temperature measurement and control instrumentation.

In thermocouple temperature measurement applications, outstanding features such as low drift, high noise rejection, and 1500V isolation make the 2B50 an ideal choice for systems used in harsh industrial environments.

DESIGN FEATURES AND USER BENEFITS
High Reliability: To assure high reliability and provide isolation protection to electronic instrumentation, the 2B50 has been conservatively designed to meet the IEEE Standard for transient voltage protection (472-1974: SWC) and provide 220V rms differential input protection.

High Noise Rejection: The 2B50 features internal filtering circuitry for elimination of errors caused by RFI/EMI, series mode noise, and 50Hz/60Hz pickup.

Ease of Use: Internal compensation enables the 2B50 to be used with seven different thermocouple types. Unique circuitry offers a choice of internal or remote reference junction temperature sensing. Thermocouple connections may be made either by screw terminals or, in applications requiring PC Board connections, by terminal pins.

Small Package: 1.5″ × 2.5″ × 0.6″ size conserves board space.

SPECIFICATIONS (typical @ +25°C and V_S = ±15V unless otherwise noted)

MODEL	2B50A	2B50B
INPUT SPECIFICATIONS		
Thermocouple Types		
Jumper Configurable Compensation	J, K, T, or B	*
Resistor Configurable Compensation	R, S, or E	*
Input Span Range	±5mV to ±100mV	*
Gain Range	50V/V to 1000V/V	*
Gain Equation	$1 + (200k\Omega/R_G)$	*
Gain Error	±0.25%	*
Gain Temperature Coefficient	±35ppm/°C max	±25ppm/°C max
Gain Nonlinearity[1]	±0.025% max	±0.01% max
Offset Voltage		
Input Offset (Adjustable to Zero)	±50µV	*
vs. Temperature	±2.5µV/°C max	±1µV/°C max
vs. Time	±1.5µV/month	*
Output Offset (Adjustable to Zero)	±10mV	*
vs. Temperature	±30µV/°C	*
Total Offset Drift	$\pm\left(2.5 + \dfrac{30}{G}\right)\mu V/°C$	$\pm\left(1 + \dfrac{30}{G}\right)\mu V/°C$
Input Noise Voltage		
0.01Hz to 100Hz, R_S = 1kΩ	1µV p-p	*
Maximum Safe Differential Input Voltage	220V rms, Continuous	*
CMV, Input to Output		
Continuous, ac or dc	±1500V pk max	*
Common Mode Rejection		
@ 60Hz, 1kΩ Source Unbalance	160dB min	*
Normal Mode Rejection @ 60Hz	70dB min	*
Bandwidth	dc to 2.5Hz (−3dB)	*
Input Impedance	100MΩ	*
Input Bias Current[2]	±5nA	*
Open Input Detection	Downscale	*
Response Time[3], G = 250	1.4sec	*
Cold Junction Compensation		
Initial Accuracy[4]	±0.5°C	*
vs. Temperature[5] (+5°C to +45°C)	±0.01°C/°C	*
OUTPUT SPECIFICATIONS		
Output Voltage Range[6]	±5V @ ±2mA	*
Output Resistance	0.1Ω	*
Output Protection	Continuous Short to Ground	*
POWER SUPPLY		
Voltage		
Output ±V_S (Rated Performance)	±15V dc ±10% @ ±0.5mA	*
(Operating)	±12V to ±18V dc max	*
Oscillator +V_{OSC} (Rated Performance)	+13V to +18V @ 15mA	*
ENVIRONMENTAL		
Temperature Range, Rated Performance	0 to +70°C	*
Operating	−25°C to +85°C	*
Storage Temperature Range	−55°C to +85°C	*
RFI Effect (5W @ 470MHz @ 3ft)		
Error	±0.5% of Span	*
PHYSICAL		
Case Size	1.5" × 2.5" × 0.6"	*

NOTES
*Specifications same as 2B50A.
[1] Gain nonlinearity is specified as a percentage of output signal span representing peak deviation from the best straight line; e.g., nonlinearity at an output span of 10V pk-pk (±5V) is ±0.01% or ±1mV.
[2] Does not include open circuit detection current of 20nA (optional by jumper connection).
[3] Open input response time is dependent upon gain.
[4] When used with internally provided CJC sensor.
[5] Compensation error contributed by ambient temperature changes at the module.
[6] Output swing of ±10V may be obtained through output scaling (Figure 5).

Specifications subject to change without notice.

OUTLINE DIMENSIONS
Dimensions shown in inches and (mm).

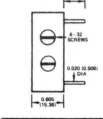

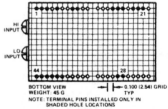

BOTTOM VIEW
WEIGHT: 45 G
NOTE: TERMINAL PINS INSTALLED ONLY IN SHADED HOLE LOCATIONS

PIN DESIGNATIONS

PIN	FUNCTION	PIN	FUNCTION
1	INPUT LO	23	
2	INPUT HI	24	
3	R_G	25	
4		26	
5	R_G COM	27	
6		28	+V OSC
7		29	OSC COM
8	+V ISO OUT	30	
9	−V ISO OUT	31	
10		32	
11		33	
12		34	
13		35	
14		36	
15	OUTPUT OFFSET ADJUST	37	
		38	
16	OUTPUT SCALE	39	OPEN INPUT DET
17	OUTPUT	40	X — T TYPE
18		41	K, T — PROGRAMMING
19	OUTPUT COM	42	J
20	+V_S	43	CJC SENSOR IN
21	−V_S	44	CJC SENSOR OUT
22			

MATING SOCKET:
AC1218

572

B DERIVATIONS OF SELECTED EQUATIONS

EQUATION (3–4)

The formula for open-loop gain in Equation (3–2) can be expressed in complex notation as

$$A_{ol} = \frac{A_{ol(mid)}}{1 + jf/f_{c(ol)}}$$

Substituting the above expression into the equation $A_{cl} = A_{ol}/(1 + BA_{ol})$, we get a formula for the total closed-loop gain.

$$A_{cl} = \frac{A_{ol(mid)}/(1 + jf/f_{c(ol)})}{1 + BA_{ol(mid)}/(1 + jf/f_{c(ol)})}$$

Multiplying the numerator and denominator by $1 + jf/f_{c(ol)}$ yields

$$A_{cl} = \frac{A_{ol(mid)}}{1 + BA_{ol(mid)} + jf/f_{c(ol)}}$$

Dividing the numerator and denominator by $1 + BA_{ol(mid)}$ gives

$$A_{cl} = \frac{A_{ol(mid)}/(1 + BA_{ol(mid)})}{1 + j[f/(f_{c(ol)}(1 + BA_{ol(mid)}))]}$$

The above expression is of the form of the first equation

$$A_{cl} = \frac{A_{cl(mid)}}{1 + jf/f_{c(cl)}}$$

where $f_{c(cl)}$ is the closed-loop critical frequency. Thus,

$$f_{c(cl)} = f_{c(ol)}(1 + BA_{ol(mid)})$$

EQUATION (5–7)

The center frequency equation is

$$f_0 = \frac{1}{2\pi\sqrt{(R_1\|R_3)R_2C_1C_2}}$$

573

Substituting C for C_1 and C_2 and rewriting $R_1 \parallel R_3$ as the product-over-sum produces

$$f_0 = \frac{1}{2\pi C \sqrt{\left(\dfrac{R_1 R_3}{R_1 + R_3}\right) R_2}}$$

Rearranging,

$$f_0 = \frac{1}{2\pi C} \sqrt{\left(\frac{R_1 + R_3}{R_1 R_2 R_3}\right)}$$

EQUATION (6–1)

$$\frac{V_{out}}{V_{in}} = \frac{R(-jX)/(R - jX)}{(R - jX) + R(-jX)/(R - jX)}$$

$$= \frac{R(-jX)}{(R - jX)^2 - jRX}$$

Multiplying the numerator and denominator by j,

$$\frac{V_{out}}{V_{in}} = \frac{RX}{j(R - jX)^2 + RX}$$

$$= \frac{RX}{RX + j(R^2 - j2RX - X^2)}$$

$$= \frac{RX}{RX + jR^2 + 2RX - jX^2}$$

$$\frac{V_{out}}{V_{in}} = \frac{RX}{3RX + j(R^2 - X^2)}$$

For a 0° phase angle there can be no j term. Recall from complex numbers in ac theory that a *nonzero* angle is associated with a complex number having a j term. Therefore, at f_r the j term is 0.

$$R^2 - X^2 = 0$$

Thus,

$$\frac{V_{out}}{V_{in}} = \frac{RX}{3RX}$$

Cancelling,

$$\frac{V_{out}}{V_{in}} = \frac{1}{3}$$

EQUATION (6–2)

From the derivation of Equation (6–1),

$$R^2 - X^2 = 0$$
$$R^2 = X^2$$
$$R = X$$

Since $X = \dfrac{1}{2\pi f_r C}$,

$$R = \frac{1}{2\pi f_r C}$$

$$f_r = \frac{1}{2\pi RC}$$

EQUATIONS (6–3) AND (6–4)

The feedback network in the phase-shift oscillator consists of three RC stages, as shown in Figure B–1. An expression for the attenuation is derived using the mesh analysis method for the loop assignment shown. All Rs are equal in value, and all Cs are equal in value.

FIGURE B–1

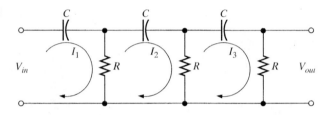

$$(R - j1/2\pi fC)I_1 - RI_2 + 0I_3 = V_{in}$$
$$-RI_1 + (2R - j1/2\pi fC)I_2 - RI_3 = 0$$
$$0I_1 - RI_2 + (2R - j1/2\pi fC)I_3 = 0$$

In order to get V_{out}, we must solve for I_3 using determinants:

$$I_3 = \frac{\begin{vmatrix} (R - j1/2\pi fC) & -R & V_{in} \\ -R & (2R - j1/2\pi fC) & 0 \\ 0 & -R & 0 \end{vmatrix}}{\begin{vmatrix} (R - j1/2\pi fC) & -R & 0 \\ -R & (2R - j1/2\pi fC) & -R \\ 0 & -R & (2R - j1/2\pi fC) \end{vmatrix}}$$

$$I_3 = \frac{R^2 V_{in}}{(R - j1/2\pi fC)(2R - j1/2\pi fC)^2 - R^2(2R - j1/2\pi fC) - R^2(R - 1/2\pi fC)}$$

$$\frac{V_{out}}{V_{in}} = \frac{RI_3}{V_{in}}$$

$$= \frac{R^3}{(R - j1/2\pi fC)(2R - j1/2\pi fC)^2 - R^3(2 - j1/2\pi fRC) - R^3(1 - 1/2\pi fRC)}$$

$$= \frac{R^3}{R^3(1 - j1/2\pi fRC)(2 - j1/2\pi fRC)^2 - R^3[(2 - j1/2\pi fRC) - (1 - j1/2\pi fRC)]}$$

$$= \frac{R^3}{R^3(1 - j1/2\pi fRC)(2 - j1/2\pi fRC)^2 - R^3(3 - j1/2\pi fRC)}$$

$$= \frac{1}{(1 - j1/2\pi fRC)(2 - j1/2\pi fRC)^2 - (3 - j1/2\pi fRC)}$$

Expanding and combining the real terms and the j terms separately,

$$\frac{V_{out}}{V_{in}} = \frac{1}{\left(1 - \dfrac{5}{4\pi^2 f^2 R^2 C^2}\right) - j\left(\dfrac{6}{2\pi f R C} - \dfrac{1}{(2\pi f)^3 R^3 C^3}\right)}$$

For oscillation in the phase-shift amplifier, the phase shift through the RC network must equal $180°$. For this condition to exist, the j term must be 0 at the frequency of oscillation f_0.

$$\frac{6}{2\pi f_0 R C} - \frac{1}{(2\pi f_0)^3 R^3 C^3} = 0$$

$$\frac{6(2\pi)^2 f_0^2 R^2 C^2 - 1}{(2\pi)^3 f_0^3 R^3 C^3} = 0$$

$$6(2\pi)^2 f_0^2 R^2 C^2 - 1 = 0$$

$$f_0^2 = \frac{1}{6(2\pi)^2 R^2 C^2}$$

$$f_0 = \frac{1}{2\pi \sqrt{6} R C}$$

Since the j term is 0,

$$\frac{V_{out}}{V_{in}} = \frac{1}{1 - \dfrac{5}{4\pi^2 f_0^2 R^2 C^2}} = \frac{1}{1 - \dfrac{5}{\left(\dfrac{1}{\sqrt{6} R C}\right)^2 R^2 C^2}}$$

$$= \frac{1}{1 - 30} = -\frac{1}{29}$$

The negative sign results from the $180°$ inversion. Thus, the value of attenuation for the feedback network is

$$B = \frac{1}{29}$$

EQUATION (8–1)

The output voltage of the upper op-amp is called V_{out1} and the output voltage of the lower op-amp is called V_{out2}. The difference in these two voltages sets up a current in the two feedback resistors, R, and R_G, given by Ohm's law.

$$i = \frac{V_{out1} - V_{out2}}{2R + R_G}$$

Because of negative feedback, ideally, the input voltage is across R_G (no voltage drop across the op-amp inputs). Applying Ohm's law again,

$$i = \frac{V_{in1} - V_{in2}}{R_G}$$

The current in the feedback resistors (R) and the gain resistor (R_G) are the same, since the op-amp inputs (ideally) draw no current. Equating the currents,

$$\frac{V_{out1} - V_{out2}}{R_G + 2R} = \frac{V_{in1} - V_{in2}}{R_G}$$

The third op-amp is set up as a unity-gain differential amplifier. Its output is

$$V_{out} = -(V_{out1} - V_{out2})$$

Substituting this result into the previous equation,

$$\frac{-V_{out}}{R_G + 2R} = \frac{V_{in1} - V_{in2}}{R_G}$$

Rearranging, changing signs, and simplifying,

$$V_{out} = \left(1 + \frac{2R}{R_G}\right)(V_{in2} - V_{in1})$$

ANSWERS TO SELF-TESTS

Chapter 1

1. (a)   **2.** (b)   **3.** (d)   **4.** (c)   **5.** (a)
6. (c)   **7.** (b)   **8.** (b)   **9.** (a)   **10.** (d)
11. (a)   **12.** (b)   **13.** (d)   **14.** (c)   **15.** (d)
16. (e)

Chapter 2

1. (c)   **2.** (b)   **3.** (d)   **4.** (b)   **5.** (a)
6. (c)   **7.** (b)   **8.** (a)   **9.** (d)   **10.** (c)
11. (d)   **12.** (a)   **13.** (b)   **14.** (c)   **15.** (d)
16. (b)   **17.** (c)   **18.** (a)   **19.** (c)   **20.** (d)

Chapter 3

1. (c)   **2.** (b)   **3.** (a)   **4.** (b)   **5.** (d)
6. (a)   **7.** (d)   **8.** (c)   **9.** (b)   **10.** (d)
11. (d)   **12.** (d)   **13.** (b)   **14.** (c)   **15.** (b)

Chapter 4

1. (c)   **2.** (a)   **3.** (c)   **4.** (e)   **5.** (b)
6. (d)   **7.** (c)   **8.** (a)   **9.** (c)   **10.** (a)
11. (b)   **12.** (c)   **13.** (b)   **14.** (d)   **15.** (d)
16. (a)   **17.** (d)   **18.** (c)

Chapter 5

1. (c)   **2.** (d)   **3.** (a)   **4.** (b)   **5.** (c)
6. (c)   **7.** (b)   **8.** (a)   **9.** (d)   **10.** (b)
11. (a)   **12.** (c)   **13.** (b)   **14.** (d)

Chapter 6

1. (b)   **2.** (a)   **3.** (c)   **4.** (b)   **5.** (d)
6. (c)   **7.** (b)   **8.** (d)   **9.** (a)   **10.** (b)
11. (c)   **12.** (e)   **13.** (c)   **14.** (d)   **15.** (d)

Chapter 7

1. (c)   **2.** (d)   **3.** (c)   **4.** (b)   **5.** (d)
6. (a)   **7.** (c)   **8.** (a)   **9.** (g)   **10.** (c)

Chapter 8

1. (d)   **2.** (b)   **3.** (a)   **4.** (e)   **5.** (c)
6. (b)   **7.** (a)   **8.** (c)   **9.** (d)   **10.** (c)
11. (a)   **12.** (b)   **13.** (f)   **14.** (b)   **15.** (b)

Chapter 9

1. (b)   **2.** (c)   **3.** (d)   **4.** (d)   **5.** (c)
6. (a)   **7.** (b)   **8.** (c)   **9.** (b)   **10.** (e)
11. (c)   **12.** (b)   **13.** (c)   **14.** (c)   **15.** (a)

Chapter 10

1. (b)   **2.** (c)   **3.** (d)   **4.** (e)   **5.** (c)
6. (c)   **7.** (c)   **8.** (b)   **9.** (b)   **10.** (d)
11. (a)   **12.** (d)   **13.** (b)   **14.** (c)   **15.** (d)
16. (a)

Chapter 11

1. (b)   **2.** (f)   **3.** (a)   **4.** (b)   **5.** (c)
6. (c)   **7.** (d)   **8.** (d)   **9.** (e)   **10.** (a)
11. (b)   **12.** (c)   **13.** (e)   **14.** (b)   **15.** (b)
16. (c)   **17.** (a)   **18.** (d)   **19.** (b)   **20.** (c)

ANSWERS TO ODD-NUMBERED PROBLEMS

Chapter 1

1. 45.4 mS

3. $\dfrac{\Delta V}{\Delta I} = \dfrac{0.75 - 0.65 \text{ V}}{8 - 3.2 \text{ mA}} = 21 \ \Omega$

5. **(a)** $V_p = 100$ V, $V_{avg} = 63.7$ V, $\omega = 200$ rad/s
   (b) 79.6 V

7. 37 kHz

9. 1.11

11. Odd harmonics

13. Voltage across 1.0 kΩ load = 1.65;
   voltage across 2.7 kΩ load = 3.25 V;
   Voltage across 3.6 kΩ load = 3.79 V

15. See Figure ANS–1.

FIGURE ANS–1 I (μA)

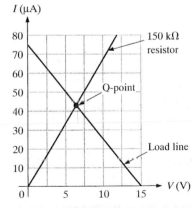

17. See Figure ANS–2.

19. 4.0 V

21. 51 dB

FIGURE ANS–2

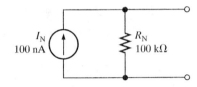

23. −60 dB

25. **(a)** −10 dB   **(b)** 10 V

27. The supply is common to both channels, so it is not the problem. Start by reversing the channels at the input of the amplifier. If the Channel 2 is still bad, the problem is most likely the amplifier or the Ch-2 speaker. Speakers can be tested by reversing them.

   If the problem changes channels when the first test is done, the problem is before the amplifier inputs and could be the A_2 microphone or a problem in wiring including the battery lead at the microphone. Test by changing SW to the B microphones. If this corrects the problem, check the A_2 microphone; otherwise look for continuity to the switch and check the switch itself.

29. Use a static-safe wrist strap (and static free work station, if possible).

Chapter 2

1. *Practical op-amp:* High open-loop gain, high input impedance, low output impedance, large bandwidth, high CMRR.
   Ideal op-amp: Infinite open-loop gain, infinite input impedance, zero output impedance, infinite bandwidth, infinite CMRR.

3. **(a)** Single-ended input; differential output
   (b) Single-ended input; single-ended output
   (c) Differential input; single-ended output
   (d) Differential input; differential output

5. V_1: differential output voltage
V_2: noninverting input voltage
V_3: single-ended output voltage
V_4: differential input voltage
I_1: bias current

7. 8.1 μA

9. 107.96 dB

11. 0.3

13. 40 μs

15. V_f = 49.5 mV, B = 0.0099

17. (a) 11   (b) 101   (c) 47.81   (d) 23

19. (a) 1.0   (b) −1.0   (c) 22.3   (d) −10

21. (a) 0.45 mA   (b) 0.45 mA
(c) −10 V   (d) −10

23. (a) $Z_{in(VF)}$ = 1.32 × 10^{12} Ω; $Z_{out(VF)}$ = 0.455 mΩ
(b) $Z_{in(VF)}$ = 5 × 10^{11} Ω; $Z_{out(VF)}$ = 0.6 mΩ
(c) $Z_{in(VF)}$ = 40,000 MΩ; $Z_{out(VF)}$ = 1.5 mΩ

25. (a) R_1 open or op-amp faulty   (b) R_2 open

27. The closed-loop gain will become a fixed −100.

Chapter 3

1. 70 dB

3. 1.67 kΩ

5. (a) 79,603   (b) 56,569
(c) 7960   (d) 80

7. (a) −0.67°   (b) −2.69°
(c) −5.71°   (d) −45°
(e) −71.22°   (f) −84.29°

9. (a) 0 dB/decade   (b) −20 dB/decade
(c) −40 dB/decade   (d) −60 dB/decade

11. 4.05 MHz

13. 21.14 MHz

15. Circuit (b) has smaller BW (97.5 kHz).

17. (a) 150°   (b) 120°   (c) 60°
(d) 0°   (e) −30°

19. (a) Unstable   (b) Stable
(c) Marginally stable

21. 25 Hz

Chapter 4

1. 24 V, with distortion

3. V_{UTP} = +2.77 V; V_{LTP} = −2.77 V

5. See Figure ANS–3.

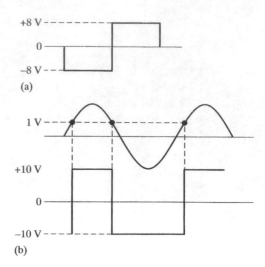

(a)

(b)

FIGURE ANS–3

7. +8.57 V and −0.968 V

9. (a) −2.5 V   (b) −3.52 V

11. 110 kΩ

13. V_{OUT} = −3.57 V; I_f = 357 μA

15. −4.46 mV/μs

17. 1 mA

19. See Figure ANS–4.

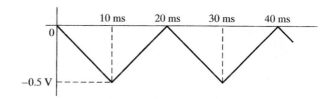

FIGURE ANS–4

21. See Figure ANS–5.

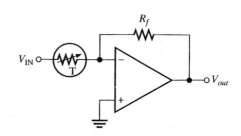

FIGURE ANS–5

23. The output is not correct because the output should also be high when the input goes below $+2$ V. Possible faults: Op-amp2 bad, diode D_2 open, noninverting $(+)$ input of op-amp2 not properly set at $+2$ V, or V_{in} is not reaching inverting input.

25. Output is not correct. R_2 is open.

Chapter 5

1. (a) Band pass   **(b)** High pass
    (c) Low pass   **(d)** Band stop

3. 48.2 kHz; No

5. 700 Hz, 5.05

7. (a) 1, not Butterworth
    (b) 1.44, approximate Butterworth
    (c) 1st stage: 1.67
        2nd stage: 1.67
        Not Butterworth

9. (a) Chebyshev **(b)** Butterworth
    (c) Bessel   **(d)** Butterworth

11. 190 Hz

13. Add another identical stage and change the ratio of the feedback resistors to 0.068 for first stage, 0.586 for second stage, and 1.482 for third stage.

15. Exchange positions of resistors and capacitors in the filter network.

17. (a) Decrease R_1 and R_2 or C_1 and C_2.
    (b) Increase R_3 or decrease R_4.

19. (a) $f_0 = 4.95$ kHz, $BW = 3.84$ kHz
    (b) $f_0 = 449$ Hz, $BW = 96.5$ Hz
    (c) $f_0 = 15.9$ kHz, $BW = 838$ Hz

21. Sum the low-pass and high-pass outputs with a two-input summer.

Chapter 6

1. An oscillator requires no input (other than dc power).

3. $1/75$

5. 733 mV

7. 50 kΩ

9. 7.5 V; 3.94

11. 136 kΩ; 691 Hz

13. Change R_1 to 3.54 kΩ

15. $R_4 = 65.8$ kΩ, $R_5 = 47$ kΩ

17. 3.33 V; 6.67 V

19. 0.0076 μF

21. 13.6 ms

23. 0.01 μF; 9.1 kΩ

Chapter 7

1. 0.033%

3. 1.01%

5. See Figure ANS–6.

7. 8.5 V

9. 9.57 V

11. 500 mA

13. 10 mA

15. $I_{L(max)} = 250$ mA, $P_{R1} = 6.25$ W

17. 40%

19. Increases

21. 14.25 V

23. 1.3 mA

25. 2.8 Ω

27. $R_{lim} = 0.35$ Ω

29. See Figure ANS–7.

FIGURE ANS–6

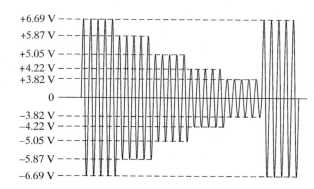

FIGURE ANS–7

Chapter 8

1. $A_{v(1)} = A_{v(2)} = 101$

3. 1.005 V

5. 9.1

7. Change R_G to 1.8 kΩ

9. 225

11. Change the 18 kΩ resistor to 270 kΩ.

13. Connect output (pin 22) directly to pin 23, and connect pin 38 directly to pin 40 to make $R_F = 0$.

15. 500 μA, 5 V

17. $A_v \cong 11.5$

19. See Figure ANS–8.

FIGURE ANS–8

21. See Figure ANS–9.

23. (a) −0.301   (b) 0.301
    (c) 1.699   (d) 2.114

25. The output of a log amplifier is limited to 0.7 V because of the transistor's *pn* junction.

27. −157 mV

29. $V_{out(max)} = -147$ mV, $V_{out(min)} = -89.2$ mV; the 1 V input peak is reduced 85% whereas the 100 mV input peak is reduced only 10%.

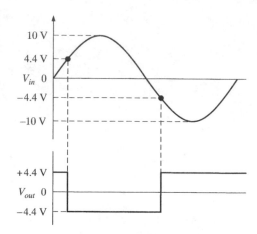

FIGURE ANS–9

31. Probe 1: ≈ 0 V
    Probe 2: ≈ 0 V
    Probe 3: 20 mV @ 1 kHz
    Probe 4: +12 V
    Probe 5: ≈ 0 V

Chapter 9

1. See Figure ANS–10.

3. 1135 kHz

5. RF: 91.2 MHz, IF: 10.7 MHz

7. 739 μA

9. −8.12 V

11. (a) 0.28 V   (b) 1.024 V
    (c) 2.07 V   (d) 2.49 V

13. $f_{diff} = 8$ kHz, $f_{sum} = 10$ kHz

15. $f_{diff} = 1.7$ MHz, $f_{sum} = 1.9$ MHz, $f_1 = 1.8$ MHz

17. $f_c = 850$ kHz, $f_m = 3$ kHz

19. $V_{out} = 15$ mV cos[(1100 kHz)2πt] − 15 mV cos[(5500 kHz)2πt]

21. See Figure ANS–11.

23. See Figure ANS–12.

25. The IF amplifier has a 450 kHz–460 kHz passband. The audio/power amplifiers have a 10 Hz–5 kHz passband.

FIGURE ANS–10

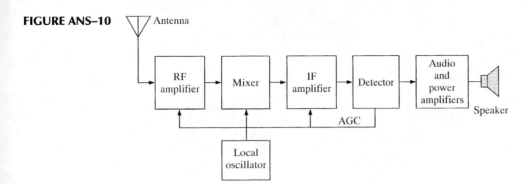

FIGURE ANS–12

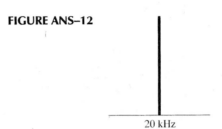

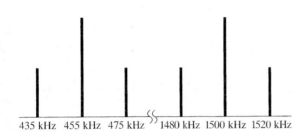

FIGURE ANS–11

27. The modulating input signal is applied to the control terminal of the VCO. As the input signal amplitude varies, the output frequency of the VCO will vary proportionally.

29. Varactor

31. (a) 10 MHz   **(b)** 48.3 mV

33. 1005 Hz

35. $f_o = 233$ kHz, $f_{lock} = \pm 103.6$ kHz, $f_{cap} \cong \pm 4.56$ kHz

Chapter 10

1. See Figure ANS–13.

3. See Figure ANS–14.

5. (a) 1   **(b)** 11

7. See Figure ANS–15.

9. 5 kΩ, 2.5 kΩ, 1.25 kΩ

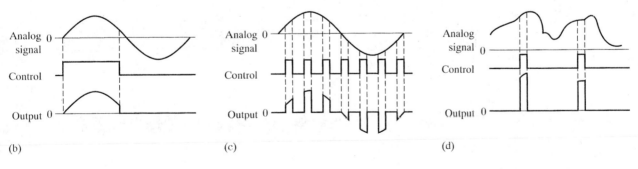

(b)   (c)   (d)

FIGURE ANS–13

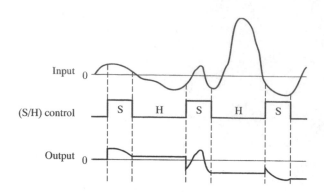

Input 0

(S/H) control — S H S H S

Output 0

FIGURE ANS–14

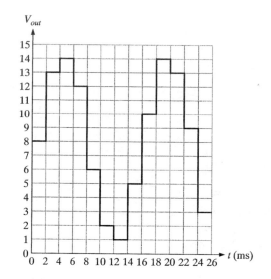

FIGURE ANS–15

13. **(a)** 16   **(b)** 32   **(c)** 256   **(d)** 65,536

15. 1 mV

17.

Sampling Time (μs)	Binary Output
0	000
10	000
20	001
30	100
40	110
50	101
60	100
70	011
80	010
90	001
100	001
110	011
120	110
130	111
140	111
150	111
160	111
170	111
180	111
190	111
200	100

11.

D_3	D_2	D_1	D_0	V_{out}
0	0	0	0	0 V
0	0	1	1	−0.50 V + (−0.25 V) = −0.75 V
1	0	0	0	−2.00 V
1	1	1	1	−2.00 V + (−1.00 V) + (−0.50 V) + (−0.25 V) = −3.75 V
1	1	1	0	−2.00 V + (−1.00 V) + (−0.50 V) = −3.50 V
0	1	0	0	−1.00 V
0	0	0	0	0 V
0	0	0	1	−0.25 V
1	0	1	1	−2.00 V + (−0.50 V) + (−0.25 V) = −2.75 V
1	1	1	0	−2.00 V + (−1.00 V) + (−0.50 V) = −3.50 V
1	1	0	1	−2.00 V + (−1.00 V) + (−0.25 V) = −3.25 V
0	1	0	0	−1.00 V
1	0	1	1	−2.00 V + (−0.50 V) + (−0.25 V) = −2.75 V
0	0	0	1	−0.25 V
0	0	1	1	−0.50 + (−0.25 V) = −0.75 V

19. f_{out} increases.

21. 691 pF (Use standard 680 pF).

23. $f_{out(min)} = 26.2$ kHz, $f_{out(max)} = 80.9$ kHz

25. The D_0 (LSB) is stuck high and the D_2 is stuck low.

Chapter 11

1. 5 V

3. 189.84°

5. $\overline{\text{Enable } M}$ and $\overline{\text{Enable } L}$ enable the tri-state output buffers.

7. Thermocouple C

9. −4.36 V

11. 590

13. 540 Ω

15. The effects of the wire resistances are cancelled in the 3-wire bridge.

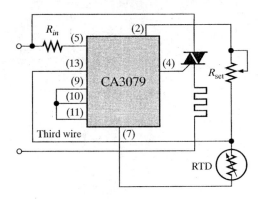

FIGURE ANS–16

17. 31

19. Pressure is measured by a strain gage bonded to a flexible diaphragm.

21. See Figure ANS–16.

GLOSSARY

Accuracy In relation to DACs or ADCs, a comparison of the actual output with the expected output, expressed as a percentage.

Acquisition time In an analog switch, the time required for the device to reach its final value when switched from hold to sample.

AC resistance The ratio of a small change in voltage divided by a corresponding change in current for a given device; also called *dynamic, small-signal,* or *bulk resistance.*

Active filter A frequency-selective circuit consisting of active devices such as transistors or op-amps coupled with reactive components.

A/D conversion A process whereby information in analog form is converted into digital form.

Amplification The process of producing a larger voltage, current, or power using a smaller input signal as a "pattern."

Amplifier An electronic circuit having the capability of amplification and designed specifically for that purpose.

Amplitude modulation (AM) A communication method in which a lower-frequency signal modulates (varies) the amplitude of a higher-frequency signal (carrier).

Analog signal A signal that can take on a continuous range of values within certain limits.

Analog-to-digital converter (ADC) A device used to convert an analog signal to a sequence of digital codes.

Antilog The number corresponding to a given logarithm.

Aperture jitter In an analog switch, the uncertainty in the aperture time.

Aperture time In an analog switch, the time to fully open after being switched from sample to hold.

Astable Characterized by having no stable states; a type of oscillator.

Attenuation The reduction in the level of power, current, or voltage.

Audio Related to the range of frequencies that can be heard by the human ear and generally considered to be in the 200 Hz to 20 kHz range.

Automatic gain control (AGC) A feedback system that reduces the gain for larger signals and increases the gain for smaller signals.

Balanced modulation A form of amplitude modulation in which the carrier is suppressed; sometimes known as *suppressed-carrier modulation.*

Band-pass filter A type of filter that passes a range of frequencies lying between a certain lower frequency and a certain higher frequency.

Band-stop filter A type of filter that blocks or rejects a range of frequencies lying between a certain lower frequency and a certain higher frequency.

Bandwidth The characteristic of certain types of electronic circuits that specifies the usable range of frequencies that pass from input to output. It is the upper critical frequency minus the lower critical frequency.

Bessel A type of filter response having a linear phase characteristic and less than −20 dB/decade/pole roll-off.

Bounding The process of limiting the output range of an amplifier or other circuit.

Butterworth A type of filter response characterized by flatness in the passband and a −20 dB/decade/pole roll-off.

Carrier The high frequency (RF) signal that carries modulated information in AM, FM, and other communications systems.

Characteristic curve A plot which shows the relationship between two variable properties of a device. For most electronic devices, a characteristic curve refers to a plot of the current, I, plotted as a function of voltage, V.

Chebyshev A type of filter response characterized by ripples in the passband and a greater than -20 dB/decade/pole roll-off.

Closed-loop An op-amp configuration in which the output is connected back to the input through a feedback circuit.

Closed-loop voltage gain The net voltage gain of an amplifier when negative feedback is included.

Cold junction A reference thermocouple held at a fixed temperature and used for compensation in thermocouple circuits.

Common mode A condition characterized by the presence of the same signal on both op-amp inputs.

Common-mode rejection ratio (CMRR) The ratio of open-loop gain to common-mode gain; a measure of an op-amp's ability to reject common-mode signals.

Comparator A circuit which compares two input voltages and produces an output in either of two states indicating the greater than or less than relationship of the inputs.

Compensation The process of modifying the roll-off rate of an amplifier to ensure stability.

Critical frequency The frequency that defines the end of the passband of a filter; also called *cutoff frequency*.

Cycle The complete sequence of values that a waveform exhibits before another identical pattern occurs.

D/A conversion The process of converting a sequence of digital codes to an analog form.

Digital-to-analog converter (DAC) A device in which information in digital form is converted to an analog form.

Damping factor (DF) A filter characteristic that determines the type of response.

dBm Decibel power level when the reference is understood to be 1 mW (see Decibel).

Decibel A dimensionless quantity that is 10 times the logarithm of a power ratio or 20 times the logarithm of a voltage ratio.

Demodulation The process in which the information signal is recovered from the IF carrier signal; the reverse of modulation.

Differential amplifier (diff-amp) An amplifier that produces an output voltage proportional to the difference of the two input voltages.

Differentiator A circuit that produces an inverted output which approximates the rate of change of the input function.

Digital signal A noncontinuous signal that has discrete numerical values assigned to the specific steps.

Discriminator A type of FM demodulator.

Domain The values assigned to the independent variable. For example, frequency or time are typically used as the independent variable for plotting signals.

Droop In an analog switch, the change in the sampled value during the hold interval.

Feedback oscillator A type of oscillator that returns a fraction of output signal to the input with no net phase shift resulting in a reinforcement of the output signal.

Feedforward A method of frequency compensation in op-amp circuits.

Feedthrough In an analog switch, the component of the output voltage which follows the input voltage after the switch opens.

Filter A type of electrical circuit that passes certain frequencies and rejects all others.

Flash A method of A/D conversion.

Fold-back current limiting A method of current limiting in voltage regulators.

Frequency The number of repetitions per unit of time for a periodic waveform.

Frequency modulation (FM) A communication method in which a lower-frequency intelligence-carrying signal modulates (varies) the frequency of a higher-frequency signal.

Gage factor (GF) The ratio of the fractional change in resistance to the fractional change in length along the axis of the gage.

Gain The amount of amplification. Gain is a ratio of an output quantity to an input quantity (e.g., voltage gain is the ratio of the output voltage to the input voltage).

Harmonics Higher-frequency sinusoidal waves that are integer multiples of a fundamental frequency.

High-pass filter A type of filter that passes frequencies above a certain frequency while rejecting lower frequencies.

Hysteresis Characteristic of a circuit in which two different trigger levels create an effect or lag in the switching action.

Instrumentation amplifier A differential voltage-gain device that amplifies the difference between the voltage existing at its two input terminals.

Integrator A circuit that produces an inverted output which approximates the area under the curve of the input function.

Inverting amplifier An op-amp closed-loop configuration in which the input signal is applied to the inverting input.

Linear component A component in which an increase in current is proportional to the applied voltage.

Linearity A straight-line relationship. A linear error is a deviation from the ideal straight-line output of a DAC.

Line regulation The change in output voltage for a given change in line (input) voltage, normally expressed as a percentage.

Load line A straight line plotted on a current versus voltage plot that represents all possible operating points for an external circuit.

Load regulation The change in output voltage for a given change in load current, normally expressed as a percentage.

Logarithm An exponent; the logarithm of a quantity is the exponent or power to which a given number called the base must be raised in order to equal the quantity.

Loop gain An op-amp's open-loop voltage gain times the attenuation of the feedback network.

Low-pass filter A type of filter that passes frequencies below a certain frequency while rejecting higher frequencies.

Mean Average value.

Mixer A nonlinear circuit that combines two signals and produces the sum and difference frequencies; a device for down-converting frequencies in a receiver system.

Modem A device that converts signals produced by one type of device to a form compatible with another; *mo*dulator/*dem*odulator.

Modulation The process in which a signal containing information is used to modify the amplitude, frequency, or phase of a much higher-frequency signal called the carrier.

Monostable Characterized by having one stable state.

Monotonicity In relation to DACs, the presence of all steps in the output when sequenced over the entire range of input bits.

Multiplier A linear device that produces an output voltage proportional to the product of two input voltages.

Multivibrator A type of circuit that can operate as an oscillator or as a one-shot.

Natural logarithm The exponent to which the base e ($e = 2.71828$) must be raised in order to equal a given quantity.

Negative feedback The process of returning a portion of the output back to the input in a manner to cancel changes that may occur at the input.

Noise An unwanted voltage or current fluctuation.

Noninverting amplifier An op-amp closed-loop configuration in which the input signal is applied to the noninverting input.

Nonmonotonicity In relation to DACs, a step reversal or missing step in the output when sequenced over the entire range of input bits.

Norton's theorem An equivalent circuit that replaces a complicated two-terminal linear network with a single current source and a parallel resistance.

Nyquist rate In sampling theory, the minimum rate at which an analog voltage can be sampled for A/D conversion. The sample rate must be more than twice the maximum frequency component of the input signal.

One-shot A monostable multivibrator.

Open-loop A condition in which an op-amp has no feedback.

Open-loop voltage gain The voltage gain of an amplifier without external feedback.

Operational amplifier (op-amp) A type of amplifier that has very high voltage gain, very high input impedance, very low output impedance, and good rejection of common-mode signals.

Oscillator An electronic circuit that operates with positive feedback and produces a time-varying output signal without an external input signal.

Passband The region of frequencies that are allowed to pass through a filter with minimum attenuation.

Period (T) The time for one cycle of a repeating wave.

Periodic A waveform that repeats at regular intervals.

Phase shift The relative angular displacement of a time-varying function relative to a reference.

Phase angle (in radians) The fraction of a cycle that a waveform is shifted from a reference waveform of the same frequency.

Phase-locked loop (PLL) A device for locking onto and tracking the frequency of an incoming signal.

Phase margin The difference between the total phase shift through an amplifier and $180°$, the additional amount of phase shift that can be allowed before instability occurs.

Pole A network containing one resistor and one capacitor that contributes -20 dB/decade to a filter's roll-off rate.

Positive feedback The return of a portion of the output signal to the input such that it reinforces the output.

Quality factor (Q) The ratio of a band-pass filter's center frequency to its bandwidth.

Quantization The determination of a value for an analog quantity.

Quantization error The error resulting from the change in the analog voltage during the A/D conversion time.

Quantizing The process of assigning numbers to sampled data.

Quiescent point The point on a load line that represents the current and voltage conditions for a circuit with no signal (also called operating or Q-point). It is the intersection of a device characteristic curve with a load line.

Radio frequency interference (RFI) High frequencies produced when high values of current and voltage are rapidly switched on and off.

Rectifier An electronic-circuit that converts ac into pulsating dc.

Regulator An electronic circuit that maintains an essentially constant output voltage with a changing input voltage or load.

Relaxation oscillator A type of oscillator that uses an *RC* timing circuit to generate a nonsinusoidal waveform.

Resistance temperature detector (RTD) A type of temperature transducer in which resistance is directly proportional to temperature.

Resolution In relation to DACs or ADCs, the number of bits involved in the conversion. Also, for DACs, the reciprocal of the maximum number of discrete steps in the output.

Resolver A type of synchro.

Resolver-to-digital converter (RDC) An electronic circuit that converts resolver voltages to a digital format which represents the angular position of the rotor shaft.

Root mean square (RMS) The value of an ac voltage that corresponds to a dc voltage that produces the same heating effect in a resistance.

Rotor The part of a synchro that is attached to the shaft and rotates. The rotor winding is located on the rotor.

Sample The process of taking the instantaneous value of a quantity at a specific point in time.

Sampling The process of breaking the analog waveform into time "slices" that approximate the original wave.

Schmitt trigger A comparator with hysteresis.

Settling time The time it takes a DAC to settle within ± ½ LSB of its final value.

Signal compression The process of scaling down the amplitude of a signal voltage.

Slew rate The rate of change of the output voltage of an op-amp in response to a step input.

Spectrum A plot of amplitude versus frequency for a signal.

Stability A condition in which an amplifier circuit does not oscillate.

Stage Each transistor in a multistage amplifier that amplifies a signal.

Stator The part of a synchro that is fixed. The stator windings are located on the stator.

Strain The expansion or compression of a material caused by stress forces acting on it.

Strain gage A transducer formed by a resistive material in which a lengthening or shortening due to stress produces a proportional change in resistance.

Successive approximation A method of A/D conversion.

Synchro An electromechanical transducer used for shaft angle measurement and control.

Synchro-to-digital converter (SDC) An electronic circuit that converts synchro voltages to a digital format which represents the angular position of the rotor shaft.

Thermal overload A condition in a rectifier where the internal power dissipation of the circuit exceeds a certain maximum due to excessive current.

Thermistor A type of temperature transducer in which resistance is inversely proportional to temperature.

Thermocouple A type of temperature transducer formed by the junction of two dissimilar metals which produces a voltage proportional to temperature.

Thevinen's theorem An equivalent circuit that replaces a complicated two-terminal linear network with a single voltage source and a series resistance.

Transducer A device that converts a physical quantity from one form to another; for example, a microphone converts sound into voltage.

Transfer curve A plot ot the output of a circuit or system for a given input.

Transducer A device that converts a physical parameter into an electrical quantity.

Trim To precisely adjust or fine tune a value.

Vector Any quantity that has both magnitude and direction.

Voltage-controlled oscillator (VCO) An oscillator for which the output frequency is dependent on a controlling input voltage.

Voltage-follower A closed-loop, noninverting op-amp with a voltage gain of one.

INDEX